International Conference on Fundamental and Applied Problems of Mechanics (FAPM-2017)

AA006956

IOP Conference Series: Materials Science and Engineering
Volume 468

Moscow, Russia
24 – 27 October 2017

ISBN: 978-1-5108-8012-2
ISSN: 1757-8981

Printed from e-media with permission by:

Curran Associates, Inc.
57 Morehouse Lane
Red Hook, NY 12571

Some format issues inherent in the e-media version may also appear in this print version.

This work is licensed under a Creative Commons Attribution 3.0 International Licence. Licence details: http://creativecommons.org/licenses/by/3.0/.

Printed by Curran Associates, Inc. (2019)

For permission requests, please contact the Institute of Physics at the address below.

Institute of Physics
Dirac House, Temple Back
Bristol BS1 6BE UK

Phone: 44 1 17 929 7481
Fax: 44 1 17 920 0979

techtracking@iop.org

Additional copies of this publication are available from:

Curran Associates, Inc.
57 Morehouse Lane
Red Hook, NY 12571 USA
Phone: 845-758-0400
Fax: 845-758-2633
Email: curran@proceedings.com
Web: www.proceedings.com

Table of contents

Volume 468

Fundamental and Applied Problems of Mechanics - 2017

24–27 October 2017, Bauman Moscow State Technical University, Moscow, Russian Federation

Accepted papers received: 1 November 2018
Published online: 14 December 2018

Preface

Fundamental and Applied Problems of Mechanics - 2017

Fundamental and Applied Problems of Mechanics - 2017

Fundamental and Applied Problems of Mechanics - 2017

Fundamental and Applied Problems of Mechanics - 2017

Peer review statement

Papers

Time variation of the attainability domain boundary for systems of third order

D I Bugrov and A M Formalskii.....1

Aerodynamic forces exerted on a body in viscous incompressible fluid resulting from vorticity generation on the body surface

G Dynnikova.....8

Footholds admissible areas structure of a two-legged walking robot on an inclined cylinder

Yu F Golubev and E V Melkumova.....15

Confidence intervals for output variable in fuzzy linear regression model

O M Poleshchuk.....24

Continental drift of the rotation pole: observation and theory

C Bizouard.....30

Modeling of Earth rotation in the gravitational field of Sun and Moon

M Yu Barkin, Yu G Markov, V V Perepelkin and A S Filippova.....41

Modelling of the microparticle spraying process in a heated gas stream

N G Bourago, A D Nikitin, I S Nikitin and B A Stratula.....45

Free vibrations of a two-layer liquid in an elastic vessel

D A Goncharov and A A Pozhalostin.....51

On the high-accuracy approach to flow simulation aroundthe airfoils by using vortex method

K S Kuzmina, I K Marchevsky and V S Moreva.....57

Theory and Practice of Harmonic Drive Mechanisms

G A Timofeyev, Yu V Kostikov, A V Yaminsky and Ye O Podchasov.....70

Features of forming a parallel staging launch vehicle mathematical model using the method of d'Alembert-Kane

L V Dokuchaev.....77

About islands of stability and limiting mass of the atomic nuclei

V S Okunev.....84

Analysis and comparison of the main characteristics of linear and nonlinear oscillations

N I Bondarenko, K B Obnosov and A V Panshina.....97

Wave Processes Regulators Optimization in Hydraulic Systems

D N Popov, N G Sosnovskiy and M V Siukhin.....105

Joint inversion of geophysical (seismic and selenodetic) and geochemical data for internal structure and composition of the Moon

E. Kronrod, K. Matsumoto, O. Kuskov, V. Kronrod, R. Yamada and S. Kamata.....112

The fluctuation perturbations in the model of the Chandler wobble

V V Bondarenko, S S Krylov and V V Perepelkin.....118

On safe orbits of a space station tethered to a precessing asteroid by two cables

A V Rodnikov.....123

V-shaped wings with the central body in supersonic flow

F A Maksimov, N A Ostapenko and M A Zubin.....128

Preliminary evaluation of multiple atmospheric re-entries in Lunar return missions

D A Grishko, V V Leonov, M A Ayrapetyan and O S Shvirkina.....139

Motion planning of a space robot

G K Borovin and V V Lapshin.....144

The stochastic theory of the turbulence

A V Dmitrenko.....153

Computing experiment for unloading dumper truck at a sloping pad

Yu N Baryshnikov.....163

Determination and investigation of asteroid Apophis' trajectories set potentially colliding with the Earth in 2036

Peng Guo, V V Ivashkin, C A Stikhno and P M Shkapov.....170

Fundamental problems of wave dynamics stratified medium modelling

V V Bulatov and Yu V Vladimirov.....178

Effects of bolide parameters on the motion and destruction in the Earth's atmosphere

N G Syzranova and V A Andrushchenko.....186

Determination of the satellite attitude motion according to the onboard measurements data

V A Pankratov and M Yu Barkin.....195

The mathematical model of short-time forecast of the Earth axial rotation irregularity

M Yu Barkin, V V Bondarenko and V V Perepelkin.....202

Smart parallel robots for massage

S N Sajapin.....207

Nambu mechanics and its applications

A O Shishanin.....214

Microrotation Waves Propagating in a Cylindrical Waveguide

E V Murashkin, Y N Radayev and V A Kovalev.....220

Experimental and theoretical studies of oscillations of stratified fluid

Win Ko Ko and A N Temnov.....226

Multifrequency model of Earth tidal deformations within framework of three-body problem

I V Skorobogatykh.....235

Flight trajectory optimization without decomposition into separate stages

S A Kumakshev and A M Shmatkov.....240

Modeling of compensatory sustainbility on the basis of algebraic formalization of systems

V I Serdyukov, N A Serdyukova and S I Shishkina.....247

Features of numerical simulation of the unsteady vortex flow around aircraft considering airbrake

A S Epikhin and V T Kalugin.....257

A method for attitude control of telescopes making use of an inverted pendulum

H Hanada, S Tsuruta, K Asari, H Araki, K Funazaki, A Satoh, H Taniguchi and M Kikuchi.....266

Constraint stabilization of two-wheeled sleigh

I Kaspirovich.....274

To development of general theory of gearings synthesis on the basis of conical wheels formed by a tool of rack and disk type

O N Tsukanov.....280

Hybrid algorithms for optimization and diagnostics of hydromechanical systems used in food production biotechnology

I G Blagoveshchenskii, M M Blagoveshchenskay, V D Sulimov and P M Shkapov.....287

Jacobi stability and updating parameters of dynamical systems using hybrid algorithms

V D Sulimov, P M Shkapov and A V Sulimov.....296

Modelling of dynamics of mechanical systems with regard for constraint stabilization

R G Mukharlyamov.....307

Considering the Motion of the Elastic Lunar Pole

M Yu Barkin and V A Pankratov.....315

Coupled Multi-Physics Modelling in Continuum Mechanics

E.V. Murashkin and Y.N. Radayev.....324

Precise measuring-computational predictive monitoring - the conditional of self-organization of the life cycle

M I Kiselev.....330

Its own spectrum band - a property of mechanical systems

I N Ovchinnikov.....336

IOP Conf. Series: Materials Science and Engineering **468** (2018) 011001 doi:10.1088/1757-899X/468/1/011001

PREFACE

In October 24-27, 2017 Bauman Moscow State Technical University held the International Conference "Fundamental and Applied Problems of Mechanics" dedicated to the 170th anniversary of the birth of the great Russian scientist Nikolay Egorovich Zhukovsky. The primary goal of this conference is to present new research results in various branches of mechanics, carried out by Russian and international scientific and educational institutions. The anniversary of N.E. Zhukovsky suggests focusing in particular on the studies that evolve and elaborate on his scientific and teaching-oriented investigations, updating them for the world of today. The conference was attended by Dr. Christian Bizouard (Director of IERS Earth Orientation Centre, Observatoire de Paris), Prof. Dr. Shuanggen Jin (Head of Satellite Navigation and Remote Sensing Group, Chinese Academy of Sciences), Prof. V.V. Sadovnichii (Member of RAS, Rector of Lomonosov Moscow State University) and other famous scientists. The scientists showed their great potential in studying various problems in numerous fields of science and technology, as well as applying new approaches, which develop and improve methods of mechanics.

Nikolay Egorovich Zhukovsky is the great Russian scientist, founder of applied aerodynamics, "father of the Russian aviation", creator of TsAGI, the Central Aerohydrodynamic Institute, founder of Theoretical Mechanics Department, Professor of the Imperial Moscow Technical School and Moscow University.

Professor N.E. Zhukovsky's scientific work initiated a new era in development of mechanical science and technology. N.E. Zhukovsky is not only responsible for a large number of important discoveries in mechanics, but he is also a scientist who dramatically changed the role of science in engineering. One of his biggest achievements is that he was able to combine the most subtle theoretical research with experience and direct all the power of these methods to study physical and technical phenomena, to solve major engineering problems. N.E. Zhukovsky realized and showed that mechanics can develop further, only being constantly enriched by experimental data, as it was in the days of Galileo, and in such a combination it will become an instrument of unprecedented strength for studying the phenomena of nature, for the advance of technology. All future and present day scientists, engineers, and researchers should know Nikolay Yegorovich Zhukovsky's prophetic words, which he said in 1898 at the Tenth Congress of the Russian Naturalists: "A man ... will fly, not leaning on the strength of his muscles, but on strength of their reason".

Content from this work may be used under the terms of the Creative Commons Attribution 3.0 licence. Any further distribution of this work must maintain attribution to the author(s) and the title of the work, journal citation and DOI.
Published under licence by IOP Publishing Ltd

IOP Conf. Series: Materials Science and Engineering **468** (2018) 011002 doi:10.1088/1757-899X/468/1/011002

International Organising Committee of the FAPM-2017:

- **Professor A.A. Aleksandrov (Chairman), Bauman Moscow State Technical University Moscow, Russia**
- **Professor P.M. Shkapov (Co-Chairman), Bauman Moscow State Technical University Moscow, Russia**
- **Astronom C. Bizouard (Co-Chairman), Observatoire de Paris, France**
- Professor V.O. Gladyshev, Bauman Moscow State Technical University, Moscow, Russia
- Professor K. Matsumoto, RISE Project, National Astronomical Observatory of Japan, Japan
- Professor V.S. Aslanov, Samara University, Samara, Russia
- Professor B.S. Bardin, Moscow Aviation Institute (National Research University), Moscow, Russia
- Professor V.D. Bertyaev, Tula State University, Tula, Russia
- Professor S.A. Berestova, Ural Federal University, Yekaterinburg, Russia
- Professor I.G. Blagoveshchenskiy, Bauman Moscow State Technical University, Moscow, Russia
- Professor Yu.I. Dimitrienko, Bauman Moscow State Technical University, Moscow, Russia
- Professor V.V. Ivashkin, Keldysh Institute of Applied Mathematics, Russian Academy of Sciences, Moscow, Russia
- Professor V.T. Kalugin, Bauman Moscow State Technical University, Moscow, Russia
- Professor V.A. Kalinichenko, Ishlinsky Institute for Problems in Mechanics, Russian Academy of Sciences, Moscow, Russia
- Professor G.N. Kuvyrkin, Bauman Moscow State Technical University, Moscow, Russia
- Professor M.Kh. Magomedov, SAUNO Company JSC, Makhachkala, Republic of Daghestan, Russia
- Professor V.I. Mayorova, Bauman Moscow State Technical University Moscow, Russia
- Professor A.N. Morozov, Bauman Moscow State Technical University Moscow, Russia
- Professor M.Yu. Ovchinnikov, Keldysh Institute of Applied Mathematics, Russian Academy of Sciences, Moscow, Russia
- Professor S.N. Sayapin, Blagonravov Mechanical Engineering Research Institute, Russian Academy of Sciences, Moscow, Russia
- Professor N.I. Sidnyaev, Bauman Moscow State Technical University, Moscow, Russia

Content from this work may be used under the terms of the Creative Commons Attribution 3.0 licence. Any further distribution of this work must maintain attribution to the author(s) and the title of the work, journal citation and DOI.
Published under licence by IOP Publishing Ltd

- Professor V.I. Smyslov, Zhukovsky Central Aerohydrodynamic Institute, Zhukovsky, Russia
- Professor Yu.P. Ulybyshev, S. P. Korolev Rocket and Space Corporation Energia, Korolyov, Russia
- Professor V.N. Chubarikov, Lomonosov Moscow State University , Moscow, Russia
- Professor A.O. Shimanovskiy, Belarusian State University of Transport, Gomel, the Republic of Belarus
- **Academic Secretary - Ph.D. N.D. Tyannikova**

IOP Conf. Series: Materials Science and Engineering **468** (2018) 011003 doi:10.1088/1757-899X/468/1/011003

Academician Committee of FAPM-2017 (International Advisory Council)

- **Academician I.B. Fedorov (Chairman), Bauman Moscow State Technical University, Moscow, Russia**
- **Professor V.N. Zimin (Co-Chairman), Bauman Moscow State Technical University, Moscow, Russia**
- **Professor V.O. Gladyshev (Co-Chairman), Bauman Moscow State Technical University, Moscow, Russia**
- **Professor H. Hanada (Co-Chairman), director of RISE Project, NAO of Japan, Japan**
- **Astronom C. Bizouard (Co-Chairman), Observatoire de Paris, France**
- Professor Shuanggen Jin, Chinese Academy of Sciences, China
- Academician I.G. Goryacheva, Russian National Committee on Theoretical and Applied Mechanics, Moscow, Russia
- Academician V.F. Zhuravlev, Ishlinsky Institute for Problems in Mechanics, Russian Academy of Sciences, Moscow, Russia
- Academician S.T. Surzhikov, Ishlinsky Institute for Problems in Mechanics, Russian Academy of Sciences, Moscow, Russia
- Academician D.V. Treshchev, Lomonosov Moscow State University, Moscow, Russia
- Academician S.L. Chernyshev, Zhukovsky Central Aerohydrodynamic Institute, Russia
- Corresponding Member of the Russian Academy of Sciences A.P. Krishchenko, Bauman Moscow State Technical University , Moscow, Russia
- Corresponding Member of the Russian Academy of Sciences O.S. Naraykin, Bauman Moscow State Technical University, Moscow, Russia
- Corresponding Member of the Russian Academy of Sciences V.A. Solovev, Bauman Moscow State Technical University, Moscow, Russia
- Professor M.M. Blagoveshchenskaya, Moscow State University of Food Production, Moscow, Russia
- Professor V.S. Zarubin, Bauman Moscow State Technical University Moscow, Russia
- Professor T.A. Ismailov, Daghestan State Technical University, Makhachkala, Republic of Daghestan, Russia
- Professor M.I. Kiselev, Bauman Moscow State Technical University Moscow, Russia

Content from this work may be used under the terms of the Creative Commons Attribution 3.0 licence. Any further distribution of this work must maintain attribution to the author(s) and the title of the work, journal citation and DOI.
Published under licence by IOP Publishing Ltd

- Professor V.V. Sazonov, Keldysh Institute of Applied Mathematics, Russian Academy of Sciences, Moscow, Russia
- Professor V.A. Samsonov, Lomonosov Moscow State University, Moscow, Russia
- Professor A.M. Tsirlin, Aylamazyan Program Systems Institute, Russian Academy of Sciences, Pereslavl-Zalessky, Russia
- Professor Yu.D. Chashechkin, Ishlinsky Institute for Problems in Mechanics, Russian Academy of Sciences, Moscow, Russia
- Professor A.M. Cherepashchuk, Sternberg Astronomical Institute, Moscow, Russia
- Professor V.I. Shcherbakov, A.F. Mozhaysky Military-Space Academy, Saint Petersburg, Russia
- **Academic Secretary - Ph.D. M.Yu. Barkin**

Content from this work may be used under the terms of the Creative Commons Attribution 3.0 licence. Any further distribution of this work must maintain attribution to the author(s) and the title of the work, journal citation and DOI.
Published under licence by IOP Publishing Ltd

IOP Conf. Series: Materials Science and Engineering **468** (2018) 011005 doi:10.1088/1757-899X/468/1/011005

Peer review statement

All papers published in this volume of *IOP Conference Series: Materials Science and Engineering* have been peer reviewed through processes administered by the proceedings Editors. Reviews were conducted by expert referees to the professional and scientific standards expected of a proceedings journal published by IOP Publishing.

Content from this work may be used under the terms of the Creative Commons Attribution 3.0 licence. Any further distribution of this work must maintain attribution to the author(s) and the title of the work, journal citation and DOI.
Published under licence by IOP Publishing Ltd

IOP Conf. Series: Materials Science and Engineering **468** (2018) 012001 doi:10.1088/1757-899X/468/1/012001

Time variation of the attainability domain boundary for systems of third order

D I Bugrov [1] and A M Formalskii [2]

[1]Lomonosov Moscow State University, Moscow, Russia
[2]Institute of Mechanics, Lomonosov Moscow State University, Moscow, Russia
E-mail: d.bugrov@mech.math.msu.su

Abstract. We consider linear steady-state systems of third order with one controlling (perturbing) action. The control action is assumed to be limited in absolute value and to be contained in the set of piecewise continuous functions. The case is investigated when our system has two complex conjugated and one real eigenvalues. Variation of the attainability region as time increases is studied. We show analytically that the changing of the attainability region with time variation depends on the correlation between the real parts of the eigenvalues.

1. Introduction

The study of the reachability domains of controlled (perturbed) systems with bounded controlling action is of interest in solving different problems of the optimal control and of the differential games. Many examples of the reachability regions using and constructing may be found in the publications of 1950s – 1970s [1–5]. In many investigations, it is suggested to approximate these domains by the ellipsoids [6–9]. Some problems on the constructing of the reachability regions for the discrete systems are considered in [10,11]. Problems on the smoothness of the reachability regions and the correlation between the smoothness and the system's observability are considered in [12,13]. In monograph [14], the attainability domains are constructed for some unstable objects. Using the found domains became possible to design feedback control in order to stabilize desired motions of these objects with large basin of attraction.

It is known fact [12] that for a linear steady-state system of arbitrary order having one bounded controlling action and at least one real eigenvalue, exactly two conical corner points on the boundary of reachability region exist. A similar case may be for time-variant objects. Two conical corner points are joined one to another by two ribs in a case of a time-invariant third order system with three real eigenvalues [12]. The reachability region does not have conical corner points, if a time-invariant system (of arbitrary order) has only complex eigenvalues and time value is sufficiently large [12]. Variation with time of the reachability region near a conical corner point for a third order steady-state system with two complex eigenvalues is studied in this paper. We prove that the sort of the reachability domain near a conical corner point depends on the difference between the real parts of the eigenvalues.

Content from this work may be used under the terms of the Creative Commons Attribution 3.0 licence. Any further distribution of this work must maintain attribution to the author(s) and the title of the work, journal citation and DOI.
Published under licence by IOP Publishing Ltd

2. Reachability domain

Let us consider steady-state system of linear equations in matrix form

$$\frac{d\boldsymbol{x}}{dt} = A\boldsymbol{x} + \boldsymbol{b}u, \quad |u(t)| \le 1; \quad A = \begin{Vmatrix} \lambda & 0 & 0 \\ 0 & \varepsilon & -1 \\ 0 & 1 & \varepsilon \end{Vmatrix}, \quad \boldsymbol{b} = \begin{Vmatrix} b_1 \\ b_2 \\ b_3 \end{Vmatrix} \tag{1}$$

Here $\boldsymbol{x}$, A and $\boldsymbol{b}$ are 3×1, 3×3 and 3×1 matrixes respectively, $u = u(t)$ is a scalar control (perturbation) assumed to be piecewise-continuous function of a time and to be bounded in magnitude. Any third order time-invariant system with one real and two complex conjugate eigenvalues may be written in form (1) after substitution of the parameters and variables. System (1) is assumed to be completely controllable system in Kalman sense. It means that $b_1 \ne 0$ and $b_2^2 + b_3^2 \ne 0$.

Fundamental matrix of the solutions to homogeneous system (1) (with $u(t) \equiv 0$) has the form

$$e^{At} = \begin{Vmatrix} e^{\lambda t} & 0 & 0 \\ 0 & e^{\varepsilon t}\cos t & -e^{\varepsilon t}\sin t \\ 0 & e^{\varepsilon t}\sin t & e^{\varepsilon t}\cos t \end{Vmatrix} \tag{2}$$

Set Ω of admissible control actions is described as

$$\Omega = \{u(t): \ |u(t)| \le 1\} \tag{3}$$

We assume that at instant $t = 0$ system (1) is placed at the origin of the coordinates, so $\boldsymbol{x}(0) = 0$. Then its solutions at instant $t = T$ is calculated by Cauchy formula

$$\boldsymbol{x}(T) = \int_0^T e^{A(T-\zeta)}\boldsymbol{b}u(\zeta)d\zeta = \int_0^T e^{A\xi}\boldsymbol{b}u(T-\xi)d\xi \tag{4}$$

Reachability domain $Q(T)$ may be determined as

$$Q(T) = \{\boldsymbol{x}(T): \ u(t) \in \Omega\} \tag{5}$$

It is well known [1–5], that in linear systems, for each finite instant T, this domain $Q(T)$ is a closed, bounded, convex, and symmetric about origin $\boldsymbol{x} = 0$. This region becomes larger as time T increases, that is $Q(T_2) \supset Q(T_1)$, if $T_2 > T_1$.

We introduce the spherical coordinates: angles φ and θ. Using these angles, we determine orientation of unit vector $\boldsymbol{\eta}$

$$\eta_1 = \sin\theta\cos\varphi, \quad \eta_2 = \sin\theta\sin\varphi, \quad \eta_3 = \cos\theta \tag{6}$$

All points of a unit sphere S are described by the following inequalities

$$0 \le \theta \le \pi, \quad 0 \le \varphi < 2\pi \tag{7}$$

We denote by $\Pi(\boldsymbol{\eta})$ the supporting plane (orthogonal to vector $\boldsymbol{\eta} \in S$) to domain $Q(T)$. Then the boundary of domain $Q(T)$ is the envelope of all supporting planes $\Pi(\boldsymbol{\eta})$, $\boldsymbol{\eta} \in S$. Consequently, those, and only those, points $\boldsymbol{x}$, that satisfy all inequalities

$$|\boldsymbol{\eta}\boldsymbol{x}| \le \rho(\boldsymbol{\eta}), \quad \forall \boldsymbol{\eta} \in S, \quad S = \{\boldsymbol{\eta}: \ \|\boldsymbol{\eta}\| = \eta_1^2 + \eta_2^2 + \eta_3^2 = 1\} \tag{8}$$

belong to domain $Q(T)$. Here $\rho(\boldsymbol{\eta})$ is the distance between the origin $\boldsymbol{x} = 0$ and supporting plane $\Pi(\boldsymbol{\eta})$ (in relations (8) $\boldsymbol{\eta}$ is a row matrix):

$$\rho(\boldsymbol{\eta}) = \max_{\boldsymbol{x}(T)\in Q(T)} (\boldsymbol{\eta}\boldsymbol{x}(T)) = \max_{u(T-\xi)\in\Omega} \int_0^T \boldsymbol{\eta} e^{A\xi}\boldsymbol{b}u(T-\xi)d\xi = \int_0^T |\boldsymbol{\eta} e^{A\xi}\boldsymbol{b}|d\xi \tag{9}$$

Admissible control $u(T-\xi)$ producing a maximum of integral (9) has a form

$$u(T-\xi)=\operatorname{sign}\left(\boldsymbol{\eta}e^{A\xi}\boldsymbol{b}\right)=\operatorname{sign}\left\{e^{\lambda\xi}\eta_1 b_1+e^{\varepsilon\xi}\left[\cos\xi\left(\eta_2 b_2+\eta_3 b_3\right)+\sin\xi\left(\eta_3 b_2-\eta_2 b_3\right)\right]\right\} \tag{10}$$

We assume without any loss in generality that $b_1>0$. Then denoting

$$r_2=b_2/b_1\,,\quad r_3=b_3/b_1\,,\quad \tilde{r}=\left(r_2^2+r_3^2\right)^{1/2},\quad \sin\alpha=r_2/\tilde{r}\,,\quad \cos\alpha=r_3/\tilde{r}$$

one can write expression (10) as:

$$u(T-\xi)=\operatorname{sign}\chi(\xi)\,,\quad \chi(\xi)=e^{(\lambda-\varepsilon)\xi}\eta_1+\tilde{r}\left[\eta_2\sin(\alpha-\xi)+\eta_3\cos(\alpha-\xi)\right] \tag{11}$$

Equations (4) and (11) have two parameters θ and φ. Angles θ and φ are the parameters varied within the ranges (7). Points on the boundary of domain $Q(T)$ can be calculated also by substituting in differential equation (1) function

$$u(t)=\operatorname{sign}\chi(T-t) \tag{12}$$

3. Corner points of the reachability domain

If

$$\theta=\pi/2\,,\quad \varphi=0 \tag{13}$$

then $\boldsymbol{\eta}=(\eta_1\ \eta_2\ \eta_3)=(1\ 0\ 0)$ and

$$u(T-\xi)=1 \tag{14}$$

for all values of ξ and T. Let $\boldsymbol{x}(T)$ be a point on the boundary of the reachability domain calculated as the control (14) substitutes into formula (4). Components η_1, η_2, η_3 of vector $\boldsymbol{\eta}\in S$ depend continuously on the angles θ and φ. Therefore, for any value $\xi\geq 0$ there exists an area near point (13) in the plane of variables θ and φ such that quantity $\chi(\xi)$ (11) is positive in this area, that is

$$e^{(\lambda-\varepsilon)\xi}\eta_1+\tilde{r}\left[\eta_2\sin(\alpha-\xi)+\eta_3\cos(\alpha-\xi)\right]>0 \tag{15}$$

It means that for each value $\xi\geq 0$ at the sphere S a neighborhood of vector $\boldsymbol{\eta}=(1\ 0\ 0)$ exists and inequality (15) holds on for all vectors placed in this neighborhood. We denote by $S(\xi)$ the whole subset of vectors $\boldsymbol{\eta}\in S$, for all of them inequality (15) holds at specified value of $\xi\geq 0$. The subset $S(\xi)$ for each value of ξ includes vector $\boldsymbol{\eta}=(1\ 0\ 0)$ and these subsets have a complete measure on the set S. The left-hand side of inequality (15) depends linearly on components η_1, η_2, η_3 of vector $\boldsymbol{\eta}$, consequently the subset $S(\xi)$ is convex cone.

Let $S(0,T)$ be the set of all the vectors $\boldsymbol{\eta}\in S$ for each of which inequality (15) holds for all values $0\leq\xi\leq T$. This set $S(0,T)$ is the intersection of $S(\xi)$ sets for all values $0\leq\xi\leq T$:

$$S(0,T)=\bigcap_{0\leq\xi\leq T}S(\xi)$$

Set $S(0,T)$ is the intersection of convex cones, therefore, it is a convex cone.

For each vector $\boldsymbol{\eta}\in S(0,T)$ control (11) is the one and the same constant quantity for all values $0\leq\xi\leq T$. Therefore, all supporting planes $\Pi(\boldsymbol{\eta})$ orthogonal to vectors $\boldsymbol{\eta}\in S(0,T)$ contain the point $\boldsymbol{x}(T)\in Q(T)$ obtained in the case of control (14). The intersection of these supporting planes forms a conical surface with the vertex at the point $\boldsymbol{x}(T)$. It is natural to call this point as the conical corner point of the region [12]. Set $S(0,T)$ of the unit vectors $\boldsymbol{\eta}$ is convex, so the intersection of supporting planes $\Pi(\boldsymbol{\eta})$ orthogonal to vectors $\boldsymbol{\eta}\in S(0,T)$ forms convex cone surface. This conclusion follows

IOP Conf. Series: Materials Science and Engineering **468** (2018) 012001 doi:10.1088/1757-899X/468/1/012001

also from convexity of the reachability region $Q(T)$ whose boundary being the envelope of the supporting planes set.
As the reachability region is symmetric about the origin of the coordinates, then it has once more conical corner point that is symmetric to the first one. This second point is the result of the control $u(T-\xi)=-1$. These symmetric points are named in [3] as vertex points. There are no other conical cone points on the boundary of the reachability region except the points obtained under control signals $u(T-\xi)=\pm 1$.

Inequality (15) shows that set $S(0,T)$ becomes less or remains the same as time T increases. Consequently as time T increases then the cone covering the conical corner point and the whole region $Q(T)$, either remains unchanged or becomes larger. A similar result holds to objects of arbitrary (not only the third) order.

4. Correlation between cone apex angle, difference $\boldsymbol{\lambda}-\boldsymbol{\varepsilon}$ and time T

Let us consider three following situations: the difference $\lambda-\varepsilon<0$, $\lambda-\varepsilon=0$ and $\lambda-\varepsilon>0$.
Initially we study the case when the following inequality holds

$$\lambda-\varepsilon<0 \tag{16}$$

Relation (15) is satisfied for row $\boldsymbol{\eta}=(1\ 0\ 0)$ at all values $\xi\geq 0$. Note the left-hand side of inequality (15) is the scalar multiplication of vectors

$$(\eta_1, \eta_2, \eta_3) \quad \text{and} \quad \left(e^{(\lambda-\varepsilon)\xi},\ \tilde{r}\sin(\alpha-\xi),\ \tilde{r}\cos(\alpha-\xi)\right) \tag{17}$$

It follows from expression (15) that the angle between vectors (17) is less than $\pi/2$. Set $S(0,T)$ of vectors (η_1, η_2, η_3) fulfilled expression (15) becomes smaller when time T increases and this set works towards row $\boldsymbol{\eta}=(1\ 0\ 0)$ as $T\to\infty$. In this case, the set of supporting planes at the corner point tends to the single supporting plane $\Pi(\boldsymbol{\eta})$ orthogonal to vector $\boldsymbol{\eta}=(1\ 0\ 0)$. Consequently, under inequality (16) the apex angle of the cone within the area of the conical corner point becomes larger as time T enlarges. Otherwise speaking, the conical corner point becomes less pointed and disappears as $T\to\infty$ (Fig. 1). There are no conical corner points in limit set $Q(\infty)$ corresponding to $T\to\infty$. It follows from said above that in the considered case (16) the region $Q(T)$ approximation by an ellipsoid having the smooth boundary, becomes more proper as time T enlarges.
Set $Q(\infty)$ is bounded in the phase space X only when, $\lambda<0$ and $\varepsilon<0$ [5]. If $\lambda<0$ and $\varepsilon\geq 0$, then $Q(\infty)$ is the band $|x_1|<|\lambda/b_1|$ [5]. Of course, this band does not have corner points.
Now we consider the case when the following equality holds

$$\lambda-\varepsilon=0 \tag{18}$$

Substituting equality (18) transforms expression (15) into

$$\eta_1+\tilde{r}\left[\eta_2\sin(\alpha-\xi)+\eta_3\cos(\alpha-\xi)\right]>0 \tag{19}$$

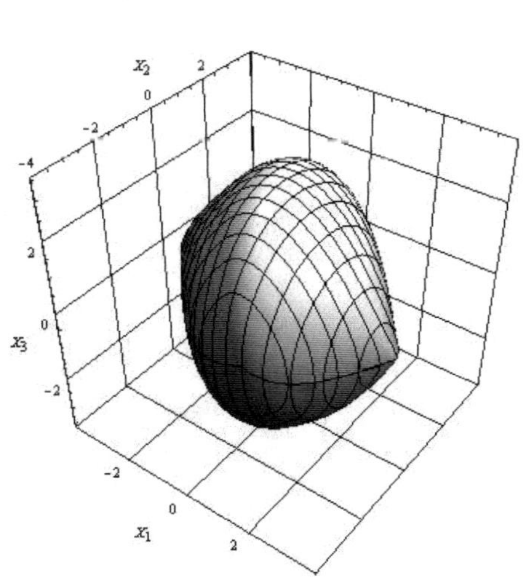

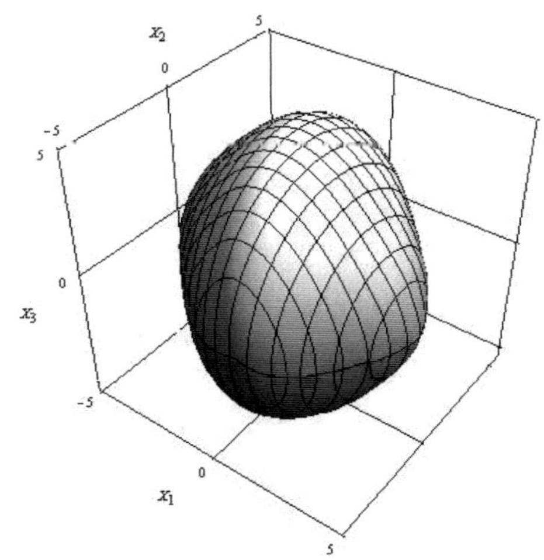

Figure 1. The domain $Q(T)$ for $\lambda = -0.2$, $\varepsilon = -0.1$, $b_1 = b_3 = 1$, $b_2 = 0$; in the left side $T = 2.5\pi$, in the right side $T = 6\pi$.

If values η_2 and η_3 of row $\boldsymbol{\eta}=(\eta_1, \eta_2, \eta_3)$ are small enough, then the second term in the left-hand side (19) is small enough too. Consequently, if vector (η_1, η_2, η_3) is located close to vector $(1, 0, 0)$, then inequality (19) holds for all values $\xi \geq 0$.

Left-hand side of inequality (19) is the scalar multiplication of vectors

$$(\eta_1, \eta_2, \eta_3) \quad \text{and} \quad (1,\ \tilde{r}\sin(\alpha-\xi),\ \tilde{r}\cos(\alpha-\xi)) \tag{20}$$

We conclude from inequality (19) that the angle between vectors (20) is less than $\pi/2$. The second vector in (20) may be represented as the sum of two vectors

$$(1,\ 0,\ 0) \quad \text{and} \quad \tilde{r}(0,\ \sin(\alpha-\xi),\ \cos(\alpha-\xi)) \tag{21}$$

It may be shown, that set $S(0,T)$ remains the same starting from the moment $T = 2\pi$. This unchanged set has a form of spherical (globular) sector or, otherwise speaking, a circular cone with the axis being the vector $(1, 0, 0)$. The apex angle of $S(0,T)$ is equal to $\pi - 2\operatorname{arctg}\tilde{r}$. So, it is a function of r_2, r_3 values, but it does not depend on time *T*. The ends of vectors η from sector $S(0,T)$ lie on a surface like a circular cap. The conical corner point together with the whole reachability domain $Q(T)$ is placed inside the circular cone conjugated to sector $S(0,T)$. The apex angle of this circular cone equals $2\operatorname{arctg}\tilde{r}$. Consequently, the larger value $\tilde{r} = \left(r_2^2 + r_3^2\right)^{1/2}$, then the larger this angle and less pointed the conical corner point.

Under condition (18), two symmetrical with respect to origin $\boldsymbol{x} = 0$ conical corner points exist for all values $0 < T < \infty$. Hence, in contrast with case (16) the conical corner points of domain $Q(T)$ do not disappear as time *T* enlarges, moreover, their pointedness remains the same (Fig. 2). Limit set $Q(\infty)$ is bounded only in the case of $\lambda, \varepsilon < 0$. It follows from said above, this bounded set has conical corner points in the case (18). Limit set $Q(\infty)$ covers the whole space *X* in case of (18) and under conditions $\lambda, \varepsilon \geq 0$. So, under conditions (18) and $\lambda, \varepsilon \geq 0$ set $Q(\infty)$ is not bounded.

IOP Conf. Series: Materials Science and Engineering **468** (2018) 012001 doi:10.1088/1757-899X/468/1/012001

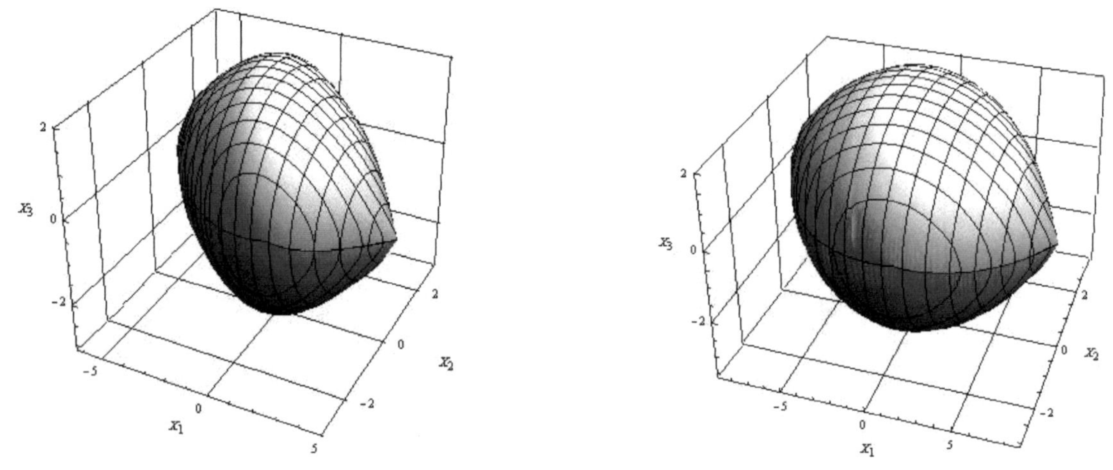

Figure 2. The domain $Q(T)$ for $\lambda=-0.2$, $\varepsilon=-0.2$, $b_1=b_3=1$, $b_2=0$; in the left side $T=2\pi$, in the right side $T=8\pi$.

Let us consider last case when the following inequality holds

$$\lambda-\varepsilon>0 \qquad (22)$$

We turn to inequality (15) review subject to this condition. In contrast with the cases (16) and (18), under condition (22) the exponent $e^{(\lambda-\varepsilon)\xi}$ arises indefinitely as the value ξ increases. As time T increases set $S(0,T)$ remains unchanged starting from some instant T^0. Thus, starting from this moment T^0 the apex angle of the conical surface encompassing the conical corner point also remains the same. It may be noted similar to case (18), the conical corner points of region $Q(T)$ do not disappear as the time T enlarges (Fig. 3).

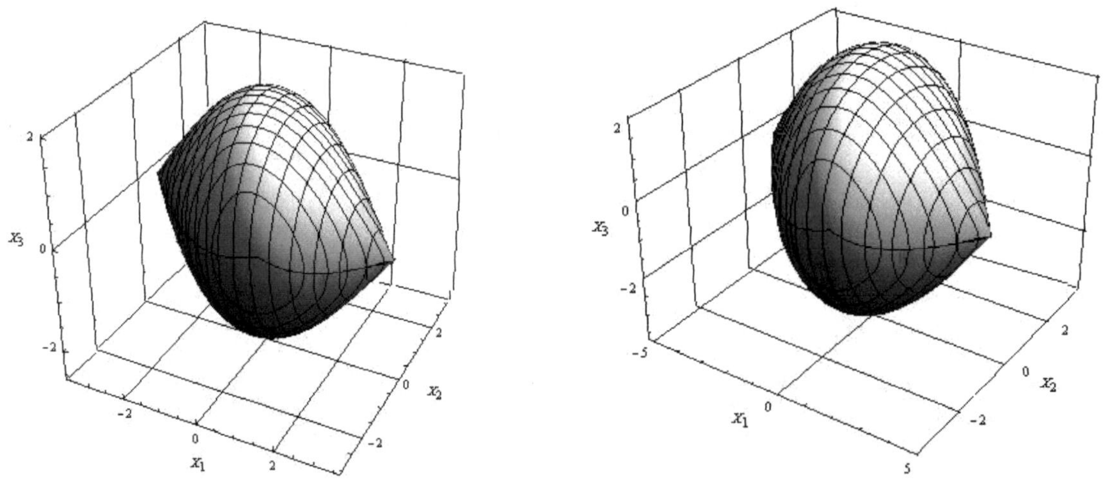

Figure 3. The domain $Q(T)$ for $\lambda=-0.1$, $\varepsilon=-0.2$, $b_1=b_3=1$, $b_2=0$; in the left side $T=3\pi$, in the right side $T=7\pi$.

The attainability domains $Q(T)$ are constructed in [15] for other parameter values, than here in Figures 1 - 3.

Under inequality $\varepsilon < 0$ the limit domain $Q(\infty)$ is bounded, if $\lambda < 0$. It follows that in case (22) this domain has conical corner points. This limit region is a cylinder as $\lambda \geq 0$ [5]. Cylinder guide is an axis x_1 ($x_2 = x_3 = 0$). The intersection of this cylinder and the plane $x_1 = 0$ is the limit reachability region of the system involving the second and third scalar equations of system (1). This region constructed in the variables plane x_2, x_3 ($x_1 = 0$) is bounded by the limit cycle. And this cycle consists of two integral curves. The integral curves are symmetric with respect to point $x_2 = x_3 = 0$ and correspond to controls $u = 1$ and $u = -1$ [5]. This plane region does not have conical corner points. Thus, the boundary of the limit reachability domain $Q(\infty)$ for system (1) is smooth, if $\lambda \geq 0$, $\varepsilon < 0$. It may be marked that for any finite time T region $Q(T)$ has two conical corner points. Related to these points cone apex angles do not "gravitate" to π under $T \to \infty$.

References

[1] Bellman R 1957 *Dynamic Programming* (Princeton, NJ: Princeton Univ. Press)
[2] Lee E B and Markus L 1967 *Foundations of Optimal Control Theory* (N. Y. etc.: Wiley)
[3] Krasovskii N N 1968 *The Theory of Motion Control* (Moscow: Nauka)
[4] Boltyanskii V G 1971 *Mathematical Methods of Optimal Control* (N. Y.: Holt, Rinehart & Winston)
[5] Formalskii A M 1974 *Controllability and Stability of Systems with Restricted Resources* (Moscow: Nauka)
[6] Chernousko F L 1994 *State Estimation of Dynamic Systems* (Boca Raton, FL: SRC)
[7] Kurzhanski A B and Valyi I 1997 *Ellipsoidal Calculus for Estimation and Control* (Boston: Birkhauser)
[8] Ovseevich A I and Chernousko F L 2008 *Methods of ellipsoidal estimation for linear control systems* Proc. 17th World Congress IFAC, Seoul, Korea, July 6–11 pp 15345-48
[9] Polyak B T , Khlebnikov M V and Shcherbakov P S 2014 *The Control of Linear Systems with External Perturbations* (Moscow: URSS)
[10] Formalsky A M and Sirotin A N 2004 On the geometric properties of reachable and controllable sets for linear discrete systems *J. Optimiz. Theory Appl.* **122** (2) pp 257-84
[11] Sirotin A N and Formalskii A M 2003 Attainability and controllability of discrete-time systems under control actions bounded in magnitude and norm *Autom. and Rem. Cont.* **64** (12) pp 1844-57
[12] Formalskii A M 1983 On the corner points of regions of attainability *J. Appl. Math. Mech.* **47** (4) pp 566-74
[13] Bolotin Yu V and Morgunova S N 2007 Observability with bearing-only observations and smoothness of the attainable set *J. Math. Sci.* **147** (2) pp 6631-38
[14] Formalskii A M 2015 *Stabilization and Motion Control of Unstable Objects* (Berlin/Boston: Walter de Gruyter)
[15] Bugrov D I and Formalskii A M 2017 Time dependence of the attainability regions of third order systems *J. Appl. Math. Mech.* **81** (2) pp 106-13

Acknowledgments
The research was supported by the Russian Foundation for Basic Research (projects: 15-05-04503 and 16-01-00683).

IOP Conf. Series: Materials Science and Engineering **468** (2018) 012002 doi:10.1088/1757-899X/468/1/012002

Aerodynamic forces exerted on a body in viscous incompressible fluid resulting from vorticity generation on the body surface

G Dynnikova

Lomonosov Moscow State University, Institute of Mechanics, Russia
E-mail: dyn@imec.msu.ru

Abstract. The vorticity evolution equation following from the Navier-Stokes equations for three-dimensional viscous incompressible fluid flows is represented in the divergent form. The partial derivative of the vorticity with respect to time is expressed as the divergence of a tensor called the vorticity transfer tensor. The concept of vorticity flux generated by a surface is defined. An expression for the pressure force in terms of the generated vorticity flux is obtained under the no-slip condition for three-dimensional bodies of constant and variable shape.

1. Introduction

The close connection between the aerodynamic forces exerted on a body and the attached vorticity and its fluxes shedding from the body surface is well known. This is the subject of the well-known Kutta-Joukowski theorem on the lifting force and the Sedov and Belotserkovskii formulas. In the studies of the authors mentioned above the formulas for the forces were obtained for potential flows. In [1-2] generalizations of these formulas to include the case of unsteady vortex ideal fluid separation flow were given.

In [3-6] expressions for the aerodynamic forces in terms of the vorticity fluxes in a viscous incompressible fluid were obtained on the basis of the representation of evolution of the vorticity field as the motion of vortices in the case of unsteady plane-parallel and unswirled axisymmetric flows. Integral formulas relating the pressure distribution over the flow field to the motion of vortices were also obtained. These formulas were widely used to solve problems of unsteady flows past bodies [see 7-13 and others] and made it possible to design the effective method of solving the conjugate problems of interconnected motion of bodies and fluid [5,14,15]. In the present study the expressions for the forces in terms of the vorticity fluxes generated by the body surface is generalized to include the case of arbitrary three-dimensional flows of a viscous incompressible fluid.

In using the vortex methods, the expression for the aerodynamic forces in terms of the vorticity fluxes is most relevant. These methods are underlain by the vorticity evolution equation which can be obtained by applying the curl operator to the Navier-Stokes equation

$$\frac{\partial}{\partial t} = -\boldsymbol{\nabla}\times(\quad \times\mathbf{V} + \nu\boldsymbol{\nabla}\times \quad), \qquad (1)$$
$$= \boldsymbol{\nabla}\times\mathbf{V}, \quad \boldsymbol{\nabla}\mathbf{V} = 0$$

Content from this work may be used under the terms of the Creative Commons Attribution 3.0 licence. Any further distribution of this work must maintain attribution to the author(s) and the title of the work, journal citation and DOI.
Published under licence by IOP Publishing Ltd

where $\mathbf{V}$ and $\mathbf{\Omega}$ are the fluid velocity and vorticity, respectively, and ν is the kinematic viscosity coefficient.

There is no pressure in these equations; therefore, the expression for the forces written directly in terms of the characteristics of vortex field is of importance.

For plane-parallel and unswirled axisymmetric flows the equation (1) can be written in the divergent form [16], [17]:

$$\frac{\partial \Omega}{\partial t} = -\nabla \cdot (\mathbf{U}\Omega), \quad \mathbf{U} = \mathbf{V} + \mathbf{V}_d, \tag{2}$$

where the vector $\mathbf{V}_d$, called the diffusion velocity [18], is equal to

$$\mathbf{V}_d = \begin{cases} -\nu \dfrac{\nabla \Omega}{\Omega} & \text{plane-parallel flows} \\ -\nu \dfrac{\nabla (r\Omega)}{r\Omega} & \text{axisymmetric flows} \end{cases}$$

The vectors $\mathbf{J} = \mathbf{V}\Omega$ and $\mathbf{J}_d = \mathbf{V}_d\Omega$ represent the convective and diffusive vorticity transfer vectors.

The divergent form of writing the vorticity evolution equation was the basis for the development of the Lagrangian vortex methods of simulating two-dimensional viscous incompressible fluid flows in which the vorticity field was represented by a set of vortex particles of given circulation [18,14]. The particles move at a velocity $\mathbf{U}$, their circulation being not varied. The fluid velocity $\mathbf{V}$ can be determined from the vortex particle distribution using the Biot-Savart formula. New particles with a circulation Γ_i^{new}, where i is the node number, are generated in the nodes of contours of the surfaces in each time step. The values of Γ_i^{new} are determined from a system of linear equations representing the boundary conditions on the surface (no-flow or no-slip conditions). The value of Γ_i^{new} is connected with the diffusive vorticity flux by the relation $\Gamma_i^{new} = \mathbf{J}_d\mathbf{n}\Delta t\Delta l$, where $\mathbf{n}$ is the normal vector to the contour and Δl is the length of a contour segment. Thus, the vorticity flux generated by the surface is known in each time step.

In [3] the following expression was obtained for the pressure force $\mathbf{F}_p$ exerted on a profile under the no-slip condition:

$$\frac{\mathbf{F}_p}{\rho} = \mathbf{e}_z \times \oint \mathbf{r}(\mathbf{J}_d\mathbf{n})\,dl + 2\dot{\boldsymbol{\omega}} \times \mathbf{r}_0 S + \ddot{\mathbf{r}}_0 S. \tag{3}$$

Here, $\mathbf{r}_0$ is the coordinates of the center of gravity of the domain inside the profile, S is its area, and $\boldsymbol{\omega}$ is its angular velocity. This formula can be simplified since the integral of the vorticity over the entire space vanishes and the vorticity inside the profile rotating with the angular velocity $\boldsymbol{\omega}$ is equal to $2\boldsymbol{\omega}$. Consequently, the increment of circulation is equal to zero: $\mathbf{e}_z \oint (\mathbf{J}_d\mathbf{n})\,dl + 2\dot{\boldsymbol{\omega}} S = 0$. With regard to this equality, the expression (3) takes the form:

$$\frac{\mathbf{F}_p}{\rho} = \mathbf{e}_z \times \oint (\mathbf{r} - \mathbf{r}_0)(\mathbf{J}_d\mathbf{n})\,dl + \ddot{\mathbf{r}}_0 S \tag{4}$$

In axisymmetric flow a similar formula can be represented as follows:

$$\mathbf{F}_p = \pi \mathbf{e}_x \oint r^2 (\mathbf{J}_d\mathbf{n})\,dl + \ddot{\mathbf{r}}_0 \Lambda$$

where $\mathbf{e}_x$ is the unit vector directed along the axis of symmetry and Λ is the body volume.

2. Vorticity transfer in three-dimensional flows

In the case of an arbitrary three-dimensional flow there exists no representation of the vorticity evolution in the form of continuous motion of vortex tubes with conserved circulation. We will show that despite this fact the equation (1) can be written in the divergent form and the concept of vorticity transfer can be defined.

The curl of the vector $\mathbf{Q} = \quad \times \mathbf{V} + \nu \boldsymbol{\nabla} \times \quad$ on the right-hand side of Eq. (2) can be written in the form of divergence of the tensor **T**

$$\nabla\times\mathbf{Q}=\begin{pmatrix}\frac{\partial}{\partial y}Q_z-\frac{\partial}{\partial z}Q_y\\ \frac{\partial}{\partial z}Q_x-\frac{\partial}{\partial x}Q_z\\ \frac{\partial}{\partial x}Q_y-\frac{\partial}{\partial y}Q_x\end{pmatrix}=\nabla\cdot\mathbf{T}=\begin{pmatrix}\frac{\partial}{\partial x}T_{xx}+\frac{\partial}{\partial y}T_{yx}+\frac{\partial}{\partial z}T_{zx}\\ \frac{\partial}{\partial x}T_{xy}+\frac{\partial}{\partial y}T_{yy}+\frac{\partial}{\partial z}T_{zy}\\ \frac{\partial}{\partial x}T_{xz}+\frac{\partial}{\partial y}T_{yz}+\frac{\partial}{\partial z}T_{zz}\end{pmatrix},\quad \mathbf{T}=\begin{pmatrix}0 & -Q_z & Q_y\\ Q_z & 0 & -Q_x\\ -Q_y & Q_x & 0\end{pmatrix}$$

According to the Stokes theorem, the integral of the function $\nabla\cdot\mathbf{T}$ over a volume τ bounded by a closed surface S is equal to the integral of the product $\mathbf{n}\cdot\mathbf{T}$ over the surface S, where $\mathbf{n}$ is the outward normal to the surface S. Consequently,

$$\frac{\partial}{\partial t}\int_{\tau}\boldsymbol{\Omega}d\tau=-\oint_{s}\mathbf{n}\cdot\mathbf{T}ds$$

Hence we can see that the tensor **T** and the vector $\mathbf{n}\cdot\mathbf{T}ds$ on the right-hand side have the meanings of the vorticity transfer tensor and its flux across the elementary area ds, respectively. We can readily verify that

$$\mathbf{n}\cdot\mathbf{T}=\begin{pmatrix}0+n_yQ_z-n_zQ_y\\ -n_xQ_z+0+n_zQ_x\\ n_xQ_y-n_yQ_x+0\end{pmatrix}=\mathbf{n}\times\mathbf{Q}=\mathbf{n}\times(\mathbf{V}\times\quad)+\nu\mathbf{n}\times(\nabla\times\quad)$$

The first and second terms on the right-hand side correspond to the convective and diffusive vorticity fluxes $\mathbf{n}\cdot\mathbf{T}_v=\mathbf{n}\times(\quad\times\mathbf{V})$ and $\mathbf{n}\cdot\mathbf{T}_d=\nu\mathbf{n}\times(\boldsymbol{\nabla}\times\quad)$, respectively.

3. Expression of the pressure force acting on a body

The pressure force $\mathbf{F}_p$ exerted on an individual body is equal to the integral over its surface S:

$$F_p=-\oint_S\mathbf{n}pds, \tag{5}$$

Here, the normal vector **n** is directed outward the body.

We will now demonstrate that the equality holds:

$$-\oint_S\mathbf{n}pds=\frac{1}{2}\oint_S\mathbf{r}\times(\mathbf{n}\times\nabla p)ds. \tag{6}$$

For this purpose, we will use the integral relation $\int_S(\mathbf{n}\times\nabla)\times\mathbf{f}ds=\oint_C d\mathbf{r}\times\mathbf{f}$ known from the vector analysis. This relation holds for a continuous vector-function **f** which has continuous partial derivatives on an open surface S bounded by a contour C. From this equality there apparently follows the vanishing of the integral over the closed surface $\oint_S (\mathbf{n}\times\nabla)\times\mathbf{f}ds = 0$. Setting $\mathbf{f}=\mathbf{r}p$, we obtain

$$\oint_S (\mathbf{n}\times\nabla)\times(\mathbf{r}p)\,ds = 0 \tag{7}$$

Applying the differential operator $(\mathbf{n}\times\nabla)\times$ to the multipliers of the product $\mathbf{r}p$, we can writ

$$(\mathbf{n}\times\nabla)\times(\mathbf{r}p) = p(\mathbf{n}\times\nabla)\times\mathbf{r} - \mathbf{r}\times(\mathbf{n}\times\nabla p) \tag{8}$$

Using the identity $(\mathbf{a}\times\nabla)\times\mathbf{r} = -2\mathbf{a}$, which can readily be verified, developing the double vector product, and differentiating the components of the vector $\mathbf{r}$, we obtain

$$(\mathbf{n}\times\nabla)\times(\mathbf{r}p) = -2p\mathbf{n} - \mathbf{r}\times(\mathbf{n}\times\nabla p).$$

Integrating this equality over the surface S and taking (7) into account, we obtain

$$0 = -2\oint_S p\mathbf{n}ds - \oint_S \mathbf{r}\times(\mathbf{n}\times\nabla p)\,ds.$$

Thus, the equality (6) is proved.

We will now express ∇p from the Navier-Stokes equation $\boldsymbol{\nabla} p = -\rho\frac{D\mathbf{V}}{Dt} - \nu\rho\boldsymbol{\nabla}\times\ \ $. Under the no-slip condition the fluid velocity on the surface is equal to the velocity of the latter $\mathbf{V}_s$; consequently, in this case

$$\mathbf{F}_p = -\frac{\rho}{2}\oint_S \mathbf{r}\times(\mathbf{n}\times\dot{\mathbf{V}}_s)\,ds - \frac{\nu\rho}{2}\circ\oint_S \times(\mathbf{n}\times(\nabla\times\ \ ))\,ds \tag{9}$$

As shown above, the expression $\nu\mathbf{n}\times(\nabla\times\ \ )$ represents the diffusive vorticity flux from the surface. We will denote this flux as follows: $\mathbf{G} = \mathbf{n}\cdot\mathbf{T}_d = \nu\mathbf{n}\times(\nabla\times\ \ )$.

Consequently, *under the no-slip condition on the body surface, the pressure force exerted on a deformable body arbitrarily traveling in a viscous incompressible fluid is equal to*

$$\mathbf{F}_p = \frac{\rho}{2}\oint_S \mathbf{G}\times\mathbf{r}ds - \frac{\rho}{2}\oint_S \mathbf{r}\times(\mathbf{n}\times\dot{\mathbf{V}}_s)\,ds$$

In the case of the rigid body, the velocity of points on its surface can be expressed in terms of the translational and angular velocities

$$\mathbf{V}_s = \dot{\mathbf{r}}_0 + \omega\times(\mathbf{r}-\mathbf{r}_0),$$

where $\mathbf{r}_0$ is the center of gravity of the homogeneous body of equivalent shape

$$\mathbf{r}_0 = \frac{1}{\Lambda}\int_\tau \mathbf{r}d\tau,$$

where Λ is the body volume and τ is the spatial domain inside the body.

The acceleration $\dot{\mathbf{V}}_s$ is equal to

$$\dot{\mathbf{V}}_s = \ddot{\mathbf{r}}_0 + \dot{\omega}\times(\mathbf{r}-\mathbf{r}_0) + \omega\times(\omega\times(\mathbf{r}-\mathbf{r}_0))$$

IOP Conf. Series: Materials Science and Engineering **468** (2018) 012002 doi:10.1088/1757-899X/468/1/012002

Using the Stokes theorem, the first integral on the right-hand side of the equality (9) can be transformed into the volume integral

$$\oint_S \mathbf{r}\times\left(\mathbf{n}\times\dot{\mathbf{V}}_s\right)ds=\int_\tau \mathbf{r}\times\left(\nabla\times\dot{\mathbf{V}}_s\right)d\tau+\int_\tau\left(\dot{\mathbf{V}}_s\times\nabla\right)\times\mathbf{r}d\tau \qquad (10)$$

Using the relations $\nabla\mathbf{r}=3,\ \ (\mathbf{a}\nabla)\mathbf{r}=\mathbf{a},\ \ \nabla(\mathbf{ar})=\mathbf{a},\ \ \nabla\times\mathbf{r}=0,\ \ (\mathbf{a}\times\nabla)\times\mathbf{r}=-2\mathbf{a}$, which hold for any vector **a**, we will transform the integrands on the right-hand side of (10)

$$\nabla\times\dot{\mathbf{V}}_s=\nabla\times\left(\dot{\omega}\times\mathbf{r}+\omega\left(\omega\mathbf{r}\right)-\omega^2\mathbf{r}\right)=\dot{\omega}\left(\nabla\mathbf{r}\right)-\left(\dot{\omega}\nabla\right)\mathbf{r}-\omega\times\nabla\left(\omega\mathbf{r}\right)-\omega^2\nabla\times\mathbf{r}=2\dot{\omega}$$

$$\left(\dot{\mathbf{V}}_s\times\nabla\right)\times\mathbf{r}=-2\dot{\mathbf{V}}_s$$

Substituting these expressions in (10), we obtain

$$\oint_S \mathbf{r}\times\left(\mathbf{n}\times\dot{\mathbf{V}}_s\right)ds=2\int_\tau \mathbf{r}\times\dot{\omega}d\tau-2\int_\tau\dot{\mathbf{V}}_s d\tau=-2\Lambda\dot{\omega}\times\mathbf{r}_0-2\dot{\mathbf{r}}_0\Lambda$$

Consequently, the pressure force exerted on a rigid body under the no-slip condition on its surface is equal to

$$\frac{\mathbf{F}_p}{\rho}=\frac{1}{2}\oint_S\mathbf{G}\times\mathbf{r}ds+\dot{\omega}\times\mathbf{r}_0\Lambda+\ddot{\mathbf{r}}_0\Lambda \qquad (11)$$

In addition, since from the circulation conservation condition it follows that

$$2\dot{\omega}\Lambda+\oint_S\mathbf{G}ds=0,$$

the equality (11) takes the final form:

$$\frac{\mathbf{F}_p}{\rho}=\frac{1}{2}\oint_S\mathbf{G}\times\left(\mathbf{r}-\mathbf{r}_0\right)ds+\ddot{\mathbf{r}}_0\Lambda \qquad (12)$$

In simulating the generated vorticity by vortex loops, the surface integral can be approximated by the sum of the contributions of each loop $\mathbf{F}_i$

$$\frac{\rho}{2}\oint_S\mathbf{G}\times\left(\mathbf{r}-\mathbf{r}_0\right)ds\approx\sum_i\mathbf{F}_i,\quad \mathbf{F}_i=-\frac{\rho s_i\Gamma_i}{\Delta t}\mathbf{n}_i$$

where Γ_i is the circulation of a discrete vortex forming the loop, s_i is its area, and Δt is the step in time.

4. Conclusion

Comparing the formulas (12) and (4), we can see that these formulas have the same structure and physical meaning since both the expression $\mathbf{e}_z\left(\mathbf{J}_d\mathbf{n}\right)$ and the vector **G** represent the diffusive vorticity fluxes from the body surface. The only difference consists in the presence of the factor 1/2 on the right-hand side of (12) related to the space dimension. Thus, for both two-dimensional and arbitrary three-dimensional flows of a viscous incompressible fluid the aerodynamic force can be explicitly expressed in terms of the vorticity flux generated by the surface and the acceleration. We should note that the vorticity flux depends implicitly on the body velocity and acceleration, as well as on the spatial vorticity distribution. In the case of using the vortex methods for calculating the flows, the vorticity flux can be determined from the no-flow condition on the surface. The expressions obtained can also be useful in calculations of flow in the natural coordinates for checking the accuracy of the results obtained.

Referenccs:

[1] Dynnikova G Ya 2001 Forces exerted on a body in an unsteady vortex separation flow of an ideal incompressible fluid *Fluid Dynamics* **36**(2) pp 285

[2] Golovkin M A, Golovkin V A, and Kalyavin V.M 2009 *Vortex Fluid Mechanics* (FIZMATLIT, Moscow) p 264 (in Russian)

[3] Andronov P R, Guvernyuk S V, and Dynnikova G Y 2006 *Vortex Methods for Calculating Unsteady Hydrodynamic Loads* (Lomonosov Moscow State University, Moscow) (in Russian)

[4] Guvernyuk S V, Dynnikova G Y 2007 Modeling the flow past an oscillating airfoil by the method of viscous vortex domains *Fluid Dynamics* vol. 42 no. 1 pp 1–11

[5] Andronov P R, Grigorenko D A, Guvernyuk S V, Dynnikova G Y Numerical simulation of plate autorotation in a viscous fluid flow 2007 *Fluid Dynamics* **42** no. 5 pp 719

[6] Dynnikova G Y 2000 An analog of the Bernoulli and Cauchy-Lagrange integrals for a time-dependent vortex flow of an ideal incompressible fluid *Fluid Dynamics* vol. **35** no. 1 pp 24–32

[7] Shcheglov G A, Ermakov A V 2012 On application of vortex element method for aeroelastic airfoil dynamics simulation *(European Congress on Computational Methods in Applied Sciences and Engineering, e-Book Full Papers)* pp 6050-6063

[8] Marchevskii I K, Ivanova O I 2009 Numerical simulation of wind resonance of a circular profile by means of the vortex element method *Journal of Machinery Manufacture and Reliability* vol. **38** no. 5 pp 420-424

[9] Marchevskii I K, Shcheglov G A 2016 On the dynamic stability of an elastically fixed high-drag airfoil under vertical parametric excitations *Mechanics of Solids* vol. 51 no. 3 pp 349-356

[10] Marchevsky I K, Moreva V S, Puzikova V V 2015 The efficiency comparison of the vortex element method and the immersed boundary method for numerical simulation of airfoil's hydroelastic oscillations *COUPLED PROBLEMS 2015* (*Proceedings of the 6th International Conference on Coupled Problems in Science and Engineering Venice*) pp 800-811

[11] Kaplunov S M, Val'es, N G Samolysov A V et al. 2015 Determining the tube bundle streamlining critical parameters using the numerical experiment method *Thermal Engeneering* vol **62** no 8 pp 593-597 doi:10.1134/S0040601515080029

[12] Andronov P R, Dosaev M Z, Dynnikova G Ya, et al. 2009 Modelling of oscillating wind turbine *Journal of Machinery Manufacture and Reliability* vol **38** no 4 pp 383-387

[13] Andronov P R, Guvernyuk S V, Dynnikova G Y 2017 Free oscillations of spheroids on the elastic spring in a viscous fluid *Journal of Mechanics (SCIREA)* vol. **2** no. 1 pp 1-10

[14] Dynnikova G Ya 2011 *Vortex methods of investigation of non-stationary viscous incompressible flows* (Theses Moscow) p 269 (in Russian)

[15] Dynnikov Y A, Guvernyuk S V, Dynnikova G Ya 2015 Simulation of the flexible body moving in viscous fluid (*Proceedings of the ECCOMAS Thematic Conference on Multibody Dynamics 2015, Multibody Dynamics*) pp 960-967

[16] Golubkin V N, Sizykh G B 1987 Some general properties of plane-parallel viscous flows *Fluid Dynamics* vol. 22 no. 3, pp 479

[17] Brutyan M A Golubkin V N, Krapivskii P L 1988 On the Bernoulli equation for axisymmetric viscous fluid flows *Uchenyie zapiski TsAGI*, XIX, no. **2**, pp 98–100 (in Russian)

[18] Ogami Y, Akamatsu T 1991 Viscous flow simulation using the discrete vortex method – the diffusion velocity method *Computers & Fluids* vol. **19** no 3/4, pp 433-441

Acknowledgment
The work was carried out with financial support from the Federal Target Program of the Russian Federation Ministry of Education and Science (Agreement No. 14.576.21.0079 and Project No. RFMEFI57614X0079).

Footholds admissible areas structure of a two-legged walking robot on an inclined cylinder

Yu F Golubev[1,2] and E V Melkumova[2]

[1]Keldysh Institute of Applied Mathematics, RAS, Moscow, Russia
[2]Lomonosov Moscow State University, Moscow, Russia
E-mail: elena_v_m@mail.ru

Abstract. We consider the walking robot with n legs and m arms, where each leg and arm contacts the surface in a single foothold. Given the motion of the robot's legs and arms with respect to its body, we solve the problem of finding the reaction forces, both analytically and numerically. The first step in solving similar problem belongs to N Y Zhukovsky. We describe the robot motion in terms of the general dynamics theorems, with six different equations of the robot's dynamics from the momentum and angular momentum theorems. In the special case a robot with two legs, the existence of the solution is related to a set of straightforward inequalities. Using numerical simulations we develop the classification of footholds positions for different values of the friction coefficient. This problem equivalent to the problem of curved object grasping by the fingers of the robot-manipulator.

1. Introduction

This paper discussed the problem of robot dynamics on one-side constraint. While the general robot motion on a plane was detail analyzed in [1, 2]. Work [3] discussed model dynamics and control problems. The case of the dynamics on a curved surface is far more complicated. In the special case of a robot with eight legs whose up porting points are restricted to the inner surface of a tube, this problem was considered in [4]. Present work is devouted to the more general case of a robot with two arbitrary supporting points on a rough cylinder and on a curved surface. Equilibrium conditions for a solid on a rough plane was considered in [5]. Walking robot parameters optimization for the motion in tubes was discussed in [6].

2. The problem

Consider the walking robot with n legs and m arms. Each leg and arm contacts the surface in the one foothold. Suppose that robot legs and arms accomplish the desired motion with respect to the body of the robot. Using general dynamics theorems to describe the robot motion, we obtain six different equations for the robot dynamics from the momentum and angular momentum theorems. Among them there are three equations of the body translation with point A and another three describe body rotation about point A. For prescribed motion be realized then reaction in $n+m$ supporting points should satisfy following equations [7]:

$$\sum_i \widetilde{\mathbf{R}}_i = -\widetilde{\mathbf{\Phi}}, \quad \sum_i \tilde{\mathbf{r}}_i \times \widetilde{\mathbf{R}}_i = -\widetilde{\mathbf{M}}, \quad i = 1, \dots, n+m, \tag{1}$$

where $\widetilde{\mathbf{R}}_i$ is the reaction component, $\tilde{\mathbf{r}}_i$ corresponds to the i-th supporting point vector, $\widetilde{\mathbf{\Phi}}$ is the sum of the external active forces plus time derivative of desired momentum, and $\widetilde{\mathbf{M}}$ is the sum of external

Content from this work may be used under the terms of the Creative Commons Attribution 3.0 licence. Any further distribution of this work must maintain attribution to the author(s) and the title of the work, journal citation and DOI.
Published under licence by IOP Publishing Ltd

active forces momentum and time derivative of desired angular momentum with respect to the point O (which is defines as the origin fixed in absolute space). In two vector equations (1), the former corresponds to the momentum of the robot (and is equivalent to three scalar equations when projected onto the basis vectors), while the latter defines the desired change of the angular momentum.

Assuming that $\tilde{\mathbf{\Phi}}$ is orthogonal $\tilde{\mathbf{M}}$, we obtain: [8] that the system $\{\tilde{\mathbf{\Phi}}, \tilde{\mathbf{M}}\}$ can be also used at the point C

$$\tilde{\mathbf{r}}_C \times \tilde{\mathbf{\Phi}} = \tilde{\mathbf{M}}, \qquad \mathbf{r}_C = -\frac{\tilde{\mathbf{M}} \times \tilde{\mathbf{\Phi}}}{\tilde{\mathbf{\Phi}}^2}, \qquad \tilde{\Phi} = |\tilde{\mathbf{\Phi}}|,$$

where $\mathbf{r}_C$ is the vector $\mathbf{OC}$, and C corresponds to point at which the resultant of the reactions is acting. Further problem of reactions distribution $\tilde{\mathbf{R}}_i$ in some fixed point of time is investigated by the proposal that force $\tilde{\mathbf{\Phi}}$ is acting at the point C and force moment there is zero. Motion equations (1) for finding reactions of walking robot body arms and legs prescribed motion can be transformed [11]:

$$\sum_i \tilde{\mathbf{R}}_i = -\tilde{\mathbf{\Phi}}, \quad \sum_i \tilde{\mathbf{r}}_i \times \tilde{\mathbf{R}}_i = -\tilde{\mathbf{r}}_C \times \tilde{\mathbf{\Phi}}. \tag{2}$$

Assuming that the robot footholds are on the surface of a rough cylinder of radius ρ with a friction coefficient k, we introduce the coordinate system $Oxyz$ such that the axis Ox is directed along the cylinder axis (so that the projection of $\tilde{\mathbf{\Phi}}$ on the axis Ox is negative - see Fig 1.), the axis Oz is parallel to the vector $\tilde{\mathbf{\Phi}}$, and the angle between the cylinder axis and the vector $\tilde{\mathbf{\Phi}}$ is α.

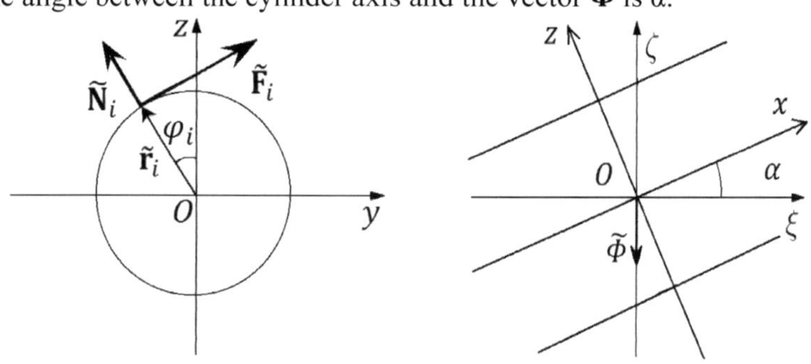

Figure 1. Cylinder.

The problem of finding the reaction forces (2) is similar to the foothold reactions distribution problem in homogeneous gravity field, when the footholds are on the external surface of a rough inclined cylinder where the axis has an angle α with respect to the vertical vector $\tilde{\mathbf{\Phi}}$. It has been considered in Ref [9], where the problem of searching the reactions components along the cylinder axis when α = 0 and the work [10] for arbitrary α was considered.

In the coordinates $Oxyz$ we define $\tilde{\mathbf{r}}_i = (\tilde{x}_i, \tilde{y}_i, \tilde{z}_i)$, $\tilde{\mathbf{R}}_i = (\tilde{R}_i^x, \tilde{R}_i^y, \tilde{R}_i^z)$, $\tilde{\mathbf{r}}_C = (\tilde{x}_C, \tilde{y}_C, \tilde{z}_C)$, and $\tilde{\mathbf{\Phi}} = (-\tilde{\Phi}\sin\alpha, 0, -\tilde{\Phi}\cos\alpha)$. In case of a one-sided surface, we have additional restrictions on normal reactions $\tilde{\mathbf{N}}_i$ [11]:

$$\tilde{N}_i = \tilde{\mathbf{R}}_i \cdot \mathbf{e}_i^\nu \geq 0, \tag{3}$$

where $\mathbf{e}_i^\nu$ is an external normal to i-th supporting point on the cylinder, while the tangential components are given by $\tilde{\mathbf{F}}_i = \tilde{\mathbf{R}}_i - \tilde{N}_i \mathbf{e}_i^\nu$. For the reactions to be in the friction cones (2), we have following inequalities [12]:

$$|\tilde{\mathbf{F}}_i| \leq k\tilde{N}_i, \; i = \overline{1, n+m}, \tag{4}$$

i.e. the tangential reactions $\tilde{\mathbf{F}}_i$ are restricted by Coulomb limiting friction value. When $|\tilde{\mathbf{F}}_i|$ exceeds this limiting value, the robot legs and arms begin to slide along a surface.

The reaction distribution problem then reduces to the solution of equations (2), and inequalities (3), (4), for reactions limited to the friction cones. The restricted motion can only be realized if the solution of (2)-(4) does exist [13].

The problem of finding the reaction forces (2) is the problem with two-sided constraint [14]. And adding by inequalities (4). For example, let n is even. And first n/2 legs are on the cylinder surface, second n/2 legs are on the inner cylinder surface. The equivalent problem is the same grasp with even supporting points.

If the kinematics of the motion is given, the supporting points reactions are not immediately defined, and we can impose additional conditions or restrictions on the system. For example, we can limit the forces acting on the surface (due e.g. the requirement to avoid its damage), which implies a restoration on $\widetilde{N}_i$.

For vector $\boldsymbol{r}_i = \tilde{\mathbf{r}}_i/\rho = (x_i,\ y_i,\ z_i)$, in the cylinder coordinate: $\boldsymbol{r}_i = (x_i, -\sin\varphi_i, \cos\varphi_i)$, $\mathbf{e}_i^y = (0, -\sin\varphi_i, \cos\varphi_i)$, $\boldsymbol{N}_i = \widetilde{\mathbf{N}}_i/\widetilde{\Phi} = (0, -N_i\sin\varphi_i, N_i\cos\varphi_i)$, where φ_i is the angles between axis Oz and cylinder normal $\mathbf{e}_i^y$. We define $\mathbf{e}_x$ as the unitary vector in the Ox axis, while the vector $\mathbf{e}_i^\tau = (0, \cos\varphi_i, \sin\varphi_i)$ as the tangential to the cylinder. Then the tangential reaction component: $\boldsymbol{F}_i = \tilde{\mathbf{F}}_i/\widetilde{\Phi} = \left(F_i^x, F_i^{yz}\cos\varphi_i, F_i^{yz}\sin\varphi_i\right)$, where $F_i^x = \boldsymbol{F}_i\cdot\mathbf{e}_x$, $F_i^{yz} = \boldsymbol{F}_i\cdot\mathbf{e}_i^\tau$, the reaction vector $\boldsymbol{R}_i = \widetilde{\mathbf{R}}_i/\widetilde{\Phi} = \left(R_i^x, R_i^y, R_i^z\right)$, $\boldsymbol{r}_C = \tilde{\mathbf{r}}_C/\rho = (x_C, y_C, z_C)$.

We further define the supporting points the coordinate differences along the cylinder, and angles difference: $\Delta x = x_2 - x_1$, $\Delta\varphi = \varphi_2 - \varphi_1$. We then project system (2) onto the axes $Oxyz$. For arbitrary surface we find that the second equation of (2) (corresponding to the angular moment) has the skew-symmetric matrix with respect to the component R_i^x [10]. These are two independent equations, while the third equation corresponds to the restriction of the point C to the plane containing the two supporting points. As a result, the system (2) yields five independent equation and a restriction:

$$\begin{gathered}
R_1^x = -R_2^x + \sin\alpha,\ R_1^y = -R_2^y,\ R_1^z = -R_2^z + \cos\alpha,\\
(z_1 - z_2)R_1^x + (x_2 - x_1)(R_1^z - \cos\alpha) = (z_C - z_2)\sin\alpha,\\
(y_2 - y_1)R_1^x - (x_2 - x_1)R_1^y = (y_2 - y_C)\sin\alpha,\\
\mathrm{tg}\alpha = \frac{(x_2 - x_1)(\sin\varphi_2 + y_C) + (x_C - x_2)(\sin\varphi_2 - \sin\varphi_1)}{z_C(\sin\varphi_2 - \sin\varphi_1) + \sin(\varphi_1 - \varphi_2) + y_C(\cos\varphi_2 - \cos\varphi_1)}.
\end{gathered} \tag{5}$$

If $p = R_2^x - R_1^x$, then from (5):

$$\begin{gathered}
R_1^x = (p + \sin\alpha)/2, \qquad R_2^x = (\sin\alpha - p)/2,\\
R_1^y = \frac{p(y_2 - y_1)}{2(x_2 - x_1)} + \frac{2(y_C - y_2) + y_2 - y_1}{2(x_2 - x_1)}\sin\alpha,\\
R_2^y = \frac{p(y_1 - y_2)}{2(x_2 - x_1)} + \frac{2(y_2 - y_C) + y_1 - y_2}{2(x_2 - x_1)}\sin\alpha,\\
R_1^x = \frac{2(x_2 - x_C)\cos\alpha + p(z_2 - z_1)}{2(x_2 - x_1)} + \frac{2(z_C - z_2) + z_2 - z_1}{2(x_2 - x_1)}\sin\alpha,\\
R_2^x = \frac{2(x_C - x_1)\cos\alpha - p(z_2 - z_1)}{2(x_2 - x_1)} + \frac{2(z_2 - z_C) + z_1 - z_2}{2(x_2 - x_1)}\sin\alpha,\\
\mathrm{tg}\alpha = \frac{(x_2 - x_1)(\sin\varphi_2 + y_C) + (x_C - x_2)(\sin\varphi_2 - \sin\varphi_1)}{z_C(\sin\varphi_2 - \sin\varphi_1) + \sin(\varphi_1 - \varphi_2) + y_C(\cos\varphi_2 - \cos\varphi_1)}.
\end{gathered} \tag{6}$$

2.1. *Two-legged Robot*

During the robot motion one-supporting and two-supporting points phases are changed. First, we consider the one-supporting phase. Let $n = 1$, then the motion existing condition is reaction is equal to force $\widetilde{\boldsymbol{\Phi}}$ and supporting point and the point C are on one line along force $\widetilde{\boldsymbol{\Phi}}$, while the angle between $\widetilde{\boldsymbol{\Phi}}$ and the normal do not exceed the friction angle.

Let $n = 2$, and $x_1 \neq x_2$. Then $p = F_2^x - F_1^x$, and from (6):

$$F_1^x = (p + \sin\alpha)/2, \qquad F_2^x = (\sin\alpha - p)/2,$$

$$N_1 = \frac{(x_2 - x_C)\cos\varphi_1 - p\sin^2\frac{\varphi_2-\varphi_1}{2}}{x_2 - x_1}\cos\alpha + N_1^{\alpha},$$
$$N_2 = \frac{(x_C - x_1)\cos\varphi_2 - p\sin^2\frac{\varphi_2-\varphi_1}{2}}{x_2 - x_1}\cos\alpha + N_2^{\alpha},$$
$$F_1^{yz} = \frac{2(x_2 - x_C)\sin\varphi_1 - p\sin(\varphi_2-\varphi_1)}{2(x_2 - x_1)}\cos\alpha + F_1^{(yz)\alpha},$$
$$F_2^{yz} = \frac{2(x_C - x_1)\sin\varphi_2 - p\sin(\varphi_2-\varphi_1)}{2(x_2 - x_1)}\cos\alpha + F_2^{(yz)\alpha},$$
$$\mathrm{tg}\alpha = \frac{(x_2 - x_1)(\sin\varphi_2 + y_C) + (x_C - x_2)(\sin\varphi_2 - \sin\varphi_1)}{z_C(\sin\varphi_2 - \sin\varphi_1) + \sin(\varphi_1 - \varphi_2) + y_C(\cos\varphi_2 - \cos\varphi_1)}. \quad (7)$$

where

$$N_1^{\alpha} = \frac{z_C\cos\varphi_1 - y_C\sin\varphi_1 - \cos^2\frac{\varphi_2-\varphi_1}{2}}{x_2 - x_1}\sin\alpha,$$
$$N_2^{\alpha} = \frac{y_C\sin\varphi_2 - z_C\cos\varphi_2 + \cos^2\frac{\varphi_2-\varphi_1}{2}}{x_2 - x_1}\sin\alpha,$$
$$F_1^{(yz)\alpha} = \frac{2(z_C\sin\varphi_1 + y_C\cos\varphi_1) + \sin(\varphi_2-\varphi_1)}{2(x_2 - x_1)}\sin\alpha,$$
$$F_2^{(yz)\alpha} = \frac{\sin(\varphi_2-\varphi_1) - 2(z_C\sin\varphi_2 + y_C\cos\varphi_2)}{2(x_2 - x_1)}\sin\alpha.$$

From the conditions (3)

$$p \le \frac{(x_2 - x_C)\cos\varphi_1\cos\alpha + (x_2 - x_1)N_1^{\alpha}}{\sin^2\frac{\varphi_2-\varphi_1}{2}},$$
$$p \le \frac{(x_C - x_1)\cos\varphi_2\cos\alpha + (x_2 - x_1)N_2^{\alpha}}{\sin^2\frac{\varphi_2-\varphi_1}{2}}. \quad (8)$$

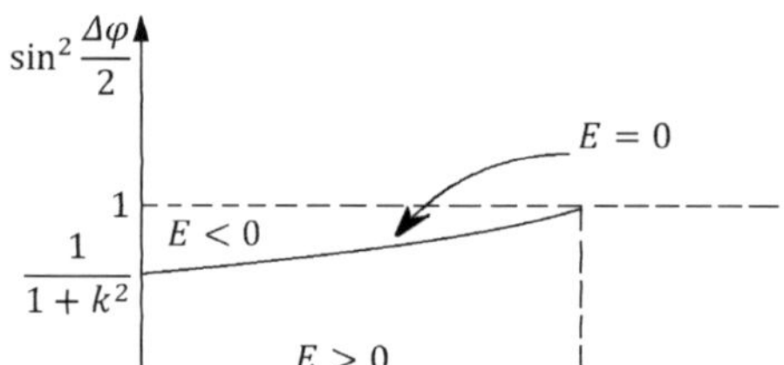

Figure 2. The parameter diagram.

We further define $s\varphi_{21} = \sin\varphi_2 - \sin\varphi_1$, $c\varphi_{21} = \cos\varphi_2 - \cos\varphi_1$. The conditions (4) can be displayed in the form:

$$Ep^2 + B_1p + C_1 \le 0, \qquad Ep^2 + B_2p + C_2 \le 0, \quad (9)$$

where

$$E = \Delta x^2 + \sin^2\Delta\varphi - 4k^2\sin^4\frac{\Delta\varphi}{2},$$

$$B_1 = 4\sin\frac{\Delta\varphi}{2}\left\{x_2\cos\alpha\left[\sin\varphi_1\ \cos\frac{\Delta\varphi}{2} - k^2\cos\varphi_1\sin\frac{\Delta\varphi_2}{2}\right] + B_1^\alpha\sin\Delta\varphi\right\},$$

$$B_2 = 4\sin\frac{\Delta\varphi}{2}\{x_1\cos\alpha\left[\sin\varphi_2\cos\frac{\Delta\varphi}{2} + k^2\cos\varphi_2\sin\frac{\Delta\varphi}{2}\right] + B_2^\alpha\sin\Delta\varphi\},$$

$$C_1 = {x_2}^2[\sin^2\varphi_1 - k^2\cos^2\varphi_1]\cos^2\alpha + (x_2 - x_c)C_{11}^\alpha\sin2\alpha + C_{12}^\alpha\sin\alpha^2,$$

$$C_2 = {x_1}^2[\sin^2\varphi_2 - k^2\cos^2\varphi_2]\cos^2\alpha + (x_c - x_1)C_{21}^\alpha\sin2\alpha + C_{22}^\alpha\sin\alpha^2,$$

$$B_1^\alpha = \cos\frac{\Delta\varphi}{2}(y_c\cos\varphi_1 + z_c\sin\varphi_1) - k^2\sin\frac{\Delta\varphi}{2}(z_c\cos\varphi_1 - y_c\sin\varphi_1 - 1) - 2\Delta x^2,$$

$$B_2^\alpha = \cos\frac{\Delta\varphi}{2}(\sin\Delta\varphi - y_c\cos\varphi_2 - z_c\sin\varphi_2) + k^2\sin\frac{\Delta\varphi}{2}(y_c\sin\varphi_2 - z_c\cos\varphi_2 + \cos\Delta\varphi),$$

$$C_{11}^\alpha = (y_c\cos\varphi_1 + z_c\sin\varphi_1)\sin\varphi_1 - k^2(z_c\cos\varphi_1 - y_c\sin\varphi_1 - 1)\cos\varphi_1,$$

$$C_{12}^\alpha = (y_c\cos\varphi_1 + z_c\sin\varphi_1)^2 - k^2(z_c\cos\varphi_1 - y_c\sin\varphi_1 - 1)^2,$$

$$C_{21}^\alpha = (\sin\Delta\varphi - z_c\sin\varphi_2 - y_c\cos\varphi_2)\sin\varphi_2 - k^2(y_c\sin\varphi_2 - z_c\cos\varphi_2 + \cos\Delta\varphi)\cos\varphi_2,$$

$$C_{22}^\alpha = (\sin\Delta\varphi - z_c\sin\varphi_2 - y_c\cos\varphi_2)^2 - k^2(y_c\sin\varphi_2 - z_c\cos\varphi_2 + \cos\Delta\varphi)^2.$$

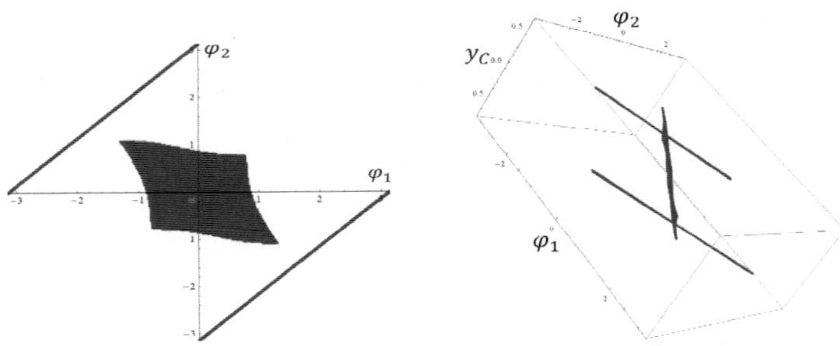

Figure 3. The case of $\alpha = 0$; $x_2 = -x_1 = \rho = k = 1$.

The boundaries between different regimes can be determined analytically. For example, in the case of significant friction when $E < 0$, the solution exists, and can be obtained analytically [10], as shown in Fig 2. Note that in this case it is limited to the range $x_c - x_1 \leq 2k\rho$. In contract to this behavior, for $E \geq 0$ there is no such restriction and an additional step is required to address the question of the existence of the solution. At the point $(0, 0)$ we find: $E = 0$, which means that two supporting points are orthogonal to the cylinder axis. Here, two possible solutions are either identical, or limited to a single diameter. In the latter case, point C and the reaction have to be in one plane, parallel to force $\widetilde{\mathbf{\Phi}}$, and the problem has a solution.

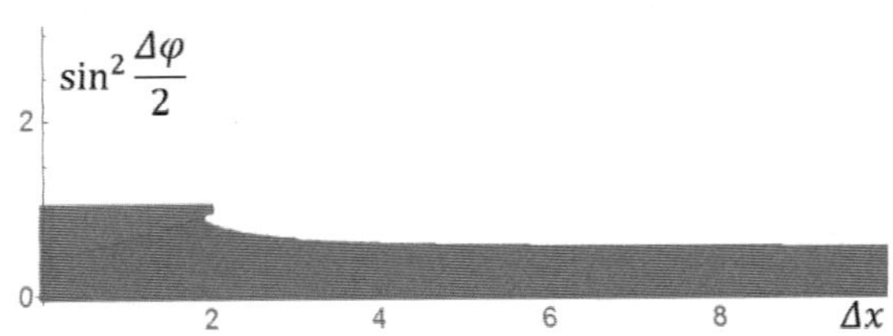

Figure 4. The case of $\boldsymbol{\alpha = 0}$; $\boldsymbol{x_2 = -x_1 = \varphi_2 = -\varphi_1}$.

For the desired legs and arms configurations and given point C, the problem can be solved numerically. In Fig 3, we present the numerical solution for the example when $\alpha = 0$, $x_2 = -x_1 = \rho = k = 1$. Note that in this case $E > 0$.

In the numerical simulations, we use a 300×300 array for the points (φ_1, φ_2), in the interval $[-\pi, \pi]$, and for each point verify the conditions (8), (9). Specifically, the condition (9) was analyzed in two cases, when $E = 0$ and $E > 0$, and when the solution of the problem does exist, the solutions were shown in the plot.

When $E = 0$, the reaction distribution problem reduces to the linear inequalities (8), (9) for the parameter p. To avoid zero division, signs of the coefficients in cycle were verified. For example if $B_1 = 0$, then necessary condition of the reaction distribution problem is $C_1 \leq 0$.

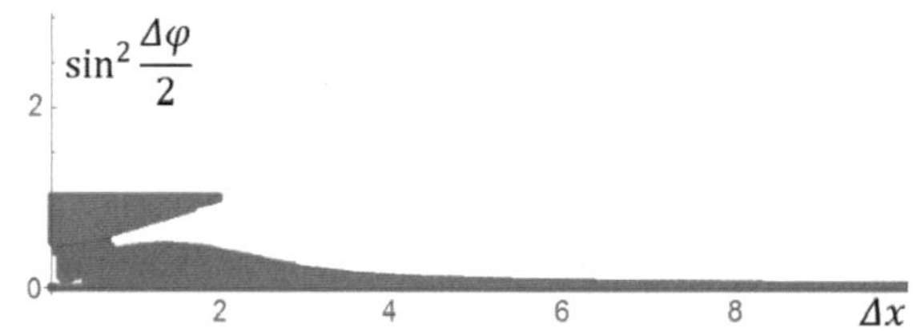

Figure 5. The case of $\boldsymbol{\alpha} = \boldsymbol{\pi/4}$; $\boldsymbol{x_2 = -x_1 = \varphi_2 = -\varphi_1}$.

For $E > 0$, we need to consider two conditions. First is the restriction on the determinants $D_i \geq 0$, while the second is the requirement of a non-empty intersection of the set of points of the intervals between the roots of quadratic equations. If the point (φ_1, φ_2) satisfies the conditions, then it shown on plot that shows the existing field of the reaction distribution problem. From this plot we see that, if two points are on one diameter, then the solution of the reaction distribution problem exists. The two lines in the plot, correspond to $\varphi_1 = \varphi_2 + \pi$ or $\varphi_1 = \varphi_2 - \pi$. The rhombus form represents the requirement on the determinants $D_i \geq 0$, while additional conditions further restrict the range. This is further illustrated in Fig 3, on the right, where the $3D$-plot shows parameter range for the existence of the reactions distribution problem.

In Figs 4, 5 and 6, we present the results for $E > 0$ and $E \leq 0$, when $x_2 = -x_1$, $\varphi_2 = -\varphi_1$. Fig 4 shows the case of $\alpha = 0$, Fig 5 corresponds to $\alpha = \pi/4$, and Fig 6 describes the case when α increased to $\pi/2$. Note that when $\alpha = \pi/2$, the solution exists only for diametrical footholds.

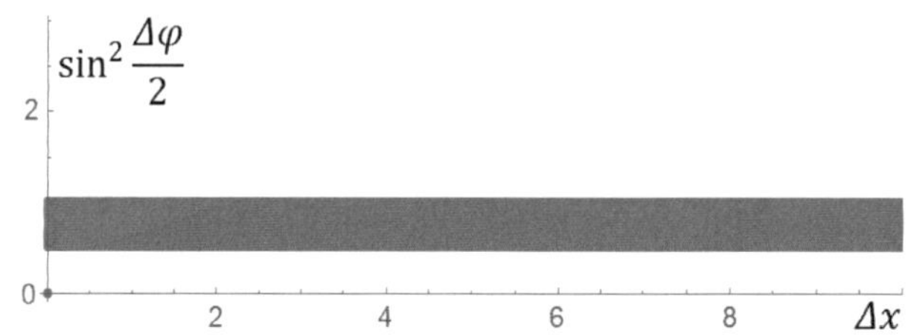

Figure 6. For $\boldsymbol{\alpha}$ increased to $\boldsymbol{\pi/2}$; $\boldsymbol{x_2 = -x_1 = \varphi_2 = -\varphi_1}$.

The boundaries between different regimes can be determined analytically. For two-legged robot when E is negative, the solution exists, and obtained analytically. Using numerical simulations we explain the reaction distribution problem existing and build this problem solution existing fields for given supporting points and point C positions. For example, for two supporting points phase, we consider symmetric, about point C, along and orthogonal cylinder axis, robot configurations. For first of these configurations examined three cases with nonnegative E coefficient, for distance x, between point C and footholds: 0,9; and 1,1 at ρ and k equal 1, α from 0 to π (in all 13 different values cylinder inclination angles). Reactions distribution problem solution existing fields constructed on the two angles plane, corresponds to supporting points projections on the cylinder base and three dimensional fields which supplement this plane by point C z-coordinate altitude. When α equals to 0, x equals to 1, the field consist of three separate situated subregions. On the angle plane each of pair parallel lines

IOP Conf. Series: Materials Science and Engineering **468** (2018) 012003 doi:10.1088/1757-899X/468/1/012003

corresponds to support on the cylinder diameter plane section contained point C. There is connected field between these lines. It contains the line segment corresponding to the angles equality, robot supported above on the line which is parallel to cylinder axis and satisfy force direction deviation restriction. The indicated segment on the plot disappear when x equals to 0,9 for α equals $\pi/4$, and at increasing x, later, for $4\pi/9$. It corresponds to the robot beginning sliding down the cylinder. When x equals to 1,1 for α equals $\pi/3$ in three-dimensional fields observed bundles of separate points, Fig 7. That means that the point C altitude position more harsh change while changing the angles.

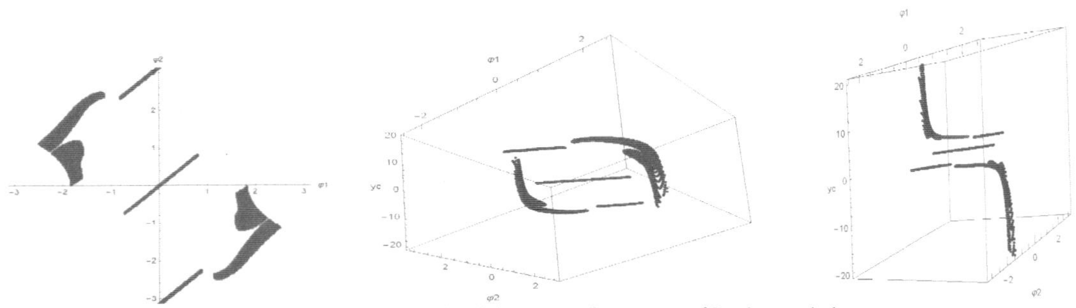

Figure 7. Admissible areas for $\alpha = \pi/3$; $\Delta x = 1{,}1$.

2.1.1. An objects grasping problem. Let consider the problem of curved object grasping by the fingers of the robot-manipulator. An object grasping problem is equivalent to the problem of the walking robot with n legs. Consider a grasp with n fingers. Each finger contacts an object in one point. Let C is the object center of mass.

An object grasp kinetostatic described by general dynamics theorems. The momentum and angular momentum theorems give six differential kinetostatic equations. Among them there are three equations of the body translation with point A and another three describe body rotation about point A. Let the point C is an origin fixed in absolute space. For a grasp of an object be realized then reaction in n contact points should satisfy following equations investigated by [1, 7, 8, 12 − 14]:

$$\sum_i \tilde{\mathbf{R}}_i = -\tilde{\mathbf{P}}, \qquad \sum_i \tilde{\mathbf{r}}_i \times \tilde{\mathbf{R}}_i = -\tilde{\mathbf{r}}_C \times \tilde{\mathbf{P}}, \qquad i = 1, \dots, n, \tag{10}$$

where $\tilde{\mathbf{R}}_i$ is reaction component, $\tilde{\mathbf{r}}_i$ - i-th finger supporting point vector, $\tilde{\mathbf{P}}$ is gravity force, $\tilde{\mathbf{r}}_C \times \tilde{\mathbf{P}}$ is gravity force momentum about point O, $\tilde{\mathbf{r}}_C - \mathbf{OC}$ is the center of mass vector. First vector equations express object momentum. It equivalents to three scalar equations in basic vectors projections. The second equation of (10) defines angular momentum.

We introduce right axes Oξηζ so that an axis Oζ be parallel $\tilde{\mathbf{P}}$.

Suppose a grasp supporting points be on a rough cylinder as for walking robot. The force may be $\tilde{\mathbf{P}}$. Or the up theory for arbitrary force $\tilde{\mathbf{\Phi}}$ and up built point C as for walking robot.

In case of the one-side surface, we have adding restrictions on normal reactions N_1 as [7, 10, 11]. For reactions be in the friction cones (2) we have inequalities (3), (4). Inequalities (3) in case of the grasp inside the object, for example, the cylinder. And opposite sign in inequalities (3) if the grasp is out the object. That is the tangential reactions $\tilde{\mathbf{F}}_i$ are restricting by Coulomb limiting friction value. When $|\tilde{\mathbf{F}}_i|$ became more then limiting value, sliding of the cylinder begun along the grasp.

The reaction distribution problem formulated so. We want to find the reaction corresponding to equations (10), inequalities (3) or opposite sign, and (4), which be in the friction cones.

For two-sided constraint does not matter in which direction. For one-sided constraint the reaction must be on the side permitted for living.

The equations (10) be added by square inequalities, for reactions be in the friction cones. And the properties of the coefficient E are the same as for walking robot in case of the grasp inside the cylinder or an another surface object.
If the solution of (10), and pointed inequalities exists than restricted motion can be realized. If not exists, than can not be realized. This problem equivalent to the problem of a walking robot on the rough cylinder.
For the cylinder the equations (10) are:

$$\sum_i (\boldsymbol{N}_i + \boldsymbol{F}_i^x + \boldsymbol{F}_i^{yz}) = -\boldsymbol{P}, \qquad \sum_i \boldsymbol{r}_i \times (\boldsymbol{N}_i + \boldsymbol{F}_i^x + \boldsymbol{F}_i^{yz}) = -\boldsymbol{r}_C \times \boldsymbol{P}, \qquad i = 1, \dots, n, \tag{11}$$

where $\boldsymbol{F}_i^x = F_i \cdot \mathbf{e}_x$, $\boldsymbol{F}_i^{yz} = F_i \cdot \mathbf{e}_i^{\tau}$.
Then projection on Oxyz the first and the second vector equations (11):

$$\begin{gathered}
\sum_i F_i^x = sin\alpha, \\
\sum_i (N_i sin\varphi_i - F_i^{yz} cos\varphi_i) = 0, \\
\sum_i (N_i cos\varphi_i + F_i^{yz} sin\varphi_i) = cos\alpha, \\
\sum_i F_i^{yz} = -y_C cos\alpha, \\
\sum_i (x_i N_i cos\varphi_i + x_i F_i^{yz} sin\varphi_i - F_i^x cos\varphi_i) = x_C cos\alpha - z_C sin\alpha, \\
\sum_i (x_i N_i sin\varphi_i - x_i F_i^{yz} cos\varphi_i - F_i^x sin\varphi_i) = y_C sin\alpha.
\end{gathered} \tag{12}$$

Robot can hold the surface object by one finger. If the finger inside the cylinder, the center mass of an object is under the finger. If the finger out the cylinder, the center mass of an object is up the finger. And the angle between $\boldsymbol{P}$ and the normal not exceed friction angle.

3. Conclusion

During the robot motion, one-foothold and two-foothold phases are changed. Reaction distribution problem have a solution in following cases.
1. One foothold phases. So the motion existing condition is reaction is equal to force $\widetilde{\boldsymbol{\Phi}}$ and foothold and the point C are on the line along $\widetilde{\boldsymbol{\Phi}}$. And the angle between $\widetilde{\boldsymbol{\Phi}}$ and the normal not exceed friction angle.
2. Two-foothold phases. The point C and the reactions have to be in the plane parallel to force $\widetilde{\boldsymbol{\Phi}}$.
2.1. If footholds are on one diameter.
2.2. When the coefficient

$$E = \Delta x^2 + \sin^2\Delta\varphi - 4k^2\sin^4\frac{\Delta\varphi}{2} < 0.$$

And in some fields with connected set of points, when $E \geq 0$.
So robot can move along the cylinder changing one and two-footholds phases.
3. And for example monkey-robot with 20 fingers or two legged human-robot with 10 arms fingers can hold the object by one and grasp by two-fingers.
Robot can hold the surface object by one finger. If the finger inside the cylinder, the center mass of an object is under the finger. If the finger out the cylinder, the center mass of an object is up the finger. And the angle between $\boldsymbol{P}$ and the normal not exceed friction angle.

References

[1] Okhotsimski D E and Golubev Yu F 1984 *Mechanics and Motion Control of Automatic Locomotion Apparatus* (Moscow: Nauka) (In Russian)

[2] Zhukovsky N 1937 Equilibrium Condition of a Rigid Body Supporting by a Some Place on Unmoving Plane, and can Moving Along this Plane with Friction. *Complete works, (M.-L., ONTI NKTP USSR, The main redaction of aviation literature)* V.**1** pp 433-449.

[3] Beletski V V, *Two-legged Locomotion. Model Dynamics and Control Problems* (Moscow: Nauka) (In Russian)

[4] Pfeiffer F, et al. 1996 Optimization of Structural Parameters and Gates of a Pipe Crawling Robot, *IUTAM Symposium on Optimization of Mechanical Systems*, Bestli D; Schiehlen W, IAF, (Kluwer Academic Publisgers) pp 231-238.

[5] Chernousko F L 1988 Equilibrium conditions for a solid on a rough plane *Izvestiya RAN, series: Solid Mechanics,* № 6, pp 6-17.

[6] Chernousko F L and Bolotnik N N 1995 Walking robot parameters optimization for the motion in tubes *Izvestiya RAN, series: Solid Mechanics*, № **6**, pp 27-41

[7] Golubev Yu F 2000 *The Fundamentals of Theoretical Mechanics* (Moscow State University) (In Russian)

[8] Golubev Yu F and Melkumova E V 1999 Static-Stability Conditions for a Walking Apparatus in Horizontal Cylinder and on Two Planes *Journal of Computer and Systems Sciences International. Robotics* (Pleiades Publishing, Ltd) volume **38** № 2 pp 278-284. (Translated from Izvestiya Akademii Nauk. Teoriya i Sistemy Upravleniya, № **2**, pp 116-122, 1999.)

[9] Golubev Yu F and Melkumova E V 2000 On Stability of Equilibrium Positions for Walking Robot in a Smooth Horizontal Tube. *Proc. of the Third International Conference on Climbing and Walking Robots*, Madrid, Spain, pp 433-440.

[10] Golubev Yu F and Melkumova E V 2010 Equilibrium Problem for a Two-Legged Robot on Rough Horizontal Cylinder Taking into Account the Reaction Components Along the Cylinder Axis (Moscow, Russia: MAKS Press) (In Russian)

[11] Golubev Yu F and Melkumova E V 2010 Trace Point Choice for a Biped Robot Motion on Rough Horizontal Cylinder. *Proceedings of the XI International Conference "Stability and Oscillations of Nonlinear Control Systems" (Pyatnitskiy's Conference)*, Moscow, Russia, pp 95-96, (1-4 of June 2010) (In Russian)

[12] Golubev Yu F and Melkumova E V 2012 Walking Robot Dynamics on a Rough Inclined Cylinder *ESMC-2012 - 8th European Solid Mechanics Conference*, Graz, Austria, July 9-13, 2012, Book of Abstracts, Editors: Holzapfel G A and Ogden R W.

[13] Golubev Yu F and Melkumova E V 2016 Walking Robot Prescribed Motion Existence on Tube Support *VII All-Russian Theoretical Mechanics Conference-Seminar, Robotics, Mechatronics Universities Departments Heads and Lecturers of the Russian Federation. // Conference materials. Under Samsonov V A edition.* (Makhachkala.: Publishing center "Master" (Didkovskaya N V)) pp 30-33. (In Russian.)

[14] Golubev Yu F and Melkumova E V 2017 Footholds Admissible Areas Structure Properties for a Two-legged Walking Robot on an Inclined Cylinder. *Proceeding of the Lomonosov Reading, 17-26 April, 2017 Mechanics Section. MSU*, pp 64-65. (In Russian.)

Acknowledgments
This work was supported by the Russian Foundation for Basic Research
(grant no. $16-01-00131\,A$).

IOP Conf. Series: Materials Science and Engineering **468** (2018) 012004 doi:10.1088/1757-899X/468/1/012004

Confidence intervals for output variable in fuzzy linear regression model

O M Poleshchuk

Bauman Moscow State Technical University, Moscow, Russia
E-mail: olga.m.pol@yandex.ru

Abstract. An analogue of confidence intervals with a given level of possibility for output variable in a fuzzy linear regression model has been developed in the paper. The methods of fuzzy regression analysis extend the methods of classical regression analysis and allow to solve different problems in conditions of fuzzy and incomplete initial information without the limits of the probabilities methods. The developed approach opens up new opportunities for predicting of output fuzzy variable.

1. Introduction

The general problem of studies dependencies within the framework of classical regression analysis can be formulated as follows: based on the values $x_{i1},...x_{ik}, y_i, i=1,n$ of variables $X_1,...X_k,Y$, construct a function $f(X_1,..,X_k)$ that would restore (in a certain sense) the values of the variable Y with respect to the values of the variables $X_1,...X_k$. In the classical linear regression model, the model can be written in the general form as:

$$y_i = a_1 x_{11} + ... + a_k x_{ik} + b + \varepsilon_i, i = \overline{1,n}.$$

A special case of this model is a linear regression model for two variables (sample regression model):

$$y_i = ax_i + b + \varepsilon_i,$$

where x_i - non-random variables, y_i, ε_i - random variables, ε_i - regression errors, standard normal random variables, the vector $(x_1, x_2, ..., x_n)$ is not collinear to the vector $(1,1,...,1)$, $M(\varepsilon_i)=0, D(\varepsilon_i)=\sigma^2$, $i=\overline{1,n}$, $M(\varepsilon_i,\varepsilon_j)=0, i \neq j$.

According to the theorem Gauss – Markov [1], under these assumptions, unknown model parameters, obtained by the least squares method, are effective, that is, have the smallest variance in the class of all unbiased estimates.

To analyze relations between qualitative characteristics and the prediction of their values the methods of fuzzy regression analysis are used, which being actively developed have already considerably expanded boundaries of application of classical regression analysis methods, i.e. they allow to construct the regression relations on the basis of fuzzy initial information. Besides, this information can be of both quantitative and qualitative nature, thus, making possible application of methods of fuzzy regression analysis in the theory of expert evaluations and ensuring practical applications in various spheres of human activity [2, 3].

Methods of fuzzy regression analysis are used to study behavior of complex engineering, ecological and other systems with output indexes depending on a great many of parameters [2]. These methods

Content from this work may be used under the terms of the Creative Commons Attribution 3.0 licence. Any further distribution of this work must maintain attribution to the author(s) and the title of the work, journal citation and DOI.
Published under licence by IOP Publishing Ltd

are applied to construct regression models not only within the limits of the fuzzy initial information, but also within the limits of the definite information. In this case the predicted output values are provided as fuzzy numbers. Such representation is explained by the fact that the real system is always more complex than any of its model not capable of combining all input variables on which the output variable depends.

The first fuzzy linear regression model [4] excited interest in contributors, thus resulting in occurrence of new fuzzy regression models based on various optimizing criteria. Today, a number of linear fuzzy regression models [5-11] is developed, and approaches to building of nonlinear fuzzy regression models [2] are outlined.

There appears to be three different approaches under the heading of «Fuzzy Regression»:

(a) Methods that were proposed by H. Tanaka [4] and further elaborated in current literature, where the coefficients of input variables are assumed to be fuzzy numbers. These fuzzy regression models are based on the possibility theory instead of the probability theory or they are based on both possibility and probability theories.

(b) Method proposed by R.J. Hathaway and J.C. Bezdek [5] where first the fuzzy clusters determined by an fuzzy c-means clustering algorithm define how many ordinary regressions are to be constructed, one for each cluster. Next each fuzzy cluster is used essentially for switching purposes to determine the most appropriate ordinary regression that is to be applied for a new input from amongst a number of ordinary regressions determined in the first place.

(c) Methods proposed by I.B.Turksen, A. Celikyilmaz, where the fuzzy functions approach to system modeling was developed. The new fuzzy functions approach augments the membership values together with their transformations to form a new input variable to find local functions. First the given system domain is fuzzy partitioned into c clusters using fuzzy c-means clustering algorithm. Then, one regression function is calculated to model the behavior of each partition. In [9] linear regression function to estimate the parameters of each function is proposed. A new fuzzy system modeling approach that identifies the fuzzy functions using support vector machines is proposed in [10]. This new approach is structurally different from the fuzzy rule base approaches and fuzzy regression methods. Method support vector machines is applied to determine the support vectors for each fuzzy cluster obtained by fuzzy c-means clustering algorithm. Original input variables, the membership values obtained from the fuzzy c-means clustering algorithm together with their transformations form a new augmented set of input variables.

A major difference between fuzzy regression and classical regression is in dealing with errors as fuzzy variables in fuzzy regression modeling, and in dealing with errors as random variables in classical regression modeling. The researchers have tried to integrate both fuzziness and randomness into regression model. As a result of this the hybrid fuzzy least-squares regressions were developed [2, 3, 11].

The problem of confidence intervals constructing for the output variable in fuzzy regression models was practically not considered, therefore the research topic is new and relevant.

2. Necessary information of fuzzy sets theory

According to [12], the set of pairs $\{(x, \mu_{\tilde{A}}(x)) : x \in X\}$ is referred to as a fuzzy set $\tilde{A}$, where $\mu_{\tilde{A}}(x) : X \to [0,1]$. Function definition area of $\mu_{\tilde{A}}(x)$ is referred to as universal set X of fuzzy set $\tilde{A}$.

A subset of universal set, defined as

$$A_\alpha = \{x \in X : \mu_{\tilde{A}}(x) \geq \alpha\}, \alpha \in [0,1]$$

is referred to as set of α - cut (α -level) of fuzzy set $\tilde{A}$ with membership function $\mu_{\tilde{A}}(x)$.

Fuzzy number $\tilde{A}$ is referred to as fuzzy subset of set of real numbers R possessing membership function

$$\mu_{\tilde{A}} : R \to [0,1].$$

The subset $S_A \subset R$ is referred to as the support of fuzzy number $\tilde{A}$, if

$$S_A = \sup p\, A = \{x : \mu_{\tilde{A}}(x) > 0\}.$$

Fuzzy number $\tilde{A}$ is referred to as positive, if $\forall x \in S_A, x > 0$, and negative, if $\forall x \in S_A, x < 0$.
A subset of real line R, defined as

$$A_\alpha = \{x \in R : \mu_{\tilde{A}}(x) \geq \alpha\}, \alpha \in [0,1]$$

is referred to as set of α - cut (α -level) of fuzzy number $\tilde{A}$ with membership function $\mu_{\tilde{A}}(x)$.
Fuzzy number $\tilde{A}$ with membership function $\mu_{\tilde{A}}(x)$ is referred to as normal, if

$$\max_x \mu_{\tilde{A}}(x) = 1, x \in R.$$

A normal triangular fuzzy number $\tilde{A} = (a, a_L, a_R)$ is fuzzy number with membership function

$$\mu_{\tilde{A}}(x) = \begin{cases} 1 - \dfrac{a-x}{a_L}, \ 0 < \dfrac{a-x}{a_L} \leq 1, \ a_L > 0 \\ 1 - \dfrac{x-a}{a_R}, \ 0 < \dfrac{x-a}{a_R} \leq 1, \ a_R > 0 \\ 1, \ x = a \\ 0, \ x < a - a_L \ \text{или} \ x > a + a_R \end{cases}.$$

In that case $A_\alpha = [A_\alpha^1, A_\alpha^2] = [a - (1-\alpha)a_L, a + (1-\alpha)a_R], \alpha \in [0,1]$.
The definition of weighted point A for a normal triangular number $\tilde{A} = (a, a_L, a_R)$ was given in [13]:

$$A = \frac{\int_0^1 \left(\dfrac{A_\alpha^1 + A_\alpha^2}{2} \right) 2\alpha d\alpha}{\int_0^1 2\alpha d\alpha} = \int_0^1 (A_\alpha^1 + A_\alpha^2)\alpha d\alpha = \int_0^1 (a - (1-\alpha)a_L + a + (1-\alpha)a_r)\alpha d\alpha = a + \frac{1}{6}(a_R - a_L).$$

The weighted interval $[A_1, A_2]$ for the normal triangular number $\tilde{A} = (a, a_L, a_R)$ was given in [13], such as $A_1 = a_1 - \frac{1}{6}a_L$, $A_2 = a_2 + \frac{1}{6}a_R$

3. Fuzzy linear regression model

Let us consider a problem of fuzzy linear regression construction. Relation between input and output data $x_i, y_i, i = \overline{1,n}$ of variables X, Y is determined as

$$Y = \tilde{a}X + \tilde{b},$$

where $\tilde{a} \equiv (a, a_L, a_R)$, $\tilde{b} \equiv (b, b_L, b_R)$ are unknown coefficients of the regression model, which are determined as normal triangular fuzzy numbers.
The method of regression's creation is based on the transformation of the input and output fuzzy numbers into weighted intervals.

If $x_i \geq 0$, then the weighted interval of $\tilde{a}x_i + \tilde{b}$ has the left border equals to $x_i\left(a - \frac{1}{6}a_L\right) + \left(b - \frac{1}{6}b_L\right)$ and has the right border equals to $x_i\left(a + \frac{1}{6}a_R\right) + \left(b + \frac{1}{6}b_R\right)$. If $x_i < 0$, then the weighted interval of $\tilde{a}x_i + \tilde{b}$ has the left border equals to $x_i\left(a + \frac{1}{6}a_R\right) + \left(b - \frac{1}{6}b_L\right)$ and has the right border equals to $x_i\left(a - \frac{1}{6}a_L\right) + \left(b + \frac{1}{6}b_R\right)$.

Let us define an affinity measure for two numbers $\tilde{A}, \tilde{B}$ with the weighed intervals $[A_1, A_2]$, $[B_1, B_2]$:

$$f(\tilde{A}, \tilde{B}) = \sqrt{(A_1 - B_1)^2 + (A_2 - B_2)^2}.$$

Let us consider a functional $F = \sum_{i=1}^{n} f^2(Y_i, \tilde{Y}_i)$, which characterizes an affinity measure between initial Y_i and model output data $\tilde{Y}_i$.

The optimization problem is set as follows:

$$F(a, a_L, a_R, b, b_L, b_R) = \sum_{i=1}^{n} f^2(Y_i, \tilde{Y}_i) \to \min,$$

$$a_L \geq 0, a_{R,} \geq 0, b_L \geq 0, b_R \geq 0.$$

If $x_i \geq 0$, then $f^2(Y_i, \tilde{Y}_i)$ equals to

$$\left(y_i - x_i\left(a - \frac{1}{6}a_L\right) - \left(b - \frac{1}{6}b_L\right)\right)^2 + \left(y_i - x_i\left(a + \frac{1}{6}a_R\right) - \left(b + \frac{1}{6}b_R\right)\right)^2, \ i = \overline{1, n}.$$

If $x_i < 0$, then $f^2(Y_i, \tilde{Y}_i)$ equals to

$$\left(y_i - x_i\left(a + \frac{1}{6}a_R\right) - \left(b - \frac{1}{6}b_L\right)\right)^2 + \left(y_i - x_i\left(a - \frac{1}{6}a_L\right) - \left(b + \frac{1}{6}b_R\right)\right)^2, \ i = \overline{1, n}.$$

Since sum of $f^2(Y_i, \tilde{Y}_i)$ is piecewise linear functions in the field $a_L \geq 0, a_R \geq 0, b_L \geq 0, b_R \geq 0$, then F is piecewise differentiable function, and solutions of an optimization problem are determined by means of known methods [14].

4. Confidence intervals for output variable in linear regression models

Let us consider a statistic $\frac{Y - (a + bX)}{\sigma_y}$, that belongs to the standard normal distribution with known σ_y, a, b. In that case we would use Laplace function for prediction output variable Y in the form of confidence intervals. Since σ_y, a, b are unknown, we will use their estimates and in particular

$$S_y^2 = \sigma^2\left(\frac{1}{n} + \frac{(X - \overline{X})^2}{\sum_{i=1}^{n}(X_i - \overline{X})^2}\right).$$

Accordingly for constructing of confidence intervals we will use Student distribution.

Since σ^2 is unknown, we will use their estimate $S^2 = \frac{1}{n-2}\sum_{i=1}^{n}\varepsilon_i^2$, then

$$S_y^2 = S^2\left(\frac{1}{n} + \frac{(X - \overline{X})^2}{\sum_{i=1}^{n}(X_i - \overline{X})^2}\right).$$

IOP Conf. Series: Materials Science and Engineering **468** (2018) 012004 doi:10.1088/1757-899X/468/1/012004

Then random variable $\frac{Y-(a+bX)}{S_y}$ has Student distribution t_{n-2} with the number of degree of freedom - $n-2$, $P(|t_{n-2}| > t_{n-2,\alpha}) = \alpha$. Then:

$$a + bX - t_{n-2,\alpha} S_y \le Y \le a + bX + t_{n-2,\alpha} S_y .$$

The developed in [13] method of regression's creation, based on the transformation of the input and output fuzzy numbers into intervals, allows to construct an analogue of confidence intervals for output variable in a fuzzy linear regression model. It should be emphasized that this does not require knowledge of the distribution of regression errors and any conditions on the initial information.

From the perspective of the possibilities theory, the membership function of a fuzzy number is a function of the possibilities distribution, that a fuzzy number takes one or another of its values. If $\mu_{\tilde{A}}(x) \ge 1-\alpha$, then $a - \alpha a_L \le x \le a + \alpha a_R$ with the possibility greater or equal to $1-\alpha$.

Based on this, a confidence interval for the output variable with the possibility $\mu_{\tilde{Y}}(x) \ge 1-\alpha$ can be determined as follows:

$$x(a-\alpha a_L) + b - \alpha b_L \le y \le x(a+\alpha a_R) + b + \alpha b_R \quad \text{with} \quad x \ge 0,$$
$$x(a+\alpha a_R) + b - \alpha b_L \le y \le x(a-\alpha a_L) + b + \alpha b_R \quad \text{with} \quad x < 0.$$

Conclusion

An analogue of confidence intervals for output variable in a fuzzy linear regression model is developed in this paper. The methods of fuzzy regression analysis extend the possibilities of classical regression analysis as these methods allow to fit a model to fuzzy data, crisp data, and their mixture.

The problem of confidence intervals constructing for the output variable in fuzzy regression models was practically not considered, therefore the research topic is new and relevant. It should be emphasized that developed constructing does not require knowledge of the distribution of regression errors and does not require any conditions on the initial information. Based on this, an analogue of confidence intervals with a given level of possibility for output variable in a fuzzy linear regression model has been developed in the paper. The developed approach opens up new opportunities for predicting of output fuzzy variable.

References

[1] Gnedenko B V 1988 *Probability theory* (M: Nauka) p 448.

[2] Poleshchuk O, Komarov E 2011 *Expert fuzzy information processing*, Studies in Fuzziness and Soft Computing, vol. **268**. pp 1-239.

[3] Poleshuk O M, Komarov E G 2008 Multiple hybrid regression for fuzzy observed data, *Annual Conference of the North American Fuzzy Information Processing Society - NAFIPS - 2008*. p 4531224.

[4] H Tanaka, S Uejima, K Asai 1982 Linear regression analysis with fuzzy model, *IEEE. Systems, Trans. Systems Man Cybernet. - SMC-2,* pp 903-907.

[5] R J Hathaway, J C Bezdek 1993 Switching regression models and fuzzy clustering, *IEEE Transactions on fuzzy systems,* vol. 1, № **3**, pp 195-203.

[6] D A Sabic, W Pedrycr 1991 Evaluation on fuzzy linear regression models, *Fuzzy Sets and Systems*, vol. **39**, pp 51 - 63.

[7] Y-H O Chang 2001 Hybrid fuzzy least-squares regression analysis and its reliabity measure, *Fuzzy Sets and Systems*, vol. **119**, pp 225-246.

[8] Y-H O Chang, B M Ayyub 2001 Fuzzy regression methods – a comparative assessment, *Fuzzy Sets and Systems*, vol. **119**, pp 187-203.

[9] I.B. Turksen 2008 Fuzzy functions with LSE, *Applied Soft Computing*, vol. 8 № **3**, pp 1178-

1188.
[10] A. Celikyilmaz, I.B. Turksen 2007 Fuzzy functions with support vector machines, *Information Sciences*, vol. **177**, pp 5163–5177.
[11] Poleshchuk, O.M., Komarov, E.G., Darwish, A. 2016 A fuzzy linear regression model with interval type-2 fuzzy coefficients, *Proceedings of the 19th International Conference on Soft Computing and Measurements, SCM - 2016*, pp 388-391.
[12] L.A. Zadeh 1975 The Concept of a linguistic variable and its application to approximate reasoning, *Information Sciences*, vol. **8** pp 199-249.
[13] Poleshuk O. M., Komarov E. G. 2008 New defuzzification method based on weighted intervals, *Annual Conference of the North American Fuzzy Information Processing Society – NAFIPS 2008*, p. 4531223.
[14] T .F. Coleman, Y. Li 1996 A reflective newton method for minimizing a quadratic function subject to bounds on some of the variables, *SIAM J. Optim*, vol. 6, pp 1040-1058.

Continental drift of the rotation pole: observation and theory

C Bizouard

Paris Observatory, SYRTE and IERS Earth Orientation Center, Paris, France
E-mail: christian.bizouard@obspm.fr

Abstract. The Earth rotation pole moves with respect to the crust. Along with a nearly seasonal wobble of about 0.2" and other sub-seasonal oscillations of smaller amplitude, it drifts at longer periods. In light of astro-geodetic observations going back to 1890, this drift is composed of a linear time varying term and of a multi-decadal wobble. Linear trend is interpreted as the visco-elastic response of the Earth to the last deglaciation, however tectonic processes with mantle downwelling or upwelling could also account for a part of it. In this respect, it provides constrains on law mantle viscosity. According to last studies, decadal change of the mean pole path mostly reflects climatic changes through variation in the continental ice coverage.

1. Introduction

Owing to the flattening of the Earth of and its diurnal rotation, the rotation pole should freely wobble with respect to the Earth at a period of 304 days. This prediction was made by Euler in the middle of eighteen century by widening Newtonian mechanics to an extended rigid body. The motion of the rotation pole was searched for as latitude variations. Indeed, if rotation axis moves with respect to continents, the astronomical colatitude, namely the angle between the rotation axis and the local vertical, should change accordingly. This phenomena was finally discovered at the end of the nineteen century, but not with the expected period of 303 days. American astronomer Chandler put forward a cycle of about 0.3" or 300 milli-arc-seconds (mas) in 430 days, accompanied by an annual oscillation of smaller amplitude. The alternative destructive and constructive interference between these two modes produces an amplitude modulation of 6.4 years, appearing in time series of the pole coordinates x_p and y_p displayed in Figure 1. Pole coordinates x_p and y_p are the cosine director of the rotation axis with respect to the axes Gx and Gy of the terrestrial frame $Gxyz$ respectively. The terrestrial frame is a crust tied Cartesian system, which the center is the center of mass of the Earth, and in which the crust presents only deformations. The Gz axis defines the geographic axis, Gx and Gy axes define the geographic equator with Gx lying in the prime meridian (the Greenwich meridian), and Gy at 90°East. At this time, astrometric pointing had a precision of $0.03''$. According to the contemporaneous interpretation, worked out by Newcomb (1992) and other prominent scientists of this epoch like Zhukovsky[1], the Chandler wobble is the free wobble, which the period is lengthened by the non-rigid feed-back of the Earth to the centrifugal variation generated by the wobble itself. Especially it reflects the quasi-elasticity of the mantle and the hydro-static deformation of the ocean at the level of 1 cm. In reason of the interest raised by this phenomenon, its implication for Earth

[1]During conference, Pr. VyacheslavIvashkin let us aware of the Zhukovsky contribution [6] and offered us a copy of it in Keldish Institute

Content from this work may be used under the terms of the Creative Commons Attribution 3.0 licence. Any further distribution of this work must maintain attribution to the author(s) and the title of the work, journal citation and DOI.
Published under licence by IOP Publishing Ltd

IOP Conf. Series: Materials Science and Engineering **468** (2018) 012005 doi:10.1088/1757-899X/468/1/012005

rheology, and possibly internal structure, observations of the astronomical latitude started to be coordinated in 1900 trough a set of 5 observatories, all located at about the same latitude ($39^{\circ}8'$ N). The International Latitude Service (ILS) was born.

Meanwhile the rotation pole lost its status of fixed point with respect to the continents, and to which the latitudes were reckoned. In this respect astronomers and geodesists defined the geographic pole as the mean pole of the epoch 1900-1906, free of its woobble. In practice, the mean pole was obtained by estimating the mean latitudes of ILS stations over a time span of 5-6 years. Yet, knowledge of this epoch was restricted to annual and 430 day wobbles, that had been related to simultaneous mass-redistribution, like the one taking place in the oceans and the atmosphere at seasonal scale. But the polar drift was still unknown.

2. Discovery of rotation pole drift

The drift of the mean pole was discovered in 1922 by American astronomer Walter D. Lambert (1879-1968) [2]. It was confirmed in the 1960's by Markowitz, E. Fedorov and Proverbio, who had longer time series at hand. So, the rotation axis slowly drifts with respect to the crust at an angular rate of 4 mas/year (equivalently the rotation pole moves by about 12 cm/year) towards Greenland. Actually, over some decades this drift is not uniform: the rate varies between 3 and 4 mas/year and longitude lies between $70^{\circ}$and 80°West (see Figure 2). Before the advent of space geodetic technique in the 1960's, this variability cannot be attributed to physical factor only, but it partly stems from the reduction of optical astrometric observation. Indeed, the latitude variation are relying on local verticals, which the direction can vary in the terrestrial frame because of miscellaneous mass transports within the Earth or at its surface in an unknown way. So, latitude variations embody the uncertainty on the direction of vertical, and this uncertainty propagate into the motion of the rotation axis in the terrestrial frame.

Until the 1960's the precision of optical determination did not increase much (30 mas). The outbreak of space geodetic techniques was accompanied by a drop of the EOP uncertainty by a factor 10 over the period 1965-1985. From 1965 the Doppler satellite tracking permitted to divide the uncertainty by a factor 2, and yielded the pole coordinates with a precision of about 50 cm (15 mas). However this technique was surpassed from 1975 by Satellite Laser Ranging (SLR), and from 1985 the Very Long Baseline radio Interferometry (VLBI) on extragalactic radio sources gave the pole coordinates with an uncertainty of about 2 mas. Since the middle of the 1980's, the optical astrometry is almost no longer used. As latitude determination fade in polar motion observations, ILS was replaced by International Polar Motion Service (IPMS) in 1967. In 1987 IPMS and Bureau International de l'Heure (BIH) were merged into the "International Earth rotation and Reference System Service" (IERS). From 1993 the Global Navigation Satellite Systems had become gradually the best technique for determining the pole coordinates, which the present accuracy is about 0.05 mas.

As the external torque that the Moon and the Sun exert on the equatorial bulge do not explain anyhow the polar drift, its cause lies in mass redistributions. A better understanding of the problem requires the lineaments of the polar motion theory, starting from a rigid Earth.

3. Dynamical Euler Equation for a rigid Earth

As Euler did, let us consider a rigid Earth and study the rotation of the principal inertia axes frame $Gxyz$, in which the inertia matrix reads

$$\underline{I} = \begin{bmatrix} A & 0 & 0 \\ 0 & B & 0 \\ 0 & 0 & C \end{bmatrix}, \quad (1)$$

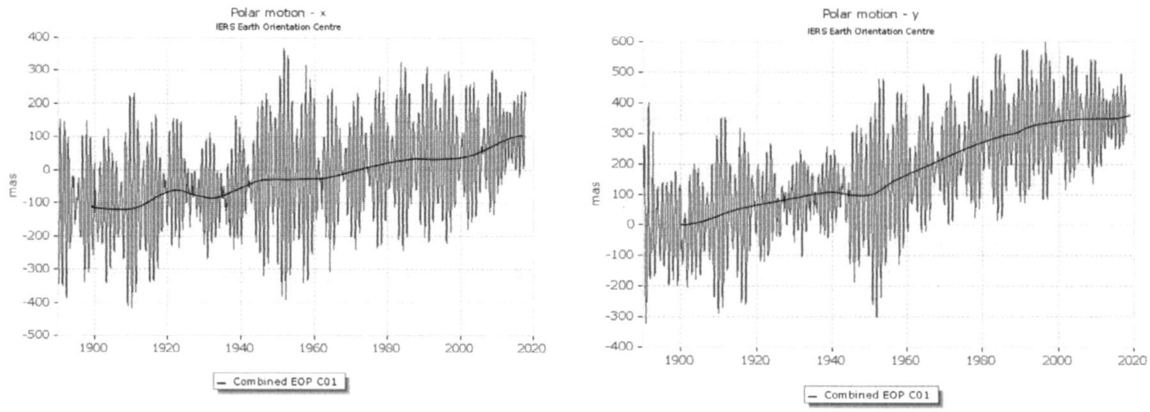

Figure 1. Time series of the pole coordinates $x = x_p$ and $y = y_p$ from 1890 to nowadays. Source: **http://iers.obspm.fr/eop-pc**

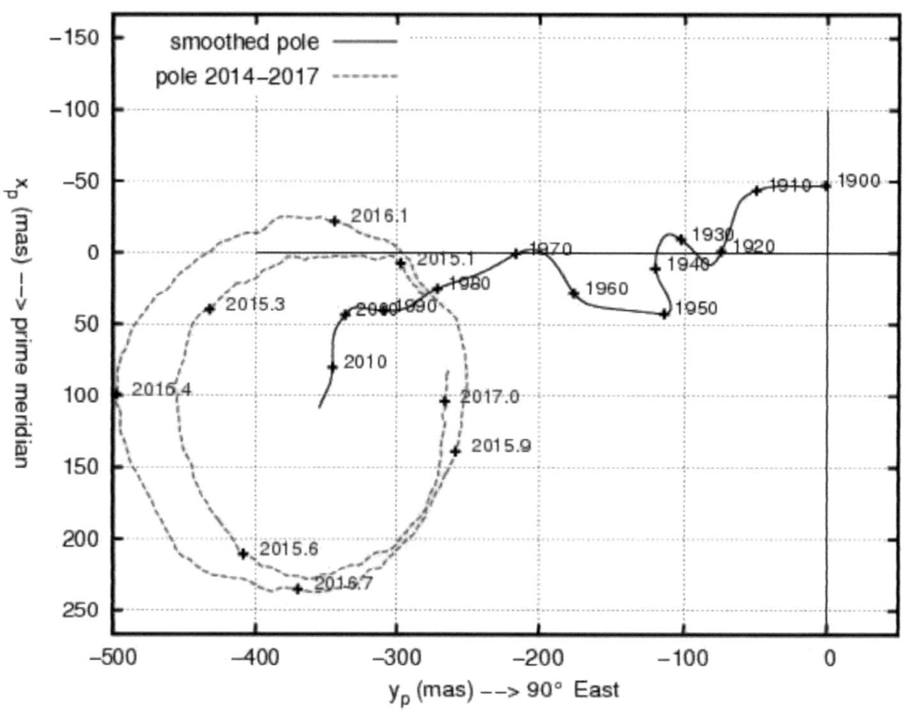

Figure 2. Polar motion over 2015-2017 superimposed to the filtered polar motion from 1900 (cut-off period of 10 years).

where C is the largest moment of inertia and $B = A$(biaxial or axi-symmetric Earth). The relative difference between axial and equatorial moment of inertia define the dynamical ellipticity
$e = (C - A) * A^{-1}$ =1/304.5.
In this geocentric frame, rotating with respect to the celestial system with the instantaneous rotation vector $\vec{\omega}$, the angular momentum balance reads

$$\frac{d\underline{\mathrm{H}}}{dt} + \underline{\omega} \wedge \underline{\mathrm{H}} = \underline{\mathrm{L}}, \tag{2}$$

where $\underline{\mathrm{H}}, \underline{\omega}, \underline{\mathrm{L}}$are column matrices of the components of the corresponding vectors in $Gxyz$. As $\underline{\mathrm{H}} = \underline{\bar{I}\omega}$, we obtain

$$\begin{aligned} A\dot{\omega}_x - (C-A)\omega_y\omega_z &= L_x \\ A\dot{\omega}_y + (C-A)\omega_x\omega_z &= L_y. \\ C\dot{\omega}_z &= L_z \end{aligned} \tag{3}$$

The free solution reads

$$\begin{aligned} \omega_z &= Constant = \Omega \\ \omega_x + i\omega_y &= \omega_0 e^{\mathrm{i}\sigma_e t}, \end{aligned} \tag{4}$$

where $\Omega = 2\pi$ rad/sidereal day is the constant diurnal rotation, and $\sigma_e = \mathrm{e}\Omega \sim \Omega/304.5$ is the so-called Euler angular frequency. The rotation axis is described by the complex equatorial coordinate $\omega_x + \mathrm{i}\omega_y$. It presents a free wobble at the period of 304.5 sidereal day, or 304 mean solar days. Whereas this simple derivation can account for the existence of a free polar motion, it does not account for the period of the main cycle in 430 days. A better theory has to include the non-rigid Earth response to polar motion, as show now.

4. Linear theory of the rotation of a non rigid Earth

Let us recall the lineaments of the polar motion theory for a real Earth, as exposed in [3]. As we do not know the terrestrial mass distribution perfectly, we do not have access to the exact orientation of the main principal axes frame with respect to the crust. Moreover this mass distributions is not constant. In practice the terrestrial frame $Gxyz$ is attached to the ground and is dissociated from the main principal axes frame. We only know that in $Gxyz$, the inertia moment matrix almost is almost diagonal, reading

$$\underline{\bar{I}}(t) = \begin{bmatrix} A + c_{11} & c_{12} & c_{13} \\ c_{21} & A + c_{22} & c_{23} \\ c_{31} & c_{32} & C + c_{33} \end{bmatrix}, \tag{5}$$

where the $c_{ij}$are variable quantities smaller than $10^{-7}A$at time scale smaller than 100 years. The winds in the atmosphere, current in oceans create a relative angular momentum thatwill be noted$\underline{h}$, and the total angular momentum of the Earth is expressed by$\underline{H} = \underline{\bar{I}\omega} + \underline{h}$. The instantaneous rotation vector$\underline{\omega}$ is now written

$$\underline{\omega} = \Omega(m_1, m_2, 1 + m_3), \tag{6}$$

where it is easy to show that $m_1 = x_p$and $m_2 = y_p$. At sub-secular scale, observations show that $m_1, m_2 \leq 10^{-6}$ and $m_3 \leq 10^{-8}$. Accounting for non-rigid effect, the basic equation (2) becomes the Liouville equation

$$\frac{d(\underline{\bar{I}}(t)\underline{\omega} + \underline{h}(t))}{dt} + \underline{\omega} \wedge \left(\underline{\bar{I}}(t)\underline{\omega} + \underline{h}(t)\right) = \underline{\mathrm{L}}. \tag{7}$$

In regard of the smallness of both c_{ij}, h_i and m_i, linearisation can be done, leading to

$$m + \frac{i}{\sigma_e}\dot{m} = \chi - \frac{i}{\Omega}\dot{\chi} + \mathrm{i}\frac{L}{\Omega^2(C-A)} \tag{8}$$

where

$$\begin{gathered} m = m_1 + \mathrm{i}m_2 \\ \chi = \frac{c_{13} + \mathrm{i}c_{23}}{C-A} = \frac{h_1 + \mathrm{i}h_2}{C-A} \\ L = L_1 + \mathrm{i}L_2 \end{gathered} \tag{9}$$

Now, we have to notice that the complex off-diagonal moment of inertia $c = c_{12} + \mathrm{i}c_{23}$ in RHS of (8) will depend on polar motion m if the Earth is non rigid. Indeed, at a given location, the distance to the axis of rotation varies, centrifugal forces change consequently and the Earth undergoes a deformation called pole tide. This deformation is commonly modelled as hydrostatic for the ocean and quasi-elastic for the solid part of the Earth. In turn, the pole tide causes a change c_{rot} of the inertia moment c, which can be shown to be linear with respect to the pole displacement m:

$$\frac{c_{rot}}{C-A} = \frac{k}{k_s}m\,, \tag{10}$$

where $k \sim 0.35$ and $k_s = 0.94$. After reporting this contribution in the LHS of (8), we obtain

$$m + \frac{\mathrm{i}}{\sigma_e}\dot{m} = \frac{k_s}{k_s - k}\chi - \frac{\mathrm{i}}{\Omega}\dot{\chi}\,, \tag{11}$$

where the equatorial external torque L has been cast aside for the considered problem of a mass transport in the mechanical system, and

$$\sigma_c = \left(1 - \frac{k}{k_s}\right)\sigma_e\,, \tag{12}$$

corresponding to a period of about 480 days for the free wobble. By considering moreover that the liquid core does not present any pole tide, one gets the observed period of 430 days. That is the essence of the polar motion theory for time scale smaller than 100 years.

According to our contemporaneous understanding, at sub-secular time scale polar motion results from transports in the surface layer of the Earth: atmosphere, ocean, in-land water and continental ice (see Figure 4). Models based upon the assimilation of observations like wind, ocean current, temperature,...allow to construct 3D carts of the state of these layers at regularly sampled times. Then integration yields induced moment of inertia c_l (the index l means load) or relative angular momentum $h = h_1 + \mathrm{i}h_2$ for each of these layers with a time resolution generally longer than 3 hours. Such time series then provide a good account for sub-decadal polar motion (period smaller than 10 year) adopting the framework of a quasi-elastic Earth resulting in Equation (11) for polar motion. For a low frequency mass load ($\sigma \leq \Omega$), (11) becomes

$$m = \frac{k_s}{k_s - k}\frac{c_l}{C-A} \sim 1.6\frac{c_l}{C-A}\,. \tag{13}$$

In c_l we include the (elastic) deformation of the solid Earth under the loading itself.

At decadal time scale, the quasi-elastic behaviour of the Earth is still valid. But the causes of polar motion are partly elucidated. Change in the land water distribution or ice coverage seem to be the major contributor. For instance, it accounts for the deflection of the rotation pole since 2000 towards Europe shown in Figure 2 [1].

Around 30 years appears a clockwise oscillation with an amplitude of 10 mas, called Markowitz wobble, and captured by the wavelet transform of Figure 3. Some study speculate on its causes. It could originate in electromagnetic coupling bewteen the iron-nickel core and the thin conducting layer located at the bottom of the mantle.

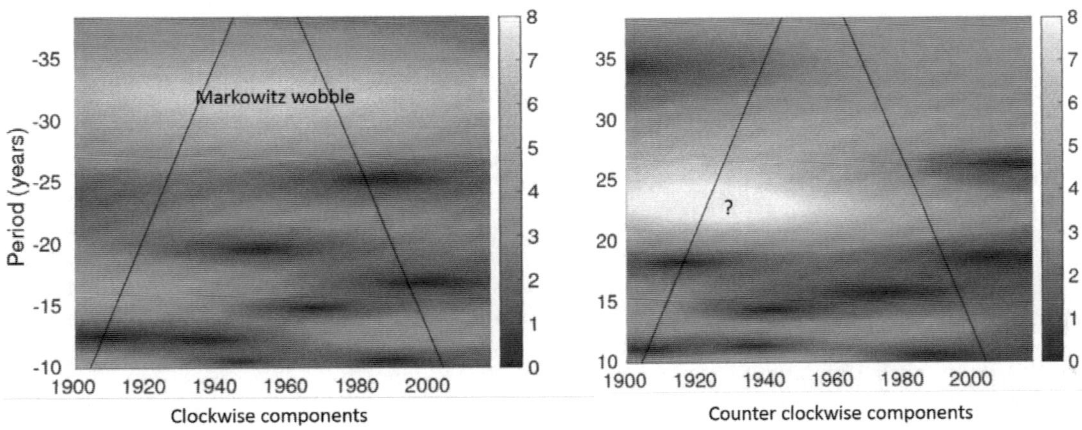

Figure 3. Wavelet transform on the 2D polar motion in the band 10 year - 40 years. Amplitude is expressed in mas.

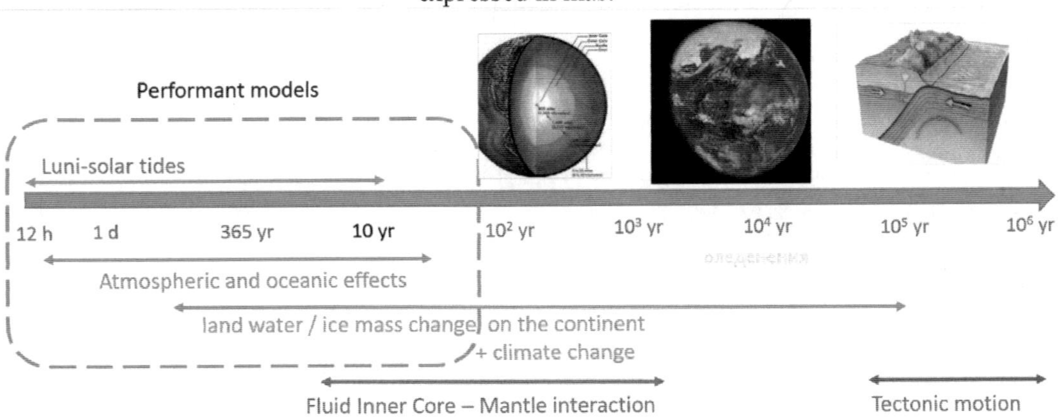

Figure 4. Synoptic of geophysical processes causing polar motion from 12 h to one million years

5. Visco-elastic Earth and secular drift

The trend observed in mass redistributions, either located in the atmosphere or in the hydrological layer, are too small to induce the observed secular drift. The most plausible explanation is an enormous moment of inertia change, almost reaching the Earth triaxiality ($A - B{\sim}10^{-5}A$), that occurred long time in the past, but which the effect continues because of the visco-elastic response of the Earth. The longest is the period of the exciting process, the largest will be delayed the deformation response. The relation (10) becomes frequency dependent:

$$c_{rot}(\sigma) = \frac{k(\sigma)}{k_s}(C - A)m(\sigma)\,. \tag{14}$$

In time domain this is equivalent to a convolution product. Rheological models yield expression having the form [4]

$$k(t) = k_0\delta(t) + \sum_{j=1}^{J} H(t)k_j\mathrm{e}^{-s_j t}\,, \tag{15}$$

where $H(t)$is the Heaviside function. The elastic/hydrostatic response is accounted by $k_0\delta(t) \approx 0.035\delta(t)$ and the viscosity is embodied in the terms $k_j e^{-s_j t}$. In frequency domain we have

$$k(\sigma) = k_0 - \sum_{j=1}^{J} \frac{k_j}{s_j + i\sigma}. \tag{16}$$

In the quasi-elastic approximation, the viscosity term $\sum \frac{k_j}{s_j+i\sigma}$ is taken as a constant complex number, independent from frequency.

As the variations of all pertaining quantities are still small in regard to the total Earth inertia moment or to the Earth angular velocity, linearity of Liouville holds. At this time scale excitation is limited to off-diagonal inertia change c in (\ref{EL1}). As the pole tide effect becomes a convolution product in time domain, it is easier to solve (\ref{EL1}) in Laplace or Fourier domain, and go back to time domain. It can be shown that the solution takes the form

$$m(t) = H(t)(\rho_0 + \sum_{1}^{J+1} \rho_j\mathrm{e}^{-\sigma_j t}) * \frac{c_l(t)}{C-A}. \tag{17}$$

with $\rho_0 \approx 1$. Considering the extension of ice coverage during the last glaciation and its thickness (up to 1000 m), models gives for c_l a maximum of 10^{32} kg m^2. For a unique mode, occurring for a mono-layered Earth for instance, we have

$$k(t) = k_0\delta(t) + H(t)k_1\mathrm{e}^{-s_1 t}\,, \tag{18}$$

and we obtain the polar motion

$$m(t) \approx H(t)\sum_{1}^{J+1} \rho_j\mathrm{e}^{-\sigma_j t} * \frac{c_l(t)}{C-A}. \tag{19}$$

To the largest mode of rheological Model A in \cite{wu_peltier_1984}, namely $s_1 \approx 1.92$ kyear^{-1} and $k_1 \approx 0.94$ kyear^{-1} corresponds the solution $\sigma_1 \approx 0.46$, $\rho_1 \approx 2.12$, $\sigma_2 \approx 1.46 - \mathrm{i}5.176\ 10^3$, $\rho_2 \approx -2.12 - \mathrm{i}7.558\ 10^3$, with all these coefficients expressed in kyear^{-1}.

Here the mode of complex frequency σ_2 models the forced Chandler wobble, and the contribution of the visco-elasticity to the secular polar motion is given by the mode of frequency σ_1, namely

$$m^{ve}(t) = H(t)\rho_1 e^{-s_1 t} * \frac{c_l(t)}{C-A} = \rho_1 \int_0^t e^{-\sigma_1 u}\frac{c_l(t-u)}{C-A}du. \tag{20}$$

At epoch t, this is the weighted sum of moment of inertia at all prior instants back to the first instant $t_0 = 0$ of the mass redistribution, which the weights decrease exponentially with offset from the present epoch. Now we see the striking contrast with a rigid or elasto-fluid Earth, where the secular pole $m(t)$ is always proportional to $c_l(t)$.

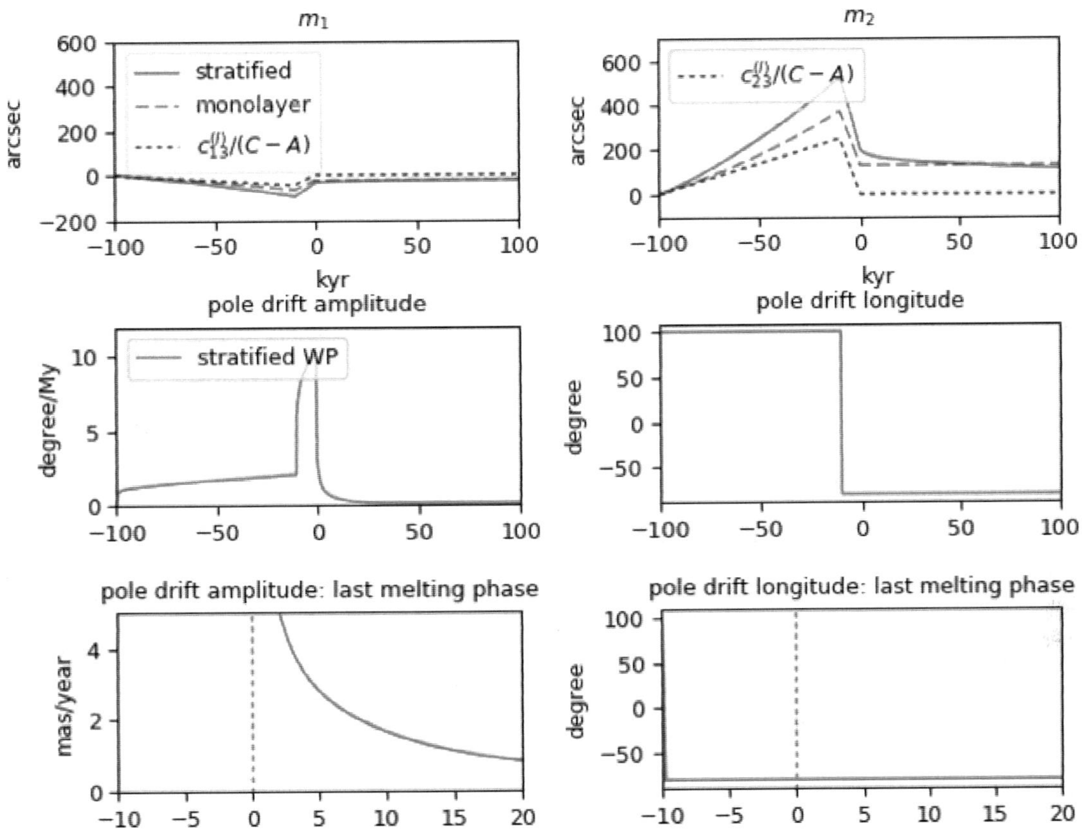

Figure 5. Pole drift resulting from one glaciation cycle lasting 100 ky

If we grossly model the evolution of c_l as a saw-like time function lasting 100 kyr, reaching its maximum 90 kyr years after the glaciation has started, and melting in 10 kyr (see top of Figure 5), then numerical evaluation of (17) gives the polar motion from -100 kyr (starting of the glaciation) to $+100$ kyr, considering both a mono-layer Earth (having an homogeneous rheology), and a refined multi-layer model. The corresponding evolution of m_1 and m_2 pole coordinates are represented at the top of Figure 5. Here $t = 0$ is for the end of the melting. Just below we show the velocity of the polar drift, indicating its amplitude in degree/Myr and its longitude. This velocity is zoomed in the bottom plots from year 0 to 20 kyr. If we assume that our epoch is $+6$ kyr after the end of the ice recession, we obtain a polar drift of about 3 mas/year towards 80° W, comparable with the contemporaneous order of magnitude. So, in this view, pole drift is the footprint of a mass load, that has totally disappeared, but which the rotational effect is still visible.

6. Qualitative description

As shown by the upper plots of Figure 5, for a visco-elastic Earth, polar drift accelerates with time, while the one estimated for an elastic Earth according to (13) increases uniformly with time as $c(t)/(C-A)$. How can we interpret this simply in light of the basic equation (8)? Let us consider successive equally spaced instant, $t_0 = 0, t_1, t_2, t_n \dots$. During each time step $\tau = [t_i, t_{i+1}]$, off-diagonal moment of inertia of the load increases by Δc. To illustrate our demonstration, let us take a positive increase of $\Delta c = \Delta c_{23} = -\int yz\, dm$ in the plane and symbolized by a trapezoidal form in Figure 6. For the sake of simplicity we assume that at time t_0 the inertia pole, the geographic pole (corresponding to the z axis of the terrestrial frame) and the Earth rotation pole coincide at the point R_0. At time $t_1 = \tau$, the Earth axis moves to the position R_1^r by $\Delta m = \Delta c_l/(C-A)$ for a rigid Earth. In

the case of a solid elastic Earth, partially covered by hydrostatic oceans, the pole tide is instantaneous. Its account - as well as a possible elastic deformation in the case of a surface loading occurring at time scales larger than 10 years-yields Euler-Liouville equation (13), and does not change this expression much. The shift is amplified, and rotation pole is now in the position R_1^e. For the next interval $[t_1, t_2]$ the same increment happens for c, the pole shift is doubled for a rigid or an elastic/fluid Earth. The new pole position are respectively R_2^r and R_2^e.
For a visco-elastic rheology, the first step $[0, t_1]$ did not present a full adjustment of the equatorial bulge around the new direction of the Earth axis. This delayed pole tide begins during the period $[t_1, t_2]$, and it causes a supplementary pole tide with the same sign than the $c(t)$. Again, this reinforces the purely elastic or hydrostatic effect, and at time t_2 rotation pole lies in the more shifted position and R_2^{ve}. Iterating this argument at instant t_3, we see that R_3^{ve} integrates the delayed pole tide effect during period $[t_1, t_2]$ and $[t_0, t_1]$ in a less measure, so that at time t_n, R_n^{ve} results not only from the pure rigid polar shift and elastic pole tide but also from the delayed pole tide over the time intervals $[t_0, t_1], [t_1, t_2], \dots, [t_{n-2}, t_{n-1}]$. So the displacement of R^{ve} is an accelerated function of t, unlike R^r or R^e.
Let us consider the simplistic rheological model (18). According to (10) the visco-elastic rotational effect on inertia moment reads then

$$\begin{aligned} c_{rot}^{(ve)}(t) &= \frac{H(t)k_1 e^{-s_1 t}}{k_s}(C-A) * m(t) \\ &= \frac{k_1}{C-A}\int_0^t e^{-s_1 u} m(t-u)du. \end{aligned} \tag{21}$$

Approximating the exponential time decrease $e^{-s_1 u}$ by a step wise function $H(\tau - u)$ dropping from 1 to zero after relaxation time $\tau = 2\pi/s_1 \sim 3$ kyears, we have

$$c_{rot}^{(ve)}(t) = \frac{k_1}{C-A}\int_0^t m(t-u)du \approx \frac{k_1}{C-A}\tau m(t-\tau). \tag{22}$$

Here τ is small with respect to the duration of the glaciation, which is 30 times longer. Put another way, the visco-elastic pole tide in the interval $\tau = [t_{i-1}, t_i]$is a constant fraction of the total polar shift $m(t_{i-1})$ at time t_{i-1}: $\Delta c_{rot}^{ve}(t_i) = \alpha m(t_{i-1})$, with $0 < \alpha \leq 1$. So, at instant $t_2$the total displacement of the visco-elastic pole R^{ve} is $R_0 R_2^{ve} = (2+\alpha)\Delta m$, whereas R^e has only moved by $2\Delta m$. At time t_3, the rotation pole R^{ve}has moved by

$$\mathrm{R_0 R_3^{ve} = R_0 R_2^{ve} + \Delta m + \alpha R_0 R_2^{ve} = (2+\alpha)\Delta m + \Delta m + 2\alpha\Delta m = (3+3\alpha)\Delta m.}$$

Iterating this argument at equally spaced instants, we find that the full displacement is $(n + \alpha(n-1)n/2)\Delta m$ at instant t_n, exceeding the rigid or the elastic one by $(n-1)n/2\, \alpha\Delta m$. In this expression n and t are proportional, so that the departure for the elastic/fluid pole R^r is quadratic function of time, while the moment of inertia c_l linearly increases with time. This qualitative approach well accounts for the parabolic acceleration of the pole shift observed in Figure 5 for the glaciation phase.

7. Summary and discussion

The low frequency polar motion, embodying periods above 10 years, reflects miscellaneous geophysical processes, and is a valuable source of information for the Earth rheology. Multi-annual oscillations are associated with changes in land water storage and continental ice coverage. The 25-30 wobble is not explained yet, whereas some study suggest the interaction between the fluid core and the mantle.Finally the secular pole drift results from the visco-elastic response of the Earth to the last glaciation cycle. Notwithstanding, the value of the secular polar drift depends on the viscosity of the

lower mantle, which the value is poorly known (between 10^{21} and 10^{22} Pas s). In turn the reconstructed pole drift can vary by an order of magnitude.
Meanwhile, we provided a qualitative description of the visco-elastic mechanism of the secular pole drift. In parallel to the elastic/fluid pole tide linearly amplifies the polar displacement of a rigid Earth, the visco-elasticy let appear a delayed tide reflecting the pole displacements at all former instants. In the case of a mass load increase or decrease in a given geographic area (like an ice cap), these visco-elastic pole tide effects are added and tend to strongly accentuate the pole shift.
Over million year, the effect of tectonic mass redistribution takes precedence over the effect of glaciations. Moreover at this time scale, the polar displacement is blurred by the continent sliding over the mantle, which is grossly reconstructed from geological observations, especially assuming that some volcano ranges (like the one of Hawai islands) reflect the trace on the Earth crust of magmatic plumes originating from the mantle, the hot spots. Yet, it is not excluded that the continents present of global rotation with respect to the mantle up to a few mas/year, then partially or totally invalidating the role of glaciation in the observed drift. For instance, if the continents globally rotate towards $180° - 80° = 100°$ East, then the rotation axis, fixed with respect to the mantle will shift in the observed direction with respect to the crust.

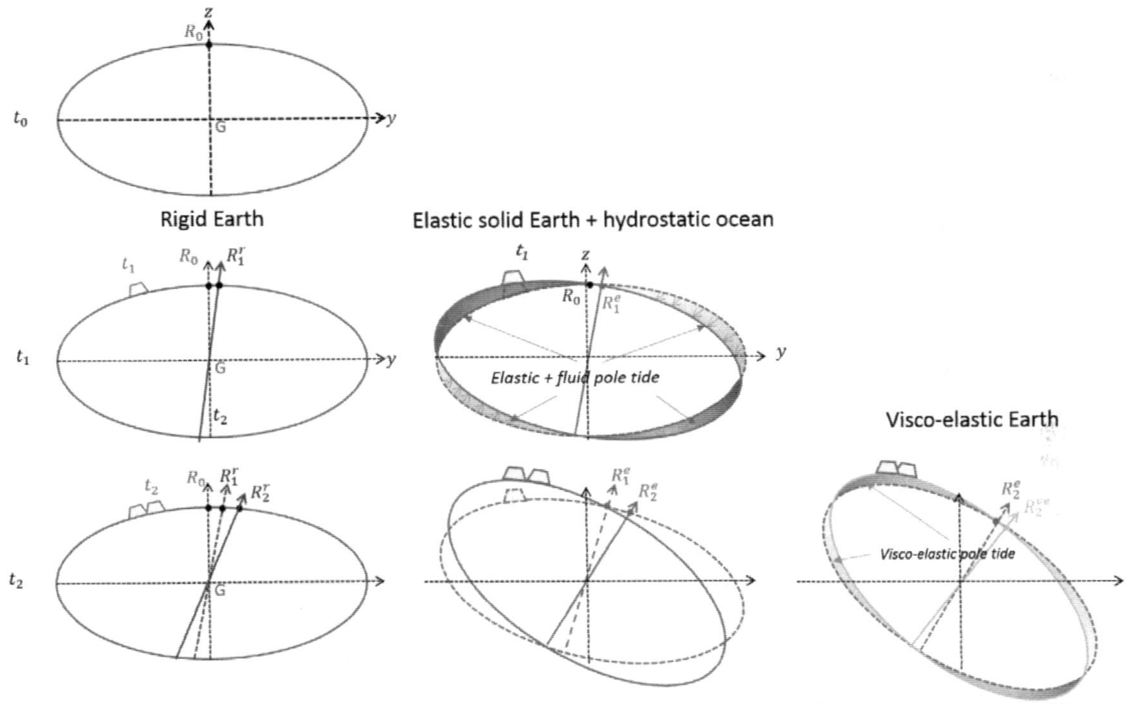

Figure 6. Schematic representation of the polar shift mechanism when off-diagonal moment of inertia c_{23} linearly increases with time. At time $t_1 = t_0 + \tau$a positive inertia moment change c_{23}(in the plane yz) is added, pictured by a trapezoidal form. A time $t_2 = t_1 + \tau$ this moment inertia variation is multiplied by 2, and so one. The subsequent displacement of the rotation pole R are represented (magnified) for three Earth models: R^r for a rigid Earth,R^e for a elastic/fluid Earth and $R^{ve}$for a visco-elastic Earth.

References

[1] Adhikari S and Ivins E R 2016 Climate-driven polar motion: 20032015, *Science Advances* Vol. **2**, no. 4, e1501693, DOI: 10.1126/sciadv.1501693
[2] Lambert W D 1922 The interpretation of apparent changes in mean latitude, *Astronomical Journal* vol. **34** iss. 804 pp 103-110

[3] Munk W H and Mac Donald G 1960 *The rotation of the Earth* (Cambridge University Press)
[4] Sabadini R , Vermeersen B and Cambiotti G 2016 *Global Dynamics of the Earth: Applications of Viscoelastic Relaxation Theory to Solid-Earth and Planetary Geophysics*, Springer, pp 335.
[5] Wu P and Peltier W R 1984 Pleistocene deglaciation and the Earth's rotation: a new analysis *Geophys. J.R. astr. Soc.* **76** pp 753-791
[6] Zhukovsky (1904), Geometricheskaya interpretsiya teorii dvizheniya polyusov vrashcheniya zemli po ee poverxnosti, Collected paper of Zukovsky, Vol. **1**, edited by A.P. Kolesnikov, Editorial Officie for Aviation Litterature (1937).

Acknowledgments
We would like to acknowledge the financial support of our journey in Moscow by the High School of Economy. We also express our gratitude to Bauman Institute for its invitation to give this talk, and Guy Berthault for encouraging us to undertake the study of rotation pole tilt in view of its coupling with marine recessions and transgressions.

Modeling of Earth rotation in the gravitational field of Sun and Moon

M Yu Barkin[1], Yu G Markov [2], V V Perepelkin[2] and A S Filippova[2]

[1]Bauman Moscow State Technical University, Moscow, Russia
[2]Moscow Aviation Institute (National Research University), Moscow, Russia
E-mail: vadimkin1@yandex.ru

Abstract. A celestial mechanics model of the deformable Earth is studied within the limits of the spatial variant of the problem "deformable Earth - Moon in the gravitational field of the Sun". An amplitude-frequency analysis of the perturbed Chandler wobble of the Earth pole due to the precessional motion of the lunar orbit is provided. Using a numerical-analytical approach, we consider the possibilities for identifying the Chandler wobble parameters and fitting the developed refined theoretical model to the real trajectory measurements of the Earth pole.

1. Perturbed Chandler oscillation of the Earth Pole

Creating a mathematical model, that is adequate to the observation data obtained by International Earth Rotation and Reference Systems Service (IERS) [1] and describes the real motion of the instantaneous axis of rotation in reference frame associated with Earth, is an interesting and complex problem in the fields of theoretical and celestial mechanics. It is significant for many practical problems, including the problems of improving the precision characteristics of the spacecraft coordinate-time and navigation support.

This formulation of the dynamical problem of the Chandler term of the pole motion is based on the spatial variant of the problem: deformable Earth-Moon motion in the Sun gravitational field [2, 3]. It is assumed that the viscoelastic planet (Earth) and the point satellite (Moon) perform a mutual translational-rotational motion around their common center of mass (barycenter).

The most convenient generalized coordinates for qualitative describing the Earth rotation around its center of mass are canonical action-angle coordinates I_i, w_i (i=1, 2, 3). Perturbed Raus functional is [4, 5]:

$$R = R_0(I_1, I_2) + \varepsilon R_1(I_1, I_2, w_1, w_2, \mathbf{u}, \dot{\mathbf{u}}) + \varepsilon^2 ..., \tag{1}$$

where $\varepsilon > 0$ is small dimensionless parameter, I_2 is Earth angular momentum module, I_1 – the angular momentum projection on the figure axis, $\mathbf{u}$ – displacement vector of the viscoelastic environment points (the mantle); εR_1 is the perturbed funclional due to the gravitational tidal forces from the Moon and the Sun. The structure of the perturbed functional εR_1 in (1) is the subject of the further analysis and simplification on the basis of the averaging method.

Content from this work may be used under the terms of the Creative Commons Attribution 3.0 licence. Any further distribution of this work must maintain attribution to the author(s) and the title of the work, journal citation and DOI.
Published under licence by IOP Publishing Ltd

The perturbed Raus functional of the studied problem is

$$R_0 = \frac{I_2^2}{2A^*}\left(1 - \frac{C^* - A^*}{C^*}\frac{\kappa^2}{\kappa^2+\lambda^2}\right) + \mu_1 R_{21}^{-3}\int_\Omega \left[(\mathbf{r},\mathbf{u}) - 3(O^{-1}\mathbf{R}_{21}^0,\mathbf{r})(O^{-1}\mathbf{R}_{21}^0,\mathbf{u})\right]\rho dv,$$
$$\kappa^2 = \frac{C^*(A^*-B^*)}{A^*(B^*-C^*)}, \quad \lambda^2 = \kappa^2\frac{2EC^* - I_2^2}{I_2^2 - 2EA^*}, \quad C^* > B^* > A^*. \tag{2}$$

Here κ^2, λ^2 are the main dynamical parameters; for the Earth $0 < \lambda^2 < 1$; E is the constant of the energy integral. Values A^*, B^*, C^* are the main central inertia moments of the deformable Earth under the centrifugal forces caused by the Earth's rotation and the gravitational field of the Moon. The first term in (2) describes the unperturbed motion of the Earth around its center of mass. The second term there are designations: $\mathbf{r}$ is the Earth mantle points radius-vector; $\mathbf{u}$ is the displacement vector of the Earth mantle points; $\mathbf{R}_{21}^0$ is a unit vector describing the direction from the Earth to the Moon; O is a transfer matrix from the fixed axes to the inertial ones; μ_1 is the Moon gravitation parameter; ρ is the density; dv is the elementary volume.

Omitting the intermediate calculations for the Earth pole coordinates, we obtain the following expressions:

$$x_p \approx \frac{C^*}{A^*}\frac{\lambda}{\kappa}\cos w_1 + \frac{A^*}{C^*}\left(\frac{\lambda}{\kappa}\right)^2\frac{2+\kappa^2}{2I_2\kappa_1^2}\left[-\frac{a_2^c}{2(n_1-\dot{h})}\cos 2(w_1-h) - \frac{a_2^c}{2(n_1+\dot{h})}\cos 2(w_1+h) + \frac{a_2^s}{2(n_1-\dot{h})}\sin 2(w_1-h) - \frac{a_2^s}{2(n_1+\dot{h})}\sin 2(w_1+h) - \frac{a_1}{2n_1-\dot{h}}\cos(2w_1-h-w_3) - \frac{a_1}{2n_1+\dot{h}}\cos(2w_1+h+w_3)\right]\cos w_1,$$

$$y_p \approx \frac{C^*}{B^*}\frac{\lambda}{\kappa}\kappa_*\sin w_1 + \frac{B^*}{C^*}\left(\frac{\lambda}{\kappa}\right)^2\frac{2+\kappa^2}{2I_2\kappa_1^2}\left[-\frac{a_2^c}{2(n_1-\dot{h})}\cos 2(w_1-h) - \frac{a_2^c}{2(n_1+\dot{h})}\cos 2(w_1+h) + \frac{a_2^s}{2(n_1-\dot{h})}\sin 2(w_1-h) - \frac{a_2^s}{2(n_1+\dot{h})}\sin 2(w_1+h) - \frac{a_1}{2n_1-\dot{h}}\cos(2w_1-h-w_3) - \frac{a_1}{2n_1+\dot{h}}\cos(2w_1+h+w_3)\right]\sin w_1. \tag{3}$$

2. Chandler oscillations of the Earth Pole according to the Observation Data

The dynamics of the perturbed motion of the instantaneous axis is associated, in particular, with the variability of the angle δ_2, which determines the amplitude of the Chandler wobble:

$$\dot{\delta}_2 \approx -\left(\frac{\lambda}{\kappa}\right)^3\frac{2+\kappa^2}{I_2\kappa_*}\left[a_1\cos(h+w_3) + a_2\cos(2h+\alpha)\right]\{\varepsilon\sin w_1 + \mu\}, \tag{4}$$

where h is the longitude of the ascending node of the lunar orbit.

The coefficients a_2^c, a_2^s, a_1 in equations (3) are determined by processing the long series of IERS observations for the pole coordinates (x_p, y_p).

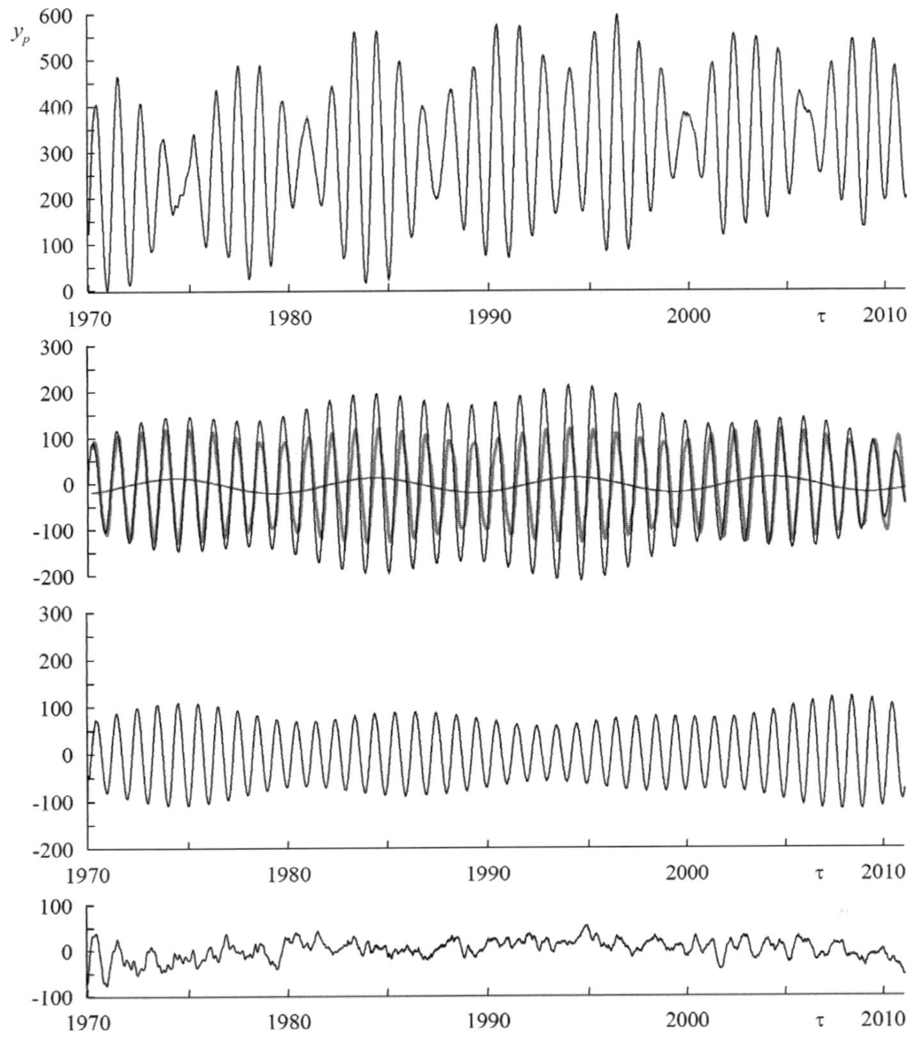

Figure 1. Decomposition of the Earth pole oscillations in coordinate y_p into their main components. From top to bottom: the IERS observations, the Chandler wobble (gray line) in comparison with the theoretical model y_p (contrasting red line) and the amplitude variation (smooth line), the annual oscillation, and the small-scale fluctuations.

Figure 1 gives a decomposition of the Earth pole oscillation process into components for the coordinate y_p. The resultant oscillation is presented at the top; this is followed by the Chandler, annual components, and the residual. A clearly distinguishable variation of the amplitude with a frequency that is twice the frequency of the precessional motion of the lunar orbit (twice the revolution frequency of the node of the lunar orbit) can be noted in the Chandler component of the oscillation process.

References

[1] International Earth Rotation and Reference Systems Service - IERS Annual Reports, http://www.iers.org.
[2] Zlenko A A 2015 *Astron. Rep.* **59** (1) pp 72 - 87
[3] Munk H, MacDonald G 1960 *The rotation of the Earth* (University Press) p 323
[4] Markov Yu G, Perepelkin V V, Filippova A S 2017 *Dokl. Phys.* **62** (6) pp 318-322
[5] Akulenko L D, Markov Yu G, Rykhlova L V 2001 *Dokl. Phys.* **46** (4) pp 261-263

Acknowledgements
The present study is financially supported by the Ministry of Science and Higher Education of the Russian Federation (State Project Number 9.7555.2017/BP).

Modelling of the microparticle spraying process in a heated gas stream

N G Bourago[1], A D Nikitin[2], I S Nikitin[2] and B A Stratula[2,3]

[1]Ishlinsky Institute for Problems in Mechanics of RAS, Moscow, Russia
[2]Institute of Computer Aided Design of RAS, Mechanics of RAS, Moscow, Russia
[3]Moscow Aviation Institute (National Research University), Moscow, Russia
E-mail: i_nikitin@list.ru

Abstract. Approximate analytical estimates for the final thickness of a deformed hot particle incident on an undeformed substrate, its radius and the collision time as a function of the speed are obtained for the law of hardening close to linear by the perturbation method. The system of ordinary differential equations of the approximate model under consideration is numerically and analytically solved. A comparison of the numerical and analytical results in the one-dimensional approximation with the results of multidimensional modelling of the unsteady collision process using finite-element schemes is carried out.

1. Introduction

The result of a building-up process such as layers thickness and mechanical properties are determined by thickness of almost melted particles after collisions. The Tailor model for single-dimensional collision of a pure plastic bar with an undeformed substrate was applied to model the process. In case of a hardening process similar to a linear one the approximate analytical evaluation of particle's final thickness, radius, and impact time as function of particle' speed before the collision were conducted using the perturbation method. This approach was approved by comparison with a multidimensional FEM calculation.

2. The Tailor model

Let us consider a dynamic deformation pattern that corresponds to the classical Tailor's model and was given in [1-3]. A bar of length h_0, radius r_0, and made of pure plastic material is moving at a velocity of v_0 (the velocity vector and bar's side with given length are parallel) towards an undeformed substrate. At $t=0$ the bar hits the substrate that is normal to the velocity vector (Fig. 1). We consider a movement of the bar as a quasi-single-dimensional; speed and strain distributions are homogenous within an every cross-section of the bar.

Content from this work may be used under the terms of the Creative Commons Attribution 3.0 licence. Any further distribution of this work must maintain attribution to the author(s) and the title of the work, journal citation and DOI.
Published under licence by IOP Publishing Ltd

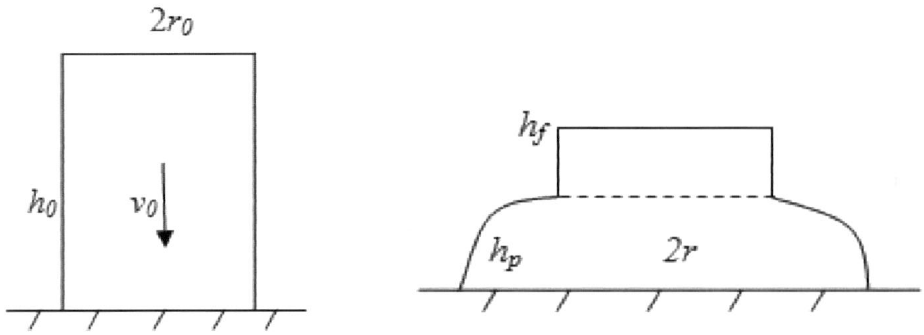

Figure 1. The Tailor collision model

A system of equations involves a relation for deformation behind a plastic wave, the conservation of momentum at the plastic wave front, a kinematic relation for a height of bar's intact part, the Newton's law for the intact part, and a plastic deformation law:

$$\varepsilon = v/(v+w)$$

$$\rho(v+w)v = \sigma - \sigma_s$$

$$dh/dt = -(v+w)$$

$$\rho h dv/dt = -\sigma_s$$

$$\sigma = \sigma_s + E_p \varepsilon^m$$

where $v(t)$ is intact part's velocity; $w(t)$ is plastic wave speed; $\sigma(t)$ is stress behind the shock wave; σ_s is a yield stress; ρ is a density; E_p is a strain hardening modulus; m is a power of hardening. Initial conditions are:

$$t=0:\ v=v_0,\ h=h_0$$

Conditions to determine a stopping time $t=t_f$ and the intact part's height h_f are:

$$v=0,\ w=0$$

One can transform the considered system of equations to:

$$dh/dt = -(v+w)$$

$$dv/dt = -\sigma_s/(\rho h)$$

$$w = \left(E_p/\rho\right)^{1/(m+1)} v^{(m-1)/(m+1)} - v$$

It seemed to be impossible to solve the system of equations for the arbitrary m. Therefore an approximate solution was made with the principal terms obtained at the point with $m = 1+\gamma > 1$ where $\gamma << 1$. It was assumed the $\sigma - \varepsilon$ locus within a plastic domain close to linear ($E_p << E$ where E is a Young's modulus).

The result of the approximate solution is:

$$h(v) = h_0 \exp((v - v_0) / v_*)$$

$$v(t) = v_0 + v_* \ln(1 - t / t_0)$$

$$w(t) = v_c((v / v_c)^{\gamma/2} - v / v_c)$$

$$h_f = h_0 \exp(-v_0 / v_*)$$

$$t_f = t_0(1 - \exp(-v_0 / v_*))$$

where $v_c = \left(E_p / \rho\right)^{1/2}$, $v_* = \sigma_s / \left(\rho E_p\right)^{1/2}$, $t_0 = h_0 / v_c$.

One can obtain the plastic zone's height h_p using the plastic wave speed $w(t)$:

$$h_p = \int_0^{t_f} w(t)dt = h_0\left(\left(\frac{v_0}{v_c}\right)^{\gamma/2}\left[1-\exp\left(\frac{-v_0}{v_*}\right)\right]-\frac{v_0}{v_c}+\frac{v_*}{v_c}\left[1-\exp\left(\frac{-v_0}{v_*}\right)\right]\right)$$

Total bar's height is $h_l = h_p + h_f$. It is important to notice that bar plastic part's shape is implicit according to the approach; therefore the shape is based on an assumption of incompressibilityof bar's material.

Using the primary formulas for the numerical evaluation of the shape of the bar a number of initial conditions have been analyzed. Listed below parameters were used: particle's material is titanium; height $h_0 = 30 * 10^{-6}$ m; $r_0 = h_0/2$; density $\rho = 4500$ kg/m^3; Young's modulus $E = 116 * 10^9$ Pa; Poisson' ratio $\nu = 0.32$; yield stress at normal conditions $\sigma_{s0} = 3 * 10^8$ Pa; initial velocity $v_0 = 20$ m/s. The bar's temperature is within the range $T = 1600$ - 1660 °C, whereas titanium's melting point T_m is 1665 °C. Defining yield stress at the temperature about the melting point is rather uncertain task; thus a linear approach $\sigma_s = \max\left(\sigma_{s0}\left(T_m - T\right)/T_m, 0\right)$ was used.

Fig. 2 and 3 shows the shape after collisions at different stream's temperatures, which are close to the melting point.

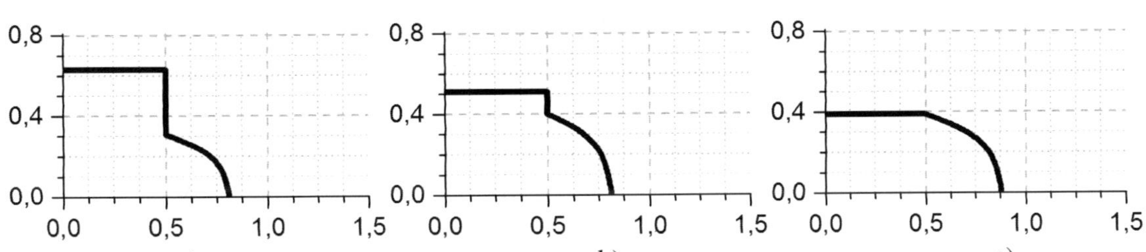

Figure 2. Particle's shape at $v_0 = 20$ m/s, $E_p = 2.5 * 10^7$ Pa, $m = 1.2$;
a) $T = 1600$; b) $T = 1620$; c) $T = 1633$.

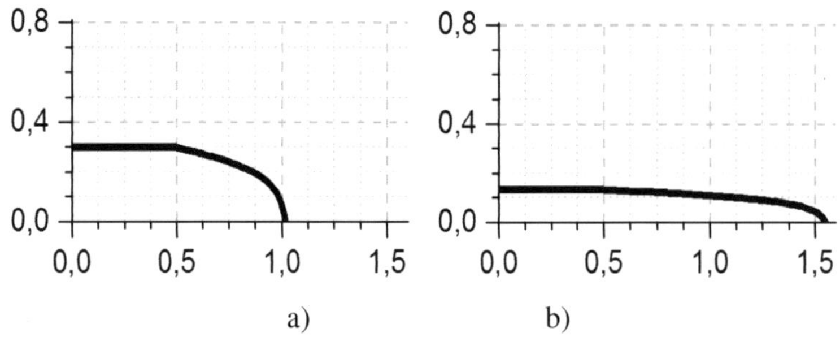

Figure 3. Particle's shape at $v_0 = 20$ m/s, $E_p = 2.5 * 10^7$ Pa, $m = 1.2$;
a) $T = 1640$; b) $T = 1650$.

Fig. 4 shows the shape after collisions at different stream's speeds. Fig. 5 shows an impact of the strain hardening modulus E_p and the power of hardening m on the shape.

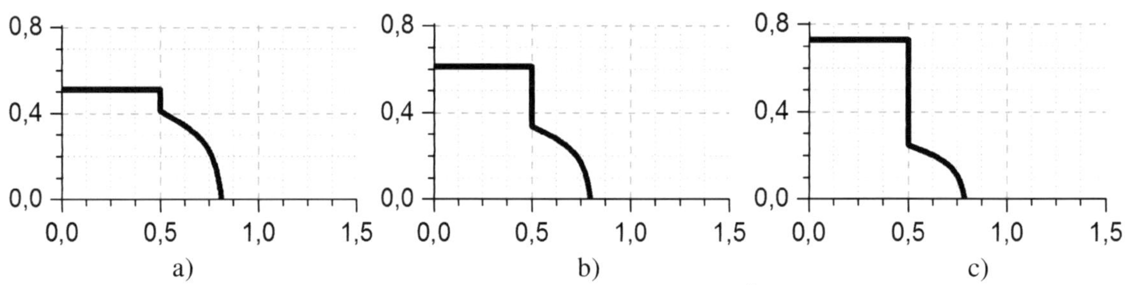

Figure 4. Particle's shape at $T = 1620$, $E_p = 2.5 * 10^7$ Pa, $m = 1.2$;
a) $v_0 = 20$ m/s; b) $v_0 = 15$ m/s, c) $v_0 = 10$ m/s.

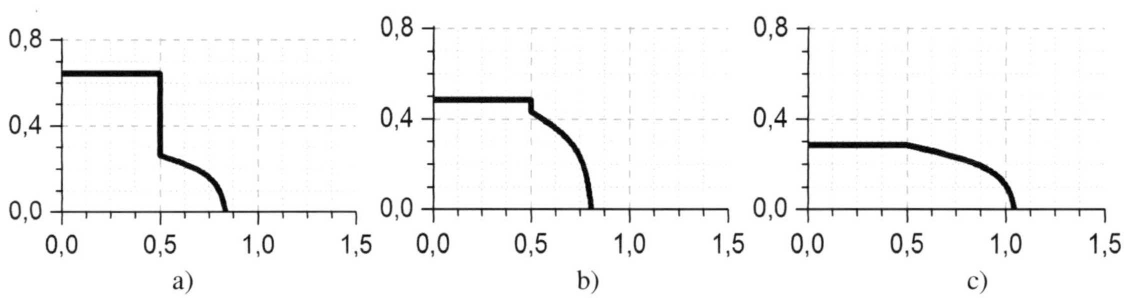

Figure 5. Particle's shape at $v_0 = 20$ m/s, $T = 1620$;
a) $E_p = 2.5 * 10^7$ Pa, $m = 1.8$; b) $E_p = 1 * 10^7$ Pa, $m = 1.8$; c) $E_p = 1 * 10^7$ Pa, $m = 1.2$.

3. Multidimensional modelling

In order to compare the above results a multidimensional modelling based on the numerical method from [5] was performed. Fig. 6 and 7 show ultimate bar's shapes with the same as Fig. 2 initial conditions.

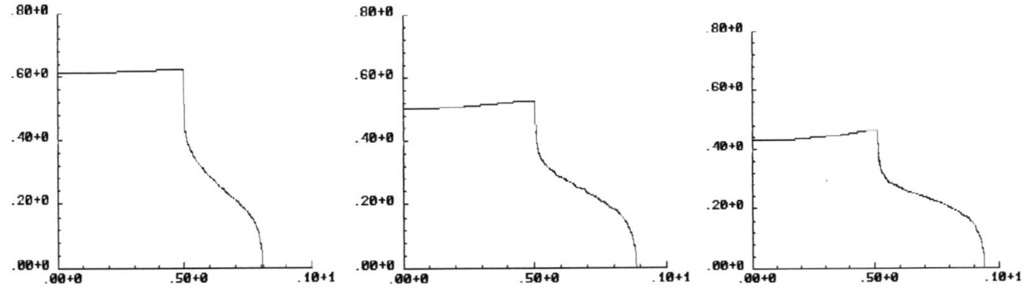

a) b) c)

Figure 6. Particle's shape at v_0 = 20 m/s;
a) T = 1600; b) T = 1620; c) T = 1633.

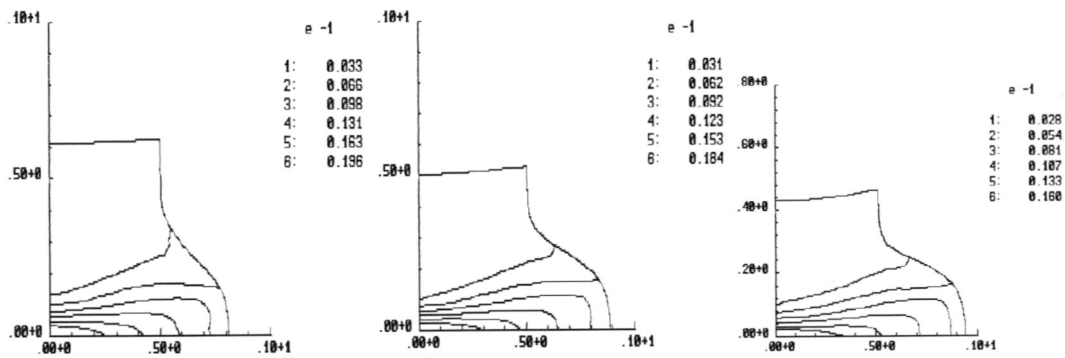

a) b) c)

Figure 7. Contour plot of the plastic work at v_0 = 20 m/s;
a) T = 1600; b) T = 1620; c) T = 1633.

Comparing Fig. 2 and Fig. 6 one can see the one-dimensional (bar-shaped) model gets ultimate thickness on the plastic part rather well within studied boundaries comparing to the multidimensional one, although it is important to notice bar's height and width are the same so this conclusion is not clear.

4. Conclusions

Approximate analytical estimates for the finite thickness of a deformed hot particle incident on an undeformed substrate, its radius and the collision time as a function of the speed are obtained for the law of hardening close to linear, by the perturbation method. The system of ordinary differential equations of the approximate model under consideration is numerically and analytically solved. A comparison of the numerical and analytical results in the one-dimensional approximation with the results of multivariate modelling of the unsteady collision process using finite-element schemes is carried out. The efficiency of the obtained formulas for estimating the thickness of the build-up layers as a function of particle velocity and flow temperature is shown.

References

[1] Zukas J A, Nicholas T and Swift H F 1982 *Impact Dynamics* (Skokie: John Wiley&Sons) p 452
[2] Goldsmith W 1965 *Impact* (Moscow: Stroyizdat) p 448
[3] Barvinok V A and Bogdanovich V I 2012 Physical and mathematical simulation of the formation of mesostructure-ordered plasma coatings *Technical Physics* **57** pp 262-269
[4] Miklyaev P G 1994 *Mechanical properties of light alloys at different temperatures and rates of pressure treatment* (Moscow: Metalurgia) p 279
[5] Burago N G, Nikitin I S and Yakushev V L 2016 Hybrid numerical method with adaptive overlapping meshes for solving nonstationary problems in continuum mechanics *Computational Mathematics and Mathematical Physics* **56** pp 1065-74

IOP Conf. Series: Materials Science and Engineering **468** (2018) 012008 doi:10.1088/1757-899X/468/1/012008

Free vibrations of a two-layer liquid in an elastic vessel

D A Goncharov* and A A Pozhalostin**

Bauman Moscow state Technical University, 105005, 2-ya Baumanskaya str., Moscow, Russia

E-mail: *goncharov@bmstu.ru, **a.pozhalostin@mail.ru

Abstract. In the article we consider small motions of an ideal, non-compressible two-layer fluid with, divided by rigid permeable flat septum. Fluid fills the cylindrical tank with the sloping spherical bottom. The fluid motions we consider as potential. The velocity potential satisfies the Laplace equation. The resistance of the septum we assume linear with a constant coefficient of resistance. The effect of a permeable partition dividing the liquid column in the vessel is taken into account as a certain linearly viscous resistance with a reduced resistance coefficient specified as a parameter. The elasticity of the bottom of the vessel is taken into account by introducing the Airy function in the right-hand side of the differential equation corresponding to the movements of the elastic bottom.

1. Introduction

The task arises from practice. In rocket technology, various intra-tank structures are widely used to launch a propulsion system in space flight conditions [1]. These designs significantly complicate the frequency analysis. At the same time, frequency analysis is very important, since the elastic tank, the propulsion path of the rocket engine and the engine itself form a closed system with feedback in which self-oscillations are possible [2]. Given the importance of these phenomena and their influence on the performance determining spacecraft and launch vehicle study symmetric oscillations of the liquid in the vessels of various shapes and configurations devoted considerable number of studies, such as [3], [4], [5], [6], [7], [8]. Some generalizations are given in the monograph [9]. Some experimental studies and simulations are also present in the article [10].

2. Formulation of the problem

We consider small axisymmetric oscillations of an ideal fluid filling a vessel of cylindrical shape of radius R with rigid walls. We introduce the cylindrical coordinates $O_1x_1\eta r$ and $O_2x_2\eta r$ associated with the centres of the partition and the bottom, respectively. Let the liquid fills the areas ϑ_1 and ϑ_2. The fluid motion is assumed to be potential with a velocity potential $\Phi_j = \Phi_j(x_j, r, t)$. We denote the velocity potential Φ_j for the j layer. In areas ϑ_1 and ϑ_2 the Laplace equation will be fair

$$\Delta\Phi_j = 0, \quad (1)$$

where Δ is a Laplace-operator. On the free surface we take a simplified form of the linearized boundary condition

Content from this work may be used under the terms of the Creative Commons Attribution 3.0 licence. Any further distribution of this work must maintain attribution to the author(s) and the title of the work, journal citation and DOI.
Published under licence by IOP Publishing Ltd

$$\Phi_1\big|_{x_1=h_1}=0. \tag{2}$$

The non-flow conditions on the tank wall also are satisfied

$$\left.\frac{\partial\Phi_j}{\partial r}\right|_{r=R}=0. \tag{3}$$

The condition on the septum takes the next form:

$$\rho\left.\frac{\partial\Phi_1}{\partial t}\right|_{x_1=0}+\rho\left.\frac{\partial\Phi_2}{\partial t}\right|_{x_2=h_2}-\gamma\left.\frac{\partial\Phi_1}{\partial x_1}\right|_{x_1=0}=0, \tag{4}$$

where ρ is a fluid density, γ is a reduced coefficient of separator resistance. The bottom condition is

$$\nabla^4\dot{w}+k^4\dot{w}+\rho_0\frac{R^4}{D}\delta\ddot{w}=\frac{R^4}{D}\dot{q}+\frac{4R^2}{DR_1}\left(\tilde{C}_2+C_4\ln\xi\right), \tag{5}$$

where R_1 is a radius of curvature of the bottom, ρ_0 is a bottom material density, δ is a shell thickness, $D=\dfrac{E\delta^3}{12(1-\nu^2)}$ is a bending stiffness of the shell, ν is a Poisson's ratio, E is an elastic modulus, $k^4=\dfrac{E\delta R^4}{DR_1^2}$, $\dot{q}=-\rho\dot{\Phi}_2+\rho g\dot{w}$, $\xi=\dfrac{r}{R}$, and an arbitrary constant $\tilde{C}_2=C_2+C_4$. The kinematic condition on the bottom has the next form

$$\dot{w}=-\left.\frac{\partial\Phi_2}{\partial x_2}\right|_{x_2=0}. \tag{6}$$

The kinematic condition on the separator takes the form

$$\left.\frac{\partial\Phi_2}{\partial x_2}\right|_{x_2=h_2}=\left.\frac{\partial\Phi_1}{\partial x_1}\right|_{x_1=0}. \tag{7}$$

3. Solution of the boundary-value problem

Expressions for the term of the series in the expansion of the velocity potentials take the form

$$\Phi^{(k)}{}_1=\left[C_{10}+C_{20}\frac{x_1}{R}+J_0\left(\xi_k\frac{r}{R}\right)\left(C_{1k}\mathrm{ch}\left(\xi_k\frac{x_1}{R}\right)+C_{2k}\mathrm{sh}\left(\xi_k\frac{x_1}{R}\right)\right)\right]e^{i\omega t}, \tag{8}$$

$$\Phi^{(k)}{}_2=\left[A_0+B_0\frac{x_2}{R}+J_0\left(\xi_k\frac{r}{R}\right)\left(A_{1k}\mathrm{ch}\left(\xi_k\frac{x_2}{R}\right)+B_{2k}\mathrm{sh}\left(\xi_k\frac{x_2}{R}\right)\right)\right]e^{i\omega t}, \tag{9}$$

where ξ_k are the roots of the transcendental equation $J_1(r)=0$. In further investigations we confine ourselves to the first terms of the expansion. From the (2) we find

$$C_{10} = -C_{20}\frac{h_1}{R}, \tag{10}$$

and also

$$C_2 = -C_1 \mathrm{cth}\left(\xi_1 \frac{h_1}{R}\right). \tag{11}$$

The equation of motion for a shallow spherical shell has the form:

$$\nabla^4 \dot{w} - \beta^4 \dot{w} = \frac{4R^2}{DR_1}\left(\tilde{C}_2 + C_4 \ln r\right) - \frac{\rho\omega^2 R}{D_1}\left[A_1 J_0\left(\xi_1 \frac{r}{R}\right) + A_0\right] e^{i\omega t}, \tag{12}$$

where $\beta = \frac{R^4}{D} E\delta\left(\frac{\rho_0\omega^2}{E} - \frac{1}{RR_1}\right)$. We set the constant C_4 to zero to exclude the singularity at $r = 0$. The general integral of the expression (12) has the form:

$$\dot{w} = C_5 J_0(\beta r) + C_6 Y_0(\beta r) + C_7 I_0(\beta r) + C_8 K_0(\beta r) - \frac{4R^2}{DR_1\beta^4}, \tag{13}$$

where $J_0(\beta r)$ is a Bessel function zero order, 1st kind, $Y_0(\beta r)$ is a Neumann function zero order, $I_0(\beta r)$ is a Infeld function and $K_0(\beta r)$ is a McDonald function. We also set the constants C_6 and C_8 to zero to exclude the singularity at $r = 0$.

We now find a particular solution. We introduce the notation

$$\dot{w}^* = D_1 J_0\left(\xi_1 \frac{r}{R}\right) + D_2, \tag{14}$$

where

$$D_1 = -\frac{\rho\omega^2 R^4}{D\left(\beta_1^4 - \beta^4\right)}, \tag{15}$$

and

$$D_2 = \frac{\rho\omega^2 R^4}{D\beta^2} A_0. \tag{16}$$

Finally, we find [11]

$$\dot{w} = \left\{-\frac{4R^2}{DR_1\beta^4}\tilde{C}_2 - \frac{\rho\omega^2 R^4}{D}\left[\frac{A_1 J_0(\beta_1 r)}{\left(\beta_1^4 - \beta^4\right)}\right] + C_5 J_0(\beta r) + C_7 I_0(\beta r)\right\} e^{i\omega t}. \tag{17}$$

Next we introduce an auxiliary function:

$$\dot{F} = -\frac{E\delta R^2}{R_1}\nabla^2\dot{\varphi}, \tag{18}$$

or

$$\dot{F} = -\frac{E\delta R^2}{R_1}\left[-\frac{R^2}{DR_1\beta^4}\tilde{C}_2 r^2 + A_0\frac{\rho\omega^2 R^4}{D}\frac{r^2}{4} + C_5\left(-\frac{J_0(\beta r)}{\beta^2}\right) - C_7\frac{I_0(\beta r)}{\beta^2}\right],$$

and also for the function $\dot{F}$ the following condition is satisfied:

$$\dot{F}'' - \frac{R}{r}\dot{F}' = 0. \tag{19}$$

Thus, the fulfillment of the boundary conditions (2) – (4) leads to the following group of equations closed with respect to the constants $\tilde{C}_2$ and C_5:

$$-\frac{4R^2}{DR_1\beta}\tilde{C}_2 + \frac{\rho\omega^2 R^4}{D}\chi\left[-\frac{4R^2}{DR_1\beta^4}\tilde{C}_2 + C_5\alpha_{30} + C_7\alpha_{20}\right] + C_5 J_0(\beta R) + C_7 I_0(\beta R) = 0, \tag{20}$$

where $\chi = 1 + \frac{\rho_0\delta}{h_2\rho}$,

$$C_5 J_0'(\beta R) + C_7 I_0'(\beta R) = 0, \tag{21}$$

$$\frac{E\delta R^2}{R_1}\left[-\frac{2R^2}{DR_1\beta^4}\tilde{C}_2 + \frac{\rho\omega^2 R^4}{2D}A_0\right] = 0, \tag{22}$$

4. Conclusion

Thus, expressions 20-22 close the system of constants into two quantities: $\tilde{C}_2$ and C_5. Constructing from them a determinant in the form

$$\begin{vmatrix} \tilde{C}_2 a_{11}(\omega) & C_5 a_{12}(\omega) \\ \tilde{C}_2 a_{21}(\omega) & C_5 a_{22}(\omega) \end{vmatrix} = 0, \tag{23}$$

where $\omega = \alpha + i\beta$ and revealing it, we obtain a system of two equations

$$\begin{cases} \Theta_1(\alpha,\beta,\gamma) = 0, \\ \Theta_2(\alpha,\beta,\gamma) = 0, \end{cases} \tag{24}$$

that allows, for a given γ, to determine α and β. In the case under consideration

$$\tilde{U}(\tilde{\alpha},\tilde{\beta},\tilde{\gamma}) + i\tilde{V}(\tilde{\alpha},\tilde{\beta},\tilde{\gamma}) = 0, \tag{25}$$

where

$$\tilde{U}(\tilde{\alpha},\tilde{\beta},\tilde{\gamma}) = b_1\tilde{\alpha}^3 - 3b_1\tilde{\alpha}\tilde{\beta} - 2b_2\tilde{\alpha}\tilde{\beta} + b_3\tilde{\alpha} + b_4\tilde{\alpha}\tilde{\gamma}$$

$$\tilde{V}(\tilde{\alpha},\tilde{\beta},\tilde{\gamma}) = 3b_1\tilde{\alpha}^2\tilde{\beta} - b_1\tilde{\beta}^3 + b_2\tilde{\alpha}^2 + b_3\tilde{\beta} + b_4\tilde{\beta}\tilde{\gamma} + b_0,$$

$$b_1 = \frac{R^9\rho\kappa\left[\alpha_{30}I_0'(\beta R) - \alpha_{20}J_0'(\beta R)\right]}{D^2R_1\beta^4 I_0'(\beta R)},$$

$$b_2 = \frac{-1}{DR_1\beta^4 I_0'(\beta R)}\left[-iR^4\rho H_2\alpha_{20}J_0'(\beta R) - iR^4\rho H_1\alpha_{30}I_0'(\beta R) - \right.$$
$$-iR^4\rho H_1 J_0(\beta R)I_0'(\beta R) - iR^4\rho H_2 J_0'(\beta R)I_0(\beta R) + iR^4\rho H_2 J_0(\beta R)I_0'(\beta R) + iR^4\rho H_1\alpha_{20}J_0'(\beta R) +$$
$$\left. +iR^4\rho H_1 I_0(\beta R)J_0'(\beta R) + iR^4\rho H_2\alpha_{30}I_0'(\beta R)\right],$$

$$b_3 = \frac{R^4}{DR_1\beta^4 I_0'(\beta R)}\left[\alpha_{30}I_0'(\beta R) - \alpha_{20}J_0'(\beta R) - I_0(\beta R)J_0'(\beta R) + J_0(\beta R)I_0'(\beta R)\right],$$

$$b_4 = \frac{-\alpha_{20}J_0'(\beta R) - I_0(\beta R)J_0'(\beta R) + \alpha_{30}I_0'(\beta R) + J_0(\beta R)I_0'(\beta R)}{R_1\beta^4 I_0'(\beta R)},$$

$$\beta = \frac{R^4}{D}E\delta\left(\frac{\rho_0\omega^2}{E} - \frac{1}{RR_1}\right),$$

$$\alpha_{3i} = \int_0^R rJ_0(\beta r)J_0(\beta_i r)dr = -\frac{R}{\beta^2 - \beta_i^2}\left[\beta J_0'(\beta R)J_0(\beta_i R) - \beta_i J_0'(\beta_i R)J_0(\beta R)\right],$$

$$\alpha_{2i} = \int_0^R rI_0(\beta r)J_0(\beta_i r)dr = -\frac{R}{\beta^2 - \beta_i^2}\left[\beta I_0'(\beta_i R)J_0(\beta R) - \beta_i J_0'(\beta_i R)I_0(\beta R)\right], \beta_i = \frac{\xi_i}{R},$$

connecting the coefficient of damping with the reduced coefficient determined by the characteristics of the porous medium and its interaction with the liquid.

5. References

[1] Bagrov V V, Kurpatenkov A V and Polyaev V M 1997 *Capillary systems for Fluid Extraction from Tanks of Spacecrafts* (Moscow: UNPC Energomash) [in Russian]
[2] Mikishev G N and Rabinovich B I 1971 *Dynamics of thin-walled structures with compartments containing liquid* (Moscow: Mashinostroyenie) [in Russian]
[3] Okhotsimsky D Y 1956 Theory of motion of a body with cavities partially filled with liquid *Prikl. Mat. Mekh.* **20** (1) 3 - 20
[4] Chernousko F L 1966 On free oscillations of a viscous fluid in a vessel *J. Appl. Math. Mech.* **30** (5) 990–1003
[5] Bauer H F 1981 Hydroelastische schwingungen einer oberflächenspannungsstruktur in einem sateliitenbehälter *Z. Flugwiss. Weltraumforsch.* **5** (5) 303-13
[6] Kononov Yu N and Tatarenko E A 2006 Free vibrations of elastic membranes that separate a multilayer liquid in a cylindrical vessel with an elastic bottom *Dinamich. sist.* No 21 7-13
[7] Kravtsov A V and Sekerzh-Zen'kovich S Ya 2004 On free vibrations of a low-viscosity liquid in a vessel partially filled with a porous medium *Zh. Vych. Mat. Mat. Fiz.* **44** (4) 746-51
[8] Bukreev V I, Sturova I V and Chebotnikov A V 2014 Seiche oscillations in a reservoir filled with a doublelayer fluid *Fluid Dynamics* **49** (3) pp 395 – 402
[9] Ibrahim R A 2005 *Liquid sloshing dynamics: Theory and Applications.* (Cambridge: Univ. Press)

[10] Gribkov V A, Adamenko R A 2017 Two-dimensional fluid model for calculating the natural vibration frequencies of axially symmetric hydro-shell systems *Engineering Journal: Science and Innovation* **63** (3) DOI: 10.18698/2308-6033-2017-3-1593
[11] Pozhalostin A A and Goncharov D A 2015 Free axisymmetric oscillations of a two-layer liquid with an elastic separator between layers *Russ. Aeronaut.* **58** (1) 37 - 41

IOP Conf. Series: Materials Science and Engineering **468** (2018) 012009 doi:10.1088/1757-899X/468/1/012009

On the high-accuracy approach to flow simulation aroundthe airfoils by using vortex method

K S Kuzmina[1,2], **I K Marchevsky**[1,2] **and V S Moreva**[1,2]

[1] Bauman Moscow State Technical University, Moscow, Russia
[2] Ivannikov Institute for System Programming of the Russian Academy of Sciences, Moscow, Russia

Abstract. Some problems connected to vortex methods development for 2D incompressible flow simulation around airfoils are discussed. In the numerical schemes and algorithms which are normally used in the vortex methods, the airfoil is approximated by a polygon consists of rectilinear panels, and the the vortex sheet intensity, which is placed on the airfoil surface line and simulates the airfoil's influence the flow, is assumed to be piecewise-constant or piecewise-linear function. The most accurate mathematical models and numerical algorithms, based on such approaches, provide the estimation for the order of the error between $O(h)$ and $O(h^2)$ for the average vortex sheet intensity over the panels ($h$is the maximal panel length). In the present research a new approach is developed, where the vortex sheet intensity is assumed to be discontinuous piecewise-linear or piecewise-quadratic function and the curvature of the panels is taken into account. In order to compute the coefficients of the main system of the algebraic equations, the least squares method is used instead of commonly used no-through or no-slip conditions at separate control points or over the panels in the average. It is shown, that the developed approach is much more accurate in comparison with previously known ones: for some test problems (flows around a circular airfoil, an elliptical airfoil and Zhukovsky airfoil) it permits to obtain numerical solution which has error of order $O(h^4)$.

1. Introduction

The vortex methods which refer to the wide class of the meshless Lagrangian CFD methods, are a powerful tool in number of engineering applications, for example, in coupled hydroelastic problems, when it is possible to consider media to be incompressible, but the flow domain changes due to motion of the airfoil which is determined, in turn, by unsteady hydrodynamic loads.

For 2D problems there are some approaches to numerical solution of the Navier — Stokes equations by using vortex methods [1–4]; the Viscous Vortex Domains (VVD) method [5] seems to be the most suitable in practice. In this method the so-called "diffusive velocity" field is introduced in order to simulate vorticity evolution in viscous incompressible flow [3, 6]. The accuracy of the flow simulation and hydrodynamic loads computation depends on number of factors:

- the accuracy of the airfoil surface line approximation;
- the accuracy of computation of the vortex sheet intensity distribution over the surface line;
- the accuracy of the vortex wake approximation and its evolution simulation.

In the present paper we consider some possible ways how to raise the accuracy of the simulation due to

Content from this work may be used under the terms of the Creative Commons Attribution 3.0 licence. Any further distribution of this work must maintain attribution to the author(s) and the title of the work, journal citation and DOI.
Published under licence by IOP Publishing Ltd

the first and the second factors, because they seem to be the most important: the vortex sheet being generated on the surface line of the airfoil, is the source for the vortex wake formation in the flow.
Normally in previously known implementations of the vortex methods developed for numerical simulation in 2D hydrodynamic and hydroelastic problems, the vortex sheet intensity is found from the singular boundary integral equation of the 1-st kind [1]. However, such approach sometimes leads to significant errors and even to qualitatively wrong solution [7, 8].
The alternative approach is also known, which makes it possible to bring the problem of the vortex sheet intensity computation to solution of the Fredholm-type integral equation of the 2-nd kind [9]. The efficiency of such method in practice has been discussed in [7, 8, 10, 11]; there some numerical schemes are also described, which are based on this approach, and it is shown, that it provides much more accurate solutions for wide class of airfoils.
In order to raise the accuracy of the flow simulation in the vortex methods, first of all, the vortex sheet intensity computation procedure should be improved. In the well-known numerical schemes it is restricted within the low quality of the shape approximation of the surface line. In the present research new algorithm is suggested and the corresponding quadrature formulae are derived, which make it possible to develop main ideas of the approach [9] taking into account the curvature of the surface line. In the framework of this method the unknown vortex sheet intensity distribution can be discontinuous piecewise-constant, piecewise-linear or piecewise-quadratic function over the curvilinear panels, which approximate the shape of the surface line (in traditional approach vortex sheet intensity normally is considered to be piecewise-constant or piecewise-linear function over rectilinear panels [12]).
The mentioned ways for the accuracy improvement at first sight seem to be similar to the approaches, which are used extensively in the so-called "panel methods" [13], but there are significant differences between them. Firstly, in the suggested method the solution is not assumed to be continuous (and differentiable even less so) over the surface line of the airfoil — it is important for correct simulation of the flow around the airfoils having sharp edges and/or angle points. So in order to compute an integrals over the curvilinear parts of the surface line, the Gaussian quadrature formulae are used instead of integrating of the power series expansions. Secondly, in order to obtain the main linear system which approximates the boundary integral equation, the ideas of the least squares method are used: the true solution should minimize the integral over the surface line from the squared difference between the flow velocity (which depends on unknown vortex sheet intensity distribution) and the velocity of the surface line of the airfoil itself.
The described approach and the corresponding numerical algorithm allow to raise the accuracy of the vortex sheet intensity computation in the vortex methods significantly.

2. The governing equations

The flow of a viscous incompressible media is described by the Navier — Stokes equations

$$\nabla \cdot \vec{V} = 0, \qquad \frac{\partial \vec{V}}{\partial t} + (\vec{V} \cdot \nabla)\vec{V} = \nu \Delta \vec{V} - \frac{\nabla p}{\rho}.$$

Here, $\vec{V}(\vec{r},t)$ is the velocity field; $p(\vec{r},t)$ is the pressure distribution in the flow; $\rho = \mathrm{const}$ is the density of the media; ν is the kinematic viscosity coefficient.
Now we consider the problem of an outer flow simulation, so we should add the no-slip boundary condition on the surface line of the airfoil K and the perturbation decay conditions on infinity:

$$\vec{V}(\vec{r},t) = \vec{V}_K(\vec{r},t), \quad \vec{r} \in K; \qquad \vec{V}(\vec{r},t) \to \vec{V}_\infty, \quad p(\vec{r},t) \to p_\infty, \quad |\vec{r}| \to \infty.$$

Here, $\vec{V}_K(\vec{r},t)$ is the velocity of the surface line of the airfoil; we consider it to be known.

The Navier — Stokes equations can be written down in Helmholtz-type form in terms of the vorticity distribution $\vec{\Omega}(\vec{r},t) = \nabla \times \vec{V}(\vec{r},t)$:

$$\frac{\partial \vec{\Omega}}{\partial t} + \nabla \times (\vec{\Omega} \times \vec{U}) = 0. \tag{1}$$

Here, $\vec{U}(\vec{r},t) = \vec{V}(\vec{r},t) + \vec{W}(\vec{r},t)$; $\vec{W}(\vec{r},t)$ is the so-called "diffusive velocity", which is proportional to the viscosity coefficient and plays the key role for correct simulation of vorticity evolution in viscous media [3, 5]:

$$\vec{W}(\vec{r},t) = \nu \frac{(\nabla \times \vec{\Omega}) \times \vec{\Omega}}{|\vec{\Omega}|^2}.$$

It follows from (1), that the vorticity, which exists in the flow domain at the initial time moment, moves with the velocity $\vec{U}$, while "new" vorticity is generated only on flow domain boundary, i.e., on the surface line of the airfoil. We consider the initial vorticity distribution $\vec{\Omega}(\vec{r},t)$ to be known.

The airfoil influences the flow in the same way as the superposition of the influences of the attached vortex sheet with intensity $\gamma_{att}(\vec{r},t)$, the attached source sheet with intensity $q_{att}(\vec{r},t)$ and the free vortex sheet of unknown intensity $\gamma(\vec{r},t)$. These sheets are placed on the surface line of the airfoil; the attached vortex and source sheets simulate the motion and deformation (if they take place) of the airfoil, so their intensities are determined by the velocities of the points on the surface line:

$$\gamma_{att}(\vec{r},t) = \vec{V}_K(\vec{r},t) \cdot \vec{\tau}(\vec{r},t), \qquad q_{att}(\vec{r},t) = \vec{V}_K(\vec{r},t) \cdot \vec{n}(\vec{r},t), \qquad \vec{r} \in K,$$

where $\vec{n}(\vec{r},t)$ and $\vec{\tau}(\vec{r},t)$ are normal and tangent unit vectors, respectively [5].

In the present paper, for simplicity we consider the airfoil to be immovable, so $\gamma_{att}(\vec{r},t) = 0$, $q_{att}(\vec{r},t) = 0$. However, such assumption is non-essential and it can be omitted. We also consider the incident flow velocity to be constant vector: $\vec{V}_\infty = \text{const}$.

When vorticity distribution in the flow is known, the velocity field can be reconstructed by using the Biot — Savart law:

$$\vec{V}(\vec{r},t) = \vec{V}_\infty + \frac{1}{2\pi} \int_S \frac{\vec{\Omega}(\vec{s},t) \times (\vec{r} - \vec{s})}{|\vec{r} - \vec{s}|^2} \mathrm{d}S + \frac{1}{2\pi} \oint_K \frac{\vec{\gamma}(\vec{s},t) \times (\vec{r} - \vec{s})}{|\vec{r} - \vec{s}|^2} \mathrm{d}l_s. \tag{2}$$

Here, S is the flow domain; K is the surface line of the airfoil; $\vec{\gamma} = \gamma \vec{k}$ and $\vec{\Omega} = \Omega \vec{k}$ are the vectors of the vortex sheet intensity on the surface line and the vorticity field distribution in the flow domain, respectively; $\vec{k}$ is the unit vector orthogonal to the 2D flow domain ($\vec{n}(\vec{r}) \times \vec{\tau}(\vec{r}) = \vec{k}$).

The vortex sheet intensity on the surface line $\gamma(\vec{s},t)$ can be found form the no-slip boundary condition on the surface line of the immovable airfoil:

$$\vec{V}(\vec{r},t) = \vec{0}, \quad \vec{r} \in K.$$

We consider the simplest model problem: without vorticity in the flow domain ($\Omega(\vec{r},t) = 0$). It is necessary to determine vortex sheet intensity distribution over the surface line of the airfoil. From mathematical point of view, this problem is equivalent to the ideal (inviscid) incompressible flow simulation around the given airfoil when total circulation of the velocity field around the airfoil is known. When the media is viscid, the vortex sheet becomes the source of vorticity in the flow domain, and all the vorticity from it becomes part of the vortex wake at every time step. Thus, the described problem of the vortex sheet intensity computation (in the presence of vorticity in the flow) is also being solved at every time step.

3. Boundary integral equation for vortex sheet intensity

Taking into account, that unknown vortex sheet intensity $\gamma(\vec{s},t)$ corresponds to the so-called "free" vorticity, which is the part of the vortex wake in the flow domain, according to (2), it is possible to show, that the limit value of the flow velocity on the airfoil is equal to (hereinafter the dependencies of all values on time are omitted)

$$\vec{V}_-(\vec{r}) = \vec{V}_\infty + \frac{1}{2\pi}\oint_K \frac{\vec{\gamma}(\vec{s})\times(\vec{r}-\vec{s})}{|\vec{r}-\vec{s}|^2} dl_s - \left(\frac{\vec{\gamma}(\vec{r})}{2}\times\vec{n}(\vec{r})\right) + \frac{1}{2\pi}\int_S \frac{\vec{\Omega}(\vec{s},t)\times(\vec{r}-\vec{s})}{|\vec{r}-\vec{s}|^2} dS, \quad \vec{r}\in K. \tag{3}$$

In the framework of the "classical" approach, which is normally used in the implementations of the vortex methods, unknown function $\gamma(\vec{r})$ can be found from the satisfaction of the boundary condition for *normal* component of the limit value of velocity on the surface line of the airfoil:

$$\vec{V}_-(\vec{r})\cdot\vec{n}(\vec{r}) = 0, \quad \vec{r}\in K. \tag{4}$$

Using (3), it is easy to obtain the boundary integral equation of the 1-st kind for unknown $\gamma(\vec{r})$, which is singular and the integral in it should be computed as the Cauchy principal value [1]. The numerical solution of such equation requires specific quadrature formulae usage and can lead to significant error, and in some cases to incorrect results in the whole.

The alternative approach is also known [9], and according to it the boundary condition (4) is equivalent to the boundary condition, written down in the form

$$\vec{V}_-(\vec{r})\cdot\vec{\tau}(\vec{r}) = 0, \quad \vec{r}\in K, \tag{5}$$

which means the equality to zero of the *tangent* component of the velocity limit value on the surface line of the airfoil.It can be shown that the boundary condition (5) leads to the Fredholm-type boundary integral equation of the 2-nd kind:

$$\frac{1}{2\pi}\oint_K \frac{\vec{n}(\vec{r})\cdot(\vec{r}-\vec{s})}{|\vec{r}-\vec{s}|^2}\gamma(\vec{s})dl_s - \frac{\gamma(\vec{r})}{2} = -\vec{V}_\infty\cdot\vec{\tau}(\vec{r}) - \frac{1}{2\pi}\int_S \frac{\vec{n}(\vec{r})\cdot(\vec{r}-\vec{s})}{|\vec{r}-\vec{s}|^2}\Omega(\vec{s},t)dS, \quad \vec{r}\in K. \tag{6}$$

The kernel of the obtained equation is a bounded function in case of smooth surface line of the airfoil:

$$\lim_{|\vec{r}-\vec{s}|\to 0}\frac{|\vec{n}(\vec{r})\cdot(\vec{r}-\vec{s})|}{|\vec{r}-\vec{s}|^2} = \frac{\rceil(\vec{r})}{2},$$

where $\rceil(\vec{r})$ is the curvature at the surface line at the corresponding point.

The boundary integral equation (6), as well as the singular integral equation following from (4), has an infinite set of solutions. In order to select the unique solution, each of them should be solved together with an additional condition, which set the value of total circulation Γ of the velocity field around the airfoil. This value is equal to total "quantity of vorticity" and usually is known from the physical problem statement:

$$\oint_K \gamma(\vec{s})dl_s = \Gamma. \tag{7}$$

For the airfoils of the simplest shape (circular, elliptical, Zhukovsky airfoils) it is possible to use the conformal mappings technique in order to solve the corresponding problem analytically and obtain the exact solution for the vortex sheet intensity distribution over the surface line [14]. These exact solutions are used for the accuracy estimation of the developed numerical schemes.

4. Numerical scheme for the vortex sheet intensity computation

Now we consider two numerical schemes for computation of the vortex sheet intensity distribution over the surface line of the airfoil.

4.1.Numerical scheme with rectilinear panels

As it was mentioned earlier, in the vortex methods the surface line of the airfoil is usually approximated with a polygon which legs (called "panels") have lengthes L_i, $i=1,\dots,N$, and the vortex sheet intensity over the panels is assumed to be constant or linear. Here we consider the numerical scheme with piecewise-constant vorticity distribution, which however provides nearly the same accuracy as piecewise-linear schemes with respect to average values of the vortex sheet intensity over the panels. The vortex wake is normally simulated with N_v separate vortex elements, which positions $\vec{r}_s^{\,v}$ and circulations Γ_s, $s=1,\dots,N_v$, are assumed to be known, so the vorticity distribution $\Omega(\vec{r})$ in the flow domain is approximated by a linear combination of the following form

$$\Omega(\vec{r}) = \sum_{s=1}^{N_v} \Gamma_s \delta(\vec{r} - \vec{r}_s^{\,v}),$$

where $\delta(\vec{r})$ is the 2D Dirac delta-function.

In the framework of these assumptions, the integral in the left-hand side in (6) can be replaced by a sum of the integrals over the individual panels, which are proportional to the corresponding intensities. The integral in the right-hand side in (6) is equal to a sum of the influences of separate vortices in the flow. The accuracy of such approach can be raised significantly if we satisfy the boundary condition (6) not only at the centers of the panels (which are usually called "control points"), but on average over the panels:

$$\frac{1}{2\pi L_i}\sum_{j=1}^{N}\gamma_j \int_{K_i}\left(\int_{K_j}\frac{\vec{n}_i\cdot(\vec{r}-\vec{s})}{|\vec{r}-\vec{s}|^2}dl_s\right)dl_r - \frac{\gamma_i}{2} = -\vec{V}_\infty\cdot\vec{\tau}_i - \frac{1}{2\pi L_i}\sum_{s=1}^{N_v}\Gamma_s\int_{K_i}\frac{\vec{n}_i\cdot(\vec{r}-\vec{r}_s^{\,v})}{|\vec{r}-\vec{r}_s^{\,v}|^2}dl_r, \quad i=1,\dots,N.$$

The coefficients of these equations and right-hand side values can be calculated analytically; the exact formulae are given in [11, 12].

The numerical experiments show that the described scheme provides the accuracy order between $O(h)$ and $O(h^2)$ for different types of airfoils. The error means the maximal difference between the average vortex sheet intensity over the i-th panel computed numerically and by integrating of the exact solution over this panel $(i=1,\dots,n)$. Here, h is the maximal panel length (figure 1).

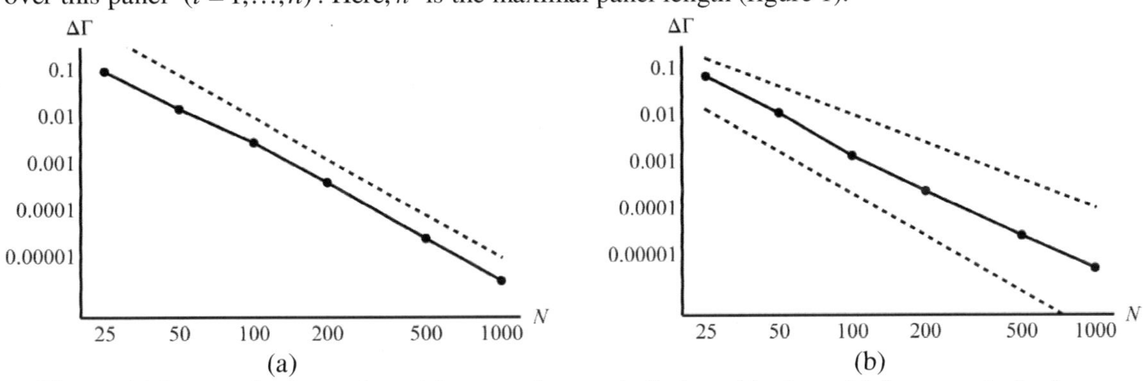

Figure 1.The error in the total vorticity over the panels (in logarithmic scale) for numerical scheme [11]:(a) — elliptical airfoil, dashed line corresponds to $O(h^2)$ error level; (b) — Zhukovsky airfoil, dashed lines correspond to $O(h)$ and $O(h^2)$ error levels.

IOP Conf. Series: Materials Science and Engineering **468** (2018) 012009 doi:10.1088/1757-899X/468/1/012009

4.2.Numerical scheme for the airfoils when the curvilinearity is taken into account

While developing of a new high-order numerical scheme, we should take into account explicitly the curvilinearity of the surface line of the airfoil, and also raise the quality of approximation of the vortex sheet intensity distribution. We assume that such distribution is linear or quadratic over each panel.

If we take into account only one of these two factors, it will provide only a slight improvement of the scheme, but the accuracy order will remain the same.

It is known [1], that vortex sheet intensity becomes unbounded in the neighborhood of "outer" angular points of the airfoil, so it is necessary to approximate smooth parts of the surface line with smooth curves when constructing high-order schemes. We consider that the shape of the surface line of the airfoil is known and it is described by the parametric piecewise-smooth dependencies $x = x(t)$, $y = y(t)$, $t \in [0, 2\pi)$, so we can exactly compute not only positions of particular points (which correspond to the endings of the panels), but also the directions of tangent lines (or their limit values) at these points.

We demand that the curve which approximates the surface line of the airfoil, passes through the chosen points and the direction of the tangent line for the curve at these points should be the same as for the original surface line.

Let's denote the beginning and the ending of the i-th chord of the surface line by C_i and C_{i+1}, respectively (these points correspond to parameter values $t = t_i$ and $t = t_{i+1}$ in the parametric equations of the surface line) and let $\vec{\tau}_i^0$ be a tangent unit vector, which direction coincides with the vector $\overrightarrow{C_i C_{i+1}}$; $\vec{n}_i^0$ is a unit vector, which is orthogonal to the vector $\vec{\tau}_i^0$ (figure 2). Hereinafter when we say "chord" we mean the rectilinear line segment between points C_i and C_{i+1}; which has in general case only common ending with the original surface lines.

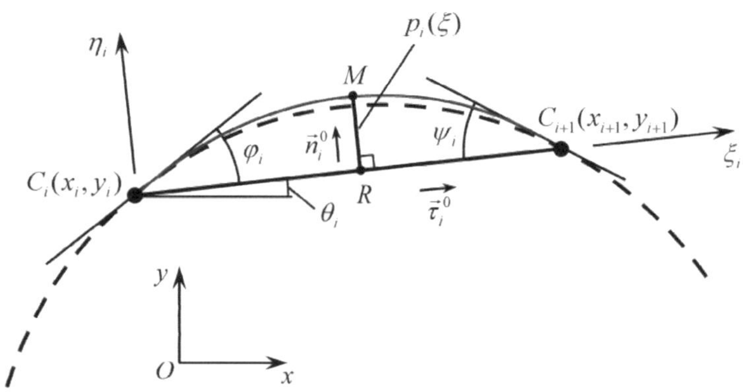

Figure 2. Piecewise-polynomial surface line approximation: dashed line — surface line of the airfoil; solid line — curvilinear panel

On each panel a local orthogonal coordinate system is introduced $C_i \xi_i \eta_i$, so that the points C_i and C_{i+1} have local coordinates $(0,0)$ and $(L_i, 0)$, respectively, where $L_i = \overrightarrow{C_i C_{i+1}}$ is the length of the i-th chord. In order to construct an interpolation curve, we firstly calculate tangents of inclination angles ϕ_i and ψ_i for the tangent lines to the panels (figure 2) by using the following formulae:

$$\tan\phi_i = -\frac{x'_+(t_i)\sin\theta_i - y'_+(t_i)\cos\theta_i}{x'_+(t_i)\cos\theta_i + y'_+(t_i)\sin\theta_i}, \quad \tan\psi_i = \frac{x'_-(t_{i+1})\sin\theta_i - y'_-(t_{i+1})\cos\theta_i}{x'_-(t_{i+1})\cos\theta_i + y'_-(t_{i+1})\sin\theta_i}.$$

Here θ_i is an angle between the i-th panel and Ox axis; x'_+, x'_-, y'_+, y'_- denote right-side and left-side derivatives of the parametric dependencies $x(t)$ and $y(t)$, being calculated at the corresponding points; ϕ_i and ψ_i can have values from the range $\left(-\frac{\pi}{2};\frac{\pi}{2}\right)$.

The equation of the interpolation curve (which is the cubic spline in the local coordinates), which approximates the surface line of the airfoil in the local coordinates over the i-th panel has the following form:

$$p_i(\xi)=\frac{\xi(\xi-L_i)}{L_i}\left(a_i+b_i\frac{\xi-L_i/2}{L_i}\right),$$

where the conditions $p_i(0)=0$, $p_i(L_i)=0$ are satisfied automatically, the values of the coefficients a_i and b_i can be found form the conditions $p_i'(0)=\tan\phi_i$ and $p_i'(L_i)=-\tan\psi_i$:

$$a_i=-\frac{1}{2}(\tan\phi_i+\tan\psi_i),\quad b_i=\tan\phi_i-\tan\psi_i.$$

The constructed interpolation curve over the i-th chord of the surface line of the airfoil we call hereinafter "curvilinear panel", or simply "panel". So the position of an arbitrary point M, which lies on the curvilinear panel and has the local coordinate (abscise) ξ, is given by a radius vector

$$\overrightarrow{OM}(\xi)=\overrightarrow{OC_i}+\xi\vec{\tau}_i^{\,0}+p_i(\xi)\vec{n}_i^{\,0}.$$

If the part of the surface line of the airfoil over the i-th panel is a smooth curve of C^4 class, then the approximation error has the order $O(L_i^4)$. This results can be easily checked by writing down the corresponding Taylor expansion.

Then we approximate the unknown vortex sheet intensity distribution over the i-th panel with a quadratic function:

$$\gamma_i(\xi)=\alpha_i+\beta_i\frac{\xi}{L_i}+\delta_i\frac{\xi^2}{L_i^2}. \tag{8}$$

The values of the coefficients α_i, β_i, δ_i, $i=1,\dots,N$, should be found from the integral equation (6) with the additional condition (7).

In order to find an approximate solution of this equation for the given dependency $\gamma_i(\xi)$, we use the least squares method and solve an unconstrained minimization problem for the following function

$$\Psi=\oint_K\left(\frac{1}{2\pi}\oint_K\frac{\vec{n}(\vec{r})\cdot(\vec{r}-\vec{s})}{|\vec{r}-\vec{s}|^2}\gamma(\vec{s})\mathrm{d}l_s-\frac{\gamma(\vec{r})}{2}+\vec{V}_\infty\cdot\vec{\tau}(\vec{r})+\frac{1}{2\pi}\sum_{s=1}^{N_v}\frac{\vec{n}(\vec{r})\cdot(\vec{r}-\vec{r}_s^{\,v})}{|\vec{r}-\vec{r}_s^{\,v}|^2}\Gamma_s\right)\mathrm{d}l_r-$$
$$-\lambda\left(\oint_K\gamma(\vec{r})\mathrm{d}l_r-\Gamma\right)\to\min \tag{9}$$

for all possible values of the parameters α_i, β_i, δ_i and the Lagrangian coefficient λ.

Both, the inner and the outer integrals, in (9) can be replaced with sums of the integrals over the curvilinear panels:

$$\Psi=\sum_{i=1}^{N}\int_{K_i}\left(\frac{1}{2\pi}\sum_{j=1}^{N}\int_{K_j}\frac{\vec{n}(\vec{r})\cdot(\vec{r}-\vec{s})}{|\vec{r}-\vec{s}|^2}\gamma(\vec{s})\mathrm{d}l_s-\frac{\gamma(\vec{r})}{2}+\vec{V}_\infty\cdot\vec{\tau}(\vec{r})+\frac{1}{2\pi}\sum_{s=1}^{N_v}\frac{\vec{n}(\vec{r})\cdot(\vec{r}-\vec{r}_s^{\,v})}{|\vec{r}-\vec{r}_s^{\,v}|^2}\Gamma_s\right)\mathrm{d}l_r-$$
$$-\lambda\left(\sum_{i=1}^{N}\int_{K_i}\gamma(\vec{r})\mathrm{d}l_r-\Gamma\right)\to\min. \tag{10}$$

All the integrals are computed in the local coordinates, so for the i-th and the j-th panels we obtain

$\vec{r} = \vec{r}_i(\xi)$ and $\vec{s} = \vec{s}_j(\zeta)$:

$$\vec{r}_i(\xi) = \overrightarrow{OC_i} + \xi\vec{\tau}_i^{\,0} + p_i(\xi)\vec{n}_i^{\,0}, \quad \vec{s}_j(\zeta) = \overrightarrow{OC_j} + \zeta\vec{\tau}_j^{\,0} + p_j(\zeta)\vec{n}_j^{\,0}.$$

The Jacobian determinant of the coordinates transformation from global to local ones has the following form:

$$J_i(\xi) = \left.\frac{\mathrm{d}l_r}{\mathrm{d}\xi}\right|_{\vec{r}=\vec{r}_i(\xi)} = \sqrt{1+(p_i'(\xi))^2},$$

and as the result, denoting for the simplicity $\gamma(\vec{r}_i(\xi)) = \gamma_i(\xi)$, $\vec{n}(\vec{r}_i(\xi)) = \vec{n}_i(\xi)$, $\vec{\tau}(\vec{r}_i(\xi)) = \vec{\tau}_i(\xi)$, the function to be minimized (10) can be written down in the following form:

$$\Psi = \sum_{i=1}^{N}\int_0^{L_i}\left(\frac{1}{2\pi}\sum_{j=1}^{N}\int_0^{L_j}\frac{\vec{n}_i(\xi)\cdot(\vec{r}_i(\xi)-\vec{s}_j(\zeta))}{|\vec{r}_i(\xi)-\vec{s}_j(\zeta)|^2}\gamma_j(\zeta)J_j(\zeta)\mathrm{d}\zeta - \frac{\gamma_i(\xi)}{2} + \vec{V}_\infty\cdot\vec{\tau}_i(\xi) + \right.$$
$$\left. + \frac{1}{2\pi}\sum_{s=1}^{N_v}\frac{\vec{n}_i(\xi)\cdot(\vec{r}_i(\xi)-\vec{r}_s^{\,v})}{|\vec{r}_i(\xi)-\vec{r}_s^{\,v}|^2}\Gamma_s\right)^2 J_i(\xi)\mathrm{d}\xi - \lambda\left(\sum_{i=1}^{N}\int_0^{L_i}\gamma_i(\xi)J_i(\xi)\mathrm{d}\xi - \Gamma\right) \to \min. \quad (11)$$

According to above mentioned assumptions, the necessary information about the shape of the surface line, which is needed for the numerical scheme construction, is limited by positions of the nodes (the panel endings) and the tangent line directions at these points. So the unit tangent vector $\vec{\tau}_i(\xi)$ in formula (11) should be calculated for the panel instead of the initial surface line:

$$\vec{\tau}_i(\xi) = \frac{\vec{\tau}_i^{\,0} + p_i'(\xi)\vec{n}_i^{\,0}}{|\vec{\tau}_i^{\,0} + p_i'(\xi)\vec{n}_i^{\,0}|}.$$

The approximation error for the tangent unit vector has the order $O(L_i^3)$ for C^4-class curves. The normal unit vector $\vec{n}_i(\xi)$ should be chosen orthogonal to the $\vec{\tau}_i(\xi)$.

After substitution of the expression (8) for unknown vortex sheet intensity distribution into (11), we obtain the function to be minimized, which depends only on unknown coefficients α_k, β_k, δ_k, $k = 1,\ldots,N$, and Lagrangian coefficient λ:

$$\Psi = \sum_{i=1}^{N}\int_0^{L_i}\left(\frac{1}{2\pi}\sum_{j=1}^{N}\left(\alpha_j\int_0^{L_j}\frac{\vec{n}_i(\xi)\cdot(\vec{r}_i(\xi)-\vec{s}_j(\zeta))}{|\vec{r}_i(\xi)-\vec{s}_j(\zeta)|^2}J_j(\zeta)\mathrm{d}\zeta + \beta_j\int_0^{L_j}\frac{\vec{n}_i(\xi)\cdot(\vec{r}_i(\xi)-\vec{s}_j(\zeta))}{|\vec{r}_i(\xi)-\vec{s}_j(\zeta)|^2}\frac{\zeta}{L_j}J_j(\zeta)\mathrm{d}\zeta + \right.\right.$$
$$\left. + \delta_j\int_0^{L_j}\frac{\vec{n}_i(\xi)\cdot(\vec{r}_i(\xi)-\vec{s}_j(\zeta))}{|\vec{r}_i(\xi)-\vec{s}_j(\zeta)|^2}\frac{\zeta^2}{L_j^2}J_j(\zeta)\mathrm{d}\zeta\right) - \frac{1}{2}\left(\alpha_i + \beta_i\frac{\xi}{L_i} + \delta_i\frac{\xi^2}{L_i^2}\right) + \vec{V}_\infty\cdot\vec{\tau}_i(\xi) + \quad (12)$$
$$\left. + \frac{1}{2\pi}\sum_{s=1}^{N_v}\frac{\vec{n}_i(\xi)\cdot(\vec{r}_i(\xi)-\vec{r}_s^{\,v})}{|\vec{r}_i(\xi)-\vec{r}_s^{\,v}|^2}\Gamma_s\right)J_i(\xi)\mathrm{d}\xi - \lambda\left(\sum_{i=1}^{N}\int_0^{L_i}\left(\alpha_i + \beta_i\frac{\xi}{L_i} + \delta_i\frac{\xi^2}{L_i^2}\right)J_i(\xi)\mathrm{d}\xi - \Gamma\right) \to \min.$$

After introducing the denotations as follows

$$\int_0^{L_j}\frac{\vec{n}_i(\xi)\cdot(\vec{r}_i(\xi)-\vec{s}_j(\zeta))}{|\vec{r}_i(\xi)-\vec{s}_j(\zeta)|^2}\frac{\zeta^r}{L_j^r}J_j(\zeta)\mathrm{d}\zeta = I_{ij}^{(r)}(\xi); \quad \int_0^{L_i}J_i(\xi)\frac{\xi^r}{L_i^r}\mathrm{d}\xi = J_i^{(r)}; \quad \frac{\vec{n}_i(\xi)\cdot(\vec{r}_i(\xi)-\vec{r}_s^{\,v})}{|\vec{r}_i(\xi)-\vec{r}_s^{\,v}|^2} = Q_s(\xi),$$

the equation (12) takes more compact form:

$$\Psi = \sum_{i=1}^{N}\int_0^{L_i}\left(\frac{1}{2\pi}\sum_{j=1}^{N}\left(\alpha_j I_{ij}^{(0)}(\xi) + \beta_j I_{ij}^{(1)}(\xi) + \delta_j I_{ij}^{(2)}(\xi)\right) - \frac{1}{2}\left(\alpha_i + \beta_i\frac{\xi}{L_i} + \delta_i\frac{\xi^2}{L_i^2}\right) + \right.$$

$$+\vec{V}_\infty \cdot \vec{\tau}_i(\xi) + \frac{1}{2\pi}\sum_{s=1}^{N_v}\Gamma_s Q_s(\xi)\Bigg)^2 J_i(\xi)d\xi - \lambda\left(\sum_{i=1}^{N}\left(\alpha_i J_i^{(0)} + \beta_i J_i^{(1)} + \delta_i J_i^{(2)}\right) - \Gamma\right) \to \min .$$

The minimal value of this function is achieved when all the partial derivatives are equal to zero:

$$\frac{\partial\Psi}{\partial\alpha_k} = \sum_{i=1}^{N}\int_0^{L_i} 2\Bigg(\frac{1}{2\pi}\sum_{j=1}^{N}\left(\alpha_j I_{ij}^{(0)}(\xi) + \beta_j I_{ij}^{(1)}(\xi) + \delta_j I_{ij}^{(2)}(\xi)\right) -$$
$$-\frac{1}{2}\left(\alpha_i + \beta_i\frac{\xi}{L_i} + \delta_i\frac{\xi^2}{L_i^2}\right) + \vec{V}_\infty \cdot \vec{\tau}_i(\xi) + \frac{1}{2\pi}\sum_{s=1}^{N_v}\Gamma_s Q_s(\xi)\Bigg)\left(\frac{I_{ik}^{(0)}(\xi)}{2\pi} - \frac{1}{2}\right)J_i(\xi)d\xi - \lambda J_k^{(0)} = 0,$$

$$\frac{\partial\Psi}{\partial\beta_k} = \sum_{i=1}^{N}\int_0^{L_i} 2\Bigg(\frac{1}{2\pi}\sum_{j=1}^{N}\left(\alpha_j I_{ij}^{(0)}(\xi) + \beta_j I_{ij}^{(1)}(\xi) + \delta_j I_{ij}^{(2)}(\xi)\right) -$$
$$-\frac{1}{2}\left(\alpha_i + \beta_i\frac{\xi}{L_i} + \delta_i\frac{\xi^2}{L_i^2}\right) + \vec{V}_\infty \cdot \vec{\tau}_i(\xi) + \frac{1}{2\pi}\sum_{s=1}^{N_v}\Gamma_s Q_s(\xi)\Bigg)\left(\frac{I_{ik}^{(1)}(\xi)}{2\pi} - \frac{1}{2}\frac{\xi}{L_k}\right)J_i(\xi)d\xi - \lambda J_k^{(1)} = 0,$$

$$\frac{\partial\Psi}{\partial\delta_k} = \sum_{i=1}^{N}\int_0^{L_i} 2\Bigg(\frac{1}{2\pi}\sum_{j=1}^{N}\left(\alpha_j I_{ij}^{(0)}(\xi) + \beta_j I_{ij}^{(1)}(\xi) + \delta_j I_{ij}^{(2)}(\xi)\right) -$$
$$-\frac{1}{2}\left(\alpha_i + \beta_i\frac{\xi}{L_i} + \delta_i\frac{\xi^2}{L_i^2}\right) + \vec{V}_\infty \cdot \vec{\tau}_i(\xi) + \frac{1}{2\pi}\sum_{s=1}^{N_v}\Gamma_s Q_s(\xi)\Bigg)\left(\frac{I_{ik}^{(2)}(\xi)}{2\pi} - \frac{1}{2}\frac{\xi^2}{L_k^2}\right)J_i(\xi)d\xi - \lambda J_k^{(2)} = 0,$$

$$\frac{\partial\Psi}{\partial\lambda} = \sum_{i=1}^{N}\left(\alpha_i J_i^{(0)} + \beta_i J_i^{(1)} + \delta_i J_i^{(2)}\right) - \Gamma = 0.$$

The derived equations can be written down in much more simple form if we additionally introduce the following designations:

$$J_{mnk}^{(p,q)} = \int_0^{L_m} I_{mn}^{(p)}(\xi) I_{mk}^{(q)}(\xi) J_m(\xi)d\xi, \qquad J_{mn}^{(p,r)} = \int_0^{L_m} I_{mn}^{(p)}(\xi)\frac{\xi^r}{L_m^r} J_m(\xi)d\xi,$$

and also denote unknown values as $\gamma_j^{(u)}$, $u = 0, 1, 2$, where

$$\gamma_j^{(0)} = \alpha_j, \quad \gamma_j^{(1)} = \beta_j, \quad \gamma_j^{(2)} = \delta_j, \quad j = 1,\ldots,N.$$

Now the derived linear system of algebraic equations can be written down in the most compact form:

$$\sum_{j=1}^{N}\sum_{u=0}^{2}\gamma_j^{(u)}\left(\frac{1}{2\pi^2}\sum_{i=1}^{N} J_{ijk}^{(u,r)} - \frac{1}{2\pi}\sum_{i=1}^{N}\left(J_{ij}^{(u,r)}\frac{L_i^r}{L_k^r}\right) - \frac{1}{2\pi}J_{jk}^{(r,u)} + \frac{1}{2}J_j^{(u+r)}\frac{L_i^r}{L_k^r}\right) - \lambda J_k^{(r)} =$$
$$= -\sum_{i=1}^{N}\int_0^{L_i}\left(\vec{V}_\infty \cdot \vec{\tau}_i(\xi) + \frac{1}{2\pi}\sum_{s=1}^{N_v}\Gamma_s Q_s(\xi)\right)\left(\frac{1}{\pi}I_{ik}^{(r)}(\xi) - \frac{\xi^r}{L_k^r}\right)d\xi, \quad k = 1,\ldots,N, \quad r = 0,1,2,$$
$$\sum_{j=1}^{N}\sum_{u=0}^{2}\gamma_j^{(u)} J_j^{(u)} = \Gamma. \qquad (13)$$

If we consider the piecewise linear distribution of the vortex sheet intensity over the panels, the variables r and u (which are the indices in the sums) in (13) will take only the values 0 and 1.
In order to compute approximately the integrals in (13), the Gaussian quadrature formulae are used for

n_{gp} nodes, and according to them

$$\int_a^b f(x)\mathrm{d}x \approx \frac{b-a}{2}\sum_{k=1}^{n_{gp}} \omega_k f\left(\frac{a+b}{2}+\frac{b-a}{2}x_k\right).$$

Here, the values of the weight coefficients ω_k and the positions of the Gaussian points x_k, $k=1,\ldots,n_{gp}$, are chosen in standard way [15].

In the present research the value $n_{gp}=7$ have been used. In the cases, when the Gaussian approximate formulae do not provide the necessary accuracy of the integrals computation (this usually occurs for the neighboring panels), an additional splitting procedure can be implemented to the chords, according to which these chords are divided into $N_{add}=4$ sub-chords of smaller length (only for numerical calculations of the integrals by using the Gaussian quadrature formulae). If it is necessary, these "small" panels should be split once again.

Numerical experiments show, that taking into account of the Jacobian determinant $J_i(\xi)$ value only slightly influences the accuracy of the solution when number of the panels N is sufficiently high, so in practice we can consider $J_i(\xi)\equiv 1$; this simplification makes the computational procedure easier.

It should be noted, that if the described numerical algorithm is implemented in some special mathematical software (MatLAB, Mathematica, etc.), built-in algorithms for numerical integration can be used, of course. It makes the development of the corresponding subroutine much easier in comparison with C++ or Fortran, but time of quadratures computation, however, becomes much higher in general case.

5. Numerical experiment

We consider three model problems of a flow simulation around the following airfoils: the circular airfoil of unit radius ($R=1$), elliptical airfoil with the semiaxes $a_1=1.0$, $b_1=0.5$ and the Zhukovsky airfoil with parameters $a=3.5$, $d=0.4$, $h=0.3$ at the angle of incidence $\beta=\pi/6$.

5.1. Exact analytical solution

Exact analytical solution for elliptical airfoils and Zhukovsky wing airfoils, which will be used for the numerical schemes accuracy estimations, can be constructed by using the conformal mappings techniques [14, 16], so for vortex sheet intensity distribution over the surface line of the airfoil the formulae can be used

$$\gamma^*(t) = \big(2V_\infty \sin(\varphi+\beta-t)+(\Gamma+W(t))/(\pi R)\big)\cdot\left|1-\frac{a^2}{(Re^{i(t-\varphi)}+H)^2}\right|^{-1}, \quad H=\mathrm{i}h-d\,\mathrm{e}^{-\mathrm{i}\varphi}. \tag{14}$$

Here, parameter $t\in[0;2\pi)$ corresponds to the position of the point on the surface line; the parametric dependencies $x(t)=\operatorname{Re} z(t)$, $y(t)=\operatorname{Im} z(t)$ determine the shape of the airfoil, where

$$z(t)=x(t)+\mathrm{i}\,y(t)=\frac{1}{2}\left(\chi(t)+\frac{a^2}{\chi(t)}\right), \quad \chi(t)=R\mathrm{e}^{\mathrm{i}(t-\varphi)}+H; \tag{15}$$

the term W expresses the influence of vortex elements (point vortices) in the flow:

$$W(t)=\sum_{s=1}^{N_v}\Gamma_s\frac{\cos(t-\theta_s)-d_s^*}{\cos(t-\theta_s)+(d_s+d_s^*)/2}.$$

In order to compute the values d_s, d_s^* and θ_s we should consider the vortices on the complex plane, then their positions will be $z_s=x_s+\mathrm{i}\,y_s$, $s=1,\ldots,N_V$,

$$d_s = |w_s - H| / R, \quad d_s^* = d_s^{-1}, \quad \theta_s = \varphi + \mathrm{Arg}(w_s - H),$$

where $w_s = z_s + (z_s^2 - a^2)^{1/2}$, and that branch of the square root function should be chosen, that provides the condition $d_s > 1$ satisfaction.
For the elliptical airfoil we should choose

$$a = (a_1^2 - b_1^2)^{1/2}, \quad R = a_1 + b_1, \quad \varphi = 0, \quad h = d = 0,$$

where a_1 and b_1 are the semiaxes of the ellipse; for the Zhukovsky airfoil

$$R = |H - a|, \quad \varphi = \arctan(h/a),$$

and the parameters a, d and h determine length, width and curvature of the airfoil, respectively.
The value of total vorticity Γ for the elliptical airfoil can be chosen arbitrary, for our test problem we assume $\Gamma = 0$ for any angle of incidence; for the Zhukovsky airfoil Γ can be chosen from the condition, that velocity at the sharp edge is bounded: its value is proportional to the incident flow velocity and depends on the shape of the airfoil and the angle of incidence, as well as on the vortex elements positions in the flow:

$$\Gamma = -2\pi V_\infty \sin(\beta + \varphi)\left((h^2 + a^2)^{1/2} + d\right) - \sum_{s=1}^{N_V} \Gamma_s \frac{\cos\theta_s - d_s^*}{\cos\theta_s - (d_s + d_s^*)/2}.$$

5.2. Numerical solution and the developed scheme accuracy estimation
Firstly, by using the developed numerical scheme with curvilinear panels, the vortex sheet intensity distribution had been computed (the values of the coefficients α_i, β_i and δ_i for all the panels). Then this distribution was integrated over the curvilinear panels, and the average value was found by dividing on the panel length:

$$\gamma_i = \int_0^{L_i} \left(\alpha_i + \beta_i \frac{\xi}{L_i} + \delta_i \frac{\xi^2}{L_i^2} \right) J_i(\xi) \mathrm{d}\xi \bigg/ \int_0^{L_i} J_i(\xi) \mathrm{d}\xi, \quad i = 1, \ldots, N.$$

The similar values γ_i^* were obtained by integration of the exact solution (14):

$$\gamma_i^* = \int_{t_i}^{t_{i+1}} \gamma^*(t) \left(x'(t)^2 + y'(t)^2 \right)^{1/2} \mathrm{d}t \bigg/ \int_{t_i}^{t_{i+1}} \left(x'(t)^2 + y'(t)^2 \right)^{1/2} \mathrm{d}t, \quad i = 1, \ldots, N.$$

where the parameter values t_i and t_{i+1} correspond to the beginning and the ending of the i-th panel, the dependencies $x(t)$ and $y(t)$ can be found from (15).
The error of the numerical solution was computed as the maximal absolute difference between the average vortex sheet intensities over the panels, computed for the exact and numerical solutions:

$$\Delta\gamma = \max_i |\gamma_i - \gamma_i^*|.$$

The results show nearly the same value of the error for the considered test problems in cases of piecewise-linear and piecewise-quadratic vorticity distributions over the panels, however, it should be noted that in case of piecewise-quadratic approximation the algebraic system (13)becomes ill-conditioned. In all cases (for the circular, elliptical and Zhukovsky wing airfoils) the error of the numerical solution has order $O(h^4)$, where h is the maximal panel length (figure 3).

6. Conclusion
A new approach and the corresponding high-order numerical algorithm are developed for the vortex sheet intensity computation on the surface line of the airfoil in incompressible flow. In order to achieve high order of accuracy, the curvature of the airfoil is taken into account as well as the piecewise-linear or

piecewise-quadratic vorticity distribution in the vortex sheet over each panel. The developed numerical scheme permits to simulate the solution discontinuities between the panels; this makes it possible to simulate airfoils with sharp edges, for example, the Zhukovsky wing airfoil. The numerical experiments proved the properties of the developed scheme and showed that it provides the 4-th order of accuracy.

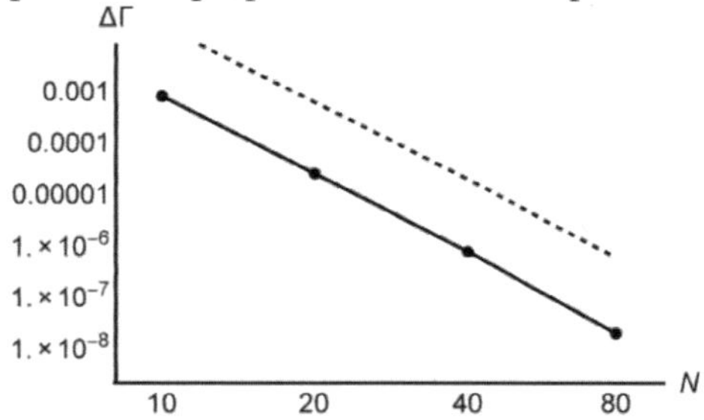

(a)

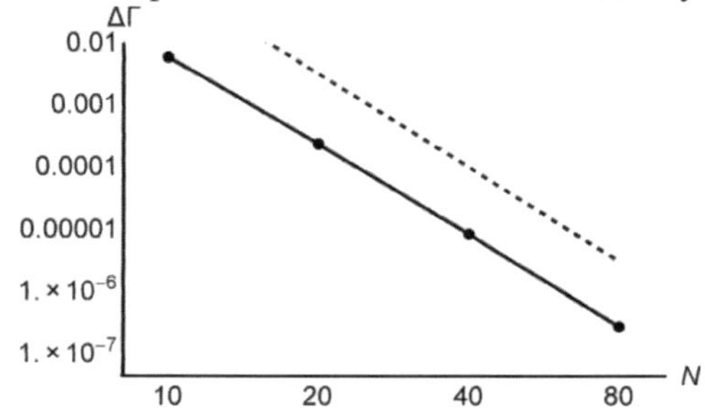

(b)

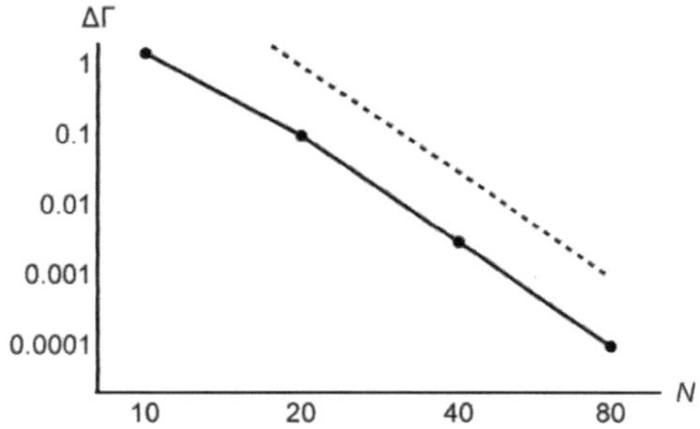

(c)

Figure 3. The error in total vorticity over the panels (in logarithmic scale) for the developed numerical scheme (13): (a) — circular airfoil; (b) — elliptical airfoil; (c) — Zhukovsky airfoil; dashed line everywhere corresponds to $O(h^4)$ error level

References

[1] Lifanov IK and Belotserkovsky SM 1992 *Methods of Discrete Vortices*(CRC Press) p 464

[2] Cottet G-H and Koumoutsakos P D 2000 *Vortex Methods: Theory and Practice* (Cambridge: Cambridge University Press) p 354

[3] Dynnikova G Ya 2003 Vortex motion in two-dimensional viscous fluid flows *Fluid Dynamics* **38:5** pp 670

[4] Kalugin VT, Mordvintcev GG and Popov VM 2011 *Modeling of the flow and the aerodynamic characteristics control for aircraft* (in Rus.)(Moscow: Bauman University Publ.) p 527

[5] Andronov P R, Guvernyuk S V and Dynnikova G Ya 2006 *Vortex Calculation Methods of Non-StationaryHydrodynamical Loadings* (in Rus.) (Moscow: Moscow University Press) p 184

[6] Ogami Y and Akamatsu T 1991 Viscous Flow Simulation Using the Discrete Vortex Model – the Diffusion Velocity Method *Comput. and Fluids* **19:3/4** pp 433

[7] Kuzmina K S and Marchevsky I K 2015 *Proc. IV Int. Conf. on Particle-Based Methods.Fundamentals and Applications PARTICLES 2015 (Barcelona)* ed E Onate,D R J Owen,T Zohdi,M Bischoffand P Wriggers (Barcelona, Spain: International Center for Numerical Methods in Engineering(CIMNE)) pp 680

[8] Kuzmina KS, Marchevsky IK and Moreva VS 2018 On vortex sheet intensity computation for airfoils with angle point in vortex methodsInt. J. of Mech. Eng. and Tech. **9:2** pp 799

[9] Kempka S N, Glass M W, Peery J S and Strickland J H 1996 Accuracy considerations for implementingvelocity boundary conditions in vorticity formulations *SANDIAREPORT* SAND96-0583 UC-700 p 52

[10] Kuzmina K S, Marchevskii I K, Moreva V S and Ryatina E P 2017 Numerical scheme of the second orderof accuracy for vortex methods for incompressible flow simulation around

airfoils *Russian Aeronautics* **60:3** pp 398

[11] Kuzmina K S, Marchevskii I K and Moreva V S 2018 Vortex sheet intensity computation in incompressibleflow simulation around airfoil by using vortex methods *Mathematical Models and Computer Simulations* **10:3** pp 276

[12] Kuzmina K S, Marchevsky I K and Ryatina E P 2017 Exact analytical formulae for linearly distributed vortexand source sheets influence computation in 2D vortex methods *J. Phys.: Conf. Series* **918** 012013

[13] Vaz G, Falcao de Campos JAC and Eca L 2013 A numerical study on low and higher-order potential based BEM for 2D inviscid flows *Computational Mechanics* **32:4-6** pp 327

[14] Kuzmina K S, Marchevsky I K and Ryatina E P 2018 Exact solutions of boundary integral equation arisingin vortex methods for incompressible flow simulation around elliptical and Zhukovsky airfoils *J. Phys.:Conf. Series* In press

[15] Abramowitz M and Stegun IA 1965 *Handbook of Mathematical Functions with Formulas, Graphs, and Mathematical Tables* (New York: Dover)

[16] Milne-Thomson L M 1962 *Theoretical hydrodynamics* (London: Macmillan & Co LTD) p 660

Acknowledgments

The research was funded by RFBR according to the research project № 18-31-00245.

Theory and Practice of Harmonic Drive Mechanisms

G A Timofeyev, Yu V Kostikov, A V Yaminsky and Ye O Podchasov

Bauman Moscow State Technical University, Moscow, Russia
E-mail: timga@bmstu.ru

Abstract. Among different mechanical drives of industrial application harmonic drive mechanisms (HDM) are distinguished by minimal materials consumption and overall dimensions, provide high efficiency, torsion rigidity and low kinematic error. Due to this they are widely used in mechanical actuators, robotics, aerial drives, supporting and turning arrangements of follow-up systems, dial indicators and other precise mechanisms. Combined design technique of different-type harmonic drive mechanisms and system concept to design process of electromechanical drives of follow-up servo is presented. Design system is made as a set of interconnected interacted computer application software which shortens design term and improves quality of the process. Drive designs based on internal and external gearing wave generating are considered, many of them are produced at the state-of-the-art level and implemented in many batch production servos.

1. Introduction

Instrument and special mechanical engineering industry requires minimal dimensions and mass of follow-up low-inertia mechanical drives for mobile devices. Multi-stage gearing drives (gear ration up to 100000 and higher) are used to reduce electrical motor high rpm (from above 10000 rpm). In these drives lesser number of reducer stages is possible due to implementation of drives distinguished by high quality characteristics of loading capability, kinematic error and efficiency. Harmonic drives are among the best in terms of these criteria [1-18]. This type drive principle of operation is based on the engagement of rigid and deformable flexible gear wheels. Deformation process is provided by wave generator. Flexible wheel is a ring or a thin shell. Variety of flexible gear geometry provides different designs of the drive which comply with different requirements in terms of configuration, motion transmission to any closed space without using any sealing units, etc. Wave generator may be positioned inside or outside relatively to the flexible wheel and deform it in one or several zones. Gear ratio of a single stage harmonic drive having steel flexible wheel ranges from 60 to 600. It provides multi-pair engagement. Total amount of teeth engaged is up to 40%. There is no engagement backlash, kinematic error decreases, efficiency and loading capability are high [1-4, 6, 8-12, 16, 17]. All this predetermines the usage of harmonic drive as the output stage of a gear train. It is also characterized by low vibration and noise [5, 13, 18].

Content from this work may be used under the terms of the Creative Commons Attribution 3.0 licence. Any further distribution of this work must maintain attribution to the author(s) and the title of the work, journal citation and DOI.
Published under licence by IOP Publishing Ltd

2. Geometrical calculation of harmonic drives

Harmonic drive geometry design and computational theory is created at Theory of Machines and Mechanisms dept. of BMSTU and is based on theory of geometry analysis of involute engagement developed by V.A. Gavrilenko and his disciples research of HDM [6, 11-13, 15-18], other proceedings.

This theory is based on the assumption that wave generator of the drive provides constant radius of curvature of deformed flexible wheel mean line within the area of engagement marked by central angle 2β in Fig. 1, a, b. Outside these areas flexible wheel has natural/loose form of deformation. Within constant radius of curvature area harmonic drive engagement is considered as internal involute engagement of rigid wheel (number of teeth is z_r) and conditional wheel which number of teeth z_c is calculated.

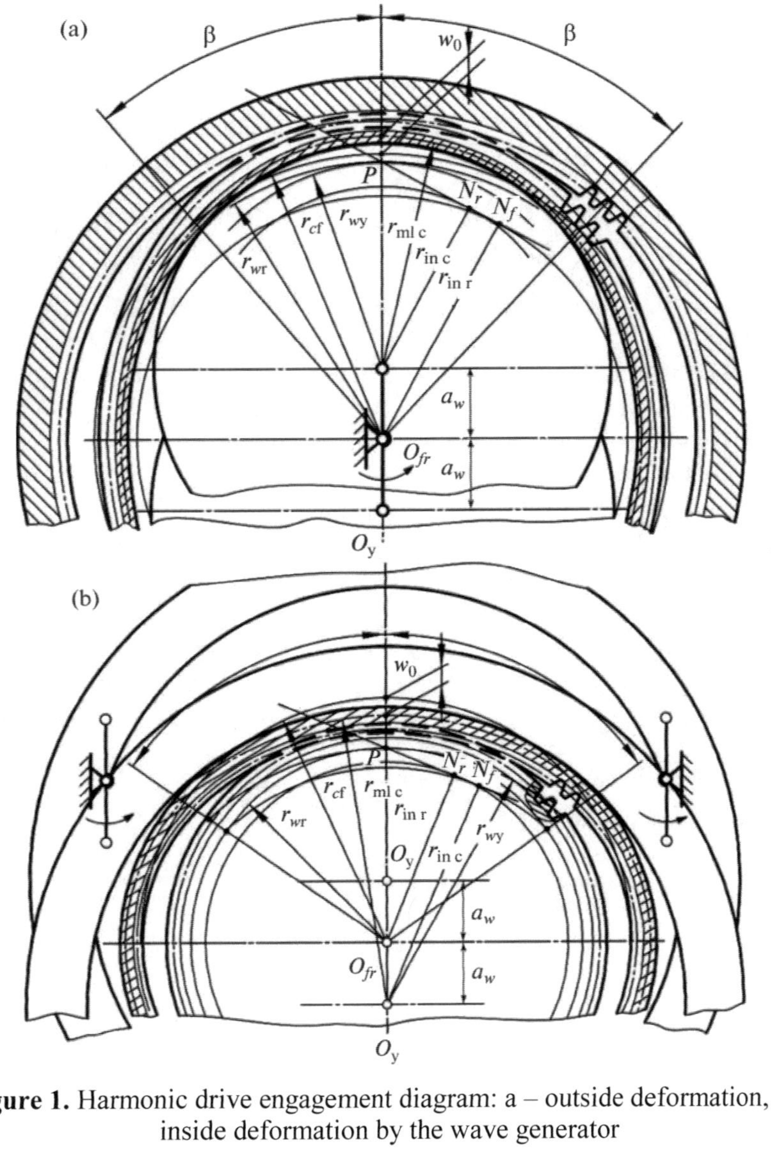

Figure 1. Harmonic drive engagement diagram: a – outside deformation, b – inside deformation by the wave generator

Calculated number of teeth is:

$$z_c = \frac{z_f}{1 \pm k_\beta \left(\frac{w_0}{r_{mlf}} \right)},$$

where w_0 – radial deformation along major axis, r_{mlf} – mean radius of the flexible wheel; $k_\beta = \frac{B}{A-B}$ - coefficient depending of the angle of wheel generator cover;

$$A = \frac{\pi}{2} - \beta - \sin\beta \cdot \cos\beta;$$

$$B = \frac{4\beta}{\pi} \cdot \sin\beta + \frac{4}{\pi} \cdot \cos\beta - 2\sin\beta \ ,$$

where β – angular coordinate of the constant radius of curvature area ($35^\circ \le \beta \le 65^\circ$).
The main input parameters for geometry characteristics calculation are: U – gear ratio of the drive; type of deformation (inside/outside); z_f, z_r – flexible and rigid wheels numbers of teeth; T, $T_{\max}$ – rated and maximal torques at the output drive shaft; n_h – rpm of the wave generator shaft; l_h – service life of the drive; ψ_{ring} – flexible wheel gear teeth tip width coefficient; flexible wheel strength parameters – hardness (HRC), endurance strength (σ_{-1}) of the material, torsion rigidity coefficient (C, optional parameter) [15].
The grater flexible wheel diameter the greater torsion rigidity and the less kinematic error. As the wheel diameter depends on the transmitted torque gear ratio of HDM should be the greatest providing compliance with overall dimensions of the unit.
Design calculation implies obtaining diameter d_{mlf} of the flexible wheel mean surface in non-deformed state in accordance with the following conditions:
1) Flexible wheel bending strength [13]:

$$d_{mlf} = 220 \cdot \sqrt[3]{k_d \cdot k_{ovl} \cdot \left(\frac{h_{ml}}{d_{mlf}} \right) \cdot \left(k_z - \frac{u}{u+1} \right) \cdot T} \ ,$$

where k_d - dynamism coefficient;

k_{ovl} - overload coefficient;

k_z - deformation form coefficient;

$\frac{h_{ml}}{d_{mlf}}$ - flexible wheel relative thickness below teeth tip;

2) Endurance calculation: $d_{mlf} = 165 \cdot \sqrt[3]{\frac{T}{(0{,}03 \cdot U - 1) \cdot \sigma_{F0}}}$,

3) Torque rigidity coefficient: $d_{mlf} = \left(1.12 - \frac{\sqrt{c}}{8000} \right) \cdot c^{\left(0.34 + \frac{\sqrt{c}}{35000}\right)}$.

The greatest among these diameters is the basis for engagement module pitch calculation which is rounded up to the closest standard value m

$$m' = \frac{d_{mlf}}{z_f}.$$

Module pitch may be readjusted.
Pitch diameters of the wheels are:

$$d_f = m \cdot z_f;$$

$$d_r = m \cdot z_r.$$

Flexible wheel rim thickness below the tooth tip:

$$h_{ml} = \left(\frac{10,6}{\sigma_{F0}} \cdot \frac{T \cdot 10^3}{d_f^3} + 0.007 \right) \cdot d_f,$$

but not greater than $0,018 \cdot d_f$.
The main variable parameter is relative radial deformation of the flexible wheel:

$$\frac{w_0}{r_{mlf}} = \left(\frac{z_r - z_f}{z_f} \right) \cdot \gamma,$$

where $\gamma = 0,9...1,2$ - coefficient of relative radial deformation.
The first expression as all the rest, containing dual arithmetic signs, is combined. The upper sign relates to inner deformation of flexible wheel by disk or cam wave generator, the lower sign – to outer deformation by ring-shaped generator.
Modification factor of the basic rack is:

$$x_f = \left(h_a^* + c^* + \frac{h_{ml}}{2m} \right) \cdot \delta,$$

where δ - coefficient of modification factor deviation.
Based on the diagram of wave engagement the radius of deformed flexible wheel mean line may be presented as

$$r_{mlc} = m \left(\frac{z_f}{2} \mp h_a^* \mp c^* \mp \frac{h_{ml}}{2m} + x_f \right).$$

δ and γ are variable parameters and are assigned as: $\delta = 1,0...1,4$, $\gamma = 0,9...1,2$ - for inside deformation, $\delta = 0,8...1,1$, $\gamma = 0,8...1,2$ - for outside deformation.
Quality of engagement may be improved considering possible ranges of input values. Goal function in this process is contact ratio coefficient.

3. Industrial harmonic drives samples

Geometry calculation algorithm is presented in [12, 13] and is applicable to HDM which gear ratio ranges from 60 to 100000. Principle schemes of these HDMs of inside and outside deformation are presented in [15].
Given below are some industrially implemented designs of drives based on HDM which geometry and kinematic characteristics were obtained based on the developed algorithm.

IOP Conf. Series: Materials Science and Engineering **468** (2018) 012010 doi:10.1088/1757-899X/468/1/012010

3.1. Two-stage gear transmission power drive

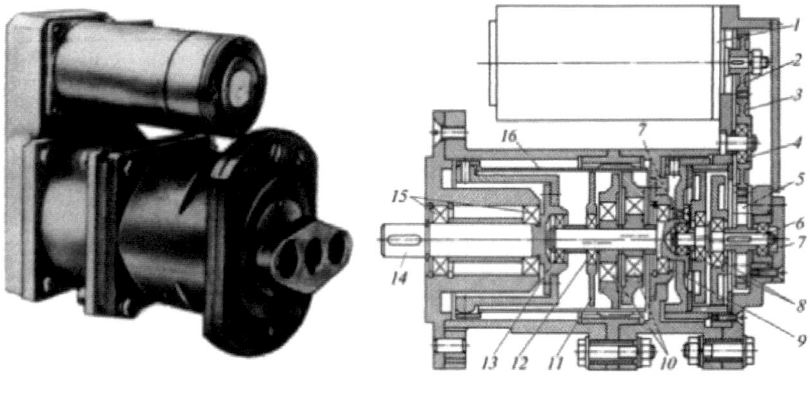

Figure 2. Two-stage gear transmission power drive

Fig. 2 presents the design of power drive based on two-stage harmonic drive. For unification purpose geometry parameters of gear engagement are the same: gear ratio is 60 or 80; module pitch is 0.4 or 0.3; total gear ratio is 4500 or 13000. Relatively to known designs overall dimensions of the drive developed are 14% less due to usage of narrow flexible wheel of intermediate stage and positioning supporting bearings 15 of the output shaft 14 inside flexible wheel 16.

Technical features:	
Output shaft load torque, N·m	80
Efficiency	0.57
Angular error of the output shaft at rated load, not greater than	0.0005
Power source voltage, V	27 ± 2
Power consumption, Wt	18
Overall dimensions, mm	$70\times120\times140$
Mass, kg	2.5

3.2. High-torque drive

Fig. 3 presents output stage of power train designed for high-torque drives of automated control. Three-disk wave generator is implemented in this design. In comparison with two-disk wave generator this design excludes sideway of the flexible wheel, twisting of its end cross-sections that improves mesh of the harmonic drive teeth.

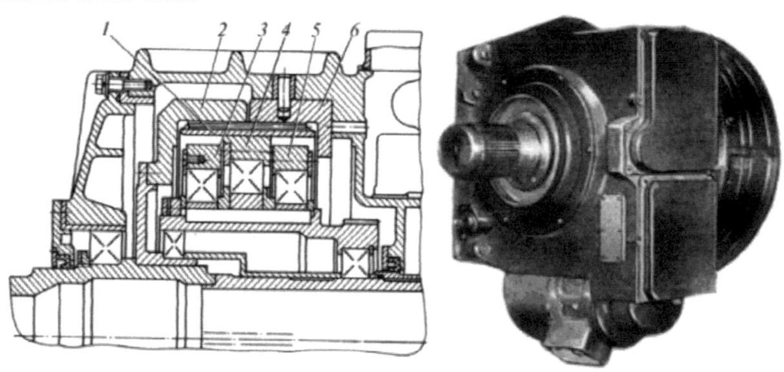

Figure 3. High-torque drive

Implementation of such design as the output stage of reducer increases its loading capability 1.5 times, efficiency – 1.4 times in comparison with the reducer containing planetary gear with five satellites as output stage.

Technical features of the drive:	
Output shaft torque, kN·m	80
Efficiency	0.63
Overall dimensions, mm	800×800×1 000
Mass, kg	2 880
Gear ratio:	
total	18 000
harmonic drive stage	120
Harmonic stage module pitch, mm	2

3.3. Discrete positioning electromechanical drive

The design of discrete positioning electromechanical drive based on HDM, presented in Fig. 4, provides new functional possibilities for transmitting continuous electrical motor shaft rotation to discrete translational motion of the output link of servo mechanism. This is three-stage drive. The first stage is cylindrical gearing of outer engagement, the second – combined harmonic drive providing discrete motion, the third, output one, is screw-and-nut drive.

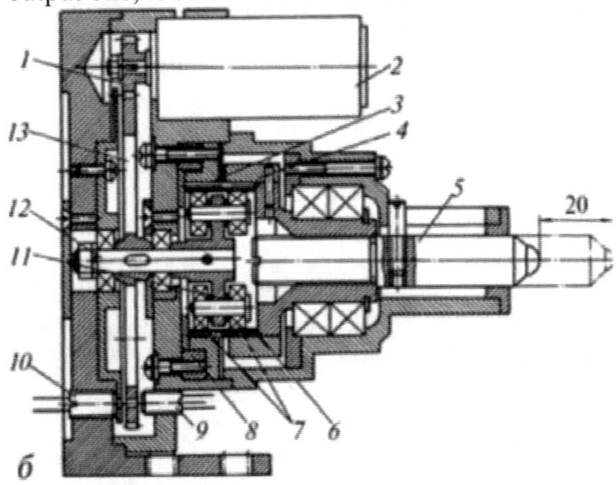

Figure 4. Discrete positioning electromechanical drive

Introduction of discrete motion harmonic drive decreased production cost of a single unit by 40% due to the 2.5 times decrease of the number of parts, metal consumption decrease (3 times) and production labor intensity reduction.

Technical features of the drive:	
Linear displacement of the output link per one step, micrometer	16
Output shaft displacement accuracy per one step, micrometer	1
Number of steps per second	25
Axial force at output shaft, N	300
Overall dimensions, mm	110x115x120
Mass, kg	1.1

Significant variety of different drive designs is presented in scientific report [12].

4. Conclusion

In some cases harmonic gear drive has no competitors. It's also clear that new HDM designs and rigid wheel gear trains are no rivals, both contribute to new decisions. HDM has some shortcomings: inability to provide low gear ratio (less than 50), rather high losses at idle, it is sensitive to lubrication and types of lubricants.

Nevertheless, the area of HDM will broaden, enriching modern servo engineering.

References

[1] Volkov D P, Krainev A F 1976 *Harmonic drives.* (Kiev, Technika) p 222.
[2] 1975, *Harmonic drives.* ed. N.I.Zeitlin, V.N.Tatischev (Moscow: Stankin) p 243.
[3] Ginzburg E G 1969 *Harmonic drives* (Leningrad, Mashinostroenie) p 169.
[4] Ivanov M N 1981 *Harmonic drives* (Moscow, High School) p 183.
[5] Kovalev N A 1979 *Transmisions with flexible gears* (Moscow, Maschinostroenie) p 200
[6] Gavrilenko V A, Skvortsova N A, Semin Yu I.et al 1978 *Works of MVTU.* vol 8. (Moscow: BMSTU) pp 22-33.
[7] Schuvalov S A 1986 *Theory and automated design of harmonic drives PhD thesises 05.02.02* (Moscow, BMSTU) p 28
[8] Krainev A F 2000 *Mechanics of mashines. Fundamental vocabulary* (Moscow: Mashinostroenie) p 904
[9] Popov P K 1997 *Computational and experimental approval of gearing accuracy D.Sc thesises 05.02.02* (Moscow, BMSTU) p 269
[10] Bleys E S, Zimin A V, Ivanov E S 1999 *Servo drives* ed. B.K.Chemodanov (Moscow: BMSTU) p 904
[11] Timofeev G A, Tarabarin W B, Yaminski A V 1988 *Construction and CAD of harmonic drives with internal and external deformation* (Moscow, VINITI) p 71
[12] Timofeev G A, Kostikov Yu V, Barbashov N N, Podchasov E O 2016 *Scientific report NIR 986.16 Development of calculation and design methods of harmonic drives for servo systems* ed. G.A. Timofeev (Moscow, BMSTU) p 78.
[13] Timofeev G A. 1997 *Development of calculation and design methods of harmonic drives for servo systems D.Sc thesises 05.02.02* (Moscow, IMASH RAN) 352 p.
[14] Timofeev G A, Kostikov Yu V, Barbashov N N 2015 *Scientific report NIR 986.15 Development of calculation and design methods of harmonic drives for servo systems* ed. G.A. Timofeev (Moscow, BMSTU) p 106.
[15] Timofeev G.A., Samoilova M.V. 2015 Comparing analysis of harmonic drives schemes for servo systems *Herald of BSTU, mashinostroenie ser.* **4** 109.
[16] Frolov K V, Popov S A, Musatov A K et al 2017 *Theory of mechanism and mechanics of machines* ed. G.A. Timofeev (Moscow, BMSTU) p 654.
[17] Kostikov Yu V, Timofeev G A, Fursiak F I 2016 The degree of influence of errors in manufacturing the details of the wave transfer on its kinematic accuracy *Machine drives and parts* **3** pp 10
[18] Timofeev G A, Kuzenkov V V 2015 Features of the servo drive dynamics with wave gearing Problems of machine *Problems of machine building and machines reliability* **6** pp 34.

Features of forming a parallel staging launch vehicle mathematical model using the method of d'Alembert-Kane

L V Dokuchaev

Central Research Institute of Machine Building, Korolev, Russia
E-mail: doklev@mail.ru

Abstract. The article discusses features of dynamic characteristics in a parallel-staging launch vehicle's elastic design. In contrast to traditional forms of body transverse vibrations in rockets with tandem staging, the parallel-staging scheme features vibrational forms of spatial character which are associated with longitudinal and torsional vibrations of side blocks of the rocket. The attention in the paper is drawn to the simplicity and clarity of drawing up general equations of dynamics in the form of d'Alambert-Kane, as well as to the features of projections of overload forces and the main engines thrust onto the related and inertial axes.
Key words: dynamic design, vibrations of an elastic body, parallel-staging rocket, launch-vehicle, natural frequencies, liquid propellant, blocks

At the beginning of the last century N E Zhukovskij considered the problem of the motion of a rigid body with cavities completely filled with a liquid. He introduced the concept of the inertia tensor of an equivalent body and obtained a number of analytic expressions for the now-known Zhukovsky potentials [1]. In connection with the instability of the flight of the first missiles in the 50s of the 20th century, a dynamic scheme of the body with cavities partially filled with liquid was developed, taking into account the wave oscillations of its free surface in the field of overload (Moiseev N N, Narimanov G S, Okhotsimsky D E, Rabinovich B I). Somewhat later, the phenomenon of unstable flight of missiles, caused by elastic oscillations of the construction, required the introduction of corresponding additional degrees of freedom into the dynamic scheme.

The hydrodynamic parameters of the dynamic scheme are determined by the solution of a homogeneous boundary value problem having a countable set of eigenvalues. Long-term experience shows that it is enough to take into account one basic tone of oscillations of the free surface of the liquid in each fuel tank of the launch vehicle (LV). This tone of fluid oscillations can be described by a pendulum model.

As a rule, the design of large-sized LV has a package layout. A feature of the LV package design is the excitation in flight of spatial flexural-longitudinal-torsional vibrations of its design. In this general case, the equations of motion of the LV, corresponding to all the control channels, are interconnected. The order of the system of equations for the study of the stability of the motion of the LV significantly increases. The presence in this system of close or multiple oscillation frequencies of carrier rocket complicates the analysis of the stability of its controlled motion.

Elastic characteristics of the dynamic scheme are calculated from the solution of the problems of elasticity theory for the construction of LV with a "frozen" liquid. At present, finite-element methods

Content from this work may be used under the terms of the Creative Commons Attribution 3.0 licence. Any further distribution of this work must maintain attribution to the author(s) and the title of the work, journal citation and DOI.
Published under licence by IOP Publishing Ltd

for determining the frequencies and forms of elastic vibrations in the construction of LV with the use of software packages Nastran, Abaqus, and others have been widely used.
In the compilation of differential equations of motion of the LV, the Lagrangian equations of the second kind are usually used. However, for this purpose it is better to use the simpler method of d'Alembert-Kane [2], in which the physical clarity of the d'Alembert principle is combined with the procedure for automatically eliminating the reactions of bonds, which is especially important in the case of nonholonomic systems.
As an illustration of this method, let us consider the procedure for constructing a dynamic scheme using the example of the design of a LV representing a package of four blocks. Each unit has an oxidizer tank, a fuel tank and a propulsion system (PS). We introduce the coupled coordinate system $Oxyz$, whose Ox axis coincides with the longitudinal axis of the LV. The axis Oy coincides with the pitch plane, the axis Oz - with the yaw plane. The longitudinal axes of the side blocks are parallel to Ox and have coordinates y_{0k}, z_{0k} $(k=1-4)$.

The associated coordinate system $Oxyz$ deviates from the trajectory coordinate system $O^*x^*y^*z^*$ by the amount of normal and lateral drift $Y(t)$ and $Z(t)$ turns to the yaw $\psi(t)$, pitch $\vartheta(t)$ and roll $\gamma(t)$ angles. In the k-th tank, the point of suspension of the pendulum has coordinates $C_k(x_C, y_{0k}, z_{0k})$. In the pendulum model of fluid oscillations, the length of the pendulum ℓ_S is represented, and the angles of its deviation from the vertical axis are equal $\left(0,\ S_{\psi_k}, -S_{\vartheta_k}\right)$. There are four engines that can be deflected in two planes with the help of 8 drives through the channels of pitch and yaw. Their rotation is determined by the coordinates $\delta_{\vartheta k}(t)$, $\delta_{\psi k}(t)$. . These motors are used for control and in the roll channel. The point of application of the tractive force P_k is located on the longitudinal axis of each block $B_k(x_B, y_{0k}, z_{0k})$. Each final element of the construction is displaced by $\Delta\vec{R}$ and rotated by a spatial angle $\Delta\vec{\Phi}$. This displacement and rotation are factorized in a series according to the forms of elastic oscillations

$$\Delta\vec{R} = \sum_{j=1}^{jq}(U_j, V_j, W_j)\, q_j(t)\ , \qquad \Delta\vec{\Phi} = \sum_{j=1}^{jq}(\phi_j, \psi_j, \chi_j) q_j(t)\ . \tag{1}$$

The speed of an arbitrary final element of the construction is composed of the velocity of the center of mass LV $\vec{V}_0 = (\dot{X},\ \dot{Y},\ \dot{Z})$; from the velocity due to the angular velocity of the LV with respect to its center of mass O

$$\Delta\vec{V} = \vec{\omega}\times\vec{R}, \quad \text{where} \quad \vec{\omega} = (\dot{\gamma},\ \dot{\psi},\ \dot{\vartheta}),\ \ \vec{R} = (x,\ y,\ z)\,.$$

In addition, there is a relative speed associated with the elastic vibrations of the construction, with fluctuations in liquid fuel, with fluctuations in the propulsion system.
The absolute velocity of the center of mass D of the rocking part of the k-th motor rotating with respect to the point B_k is

$$\begin{aligned}\vec{V}_D^k &= (\dot{X},\ \dot{Y},\ \dot{Z}) + \Big((\dot{\psi}\, z_{0k} - \dot{\vartheta}\, y_{0k}),\ (\dot{\vartheta}\, x_D - \dot{\gamma}\, z_{0k}),\ (\dot{\gamma}\, y_{0k} - \dot{\psi}\, x_D)\Big) + \\ &+ \sum_{j=1}^{i_q}(U_j^k,\ V_j^k,\ W_j^k)\Big|_{xB}\, \dot{q}_j(t) + \ell_D((\dot{\delta}^*_{\psi_k}\delta^*_{\psi_k} + \dot{\delta}^*_{\vartheta_k}\delta^*_{\vartheta_k}), -\ \dot{\delta}^*_{\vartheta_k},\ \dot{\delta}^*_{\psi_k})\ .\end{aligned} \tag{2}$$

The length of the pendulum is $\ell_D = x_B - x_D$. The angle of the absolute rotation of the pendulum is represented as the sum

$$\delta^*_{\vartheta k} = \delta_{\vartheta_k} + \sum_{j=1}^{jq} \chi^k_j(x_B) q_j(t)\,, \qquad \delta^*_{\psi k} = \delta_{\psi_k} + \sum_{j=1}^{jq} \psi^k_j(x_B) q_j(t). \tag{3}$$

The partial velocities, that is, the partial derivatives of the velocity (2) with relation to $\dot{\gamma}$, $\dot{\psi}$, $\dot{\vartheta}$, $\dot{Y}$, $\dot{Z}$, $\dot{q}_j$, $\dot{\delta}_{\vartheta k}$, $\dot{\delta}_{\psi k}$, have the form

$$\begin{gathered}
\frac{\partial \vec{V}^k_D}{\partial \dot{\gamma}} = -z_{0k}\vec{j} + y_{0k}\vec{k}\,, \qquad \frac{\partial \vec{V}^k_D}{\partial \dot{\psi}} = z_{0k}\vec{i} - x_D\vec{k}\,, \\
\frac{\partial \vec{V}^k_D}{\partial \dot{\vartheta}} = -y_{0k}\vec{i} + x_D\vec{j}\,, \qquad \frac{\partial \vec{V}_{si}}{\partial \dot{Y}} = \vec{j}\,, \qquad \frac{\partial \vec{V}_{si}}{\partial \dot{Z}} = \vec{k}\,, \\
\frac{\partial \vec{V}^k_D}{\partial \dot{q}_j} = \left(U^k_j\vec{i} + (V^k_j - \ell_D\chi^k_j)\vec{j} + (W^k_j + \ell_D\psi^k_j)\vec{k}\right)\Big|_{x_B}\,, \\
\frac{\partial \vec{V}^k_D}{\partial \dot{\delta}_{\vartheta_k}} = (\vec{i}\,\delta^*_{\vartheta k} - \vec{j})\ell_D, \qquad \frac{\partial \vec{V}^k_D}{\partial \dot{\delta}_{\psi k}} = (\vec{i}\,\delta^*_{\psi k} + \vec{k})\ell_D.
\end{gathered} \tag{4}$$

Acceleration of the center of mass of the *k*-th engine will be equal to

$$\begin{aligned}
\vec{a}^k_D = (\ddot{X}, \ddot{Y}, \ddot{Z}) &+ \left((\ddot{\psi}\, z_{0k} - \ddot{\vartheta}\, y_{0k}),\, (\ddot{\vartheta}\, x_D - \ddot{\gamma}\, z_{0k}),\, (\ddot{\gamma}\, y_{0k} - \ddot{\psi}\, x_D)\right) + \\
&+ \sum_{j=1}^{jq} (U^k_j, V^k_j, W^k_j)\Big|_{x_B} \ddot{q}_j(t) + (0, -\ell_D\ddot{\delta}_{\vartheta k}, \ell_D\ddot{\delta}_{\psi k}) + \\
&+ \sum_{j=1}^{jq} \left(0, -\ell_D\chi^k_j, \ell_D\psi^k_j\right)\Big|_{x_B} \ddot{q}_j(t)\,.
\end{aligned} \tag{5}$$

Similarly, it is possible to obtain expressions for the partial velocities and for accelerating the pendulum corresponding to the movable part of the liquid mass in the *k*-th tank, which will have a similar structure.

The main force acting on the LV is the thrust $\vec{P}_k$ created by the engines on each block. It is simultaneously a tracking and controlling force. In the stabilization planes, the vector of thrust directed from the tail to the spout of the LV is, with the axis Ox^*, the angles

$$\vartheta^k_* = \vartheta + \sum_{j=1}^{jq} \chi^k_j q_j + \delta_{\vartheta_k}\,, \qquad \psi^k_* = \psi + \sum_{j=1}^{jq} \psi^k_j q_j + \delta_{\psi_k}\,.$$

The vector of force generated by the thrust of the *k*-th engine, according to the rotation matrix, will take the following form: $\vec{P}_k = P_k(1,\, \vartheta^k_*, -\psi^k_*)$.

Therefore, the projections of the generalized force created by the propulsion system LV, on the plane of pitch and on the yaw plane, will be equal

$$\begin{gathered}
F^P_{x*} = \sum_{k=1}^{4} P_k, \qquad F^P_{y*} = \vartheta \sum_{k=1}^{4} P_k + \sum_{k=1}^{4}\sum_{j=1}^{jq} P_k \chi^k_j \Big|_{B_k} q_j + \sum_{k=1}^{4} P_k \delta_{\vartheta k}\,, \\
F^P_{z*} = -\psi \sum_{k=1}^{4} P_k - \sum_{k=1}^{4}\sum_{j=1}^{jq} P_k \psi^k_j \Big|_{B_k} q_j - \sum_{k=1}^{4} P_k \delta_{\psi k}\,.
\end{gathered} \tag{6}$$

If the forces from the thrust of the engines are projected on the axes of the associated coordinate system, then in the linear approximation we obtain

$$F_x^P = \sum_{k=1}^{4} P_k, \quad F_y^P = \sum_{k=1}^{4} F_y^{P_k}, \quad F_z^P = \sum_{k=1}^{4} F_z^{P_k},$$
$$F_y^{P_k} = P_k(\delta_{\vartheta_k} + \sum_{j=1}^{j_q} \chi_j^k \big|_{xB} q_j), \quad F_z^{P_k} = -P_k(\delta_{\psi_k} + \sum_{j=1}^{j_q} \psi_j^k \big|_{xB} q_j). \tag{7}$$

According to (4), the static moment from the traction force is

$$M_x^P = \sum_{k=1}^{4} (y_{0k} F_z^{P_k} - z_{0k} F_y^{P_k}), \quad M_y^P = \sum_{k=1}^{4} ((z_{0k} + \sum_{j=1}^{j_q} W_j^k \big|_{xB} q_j) P_k - x_B F_z^{P_k}),$$
$$M_z^P = \sum_{k=1}^{4} (x_B F_y^{P_k} - (y_{0k} + \sum_{j=1}^{j_q} V_j^k \big|_{xB} q_j) P_k), \quad x_D = x_B - l_D \tag{8}$$

Here it is taken into account that the axial line of the k-th block is in the deformed state, and the value of P_k is not a small value.
In the equation of elastic vibrations, the static component of the thrust of the motors is

$$Q_j^P = \sum_{k=1}^{4} \left(P_k U_j^k \big|_{xB} + F_y^{P_k} (V_j^k - \ell_D \chi_j^k) \big|_{xB} + F_z^{P_k} (W_j^k + \ell_D \psi_j^k) \big|_{xB} \right). \tag{9}$$

The missile in flight is affected by the quasistatic force created by the overload field. Each element $d\,m$ has a force

$$d\,\vec{F}_g = -\vec{i}_* n_x g\, d\,m\ ,$$

which is directed along the axis Ox^* of the trajectory coordinate system. The projections of this force on the axis of the coupled coordinate system will be equal to

$$d\,\vec{F}_g = -n_x g\, d\,m\,(1, -\vartheta, \psi). \tag{10}$$

Since the equations of forces are projected on the axis of the trajectory coordinate system, the projections of this force on the axis $O^* y^*$ and $O^* z^*$ are equal to zero. These forces are of decisive importance for the frequency of fluid vibrations, the approximation of which is represented by a mathematical pendulum. The point of suspension of the pendulum has coordinates $C(x_i + \ell_{S_i}, y_{0_k}, z_{0_k})$. Taking into account the deformation of the LV housing, the pendulum rotation relative to the overload vector will be determined by the angles

$$S_{\vartheta_i}^* = S_{\vartheta_i} - \sum_{j=1}^{j_q} \chi_j \big|_{x_C} q_j - \vartheta, \qquad S_{\psi_i}^* = S_{\psi_i} + \sum_{j=1}^{j_q} \psi_j \big|_{x_C} q_j + \psi\ . \tag{11}$$

By analogy with (2) - (4), the partial velocities for fluid oscillators are partial derivatives of its velocity.

$$\frac{\partial \vec{V}_{S_i}}{\partial \dot{\gamma}} = -(z_{0_k} + \ell_{S_i} S_{\psi_i}^*)\vec{j} + (y_{0_k} + \ell_{S_i} S_{\vartheta_i}^*)\vec{k}\,,$$
$$\frac{\partial \vec{V}_{S_i}}{\partial \dot{\psi}} = (z_{0_k} + \ell_{S_i} S_{\psi_i}^*)\vec{i} - (x_{c_i} - \ell_{S_i})\vec{k}\,,$$
$$\frac{\partial \vec{V}_{S_i}}{\partial \dot{\vartheta}} = -(y_{0_k} + \ell_{S_i} S_{\vartheta_i}^*)\vec{i} + (x_{c_i} - \ell_{S_i})\vec{j}\,,$$

$$\frac{\partial \vec{V}_{S_i}}{\partial \dot{q}_j} = \vec{i}\left(U_j + \ell_{s_i}(\psi_j S^*_{\psi_i} - \chi_j S^*_{\vartheta_i})\right) + \vec{j}(V_j - \ell_{S_i}\chi_j) + \vec{k}(W_j + \ell_{S_i}\psi_j), \tag{12}$$

$$\frac{\partial \vec{V}_{S_i}}{\partial \dot{S}_{\vartheta_i}} = \vec{i}\,\ell_{S_i} S^*_{\vartheta_i} + \vec{j}\,\ell_{S_i}, \qquad \frac{\partial \vec{V}_{S_i}}{\partial \dot{S}_{\psi_i}} = \vec{i}\,\ell_{S_i} S^*_{\psi_i} + \vec{k}\,\ell_{S_i}.$$

The static moment due to the overload created by the i-th oscillator is determined by multiplying the refined partial velocity (12) by the vector of the overload force (10)

$$\Delta M^c_x = 0, \qquad \Delta M^c_y = -n_x g \sum_{i=1}^{i_S} m_i (z_{0_k} + \ell_{S_i} S^*_{\psi_i}),$$
$$\Delta M^c_z = n_x g \sum_{i=1}^{i_S} m_i (y_{0_k} + \ell_{S_i} S^*_{\vartheta_i}). \tag{13}$$

In the equations of elastic vibrations, static additives will appear

$$\Delta Q^c_j = -n_x g \sum_{i=1}^{i_S} m_i \left(U_j + \ell_{s_i}(\psi_j\big|_{x_i} S^*_{\psi_i} - \chi_j\big|_{x_i} S^*_{\vartheta_i})\right). \tag{14}$$

Similar additions will appear in the equations of fluid oscillations

$$\Delta S^c_{\vartheta_i} = -n_x g\, m_i \ell_{s_i} (S_{\vartheta_i} - \sum_{j=1}^{j_q} \chi_j q_j - \vartheta),$$
$$\Delta S^c_{\psi_i} = -n_x g\, m_i \ell_{s_i} (S_{\psi_i} + \sum_{j=1}^{j_q} \psi_j q_j + \psi). \tag{15}$$

To determine the inertial forces, multiply the acceleration vector (5) of the element dm by the first partial velocity (4), which is equal to the unit vector $\vec{j}$, and integrate over the entire volume of the LV structure. Then multiply by the partial velocity equal to the unit vector $\vec{k}$, and also integrate over the volume of LV. Carrying out the addition of inertial forces with the active forces of thrust PS, static forces of the overload, as well as implicitly defined aerodynamic forces (index "a"), we obtain the equation of forces along the pitch channel

$$-(m\ddot{Y} + \sum_{i=1}^{i_S} m_i \ell_{S_i} \ddot{S}_{\vartheta_i}) + \vartheta \sum_{k=1}^{4} P_k + \sum_{k=1}^{4}\sum_{j=1}^{j_q} P_k \chi^k_j\big|_{xB}\, q_j +$$
$$+ \sum_{k=1}^{4} P_k \delta_{\vartheta_k} + \sum_{k=1}^{4} m^k_D \ell_D \ddot{\delta}_{\vartheta_k} + F_{ya} + F_{y0} = 0. \tag{16}$$

A similar equation of forces is obtained in the yaw channel

$$-(m\ddot{Z} + \sum_{i=1}^{i_S} m_i \ell_{S_i} \ddot{S}_{\psi_i}) - \psi \sum_{k=0}^{4} P_k - \sum_{k=0}^{4}\sum_{j=1}^{j_q} P_k \psi^k_j\big|_{xB}\, q_j -$$
$$- \sum_{k=0}^{4} P_k \delta_{\psi_k} - \sum_{k=0}^{4} m^k_D \ell_D \ddot{\delta}_{\psi_k} + F_{za} + F_{z0} = 0. \tag{17}$$

To obtain the moment equation over the roll channel, we must use the first partial velocity (4) with respect to the angle γ, by multiplying it by the acceleration of all the elements of the LV, integrating in volume and adding the moments of the active forces.

$$-\left(J_x\ddot{\gamma}+\sum_{i=1}^{8}m_i\,\ell_{s_i}(-z_{0_i}\ddot{S}_{\vartheta_i}+y_{0_i}\ddot{S}_{\psi_i})\right)-\sum_{k=1}^{4}m_D^k\ell_D(z_{0_i}\ddot{\delta}_{\vartheta_k}+y_{0_i}\ddot{\delta}_{\psi k})+$$
$$+\sum_{k=1}^{4}\left(-z_{0_k}P_k(\delta_{\vartheta_k}+\sum_{j=1}^{j_q}\chi_j^k\Big|_{xB}q_j)-y_{0_k}P_k(\delta_{\psi k}+\sum_{j=1}^{j_q}\psi_{jB}^k q_j)\right)+M_{x_a}+M_{x0}=0 \quad (18)$$

Similarly, it is possible to write the equations of moments in the yaw channel.

$$-\left(J_y\ddot{\psi}+\sum_{i=1}^{i_s}m_i\,\ell_{s_i}(-x_i)\ddot{S}_{\psi_i}\right)+\sum_{k=1}^{4}m_D^k\ell_D(x_B-\ell_D)\ddot{\delta}_{\psi k})+$$
$$+\sum_{k=1}^{4}\left((z_{0_k}+\sum_{j=1}^{j_q}W_j^k\Big|_{xB}q_j)P_k+x_BP_k(\delta_{\psi k}+\sum_{j=1}^{j_q}\psi_j^k\Big|_{xB}q_j)\right)- \quad (19)$$
$$-n_xg\sum_{i=1}^{i_s}m_i(z_{0_k}+\ell_{s_i}S_{\psi_i})-n_xg\sum_{k=1}^{4}m_D^k(z_{0_k}+\ell_D\,\delta_{\psi k})+M_{y_a}+M_{y0}=0.$$

The equation of moments in the pitch channel has a similar form

$$-\left(J_z\ddot{\vartheta}+\sum_{i=1}^{i_s}m_i\,\ell_{s_i}x_i\,\ddot{S}_{\vartheta_i}\right)+\sum_{k=1}^{4}m_D^k\ell_D(x_B-\ell_D)\ddot{\delta}_{\vartheta_k})+$$
$$+\sum_{k=1}^{4}\left(x_BP_k(\delta_{\vartheta_k}+\sum_{j=1}^{j_q}\chi_j^k\Big|_{xB}q_j)-(y_{0_k}+\sum_{j=1}^{j_q}V_j^k\Big|_{xB}q_j)P_k\right)+ \quad (20)$$
$$+n_xg\sum_{i=1}^{i_s}m_i(y_{0_k}+\ell_{s_i}S_{\vartheta_i})-n_xg\sum_{k=1}^{4}m_D^k(y_{0_k}+\ell_D\,\delta_{\vartheta_k})+M_{z_a}+M_{z0}=0.$$

The equations of elastic oscillations have a spatial form

$$-\left(\mu_j\ddot{q}_j+\sum_{i=1}^{i_s}m_i\,\ell_{s_i}(V_j^i\ddot{S}_{\vartheta_i}+W_j^i\ddot{S}_{\psi_i})-\right.$$
$$-\sum_{k=1}^{4}m_D^k\left(-\ell_D(V_j^k\Big|_{x_B}-\ell_D\chi_j^k\Big|_{x_B})\ddot{\delta}_{\vartheta_k}+\ell_D(W_j^k\Big|_{x_B}+\ell_D\psi_j^k\Big|_{x_B})\ddot{\delta}_{\psi_k}\right)+$$
$$+\sum_{k=1}^{4}\left(P_kU_j^k\Big|_{x_B}+P_k(\delta_{\vartheta_k}+\sum_{j=1}^{j_q}\chi_j^k\Big|_{x_B}q_j)(V_j^k\Big|_{x_B}+\ell_D\psi_j^k\Big|_{x_B})-\right.$$

$$\left.-P_k(\delta_{\psi_k}+\sum_{j=1}^{j_q}\psi_j^k\Big|_{x_B}q_j)(W_j^k\Big|_{x_B}-\ell_D\chi_j^k\Big|_{x_B})\right)-$$
$$-n_xg\sum_{k=1}^{4}m_D^k\left(U_j^k\Big|_{x_B}+\ell_D(\psi_j^k\delta_{\psi_k}^*+\chi_j^k\delta_{\vartheta_k}^*)\right)- \quad (21)$$
$$-n_xg\sum_{k=1}^{4}m_i\left(U_j^k\Big|_{x_B}+\ell_{s_i}(\psi_j^k\Big|_{x_i}S_{\psi_k}^*-\chi_j^k\Big|_{x_i}S_{\vartheta_k}^*)\right)+Q_j^0=0.$$

The equations of fluid oscillations are divided into two groups

$$-m_i \ell_{s_i} (\ddot{Y} + x_i \ddot{\vartheta} - z_{0_i} \ddot{\gamma} + \sum_{j=1}^{j_q} V_j^i \ddot{q}_j + \ell_{s_i} \ddot{S}_{\vartheta_i}) - n_x g \ell_{s_i} (S_{\vartheta_i} - \vartheta - \sum_{j=1}^{j_q} \chi_j^i q_j) = \Phi_{\vartheta_{i0}},$$

$$-m_i \ell_{s_i} (\ddot{Z} + y_{0_i} \ddot{\gamma} - x_i \ddot{\psi} + \sum_{j=1}^{j_q} W_j^i \ddot{q}_j + \ell_{s_i} \ddot{S}_{\psi_i}) - n_x g \ell_{s_i} (S_{\psi_i} + \psi + \sum_{j=1}^{j_q} \psi_j^i q_j) = \Phi_{\psi_{i0}}. \quad (22)$$

The dynamic scheme in the presented form (16) - (22) corresponds to the spatial motion of the LV. In the equations (21) - (22), there are connections between the angular coordinates in pitch, yaw and roll channels. In the particular case, it is possible to split the general system of equations of motion into subsystems describing isolated motions in the pitch plane, in the yawing plane, isolated rotation with respect to the longitudinal axis . This is possible in the case when the design of the LV 1) has two planes of symmetry, 2) the distribution of liquid masses in the lateral blocks, and also the stiffness characteristics, is strictly symmetric, and 3) the propulsion systems operate simultaneously with the same thrust. In this case, the spectrum of elastic vibrations of its construction breaks up into independent subspectra characterized by a certain type of deformation of the central block. Forms of elastic vibrations should be transformed to a symmetrical form. Instead of four lateral blocks, one "generalized" block is introduced in the mathematical model of the LV. For generalized coordinates of fluid oscillations and angular coordinates of control motors, a linear transformation is carried out. For example, instead of the four original coordinates, two new

$$s_\vartheta = s_{\vartheta_1} + s_{\vartheta_2} + s_{\vartheta_3} + s_{\vartheta_4}, \qquad s_{q\vartheta} = s_{\vartheta_1} - s_{\vartheta_2} + s_{\vartheta_3} - s_{\vartheta_4}$$

Then the general spatial system of equations of motion breaks up into three subsystems along the control channels (pitch, yaw, roll) [3]

References

[1] Zhukovsky N E 1949 On the motion of a solid body having cavities filled with a homogeneous droplet liquid *Coll. Work,* vol. **II** (Moscow-Leningrad: Gostechizdat)

[2] Kane T R 1961 Dynamics of Nonholonomic Systems. *Trans ASME, J Appl. Mech.* No **4**, pp 574-578.

[3] Dokuchaev L V and Sobolev O V 2005 Improvements in methods of research in dynamics of a clustered launch vehicle considering its symmetry, *Cosmonautics and Rocket Engineering (Russia),* No 2(39), pp 112-121

About islands of stability and limiting mass of the atomic nuclei

V S Okunev

Bauman Moscow State Technical University, Moscow, Russia.
E-mail: okunevvs@bmstu.ru

Abstract. The maximum magic number of neutrons for existing nuclei can be 260, protons - 126. The maximum deformation of existing super-heavy nuclei corresponds to the number of protons is about 120, the number of neutrons is about 120, 155 and 222. The islands (or shallows) of stability can be formed not only around the ^{298}Fl, but also around the ^{320}120 and ^{342}126. The islands are extended toward larger numbers of neutrons. An increase in the half-life should be observed for the ^{348}126 and ^{386}126. The half-life of ^{298}Fl is ~10^7...10^8 years. This is an understated estimate. The half-life of the longest-lived isotope of the 126th element with a mass number of 342 is ~10^6...10^7 s (or 0.03...0.33 years).

1. Introduction

1.1. A task

One of the fundamental tasks of modern nuclear physics is to determine the possibility of the existence of a second island of stability for super-heavy nuclei. The problem is closely related to the determination of the upper bound of the masses of atomic nuclei. Its solution is possible on the basis of predicting the properties of unknown nuclides.

1.2. Background. Islands of stability

The drip model predicts the instability of heavy nuclei with the divisibility parameter $Z^2/A \approx 45\ldots49$ (Z — nuclear charge, A — mass number). Such nuclei do not have a fission threshold. The shell model assumes the existence of a stabilizing effect of nuclear shells. The microscopic model suggests that filled proton and neutron shells increase the height of the fission threshold and slow down the decays of super-heavy nuclei.

A significant increase in the average lifetime of a super-heavy nucleus with a magic number of protons (Z = 114) and neutrons (N = 184) was predicted by V.M. Strutinsky in 1966 on the basis of the microscopic theory of the atomic nucleus [1, 2]. Simultaneously, W.D. Myers and W. Swiatecki hypothesized a possible increase in stability (with respect to spontaneous fission) of nuclei with magical Z = 114 and 126 and N = 184 [3]. Myers and Swiatecki developed a "semi-empirical theory of nuclear masses and deformations", which formed the basis of a microscopic model [4]. Thus, in 1966, the existence of stability islands in the region of trans-actinides (Z > 103). There can be several islands. Their centers are determined by the magic numbers N and Z. The contours of the first island

Content from this work may be used under the terms of the Creative Commons Attribution 3.0 licence. Any further distribution of this work must maintain attribution to the author(s) and the title of the work, journal citation and DOI.
Published under licence by IOP Publishing Ltd

($Z = 114$, $N = 184$) from the side of neutron-deficient nuclei are already known: in the laboratories of the world, such nuclei have been obtained.

1.3. Known theoretical and experimental facts

Theory.

(a) There is a hypothesis of 1966 about the possibility of the existence of two islands of stability formed around doubly magical super-heavy nuclei (^{298}Fl and ^{310}126).

(b) A theoretical transition from super-heavy nuclei to neutron stars is possible. To do this, it is sufficient to include the gravitational component in the Weizsacker formula. It is possible to extrapolate the optimal N/Z ratio taking into account the gravitational interaction

(c) The existence of bubble nuclei is predicted ($Z = 400$, $N = 900$).

Experiments and modern realities.

(a) Super-heavy elements (up to $Z = 118$) have been synthesized in various laboratories of the world.

(b) Traces of super-heavy elements (up to $Z = 130$) in the meteorites have been found.

2. Methods of research

2.1. Traditional methods

Traditionally, the methods for predicting the properties of yet undiscovered atomic nuclei are based on the adaptation of nuclear models to the region of unknown nuclides. However, it is difficult to confirm the applicability of these models in the region of unknown nuclei. The correct extrapolation of models to the region of super-actinides ($122 \leq Z \leq 157$) is laborious and may prove to be unjustified [5]. The use of known models for super-heavy nuclei is controversial [6].

2.2. Forecasting based on analysis of properties of known nuclides

The author uses an approach based on an analysis of the periodicity of the change in the properties of known atomic nuclei, and extrapolation of known semi-empirical relationships to the region of unknown nuclei. This requires the analysis of large amounts of information. The effectiveness of the analysis is determined by the successful choice of the way data is presented. Some simple considerations, based on well-known facts and the existing vast base of nuclear data, allow us to draw new interesting conclusions [7].

Based on the use of nuclear models, it was assumed that $Z = 108$ and $N = 162$ are magic numbers [8], but the simplest analysis showed that the increased stability of the nuclei of this region is related to the maximum deformation of the nuclei [9, 10]. The existence of the first island of stability around the nucleus of ^{298}Fl is beyond doubt. The analysis shows that the role of completely filled shells in the stabilization of nuclei is exaggerated. It is noticeable only near the optimal N/Z, which determines the equilibrium of the forces of nuclear (manifestation of strong), electromagnetic and weak interactions [8, 9]:

$$N/Z = 0.98 + 0{,}015\, A^{2/3}. \quad (1)$$

The relation (1) holds up to parity of N and Z in the nucleus. It is also characteristic of the longest-lived heavy and super-heavy nuclei [11]. There is reason to believe that not only at high energies, but also at small distances between interacting objects, all interactions behave as a single interaction. (This was discussed by V.L. Ginzburg [12].) The analysis allows doubting the existence of a doubly magic nuclide $Z = 126$ and $N = 184$ [9-11].

As a research method, extrapolation of known dependencies to unknown nuclei is used based on the analysis of existing files of estimated nuclear data. Such extrapolation may not have a physical meaning. Therefore, extrapolation results require additional analysis.

The simplest problem of extrapolation over an extreme node is to restore the values of the function f at the node x_{n+1} according to the known values of this function at the nodes $x_1, x_2, \ldots, x_n$. If the node x_{n+1} is far apart from the nodes $x_1, x_2, \ldots, x_n$, the known information about the behavior of the function f in the knots $x_1, x_2, \ldots, x_n$ is weakly used [13]. If the nodes are located close, then the role of errors in the initial information increases [13]. The choice of extrapolation nodes is not simple [13]. However, in this problem, the nodes $x_1, x_2, \ldots, x_{n+1}$ are known in advance and they can't be selected. When specifying a function by a graph, preference is given to approximation by polynomials. Often the approximating function is expediently sought in the form $f(x) \approx g(x, a_1, a_2, \ldots, a_n)$ [13, 14]. The parameters $a_1, a_2, \ldots, a_n$ are determined from the condition that $f(x)$ coincides with the approximating function at the nodes $x_1, x_2, \ldots, x_n$. The case of linear interpolation (extrapolation) is widespread. The approximation is sought in the form of a sum over i from 1 to n:

$$g(x, a_1, a_2, \ldots, a_n) = \Sigma\, \alpha_i\, \varphi_i(x),$$

where $\varphi_i(x)$ - the fixed functions, the values of the coefficients α_i are determined from the condition of coincidence with the approximating function at the nodes $x_1, x_2, \ldots, x_n$ [13, 14]. In this problem, such a method of extrapolation gives incorrect results. Linear extrapolation beyond the last node of the sequence of eight known magic numbers 2, 8, 20, 28, 50, 82, 114, 126 gives a value of 140 instead of 184. Extrapolation of the nine known magic numbers is 178, i.e. the monotonicity of a function drawn through given points is violated. Linear extrapolation of the number of neutrons (protons) on the shell for the ninth shell gives 30 instead of 58.

Nonlinear extrapolation is necessary. Interpolation (extrapolation) formulas for tables with constant step are known. Using the Bessel polynomial gives some advantages over using the second-degree polynomial [13]. In [15], an algorithm for choosing the most universal function ("universal pattern") is proposed for approximation based on a fairly simple cascade neural network model. However, much simpler models can be used to solve the problem.

If necessary, a plot of the known points is made using a pattern [14]. If the points are sparse, experienced engineers use a metal ruler placed on the edge, bending it so that the edge of the ruler passes through all points at once (flexible pattern) [14]. The flexible ruler is an elastic bar. The equation of its free equilibrium is $\varphi^{IV}(x) = 0$. Hence, in the interval between two neighboring nodes, the interpolation function is a polynomial of the third degree [14]. It is convenient to write it in the form:

$$\varphi(x) = a_i + b_i(x - x_{i-1}) + c_i(x - x_{i-1})^2 + d_i(x - x_{i-1})^3;\ x_{i-1} \le x \le x_i.$$

The method of solution is presented in [14]. It is applicable for solving the task.

To solve the problem, it is important to consider the following aspects, identified by the authors [14] for multidimensional interpolation.

(a) It is important that the function passing through the given nodes be monotonous. When extrapolating, it is desirable to preserve the monotonicity of the function.

(b) Not any number of nodes is advantageous, not every arrangement of nodes is permissible.

(c) Sequential interpolation (and hence extrapolation) has several advantages.

Moreover, it is better to approximate the discrete dependence (function) by a continuous dependence, i.e. the tabulation of the original data is transferred to the graph. It is also necessary to use the smoothing function passing through the specified nodes. Equally important is the way data is presented. Ideally, it is good to reduce the function to a linear or quasilinear dependence

3. Results

3.1. "Triangles of stability". Islands and shallows of stability

On the Z-N-diagram the triangles close in form to rectangular triangles can be distinguished (figure 1). We call them "triangles of stability" [9-11, 16]. One of the vertices of each of the triangles (at the right angle) is the double magic nucleus, distant from the equilibrium curve (1). On the legs are the magic nuclei. Hypotenuse is determined by the relation (1). Since the latter is true to the parity of N or Z, the hypotenuse is a sawtooth line. On this line there are stable or relatively long-lived nuclei. An analysis of the physical properties of nuclei in such triangles is a universal tool for predicting the properties of unknown nuclides.

The nuclei maximally deformed in the ground energy state are characterized by the largest internal quadrupole electric moment Q in absolute value. These nuclei are characterized by increased stability. As a result, these nuclei are similar to magical and double magic nuclei for some physical characteristics. Both have a maximum half-life $T_{1/2}$ in comparison with the neighboring nuclides of the Z-N diagram, the minimum cross section for interaction with hadrons, and the maximum threshold for inelastic processes (with the exception of the neutron capture neutron capture reaction). For magic nuclei Q $(N) = 0$ or Q $(Z) = 0$, in doubly magic Q $(N) = Q$ $(Z) = 0$. At the vertex at the right angle of the stability triangle, a nucleus with a maximum $|Q|$ [9, 10].

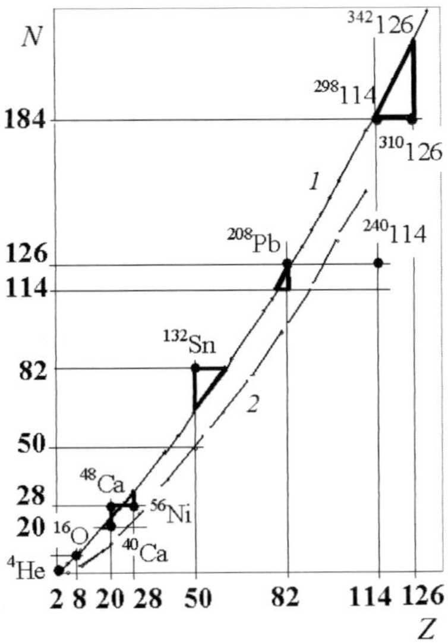

Figure 1. The arrangement of double magic nuclei and stability triangles:
1 - equilibrium curve (1),
2 – proton drip line.

In figure 2 schematically depicts the location of the island (indicated by the number *1*) and the shallows (*2*, *3*) of stability for super-heavy nuclides. Index *4* corresponds to the condition of optimal *N*/*Z*, *5* - triangle stability of trans-actinides. The shallow *2* is formed around the nucleus of the 120th element ($N = 200$, $A = 320$) with the maximum $|Q|$.

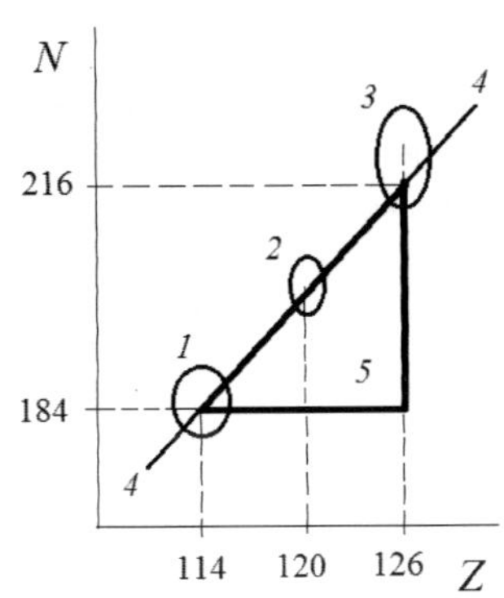

Figure 2. Diagram of the location of the triangle, the islands (shallows) stability of trans-actinides.

The mutual Coulomb repulsion of protons tends to break the nucleus. For this reason, atomic nuclei can withstand neutron overloading more easily than protons. The islands and shallows of stability should be stretched towards the large *N* (see figure 2).

The patterns of changes in the nuclear-physical properties of nuclides are the same for all stability triangles and are periodic in nature. It is possible to isolate stability triangles that contain not yet discovered nuclides (see figure 2), and predict the properties of these nuclides [9-11, 16]. So, the

IOP Conf. Series: Materials Science and Engineering **468** (2018) 012012 doi:10.1088/1757-899X/468/1/012012

double magic nucleus of the ^{298}Fl (N = 184, Z = 114) can turn out to be relatively long-lived not only due to the filling of the nuclear shells, but also due to the optimal N/Z ratio.
The double magic nucleus of the 126th element (A = 310) is far from the curve (1) and near the proton stability boundary. It can be non-existent. With Z = 126 and A = 342, the ratio $N/Z \rightarrow$ opt, which makes this nucleus the contender for the center of the shallow.

3.2. Extrapolation for the purpose of determining the numbers N and Z corresponding to the maximum stability of the nucleus in the region of super-heavy nuclides

3.2.1. General provisions. In the first approximation, especially if the distance between the shells is small, we can assume that the maximum deformation corresponds to half-filled shells, i.e. the arithmetic mean of adjacent magic numbers. The arithmetic average of the magic numbers 114 and 126 gives a value of 120; 126 and 184 gives 155. The exact value of N and Z corresponding to the maximum deformation of the nucleus can differ from 120 and 155 and is determined experimentally.
It has been experimentally confirmed that the numbers 126 and 184 are magical for neutrons. It is not known whether the magic numbers of neutrons and protons are the same for N or $Z \geq 126$. In accordance with the modern concept of nuclear shells, the magic numbers of protons and neutrons at large mass numbers must differ. It is also not known whether the values of magic numbers can change when going to bubble nuclei.
Suppose that super-heavy elements (up to Z = 130), found in trace amounts in meteorites, were formed as a result of nuclear reactions, for example, when cosmic objects collide at high velocities, and the double magic nucleus of the ^{298}Fl is the heaviest of the relatively long-lived ones. At Z > 114, in the N-Z coordinates there are only shallows (at Z about 120 and 126) containing short-lived nuclei.
The following questions remain.
1. Are there islands at Z >> 126, at the centers of which there are double magic nuclides with a half-life exceeding the half-life of ^{298}Fl or even of the nuclei composing shallows at Z about 120 or 126? Here we should distinguish two tasks for analysis.

(a) The first. Yu.T. Oganesyan suggested the possibility of a transition from traditional nuclides with a constant density of nuclear matter in the central part to exotic nuclei with a lower density of nuclear matter at the center ($Z \geq 120$) and the existence of bubble nuclei (at Z = 400 and N = 900). It should be noted that for known neutron-deficient nuclei the yield of the proton (protons) in the halo is accompanied by a significant (by orders of magnitude) increase of the half-life [9, 17]. This is due to the deceleration (stabilization) of radioactive decays by the electromagnetic interaction between the proton (protons) of the halo and the protons of the nucleus. For known nuclei with an optimal N/Z ratio, the halo does not exist. However, if a part of the neutrons and one or two protons with a close to zero binding energy in the super-heavy nucleus with the optimal N/Z evaporating will become part of the halo, and for some time they will be retained by nuclear forces, then the decay of the super-heavy nucleus electromagnetic interaction. Such nuclear structures are not known. The transition of all the nucleons of the nucleus to the halo formally corresponds to the nucleus-bubble.

(b) The second. Are there magic numbers greater than 184?

2. What are the objects located on the equilibrium curve between super-heavy (possibly, exotic) nuclei and neutron stars? It is not excluded that these are groups of nuclei connected by forces of nuclear interaction. Let's leave it for later analysis.

To study the first question, we use extrapolation. We have at our disposal a sequence of nine magic numbers (2, 8, 20, 28, 50, 82, 114, 126, 184), a sequence containing eight numbers corresponding to the distances between the nuclear shells, that is, between the magic numbers, equal to the number of nucleons of this species on the shell: 2, 6, 12, 8, 22, 32, 32, 12, 58. Sequences can be constructed for experimentally and theoretically determined maximally deformed nuclei, i.e., for N and Z for nuclei with maximal Q (elongated nuclei), minimal Q (oblate nuclei), and accordingly sequences containing N and Z located between maximally deformed nuclei.

In figure 3 (*a*) shows the dependencies of the values of *N* or *Z* on the ordinal number of the magic number. In figure 3 (*b*) shows the dependencies of intervals between magic numbers on the ordinal number of the interval. For the convenience of extrapolation, here and below, the points corresponding to the magic numbers are connected by straight lines. The discrete dependence is preliminarily approximated by a continuous dependence.

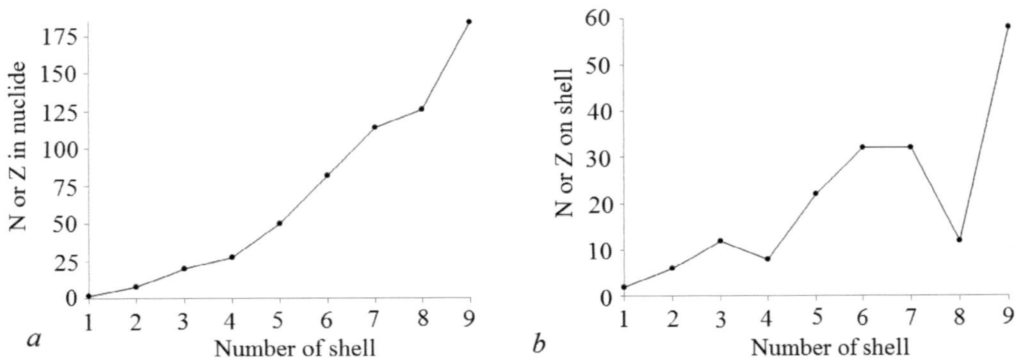

Figure 3. The magic numbers (*a*) and the number of neutrons (protons) on the shell (*b*).

Note that the magic number 28 was determined later than others. The magic number 114 was determined theoretically in the 1960s, but experimental confirmation was received only at the beginning of the 21st century. It was assumed that the numbers $Z = 108$ and 120 are also magical [8, 9]. The number 114 somewhat distorts the "perfection" of the curve in figure 3, and the number 28 distorts the curve in figure 3 (*b*). This makes extrapolation difficult. The role of different magic numbers in the stability of atomic nuclei is not the same. Analysis of *ENDF-B*/VII.1 data [18] shows that this role is less for *N* and *Z*, equal to 28. Perhaps it is less for *N* and *Z*, equal to 114. For medium atomic nuclei, the stabilization factor associated with an increase in the average specific binding energy of ε (per nucleon) is noticeable.

3.2.2. Extrapolation of magic numbers and the number of neutrons (protons) on the shell. So, we exclude from consideration the magic numbers 28 and 114, thus passing to a monotonic function (figure 4). Let's leave the magic numbers 2, 8, 20, 50, 82, 126, 184 and the intervals between them 2, 6, 12, 30, 32, 44, 58.

Extrapolation for the extreme node (184) allows us to get the next magic number - 260 and the interval between 184 and the next magic number (the number of nucleons on the shell) is 76 (see figure 4). This suggests the possibility of the existence of double magic nuclei at $Z = 126$, $N = 260$, $A = 386$ and $Z = 184$, $N = 260$, $A = 444$. The ratio of N/Z in the first case is 2.06, in the second - 1.41.

Let us determine the physical basis for such extrapolation. Because of mutual repulsion of protons in the nucleus, the average specific binding energy of nucleons in a heavy nucleus decreases with increasing mass number or number of protons. Extrapolating the known dependence of ε (*A*) in the region of super-heavy nuclei, we obtain that the specific binding energy is zero at $A \approx 500 \ldots 600$. On the other hand, according to the drop model, the nuclei with the divisibility parameter $Z^2/A \approx 45 \ldots 49$ and more should be divided spontaneously: the fission threshold is absent. The divisibility parameter of the ^{298}Fl is 43.6, of the ^{342}126 is 46.4. The divisibility parameter of the nucleus with $Z = 126$, $N = 260$, $A = 386$ is 48.64. This nucleus can exist because of magic *Z* and *N*. The divisibility parameter of the nucleus with $Z = 184$, $N = 260$, $A = 444$ is 76.25. This nucleus is at a considerable distance from the equilibrium curve (1) of the forces of nuclear, electromagnetic, and weak interaction. It nuclei does not exist. At mass number $A = 386$, the optimum values are $Z = 139$, $N = 247$, or $N/Z = 1.78$. At $A = 444$, the optimal values are $Z = 156$, $N = 288$, or $N/Z = 1.85$. The divisibility parameters for these two nuclei are 50.05 and 54.81, respectively. Such nuclei should not exist. Finally, the double magic nucleus at $Z = 184$, $N = 260$, $A = 444$ is characterized by the divisibility parameter of 76.25, which is characteristic for non-existent nuclei.

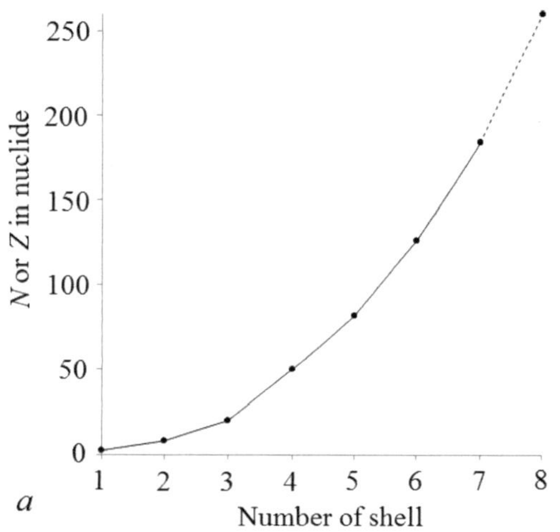

Figure 4. The results of extrapolation (dashed lines) of magic numbers (*a*) and the number of neutrons (protons) on the shell (*b*).

Stabilized can be maximally deformed super-heavy nuclei (with half-filled nuclear shells). If the magic number is 260, then the maximum strain may correspond to N or Z, equal to 222. Optimal N/Z at $Z = 126$ corresponds to $A = 342$. At $Z = 126$, $N = 222$ the mass number is $A = 348$. This nucleus is overloaded with only six neutrons. Thus, the shallows can be formed around the 126th element with a mass number of 342 and 348.

So, the simplest analysis suggests that the last magic number of neutrons can be considered 184 and, possibly, 260. If $Z = 184$ turns out to be magical, then the nuclei with such a charge should not exist. The shallows formed around the nuclei of the 126th element with a mass number of 342 and 348 are formed by the heaviest of the relatively long-lived elements.

3.2.3. Extrapolation of the numbers of neutrons (protons) corresponding to the maximum deformation of the nucleus. The maxima of Q (indicated in parentheses) and the minima of Q corresponding to the maximum difference from the spherical shape of the nucleus are attained at N or Z values equal to (1), 3, (6), 9, (12), 16, (18), ≈20 , (25), 28, (34), 44, (47), 52, (70), 83-84, (108), 128 [9]. In this case, the numbers 20 and 28 of this sequence are magical, i.e. the points corresponding to $Q = 0$ and $|Q| \rightarrow \max$ in this part of the curve $Q\ (N)$ or $Q\ (Z)$ are close.

In figure 5 the values of N (or Z) corresponding to the maximum elongated and maximally flattened nuclei are presented. Extrapolation suggests that the next largest mass of the most elongated nucleus corresponds to N (or Z), equal to 160, and maximally oblate - 184. But 184 is the magic number for neutrons. Hence, it corresponds to a stable or a long-lived nuclei.

If we consider the entire sequence N (or Z) for which $|Q| \rightarrow \max$, then the next number at which the deformation is maximal should be 152 (figure 6). Combining this sequence with magic numbers, we get the next number N (or Z) equal to 260, at which the nuclei is the most stable (see figure 6). However, as can be seen from figure 6 (*b*), extrapolation in this case is not informative, since its results are not unambiguous.

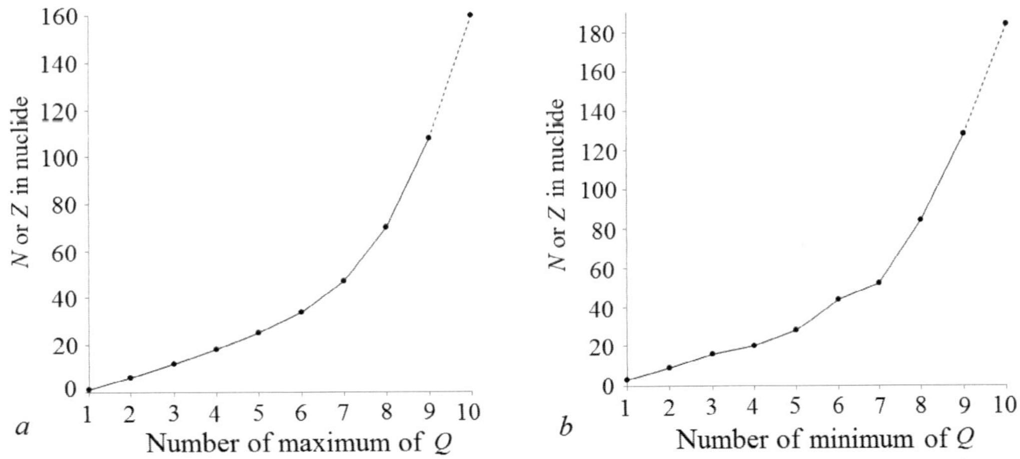

Figure 5. The results of the extrapolation (dotted line) of the number of neutrons (protons) in the nucleus corresponding to maximally elongated (*a*) and maximally flattened (*b*) nuclei.

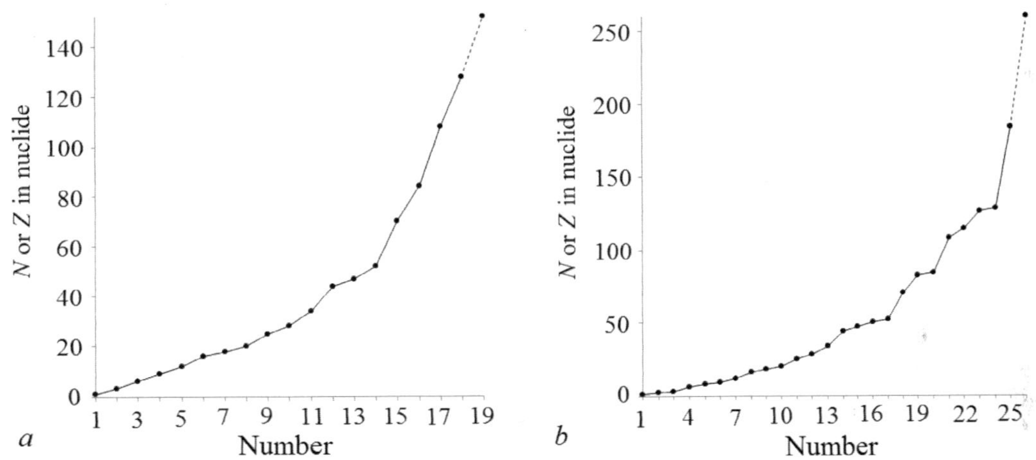

Figure 6. The results of extrapolation (dashed) of the dependence of *N* (or *Z*) on the ordinal number of the number corresponding to the maximum stability of the nucleus: without taking into account the magic nuclei (*a*), taking into account the magic nuclei (*b*)

3.3. Prediction of physical properties of super-heavy nuclei

3.3.1. Initial data. Preliminary conclusions. The method of extrapolation can be selected on the basis of an analysis of known data, for example, the corresponding region of transition from stable magic atomic nuclei of lead and bismuth to short-lived polonium isotopes. Analysis of the nuclei located in different stability triangles allows us to draw preliminary conclusions.

The half-life of a doubly magic nucleus, far removed from the equilibrium condition (1), can be noticeably smaller than the neighboring once-magic nuclides of the stability triangle. So, $T_{1/2}(^{132}Sn) < T_{1/2}(^{131}Sn)$, $T_{1/2}(^{132}Sn) < T_{1/2}(^{133}Sb)$, $T_{1/2}(^{132}Sn) < T_{1/2}(^{132}Sb)$ [9, 11]. A neutron-deficient nucleus with magic *N* can have a shorter half-life than a near-magic (*N* by 1 less), despite the odd number of neutrons in the near-magic nucleus. So, $T_{1/2}(^{215}Ac^{126}) < T_{1/2}(^{214}Ac^{125})$, $T_{1/2}(^{216}Th^{126}) < T_{1/2}(^{215}Th^{125})$ [9, 11]. As follows from the analysis of the databases [18, 19], for magical and near-magical nuclei, located at some distance from the curve (1) in the *Z-N* coordinates, the maximum of half-life (depending on *A*, *N* or *Z*) is diffuse and covers neighboring nuclides. Moreover, if the nuclei are overloaded with protons (^{215}Ac, ^{216}Th and other magic by *N*), then the isotope with a smaller *N* or an isoton with a larger *Z* will be the longest-lived one. This result is usually a much less significant reason for the stabilization of decays: a decrease in the mass number with the simultaneous transition of one of the protons of the nucleus into a neutron. The filling of nuclear shells and the parity of the number of protons and neutrons in the nucleus stabilize radioactive decays well only in the case of the optimal *N*/*Z* in the nucleus [7]. In the general case, they are less significant stabilizing factors than the equilibrium of nuclear, Coulomb, and weak forces [7, 9].

A similar pattern is observed for all nuclides.

The regularities in the change in the properties of nuclei in the transition from $Z = 114$ to 115 and 116 should be similar to the transition from $Z = 82$ to 83 and 84. In figure 7 shows the dependences $T_{1/2}(N)$ and $T_{1/2}(Z)$. For the ^{209}Bi $T_{1/2} = (1.9 \pm 0.2) \cdot 10^{19}$ years [20], the remaining data correspond to [18]. The curves in figure 7 the dependencies are characteristic for four cases: for magic by *N* (126) for $N/Z \rightarrow$ opt, for magic by *N* (114) neutron-deficient nuclei; for maximally deformed nuclei with a half-filled neutron shell ($N = 120$) and for nuclides maximally close to the proton drip line, whose properties are known ($N = 105$). For illustration, the half-lives of stable nuclides in figure 7 (*b*) are assumed to be equal to the experimentally determined $T_{1/2}$ for ^{209}Bi.

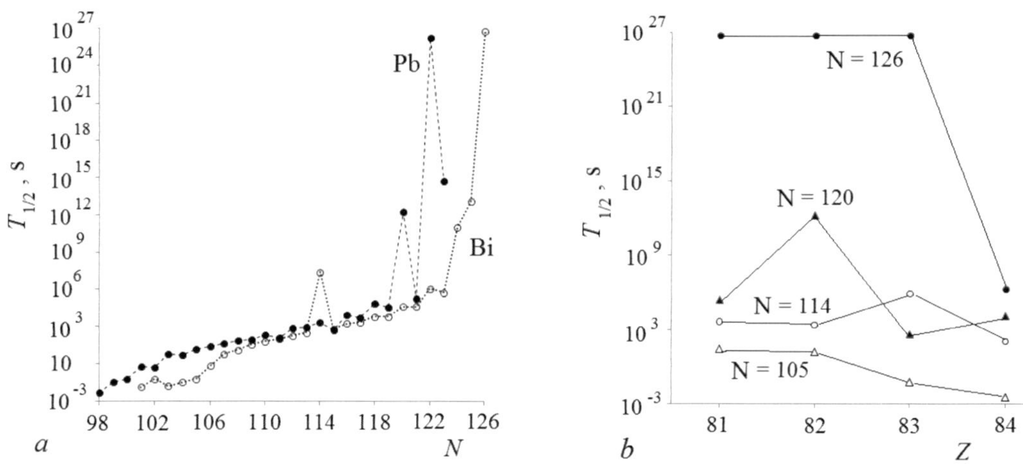

Figure 7. The dependence of the half-time on the number of nucleons in the nucleus: (*a*) $T_{1/2}$ (*N*) for neutron-deficient nuclei of Pb (●) and Bi (○); (*b*) $T_{1/2}$ (Z) for N = 105 (Δ), 114 (○), 120 (▲), 126 (●).

The first case corresponds to the transition from the heaviest stable elements to the short-lived elements. In the second case, there is an increase of $T_{1/2}$ for the magic nucleus ($N = 114$) at $Z = 83$, following the magic number 82. The superfluous proton in the nucleus has the same stabilization effect as the almost filled shell for the deformed nuclei. In the third case, the picture corresponds to the conventional one, we can say classical: the most long-lived is the magic nuclide of lead, the effect of proton pairing is clearly noticeable, as *Z* increases by 1, the half-life decreases to a greater extent than with decreasing. In the fourth case, the pairing effect is not manifested, with the filled ($Z = 82$) and almost filled ($Z = 81$) proton shell $T_{1/2}(N) \rightarrow$ max and decreases with increasing *Z* (for $Z > 82$).

Similar patterns are also characteristic of other magical and near-magical nuclei. The physical properties of the nuclides that determine their half-life in the stability triangles vary symmetrically with respect to the optimal *N*/*Z* ratio. This only confirms the universality of the approach to predicting the properties of atomic nuclei, based on the analysis of the triangles of stability.

3.3.2. Features of maximally deformed nuclei. The experimentally obtained $T_{1/2}$ (N) dependences (with respect to spontaneous fission) for certain isotopes of U ($T_{1/2} \rightarrow$ max at $N = 143$), Pu (146), Cm (146), Cf (150), Fm (152), No (152) in the vicinity of the optimal N/Z, as well as theoretical dependencies for neutron-deficient isotopes of the 106th (maximum at $N = 164$), 108th (162) and 110th (162) elements. These dependences support the assumption of nuclear deformation at $N = 152$ and 162.

Note that $N = 152$ corresponds to the maximum of $T_{1/2}$ only for the isotopes Fm and No, and $N = 162$ for the theoretical maximum for the isotopes of the 108th and 110th elements. For U, Pu, Cm and Cf, the maxima are fairly flat: three isotopes of U, four isotopes of Pu, five isotopes of Cm and two isotopes of Cf are characterized by a half-life close to the maximum. In figure 8 shows the dependence of $T_{1/2}$ on Z (*a*) and N (*b*) for some actinides with $N/Z \rightarrow$ opt and for their longest-lived isotopes (they correspond to maximally deformed nuclei). The minimum values of $T_{1/2}$ (α-decay and spontaneous fission) are given. Only some nuclei with $N/Z \rightarrow$ opt (^{231}Pa, ^{237}Np and ^{243}Am) are the longest-lived isotopes of this element.

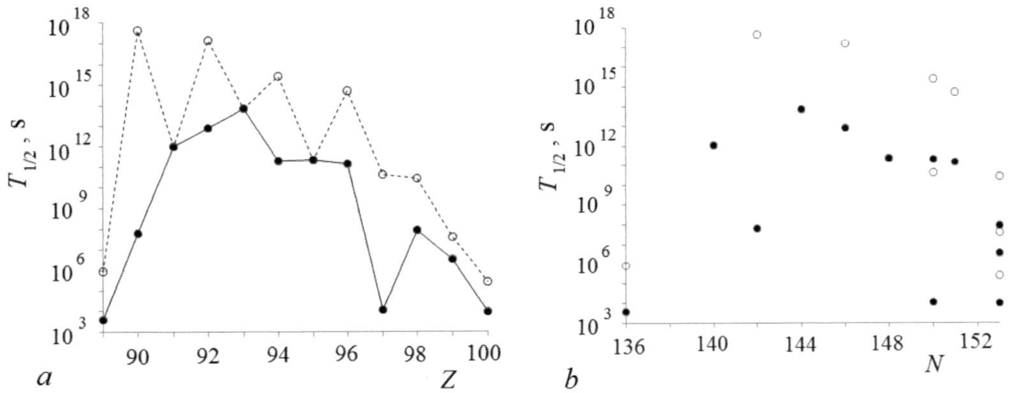

Figure 8. Dependence of the half-life on the nuclear charge (*a*) and the number of neutrons in the nucleus (*b*) for nuclides with the optimal N/Z (●) and for their longest-lived isotopes (○).

Most of the long-lived nuclei considered correspond to an odd number of neutrons $N = 153$, under which the deformation of the nucleus in the ground state is maximal. Near ($N = 150...152$) in the Z-N diagram are also relatively long-lived isotopes, i.e. $N = 152$ corresponds to the maximum of Q.

3.3.3. Estimates of the half-life of super-heavy nuclei. The patterns obtained for magical and maximally deformed heavy nuclei are characteristic of super-heavy nuclides. Their half-lives may be higher than the values of $T_{1/2}$ for deformed heavy nuclei (in the range from $Z = 90$ to 110), but lower than the values corresponding to the magic nuclei of the lead-196 and 208 in triangles of stability. According to the data of [21], the dependence $T_{1/2}$ (N) for Hs has a local ($N = 162$) and global ($N = 184$) maxima. For $_{114}$Fl, the local maximum corresponds to $N = 162$, the global $N = 182...184$. A similar phenomenon is characteristic of α-decays. According to [21], the maximum $T_{1/2}$ with respect to α-decay in the $T_{1/2}$ (N) dependence is observed at $N = 164$, the global one at $N = 184$ for Hs, and at $N = 164$ and $N = 180...184$ respectively for Fl. Obviously, the local maximum corresponds to the maximally deformed state of the nucleus ($Q \rightarrow$ max), the global maximum corresponds to the magic number $N = 114$. Q (N=114) = 0. According to estimates [21], obtained with the help of a microscopic model of nuclei, for double magic nuclides the half-life with respect to spontaneous fission can reach ~10^7 years. The lower limit of $T_{1/2}$ for super-heavy doubly magic nuclei ^{298}Fl can be estimated at ~$10^5...10^6$ years, the upper limit obtained on the basis of the theory of relativity of the limiting masses of atomic nuclei is $3 \cdot 10^8$ years [21].

In practice, interpolation and extrapolation are often used by second-degree polynomials or trigonometric polynomials. The discrete dependencies $T_{1/2}$ (N) and $T_{1/2}$ (Z) can easily be reduced to continuous differentiable functions and a uniform grid of *N* and *Z* is used. Detailed tables of $T_{1/2}$ (*N*) and $T_{1/2}$ (*Z*) dependencies allow the use of the simplest extrapolation methods that are easily performed on paper [14]. In a number of cases it is advisable to use the analysis of lg $T_{1/2}$ (*N*) or lg $T_{1/2}$ (*Z*). The form of the function depends essentially on the purpose of extrapolation. A method of linear interpolation or extrapolation with indefinite coefficients is well used [13].

For correct extrapolation, nuclides with even and odd *N* and *Z* were considered separately. In figure 9 shows the $T_{1/2}$ (*N*) and $T_{1/2}$ (*Z*) dependencies for trans-actinides (according to data [21-23]).

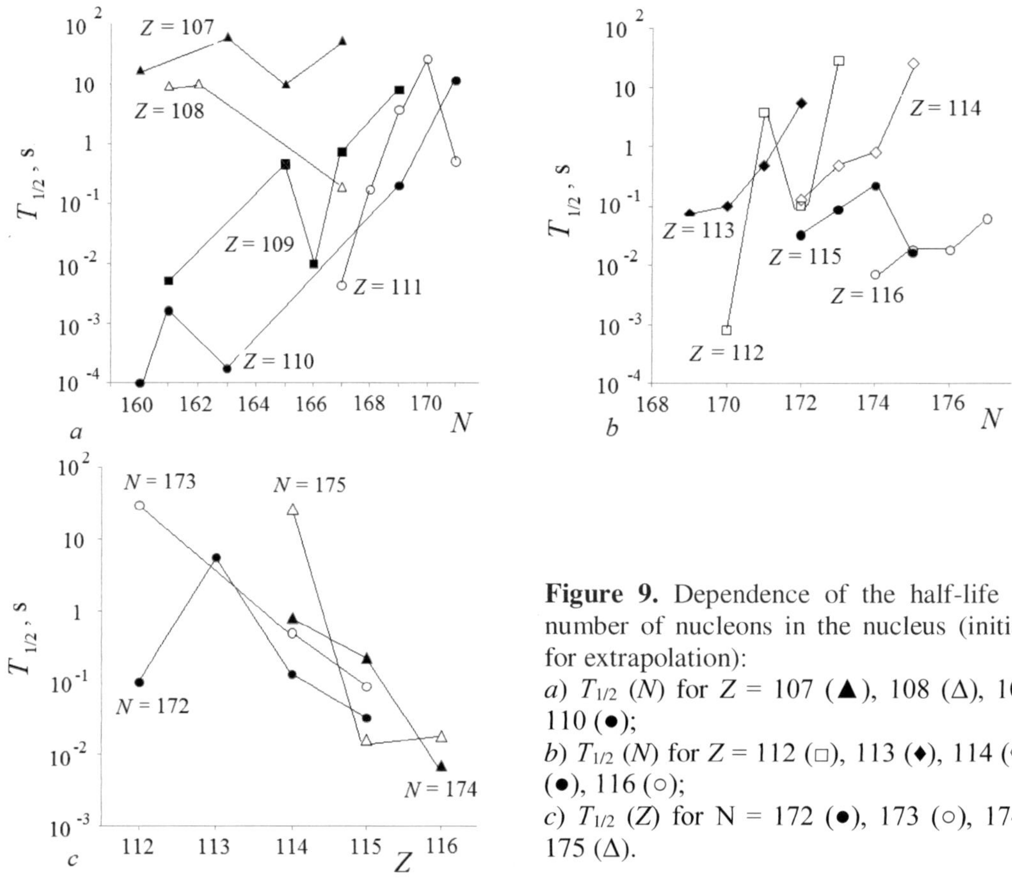

Figure 9. Dependence of the half-life on the number of nucleons in the nucleus (initial data for extrapolation):
a) $T_{1/2}$ (*N*) for *Z* = 107 (▲), 108 (Δ), 109 (■), 110 (●);
b) $T_{1/2}$ (*N*) for *Z* = 112 (□), 113 (♦), 114 (◊), 115 (●), 116 (○);
c) $T_{1/2}$ (*Z*) for N = 172 (●), 173 (○), 174 (▲), 175 (Δ).

The nature of the change in the properties of super-heavy maximum deformed or spherical nuclei as a function of *N* and *Z* is completely subject to the laws revealed for much lighter, well-studied nuclei. The simplest analysis of the triangles of stability of lead, thorium, and other actinides and their vicinities in *Z-N* coordinates, analysis of the known $T_{1/2}$ (*N*) and $T_{1/2}$ (*Z*) dependencies allow predicting the properties of new unknown elements forming a triangle of trans- and super-actinide stability with vertices at the points *Z* = 126, *N* = 184 (double magic element); *Z* = 114, *N* = 184 (^{298}Fl; *N*/*Z* → opt), *Z* = 126, *N* = 216 (Z-magic nuclide with optimal *N*/*Z*; *A* = 342).

The half-life of the nuclei varies considerably. For this reason, the most informative representation of the dependence $T_{1/2}$ (*N*) or $T_{1/2}$ (*Z*) on a semilogarithmic scale. Linear extrapolation using the logarithmic scale for $T_{1/2}$ turned out to be the most preferable from the point of view of the efficiency of predictions of the properties of strongly deformed nuclei. Such extrapolation for neutron-deficient nuclides in the region of optimal *N*/*Z* under conditions of incompleteness of the initial information (there is no data on the properties of nuclei with *N*/*Z* → opt) allowed us to estimate $T_{1/2}$ with good accuracy (table 1). The error is greatest if the longest-lived isotope of this element has a somewhat

IOP Conf. Series: Materials Science and Engineering **468** (2018) 012012 doi:10.1088/1757-899X/468/1/012012

excess *N* with respect to the optimal *N/Z* or the maximally deformed nucleus (in the ground state of energy) with respect to the spherical shape. In this case, there are two relatively long-lived nuclei.

Table 1. Estimation and exact values (*ENDF/B*-VII.1 [18]) of $T_{1/2}$ (s) for some actinides.

Nucleus	Estimated value	Exact value	Nucleus	Estimated value	Exact value
^{227}Ac	$\sim 10^{9}$	$6.87\cdot 10^{8}$	^{244}Pu	$\sim 10^{16}...10^{19}$	$2.52\cdot 10^{15}$
^{232}Th	$\sim 10^{15}...10^{18}$	$4.43\cdot 10^{17}$	^{247}Cm	$\sim 10^{14}...10^{16}$	$5.05\cdot 10^{14}$
^{236}U	$\sim 10^{17}$	$7.39\cdot 10^{14}$	^{249}Cf	$\sim 10^{10}$	$1.11\cdot 10^{10}$
^{238}U	$\sim 10^{17}...10^{20}$	$1.47\cdot 10^{17}$	^{251}Cf	$\sim 10^{10}$	$2.83\cdot 10^{10}$

For ^{208}Pb and ^{209}Bi nuclei, too low estimates are obtained (by several orders of magnitude), which is associated with a sharp change in the function $T_{1/2}$ and the derivative. For this reason, these nuclides are not represented in Table 1. If the $T_{1/2}$ estimate is acceptable for deformed nuclei, then the estimate for the doubly magical and adjacent nuclides on the *Z-N* diagram is strongly underestimated and this extrapolation is not valid. It can be expected that the use of such a simple extrapolation to estimate the half-life of the double magic nuclei of the ^{298}Fl will give a greatly underestimated estimate.

4. Findings

The ^{298}Fl nucleus will be the longest-lived trans-actinide stability triangle. The nuclei with charge $Z > 114$ will have a shorter half-life. Comparing $T_{1/2}$ (*N*) for the magic nuclei of lead and maximally deformed actinides, whose lifetime is determined by α-decay ($T_{1/2}$ with respect to spontaneous fission is large), the half-life of ^{298}Fl (for which α-activity also predominates) can be estimated in ~ $10^{7}...10^{8}$ years. This is a crude and highly underestimated estimate (one should expect $T_{1/2} >> 10^{7}...10^{8}$ years), which is in good agreement with the data [21].

The nucleus with $Z = 126$ and $N = 216$ should be relatively long-lived, but with a much smaller (approximately 10 orders of magnitude) half-life than ^{298}Fl.

Between ^{298}Fl and ^{342}126 there should be an island (or shallow) for long-lived highly deformed nuclides with a charge of about 120 and optimal or close to the optimal *N/Z* ratio (around the nucleus ^{320}120). This island (shallow) corresponds to the imposition of several triangles of stability (by analogy with actinides [10]).

An element 127 with an optimal *N/Z*, up to parity of *N* and *Z*, should have a half-life of less than ^{342}126. Next in *Z*, even an even-even nuclide with the optimal *N/Z*, may turn out to be short-lived. Extrapolation of $T_{1/2}$ and the detected traces of the 130th element allow us to make an assumption about the possible existence of nuclei $^{346}128^{218}$ and $^{350}130^{220}$.

At $N = 216$, the deformation of the nucleus should be noticeable (the location of a very shallow maximum is not yet known), which is also a stabilizing factor.

The second island of stability (or rather, the shallows) can form around the Z-magic nuclide ^{342}126 (a rough estimate yields $T_{1/2} \sim 10^{6}...10^{7}$ s) with the optimal *N/Z* ratio.

All islands and shallows should be stretched towards a larger number of neutrons. The nuclei are easier to withstand neutron overload than protons. This is due to the zero electric charge of the neutron and the absence of additional Coulomb forces destabilizing the nucleus.

5. Conclusion

A comparative analysis of the physical properties of atomic nuclei, carried out with the help of stability triangles, and the simplest extrapolation make it possible to qualitatively estimate the half-lives of the still unknown super-heavy nuclides. These are very rough estimates, since different stability triangles contain a different number of atomic nuclei. It is possible that the shallow formed around the nucleus ^{342}126, stretched towards the larger *N*, will contain relatively long-lived even-even nuclei of the 128th and 130th elements, as well as the 127th element with even N, and necessarily with a close to the optimal ratio *N/Z*.

It is possible that these are the heaviest nuclides from relatively long-lived ones. The longest-lived atomic nucleus of this shallow is its center - the nucleus $^{342}126$. The heaviest existing nucleus may be $^{386}216$. All the shallows will be stretched to the side of congestion by neutrons, since the nuclei can withstand neutron overloading more easily than protons.

Super-heavy elements with a charge $Z > 126$ can't exist with the traditional form of the nucleus, when the density of nuclear matter is maximal and approximately constant in the central part of the nucleus. The extrapolation of nuclear properties considered can precede traditional methods of predicting the properties of as yet undiscovered nuclides associated with the extrapolation of known nuclear models to the region of unknown super-heavy nuclei.

References

[1] Eisenberg M and Greiner V 1976 *Microscopic theory of the nucleus* (Moscow: Atomizdat)
[2] Sliv L *et al* 1976 *Usp. Phys. Nauk* **145 4** 553
[3] Volkov V A 1991 *Outstanding chemists of the world: biographical reference* (Moscow: High School)
[4] Myers W D and Swiatecki W 1966 *Nucl. Phys.* **81** 1
[5] Mukhin K N 1993 *Experimental Nuclear Physics: Proc. for universities* (Moscow: Energoatomizdat) vol 1.2
[6] Sobichevski A 2009 *Advances in Chemistry* **78 12** pp 1200
[7] Okunev V S 2011 *Herald of the Bauman Moscow State Tech. Univ., Nat. Sci.* **4** pp 81
[8] Dvorak J *et al* 2006 *Physical Review Letters* **97** pp 242501
[9] Okunev V S 2016 *The some paradoxes and laws of low-energy nuclear physics* (Bauman Moscow State Technical University)
[10] Okunev V S 2013 *Herald of the Bauman Moscow State Tech. Univ., Nat. Sci.* **4** pp 34
[11] Okunev V S 2012 *Herald of the Bauman Moscow State Tech. Univ., Nat. Sci.* **4** pp 13
[12] Ginzburg V L 2009 *Usp. Phys. Nauk* **179 5** pp 525
[13] Bakhvalov N S *et al* 2008 *Numerical methods* (Moscow: BINOM)
[14] Kalitkin N.N. 2011 *Numerical methods* (St. Petersburg: BHV- Petersburg)
[15] Kovalevskaya S V and Gitis V B 2008 *Artificial Intelligence* **4** 589
[16] Okunev V S 2015 *The basis of applied nuclear physics and introduction into physics of nuclear reactors. Ser.: Physics in Technical University* (Bauman Moscow State Tech. Univ.)
[17] Okunev V S 2015 *Herald of the Bauman Moscow State Tech. Univ., Nat. Sci.* **5** 69
[18] ENDF/B-VII.1 2011 (LANL, USA, htpps://t2.lanl.gov/nis/data.shtml)
[19] The JEFF Nuclear Data Library NEA Vienna 2017 (http://www.oecd-nea.org/dbdata/jeff/)
[20] Audi G *et al* 2003 *Nuclear Physics A* **729** 337
[21] Oganesyan Yu Ts 2012 *Bulletin of the International Academy of Sciences. Russian section. Ser.: Physical and technical sciences. (Moscow. Electronic Periodicals)* **2** 36
[22] Oganesyan Yu Ts and Dmitriev S N 2009 *Usp. Khimii* **78 12** pp 1165
[23] Mukhin K N and Patarakin O O 2000 *Usp. Phys. Nauk* **170 8** pp 855

IOP Conf. Series: Materials Science and Engineering **468** (2018) 012013 doi:10.1088/1757-899X/468/1/012013

Analysis and comparison of the main characteristics of linear and nonlinear oscillations

N I Bondarenko, K B Obnosov and A V Panshina

Bauman Moscow State Technical University, Moscow, Russia
E-mail: panalv@mail.ru

Abstract. In this paper, the problem of determining the positions of stable and unstable equilibrium positions of a mechanical system of the type of a planetary mechanism with one degree of freedom is considered. Analysis of possible movements of the mechanism is carried out using the phase plane. We investigate the dependence of the number of equilibrium positions on the change in the values of the parameters of the system. It is shown that with changes in the values of the parameters, the nature of the motion of the mechanism can change drastically. Some characteristics of nonlinear oscillations of the mechanism near the positions of stable equilibrium are determined. A comparative analysis of the obtained characteristics with similar characteristics for small (linear) oscillations is carried out.

1. Introduction

It is known that to ensure the long-term operation of any mechanical system in the conditions of vibrations it is necessary to conduct her dynamic calculation, that, in particular, must include determination of characteristics of her free oscillations accomplished near positions of stable equilibrium, yet on the stage of planning. Usually at the study of vibrations of the mechanical systems are limited to the consideration of those her deviations from the positions of stable equilibrium, that are described by linear differential equations with constant coefficients. The methods of decision of such equations are well worked out and the use of this method (method of linear oscillations) in many cases gives results, which mainly satisfactorily describe the most personal touches of dynamic processes in the mechanical system [1], [2]. But, as a rule, the vibrations of the mechanical systems are described by nonlinear differential equations, and their simplified description by linear equations with constant coefficients does not allow to reveal all peculiar features of the real process of oscillations. The aim of this work is to study the free nonlinear oscillations of element of the real mechanism [3], determination of some characteristics of such oscillations and comparison of them with analogical characteristics of linear oscillations.

2. Formulation of the Problem

A mechanism located in the vertical plane (Figure 1) containing a satellite 1 of radius r and of mass M_1 is considered. The satellite can roll without slipping along a fixed cylindrical surface of radius R. The axis A of the satellite is fastened with a rod 2 of length L and of mass M_2, rotating about the horizontal fixed axis O.

Content from this work may be used under the terms of the Creative Commons Attribution 3.0 licence. Any further distribution of this work must maintain attribution to the author(s) and the title of the work, journal citation and DOI.
Published under licence by IOP Publishing Ltd

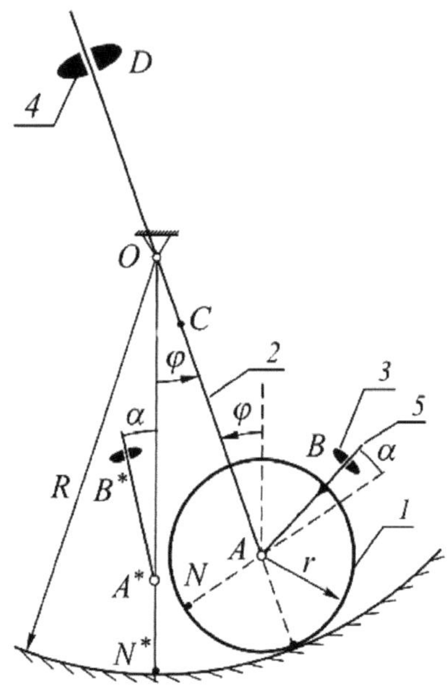

Figure 1. The diagram of the mechanism.

A weightless bar 5 with a counterweight 3 is rigidly attached to the gear 1. The counterweight of mass M_3 can be fixed on the bar 5 at a distance of $AB=l_3$. The load 4 of mass $M_4$is on the rod 2 at a distance $OD=l_4$. In the middle C of the rod 2 is located its center of mass.

A mechanism has one degree of freedom. As the generalized coordinate was selected the angle φ (radian) of deflection of the rod 2 from the vertical. The Figure 1 shoes the mechanism in current moment of time. The superscript index * indicates the position of the points N, A, B when $\varphi=0$. The position of a counterweight 3 in relation to a gear 1 determined by the angle $\alpha=const$ between the rays OA^* и A^*B^*.

In research of vibrations of mechanism we use the expressions of his potential energy $E_{П}(\varphi)$ and kinetic energy E_K, which have the form [4], [5]:

$$E_{П}(\varphi)=g\cdot\left[\left(M_1+M_2+M_3\right)\cdot\left(R-r\right)-M_2\cdot L/2-M_4 l_4\right]\cdot\left(1-\cos\varphi\right)+M_3 g l_3\cdot\left[\cos\left(\frac{R-r}{r}\varphi-\alpha\right)-\cos\alpha\right], \quad (1)$$

$$\begin{aligned}E_K=0,5\cdot\{M_1\left(R-r\right)^2+J_{1A}\frac{\left(R-r\right)^2}{r^2}+M_2\left[\frac{L^2}{12}+\left(R-r-\frac{L}{2}\right)^2\right]+\\+M_3\left(R-r\right)^2\cdot\left[1+2\cdot\frac{l_3}{r}\cos\left(\frac{R}{r}\varphi-\alpha\right)+\frac{l_3^2}{r^2}\right]+M_4 l_4^2\}\cdot\dot{\varphi}^2=\frac{1}{2}A(\varphi)\cdot\dot{\varphi}^2\end{aligned} \quad (2)$$

Where J_{1A} - a moment of inertia of the gear 1 relative to the horizontal axis A.

The analysis of possible motions of mechanism we will conduct by means of the phase plane – the Cartesian plane, along the abscissas axis of which the angular coordinate $\varphi(t)$ is plotted, and along the ordinates axis – the corresponding angular velocity $\dot{\varphi}(t)$. The trajectory of point with coordinates $\varphi(t)$, $\dot{\varphi}(t)$ (visual point) on a phase plane is called the phase trajectory [2]. Her kind depends on initial values φ_0 and $\dot{\varphi}_0$, and family of phase trajectories illustrates the possible processes in the mechanical system under consideration.

3. Research of the Movements

To obtain the equation of the phase trajectory, we use the fact that in a conservative system the mechanical energy remains constant:

$$E_k + E_{\Pi} = const = h\,, \tag{3}$$

where h is initial reserve of mechanical energy.
From (3) with an allowance (1) and (2), the equation of a phase trajectory has the form:

$$\dot{\varphi} = \pm\sqrt{\frac{2[h - E_{\Pi}(\varphi)]}{A(\varphi)}} \tag{4}$$

As can be seen, there are no phase trajectories in the areas, where $h < E_{\Pi}(\varphi)$.
The following parameter values are accepted: $M_1=0.2$ kg; $M_2=M_4=0.5$ kg; $M_3=0.4$ kg; $R=0.55$ m; $r=0.1$ m; $l_4=0.2$ m; $L=0.8$ m; $\alpha=\pi$. Note that then if $\varphi=0$ a counterweight 3 is in the lowest position. For a start, leaving the listed parameters unchanged, find out the influence of the distance l_3 a counterweight from the center C of gear on the amount of positions of equilibrium and types of motion of the mechanism.

3.1 The case 1 ($l_3=0$ m).
The Figure 2 shows for this variant the potential energy curve (Figure 2a), constructed in accordance with the equation (1), and family of phase trajectories (Figure 2b), constructed in accordance with the equation (4). As follows from a theory [2] and Figure 2a, the mechanism is in the equilibrium position at $\varphi_1=0$, $\varphi_{2,3}=\pm\pi$ rad. Moreover the angle $\varphi_1=0$ (a minimum of potential energy) is the stable equilibrium position, and the angles $\varphi_{2,3}=\pm\pi$ (a maximum of potential energy) – are unstable equilibrium positions.
The Figure 2b shows the trajectories of the visual point on a phase portrait, constructed for different values of initial reserve of mechanical energy h (Joule), which can be realized, in particular, only due to the initial angle of deflection of the rod φ_o:

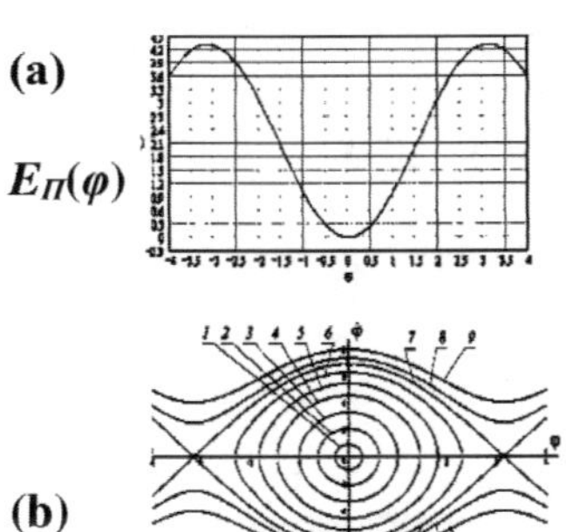

Figure 2. The characteristics of the movement mechanism when $l_3=0$.

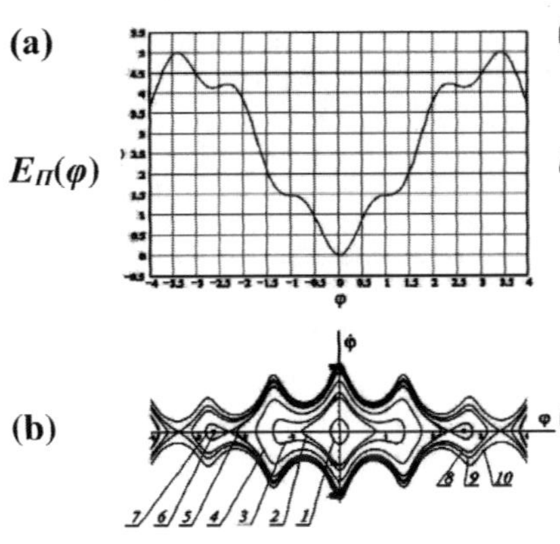

Figure 3. The characteristics of the movement mechanism when $l_3=0.1$.

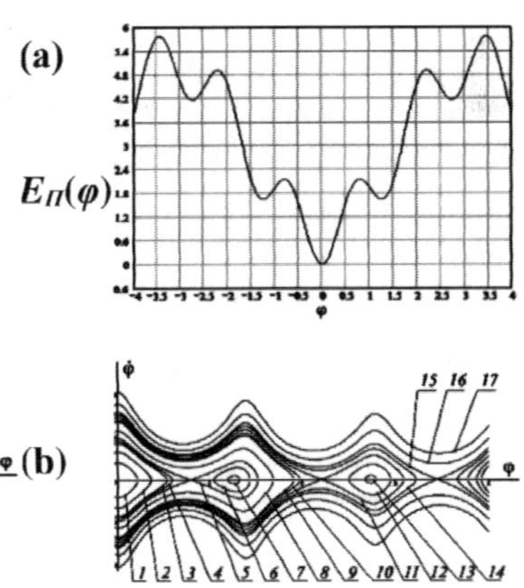

Figure 4. The characteristics of the movement mechanism when $l_3=0.2$.

$h_1=0.08$; $h_2=0.4$; $h_3=1.0$; $h_4=1.9$; $h_5=2.7$; $h_6=3.6$; $h_7=4.3162$; $h_8=4.9$; $h_9=5.8$.
Phase trajectories 1,2,3,…,6 near stable equilibrium position $\varphi_1=0$ are the disjoint closed curves similar to ellipses. It means that the motions of the rod of mechanism, described by these trajectories, will be almost harmonic motions. The unstable equilibrium positions ($\varphi_{2,3}=\pm\pi$) are described by two

intersecting trajectories 7 (separatrices) along which the visual point moves with the initial reserve of mechanical energy h_7=4.3162. Such movement is practically unrealized. Firstly, because even a small change of initial reserve of mechanical energy will lead to the exit of the visual pointfrom this trajectory, and, secondly, even with accurate observance of the necessary energy reserve, the time of motion along the separatrix tends to infinity [1].
Increasing the initial mechanical energy higher than value h_7, we pass to the new type of the motion (trajectories 8,9) – rotation of the mechanism in one direction with a variable angular velocity, which is maximal near stable equilibrium position and is minimum near unstable equilibrium position.

3.2 The case 2(l_3=0.1 m).
In the Figure 3 the results of these calculations are presented. Now the potential energy (Figure 3a) has instead of one three local minimums (φ_1=0; $\varphi_{2,3}$=±2.6583 rad) and has instead of two four local maximums ($\varphi_{4,5}$=±2.3347 rad; $\varphi_{6,7}$=±3.4157 rad). Now there are three special points of type "center" and four special points of type "saddle" [1] on a phase portrait (Figure 3b). Phase trajectories are built for the next values of h (Joule): h_1 =0.1848; h_2=1.3848; h_3 =1.7848; h_4=2.9848; h_5=4.1433; h_6 =4.1433; h_7 =4.2278; h_8= 4.7048; h_9=4.9978; h_{10} =5.5848.
The periodic oscillations of the rod are determined by the phase trajectories, covered by separatrices 7 and 9. Inside the links of the separatrix 7 there are trajectories 1,2,…,6. Trajectories 1,2,…,5 describe periodic oscillations of the rod near the basic position of stable equilibrium, determined by the angle φ_1=0, and trajectories 6 – near the additional positions of stable equilibrium, determined by the angles $\varphi_{2,3}$ =±2.6583 rad. Vibrations near the basic position of stable equilibrium (φ_1=0) practically harmonic, if the initial reserve of mechanical energy h=0.1848. The amplitude of the rotation angle of rod is somewhere about 0.2 rad (Figure 3b, trajectory 1). With increasing of the initial reserve of mechanical energy h the vibrations near this position of equilibrium become nonharmonic with varying "jerks" of angular velocity. The amplitude of the angle φ of rotation of the rod can thus reach the value somewhere about 2.18 rad (Figure 3b, trajectory 5) at an initial reserve of mechanical energy h=4.1433. From the phase portrait it follows that with fluctuations of mechanism near additional positions of stable equilibrium, for example, near φ=±2.6583 rad, the angle of rotation of the rod $2.3347<\varphi<2.8400$, that is the scope of the angle ($\Delta\varphi\approx 0.5053$) much less than, with fluctuations near the basic position of stable equilibrium.
If the mechanism is given energy $4.2278<h<4.9978$ at the initial moment of time, then the rod will also perform periodic motions near the basic stable equilibrium position (φ_1=0), but at the same time covering the positions of stable equilibrium $\varphi_{2,3}$=±2.6583 rad. The amplitude of the angle of deflection of the rod $2.8400<\varphi_0<3.4157$. If the mechanism received energy $h>4.9978$ Joules (figure 3b, trajectory 10), the rod will rotate all the time in one direction with a sharply changing angular velocity: for one turn she five times takes on values of local maximum and four times takes on values of local minimum.

3.3 The case 3 (l_3=0.2 m).
The calculations results are presented on Figure 4. The graph shows that the potential energy (Figure 4a) has now five local minimums and six local maximums, accordingly a mechanism has five stable and six unstable positions of the equilibrium. Because phase trajectories are symmetrical about the ordinate axis (for φ =0), then on the Figure 4b for comfort of analysis they are given only for φ>0. For such values we will consider the types of motions of mechanism. Three special points of type "center" (φ_1=0; φ_2=1.2583; φ_3=2.7388 (rad)) and three special points of type "saddle" (φ_4=0.7989; φ_5=2.2084; φ_6=3.4493 (rad)) are visible on the abscise axis of the phase portrait. Coordinates φ_1, φ_2, φ_3 define the first, second and third positions of the stable equilibrium. The parameter h (Joule) has following values: h_1=0.3696; h_2=0.9696;h_3=1.6696; h_4=1.8696; h_5=2.1431; h_6=1.8696; h_7=1.6696; h_8=2.9696; h_9=1.1896; h_{10}=4.4696; h_{11}=4.9115; h_{12}=4.4696; h_{13}=4.1896; h_{14} =5.3696; h_{15}=5.7711; h_{16}=7.1696; h_{17}=9.5696. As follows from a phase portrait, the rod performs periodic oscillations around the angle φ_3=*2.7388* rad(third position of stable equilibrium), if a mechanism in this position possesses a reserve

of energy $4.1666<h<4.9115$. The angle φ of the deflection of the rod 2 from the vertical line belongs to the interval $2.2084<\varphi<3.0882$ (for example, trajectories 12 and 13 on the phase portrait), that is the maximal angle of turn is equal 0.8798 rad. As seen, the maximal angle of deflection of the rod from this position of stable equilibrium depends on direction of rotation: when turning in the direction of increasing the angle φ it is equal to 0.3494 rad, and in the opposite direction – 0.5304 rad. Near the angle $\varphi_2=1.2583$ rad (second position of stable equilibrium) of the rod performs periodic oscillations, if a mechanism possesses energy of $1.6411<h<2.1431$ (Joule) in this position. In this case the angle of deflection of the rod from the vertical belongs to the interval $0.7988<\varphi<1.5140$ (for example, trajectories 6 and 7 on the phase portrait). The maximal angle of turn is equal 0.7152 rad, the angle of deflection from this position of stable equilibrium in the direction of increasing the angle is equal 0.2557 rad, and in the opposite direction – 0.4595 rad.
Near angle $\varphi_1=0$ rad (first position of stable equilibrium) the rod, depending on the initial reserve of mechanical energy, can perform three types of periodic movements:

- if $0<h<2.1431$, then amplitude of the angle of deflection of the rod from the vertical line $\varphi_0<0.7988$ rad (trajectories 1,2,3,4 on thephase portrait);
- if $2.1431<h<4.9115$, then rod vibrations cover the first and second stable equilibrium positions, the amplitude of the angle of turn belongs to the interval $1.5140<\varphi_0<2.2084$ (rad) (trajectories 8,9,10 on the phase portrait);
- if $4.9115<h<5.7711$ (Joules), then rod vibrations cover all three positions of stable equilibrium, the amplitude of the angle of turn rod belongs to the interval $3.0882<\varphi_0<3.4493$ (rad) (trajectory 14 on the phase portrait).

If to increase the energy of mechanism higher than value $h=5.7711$ Joules, then the rod passes to a non-oscillatory motion – one-way rotation (trajectories 16 and 17 on the phase portrait).
Further, supposing $l_3=0.2$ m, will define some characteristics of the free nonlinear oscillations of the mechanism and will compare them to analogical characteristics of linear oscillations. In this case the potential energy (1) and the kinetic energy (2) of mechanism are determined by next expressions:

$$E_\Pi(\varphi) = 2.943 - 2.158\cos\varphi - 0.785\cos(4.5\varphi),\ E_k = 0.5\cdot\{0.5049 - 0.324\cos(5.5\varphi)\}\cdot\dot{\varphi}^2 = 0.5A(\varphi)\cdot\dot{\varphi}^2.$$

Will analyze the motion of mechanism near the position of the stable equilibrium, determined by the angle $\varphi=0$.
If to examine the linear oscillations of the mechanism, then the simplified expressions for the kinetic and potential energies have the form:

$$E_k^* = \frac{1}{2}A(\varphi)\Big|_{\varphi=0}\cdot\dot{\varphi}^2 = \frac{1}{2}a\cdot\dot{\varphi}^2,\ a=0.1809\ \text{J(c·rad}^{-1})^2; \tag{5}$$

$$E_\Pi^* = \frac{1}{2}\left(\frac{\partial^2 E_\Pi}{\partial\varphi^2}\right)\Bigg|_{\varphi=0}\cdot\dot{\varphi}^2 = \frac{1}{2}c\cdot\dot{\varphi}^2,\ c=18.0504\ \text{J·rad}^{-2}. \tag{6}$$

Oscillation period of the mechanism

$$T = 2\pi\sqrt{a/c} = 0.629\,c \tag{7}$$

In the moment of the maximal deviation $h = E_\Pi^* = \frac{1}{2}c\cdot\varphi_0^2$. Hence the amplitude of the linear oscillations

$$\varphi_0 = \sqrt{2h/c} \tag{8}$$

If φ =0, then E_{Π}=0, and $\dot{\varphi} = \dot{\varphi}_{\max}$. In this position $h = E_k^* = \frac{1}{2} a \dot{\varphi}_{max}^2 = 0$, that is, the high angular velocity of the rod is

$$\dot{\varphi}_{\max} = \sqrt{2h/a} \tag{9}$$

Now will consider the nonlinear oscillations of the mechanism near this stable equilibrium position in detail. The Figure 5 shows the phase trajectories constructed for the different values of h. As can be seen, the family of disjoint trajectories 1-7 has a common center at the origin. It means that the considered position of equilibrium is stable, and the vibrations of the mechanism near him are periodic oscillations. The trajectories 1 and 2 are almost ellipses. It can be argued, that in these cases the mechanism, having obtained the corresponding energy reserves h_1=0.02 and h_2 =0.18, performs almost harmonic oscillations with amplitudes φ_{01}=0.047 rad (2.7^0) и φ_{02}=0.143 rad (8.2^0). Such energy reserves can be realized, in particular, by the initial deflection of the rod. With an increase in the value h (h_3=0.7; h_4=1.1; h_5=1.5; h_6=1.9; h_7=2.0 (Joule)), the trajectories become more and more diamond-shaped, and the oscillations of the rod are increasingly different from harmonic. The curve 8 in Figure 5 (separatrix) is a curve with selfintersection. The abscissas of points selfintersection correspond to the local maximums on the potential energy graph $E_{\Pi}(\varphi)$ (Figure 4a), and therefore, correspond to unstable positions of the equilibrium, that will be realized with the initial reserve of mechanical energy h_8 =2.1431. If h>h_8 (trajectory 9 on the Figure 5) the mechanism makes a different kind of movement, which differs from that considered in this part of the work.

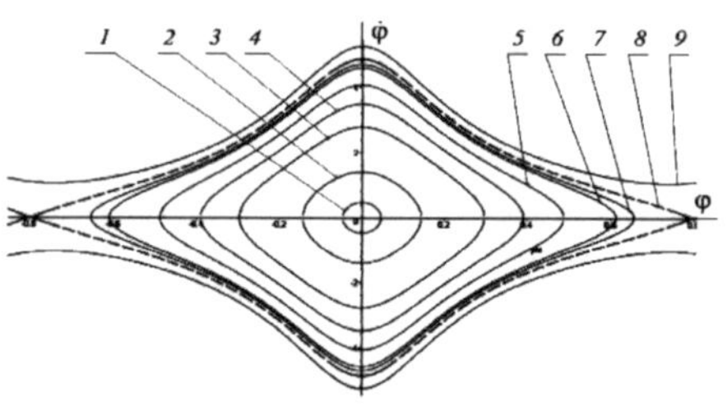

Figure 5. Phase portrait.

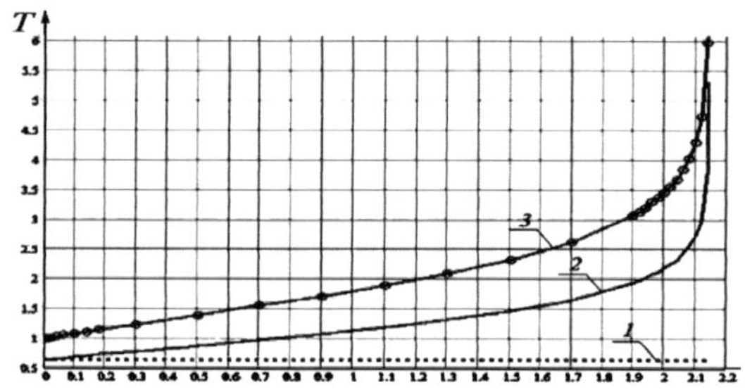

Figure 6. The dependence of the periods from h.

The important characteristic of the mechanism is the period of his free oscillations, which can be defined in the following way. If the abscissas of the points of selfintersections of any closed phase trajectory in the phase portrait (Figure 5) denote by φ_1, φ_2, then a period of nonlinear oscillations will be calculated by the formula

$$T_1 = 2\int_{\varphi_1}^{\varphi_2} \sqrt{\frac{A(\varphi)}{2\left[h - E_{\Pi}(\varphi)\right]}}\, d\varphi \tag{10}$$

The Figure 6 shows the graphic dependencies on h following parameters: 1 – period T of linear oscillations of the mechanism, calculated in accordance with a formula (7); 2 – period T_1 of nonlinear oscillations of the mechanism, calculated in accordance with (10); 3 – relation of the periods T_1 /T. The period T_1 of the nonlinear oscillations increases with increasing the value h, in contrast to the permanent period T of linear oscillations. And when approaching the value h to h_8, which characterizes the position of unstable equilibrium, T_1 tends to infinity. In practice, except the period of oscillations, it is important to know also the amplitude of the angle of deflection of the rod depending on the h.

Taking into account that at the maximal deviation of the mechanism $E_k = 0$, then the amplitude of the angle is the root of equation:

$$h = F_{\Pi}(\varphi) \tag{11}$$

The Figure 7 shows the graphic dependencies on h following parameters: 1 – amplitude φ_0 (rad) of the linear oscillations of the mechanism, calculated in accordance with a formula (8); 2 – amplitude φ_1 of nonlinear oscillations of the mechanism, calculated in accordance with (11); 3 – relation φ_1 / φ_0. Evidently, that at identical h always $\varphi_1 > \varphi_0$ (except h=0), but the relation φ_1/φ_0 although increases with increasing h, tends to the value 1.65 at $h \to h_8$. From that at $h \to h_8$ the period of nonlinear oscillations $T_1 \to \infty$, it follows that at the initial reserve of mechanical energy h_8 the mechanism practically will not be able to perform oscillations, because it requires infinitely long time.

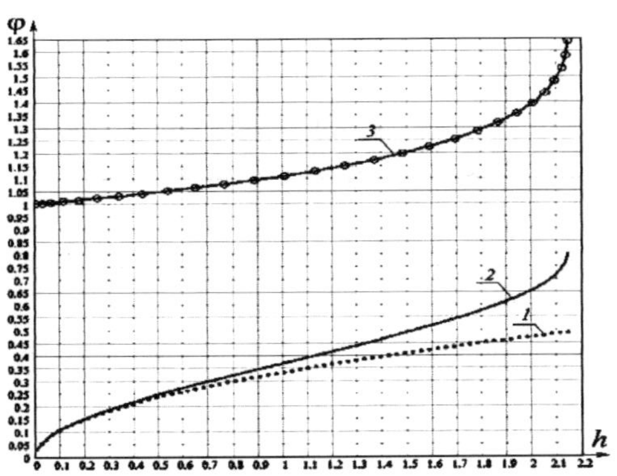

Figure 7. The dependence of oscillation amplitudes from h.

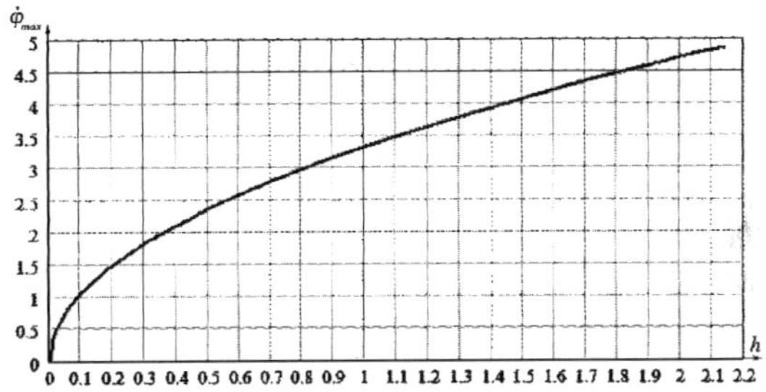

Figure 8. The dependence of the maximum angular velocity from h.

Interest is also the dependence of the top angular velocity of the rod on the initial reserve of mechanical energy $\dot{\varphi}_{max}(h)$. For linear oscillations she is determined by a formula (9), and for nonlinear – by a transformed formula (4). The maximum value of $\dot{\varphi}$ is reached at φ=0. In this position E_{Π} =0 and, in accordance with (4) and (5), $A(\varphi)=a$. Consequently, the dependences $\dot{\varphi}_{max}(h)$ for nonlinear and linear oscillations are identical and determined by the formula (9).
The Figure 8 shows the graphic dependencies $\dot{\varphi}_{max}(h)$. As expected the maximum angular velocity of the rod increases with increasing h. At h=2.14, which is close to the value of h_8, at which the positions of unstable equilibrium takes place, the angular velocity is $\dot{\varphi}$ =4.87 rad·c^{-1}.

4. Conclusion

For the real planetary mechanism, the dependence of the angular velocity of the rod on its rotation angle was determined; and the positions of equilibrium for $\alpha=\pi$, which define the location of the counterweight in the plane of the gear was determined. Using the method of phase plane the types of the movements of mechanism are found and analyzed depending on the distance l_3 of the counterbalance to the gear center. It is established that with increasing the parameter l_3 the number of stable equilibrium positions increases and can be equal to one, three or five. Depending on the initial reserve of mechanical energy the mechanism can perform free oscillations not only near separately considered stable equilibrium positions, but also at once near three or five positions of stable equilibrium. When the mechanism is oscillated near a certain stable equilibrium positions (except for the case φ_1=0), the maximum angle of deflection of the rod depends on the direction of turn: when

IOP Conf. Series: Materials Science and Engineering **468** (2018) 012013 doi:10.1088/1757-899X/468/1/012013

turning in the direction of increasing $|\varphi|$ it is less than when turning in the direction of decreasing $|\varphi|$. The angular velocity of the rod is always maximal near the stable equilibrium positions and is minimal near the unstable equilibrium positions.

For the case $\alpha=\pi$ the graphs of the amplitudes of the angle of rotation of the rod, the periods of free oscillations of the mechanism and the maximum angular velocity of the rod on the parameter h are plotted for linear and nonlinear oscillations of mechanism near the stable equilibrium position determined by the angle $\varphi=0$. The carried out analysis confirms the necessity of using methods of nonlinear mechanics to obtain more accurate results in comparison with the results obtained by linear mechanics methods.

The bifurcations observed on the phase portraits are demonstrated also on the mockup of the real mechanism. Therefore it is expedient to study some parameters of trajectories belonging to classes, bounded by separatrices and singular points. The article considers the methodology of studying of these parameters. This is the novelty of researches and of the results obtained.

References

[1] Babakov I 2004 *Theory of Oscillations*(Moscow: Drofa) [in Russian]

[2] Aldoshin G 2013 *Theory of Linear and Nonlinear Oscillations*(St. Petersburg: Lan) [in Russian]

[3] Kraynev A 1987 *Dictionary - a Directory of Mechanisms*(Moscow: Mechanical engineering) [in Russian]

[4] Obnosov K, Bondarenko N, Konyukhov M and Panshina A 2016. *Engineering Journal. With attachment* **38** pp 17

[5] Bondarenko N, Obnosov K and Panshina A 2017 *Uspekhi sovremennoy nauki* **7**(3) pp 78 – 82

Wave Processes Regulators Optimization in Hydraulic Systems

D N Popov, N G Sosnovskiy and M V Siukhin

Bauman Moscow State Technical University, Moscow, Russian Federation
E-mail: sosn@bmstu.ru

Abstract. The Article shows the methods of optimization of the regulation processes of hydraulic systems with long communication lines; describes a method providing hydraulic systems with more reliable damage protection by wave processes control via optimal regulators; describes an order of system optimization of wave processes regulation.

Systems with long communication lines between separate devices are referred to systems with distributed parameters. In such systems the oscillations may appear where the pressure amplitudes in some sections of the hydraulic lines increase significantly. It is known that resonance in systems with distributed parameters can be avoided by matching the output impedance of the system with its wave resistance [1, 2]. This position is often used when creating passive (uncontrolled) vibration dampers in hydraulic and pneumatic lines. Such dampers can be effective when the device connected to the line together with the absorber makes it possible to obtain an impedance close to the line impedance for the output section of the line. The last condition could be fulfilled only in case of using devices with simple enough transfer functions. With complicated devices passive absorbers become less effective. In this regard, began to apply the regulators of wave processes.

The method of decreasing the amplitude of pressure oscillations in systems based on the use of a regulator in hydraulic lines is considered below [3-5]. The method was tested by experimental research.

The development of mathematical models of hydraulic lines was based on the description of the nonstationary motion of a viscous compressible fluid in a pressure path, taking into account the nonstationary velocity distribution under laminar flow. Numerical values of the parameters of mathematical models were corrected by the results of experimental studies of the layout of the hydraulic system with long hydraulic lines (Figure 1).

The command sensor simulates device 1 from the servo valve. In the control loop of the servo valve, an electronic correction unit 3 with a transfer function $W_K = K_K$ is provided, which provides the required frequency characteristics of the entire device.

The hydraulic actuator plunger is represented by a device 4 consisting of a spring loaded plug in the housing and a parallel adjustable throttle. Devices 1 and 4 are connected by a hydraulic line with a length $l = 45.8$ m and a radius $r_0 = 4$ mm, in the beginning and at the end of which there are inductive pressure sensors DD1 and DD2.

Content from this work may be used under the terms of the Creative Commons Attribution 3.0 licence. Any further distribution of this work must maintain attribution to the author(s) and the title of the work, journal citation and DOI.
Published under licence by IOP Publishing Ltd

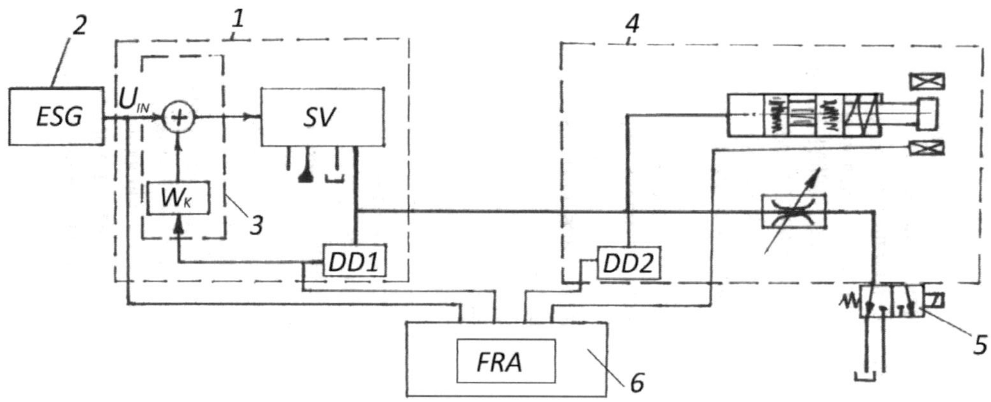

Figure 1. Schematic diagram of the hydraulic system.

The input harmonic signal U_{IN} to the hydraulic system is fed from the electrical signal generator (ESG) 2. The output signal of the device 4 can be considered the position of the plunger, which is measured by an inductive linear sensor. The sensor works in conjunction with an electronic amplifier. When adjusting the opening of the throttle, the measurement of the flow rate of the liquid flowing through the hydrolysis line is carried out in a weighted manner. Thus it is necessary to switch the valve 5.

The test results are recorded by a two-channel frequency response analyzer (FRA) 6.

The main attention was paid to checking the adequacy of the mathematical model of the hydrolysis line. In this connection, data on pressure fluctuations at the beginning and end of the hydraulic line are extracted from the hydro system test results.

The experimental and calculated Bode diagrams of the hydraulic line are given in Figure 2. The experiment was carried out with this opening of the throttle of the device 4 and the input signal at the servo valve of the device 1, so that the pressure at the beginning of the hydraulic line was 3.1 MPa, and the flow through the reactor was 0.24 l / min. The unit uses IGP-30 working fluid.

The results of testing the mathematical model of a system with long hydraulic lines of communication are practically useful for designing servo drives with throttle control.

The solution of the problem of optimization of the wave regulator in the hydraulic line is based on providing an automatic matching of the output impedance of the system with its wave resistance. This condition can be satisfied by applying the reference model of wave impedance (RMWI). The structure of the system with RMWI will depend on the way the output impedance is controlled. With the first method, a flow controller (FC) with a constant setting is turned on (Figure 3). The second one uses the adaptive regulator (ADR) (Figure 4). The following variables are indicated in the diagrams: p_2, Q_2 - pressure and flow deviations from their steady-state values at the end of the hydrolysis line; Q_2^r – Flow deviation, obtained from the output of the reference model of wave impedance; Q_ε – error on the flow; Q_p – adjustable flow deviation; Q_H – Flow deviation caused by a change in the output impedance (load) of the system; VD – the vibration damper.

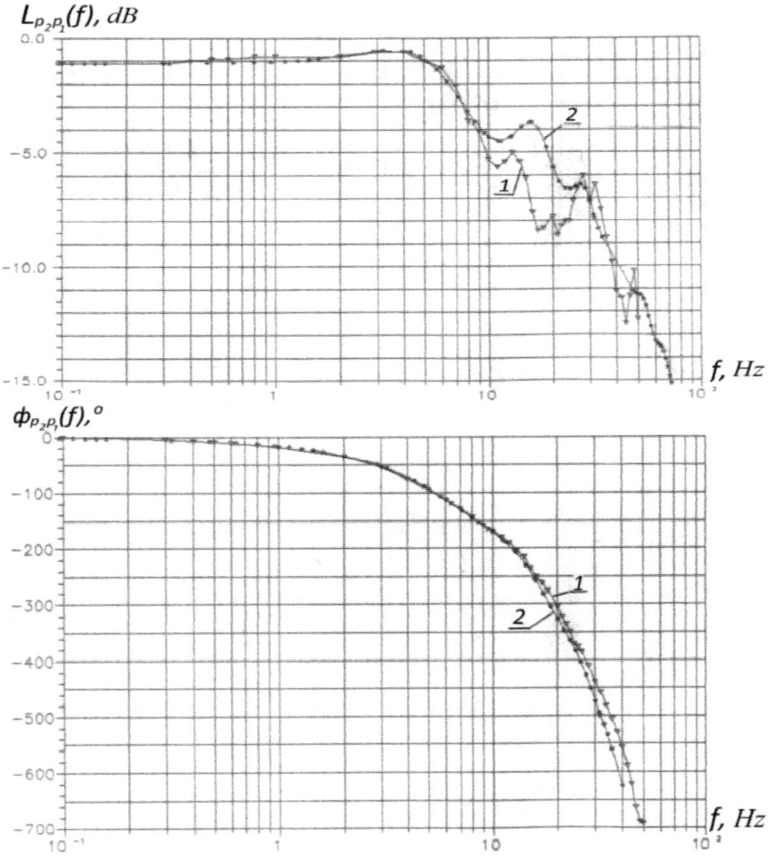

Figure 2. Bode diagrams of the hydraulic line: 1-experiment, 2-calculation.

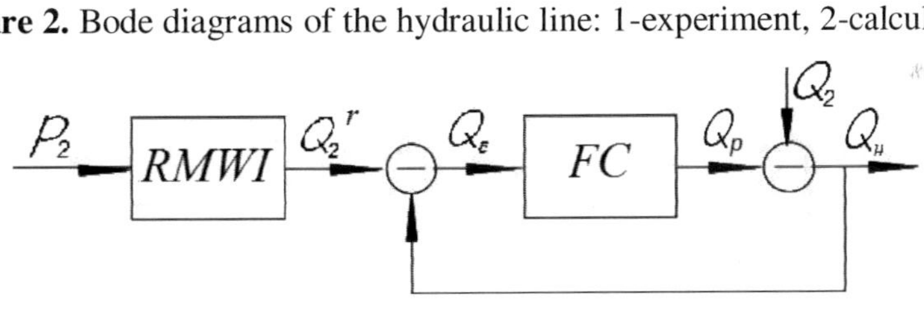

Figure 3. Diagram of a control system for oscillations in a hydraulic line with a flow regulator.

The structure of the reference model of EMWS wave resistance is determined as a result of mathematical modeling of non-stationary hydrodynamic processes in the hydrolysis [1-3]. From this description it is possible to obtain in the operator form a fairly complete mathematical model. The reference model can be substantially simplified by neglecting the resistance of the hydrolysis line. In this case we have [4]:

$$Z_l(s) = Z_0(s) = \rho c_l / F_l \,,$$

here Z_l – wave resistance of the hydraulic line; c_l – the speed of propagation of disturbances in the hydraulic line; F_l – cross-sectional area of the hydrolysis line.

In the calculation $Z_0(s)$ it should be borne in mind that the propagation velocity of the disturbance in the hydraulic line is affected by the nonstationarity of the distribution of the local velocities of the working medium and the temperatures along the cross section of the line. In order to take this influence into account, it is possible, having estimated the range of frequencies of oscillations of the working medium in the hydraulic line, to use the adjustments [1]. Assuming that after correction c_l =const, the reference model of wave impedance can be represented by a proportional link with the gain factor $K_M = F_l / \rho c_l$.

The mathematical model of the flow controller is determined by the condition that if there is no steady error during the process control, the regulator is of the type "I" (integrating). In this case for the regulator it is possible to take the transfer function

$$W_{pp}(s) = \frac{1}{T_I s}.$$

The time constant T_I must first be chosen such that the error in the rate of change of Q_ε is as small as possible.

The mathematical model of the adaptive regulator largely depends on the circuit and design of the selected vibration damper. Common, but not effective, include vibration dampers, consisting of a capacitance and a throttle, through which the vibration damper is connected to the hydrolysis line. Let us clarify the possibility of increasing the efficiency of such an absorber by means of an adaptive regulator. The processes in such a vibration damper can not be taken into account, if the inertia of the elastic filler is not taken into account. Then:

$$Q_P = K_c \frac{dp_{VD}}{dt},$$

$$Q_P = K_t (p_2 - p_{VD}),$$

here K_c – coefficient of fluid compressibility K_t – conductivity coefficient of throttle.

From these equations it is easy to obtain the transfer function

$$W_p(s) = \frac{Q_p(s)}{p_2(s)} = \frac{K_t s}{(K_c / K_t) s + 1}. \tag{1}$$

By limiting the adaptive tuning of the vibration damper to a frequency range in which the transfer function (1) can be approximately replaced by the relation 2 and equation

$$Q_p = K_t p_2. \tag{2}$$

Equation (2) shows that the damper setting can be performed by changing K_t, depending on the error Q_ε. The algorithm providing automatic tuning of the vibration damper is determined using the balance-of-expenditure equations in the algebraic adder of the circuit shown in Figure 2:

$$Q_2^r - Q_н = Q_\varepsilon,$$

$$Q_2 - Q_p = Q_н.$$

Dividing all the terms of the last two equations by p_2, you can find the corresponding impedances Z:

$$Z_0^{-1} - Z_н^{-1} = Z_\varepsilon^{-1},$$

$$Z_2^{-1} - K_t = Z_н^{-1}.$$

For $Z_0^{-1} = Z_2^{-1}$, the last equations are $K_t = Z_\varepsilon^{-1}$, which can be used in the control system diagram (Figure 4) to obtain Q_p. The structural scheme that realizes this algorithm is given in Figure 5.

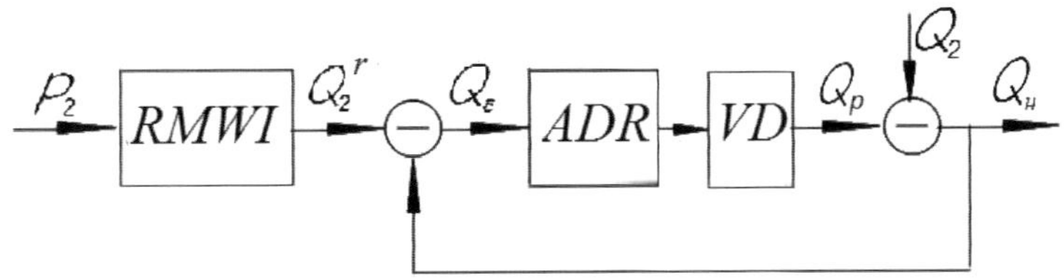

Figure 4. Diagram of a control system for oscillations in a hydraulic line with an adaptive regulator.

Systems of automatic matching of the output and wave impedances, having the structures proposed above, are themselves stable. The stability of the entire hydraulic system containing long hydraulic lines with automatic output impedance regulators depends on the dynamic characteristics of the devices connected at the ends of the lines. If these devices are represented by linear dynamic links with constant parameters, the processes caused by matching the output impedances with the wave impedances should not affect the stability of the entire system. In this case, the hydraulic lines can be considered as links of pure retardation and stability analysis of the entire system can be carried out by methods known from control theory.

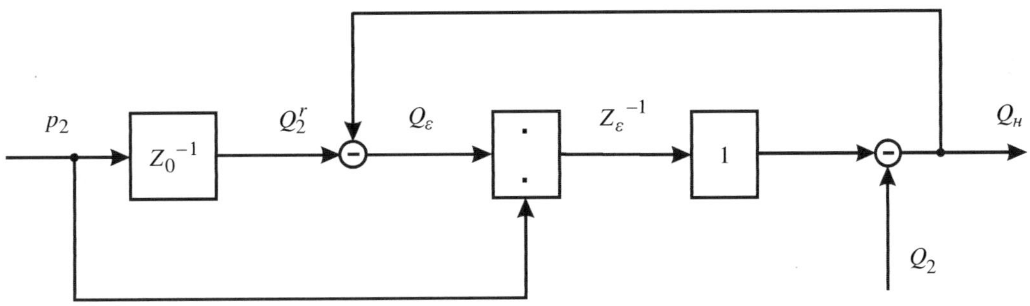

Figure 5. The algorithm is adaptive adjustment absorber fluid oscillations.

If there are nonlinear or non-stationary links in the system, the processes caused by the automatic control of the output impedances and the control processes of the entire system will affect each other. In this case, the problem of analyzing the stability of the entire system becomes so complex that its solution can be obtained only as a result of numerical simulation of the entire system on a computer.

After preliminary calculation of the parameters of the above two control systems for wave processes, to solve the design variant of the regulator it is necessary to solve the problem of multi-criteria optimization of each system [6-10]. Comparing the obtained results with each other, the designer, taking into account the factors that can not be formalized, has the opportunity to choose the most optimal variant of the regulator. The mathematical models of systems with distributed parameters required for the solution of the problem consist of partial differential equations describing non-stationary processes in hydraulic lines, and equations describing the boundary conditions. The first of these equations have the form [1]:

$$\frac{\partial v}{\partial t} = v\frac{\partial v}{\partial x} + \frac{2\tau_{0n}}{\rho r_0} = -\frac{\partial p}{\partial x},$$

$$\frac{\partial v}{\partial x} = -\frac{1}{B}(\frac{\partial p}{\partial t} + v\frac{\partial p}{\partial x}),$$

$$\tau_{0n} = A_1[v(t)],$$

$$F(B, p) = 0.$$

Here v is the velocity of the working medium, averaged over the cross-section of the hydrolysis line; $\tau_{0н}$ – non-stationary tangential stress in the place of contact of the working medium with the wall of the pressure channel (pipe); x is the coordinate measured along the hydrolysis axis; B – reduced modulus of elasticity of the working medium in the pressure channel (tube) with elastic walls; $A_1[v(t)]$ operator that determines the law of variation of τ_{0n}; $F(B,\rho)$ – a function that establishes the dependence of B on p.

Equations describing the boundary conditions should reflect those processes that occur in devices connected at the ends of the hydrolysis line. In the general case, these equations can be represented by a system of ordinary differential equations and algebraic relations. To start the line to which the controller is not connected, these equations have the form

$$\frac{dX_1}{dt} = f(X_1,t),$$

where $X_1 = (x_1^b, x_2^b, ..., x_m^b)$ is the m-dimensional vector of the state variables of the device connected at the beginning of the line.

For the end of the line, to which, in addition to the load-creating device, a regulator is connected, two sets of equations should be recorded separately. One will be similar to the above, the other refers to the regulator. The appearance of the latter depends on the selected structure and parameters of the regulator. If the system with the flow regulator is optimized, then it is possible to use the equation of a closed loop with a regulator of the form

$$T_P \frac{dQ_2}{dt} = \frac{F_l}{\rho c_l} p_2 + T_P \frac{dQ_н}{dt} - Q_н.$$

Variable $Q_н$, Q_2, p_2, which are included in this equation are both vector components X_2, where $X_2 = (x_1^e, x_2^e, ..., x_n^e)$ – n-dimensional vector of state variables unit connected to the end of the line.

In optimizing the system with an adaptive controller in accordance with the transfer function (1), equation (2), for describing the boundary conditions are the following equations:

$$\frac{K_c}{K_t}\frac{dQ_2}{dt} + Q_2 = K_c \frac{dp_2}{dt} + \frac{K_c}{K_t}\frac{dQ_н}{dt} + Q_н;$$

$$K_t = \frac{F_l}{\rho c_l} - \frac{Q_н}{p_2}.$$

When solving the optimization problem, the variable for the first scheme (Figure 3) is T_P, and for the second (Figure 4) – K_t. In addition, in the latter case, the variable parameter may be K_c. The range of variation of each variable a_i determines the inequality

$$(a_i)_{\min} \le a_i \le (a_i)_{\max}.$$

In addition to the last conditions, functional restrictions are also imposed, the form of which depends on the characteristics of the devices connected at the ends of the line and the technical requirements imposed on the hydraulic system [11].

The criteria for assessing the quality of the optimized systems are the components of the vector $K_к$. Those solutions that provide the value of $K_к$ form a subset M_0 of the optimal variants from the set M_d of admissible solutions. The components of the vector $K_к$ are also constrained in the form of inequalities that establish the boundaries in which their optimal values are found. According to the statement of the problem, the main criterion here is the deviation of the pressure from the allowable values indicated along the hydraulic line, i.e., $p = p(x)$.

Among the criteria, there can be included deviations of flow rate in different places of the system, the mass of all the elements of the system, the indicator of the energy efficiency of the system, and others.

The optimization results are tabulated in which the best values of the accepted criteria are given for each project variant. According to these tables, the optimal variant is first selected, and then the best one is selected from these two options.

As a result of the solution of the problem considered, a method is proposed to increase the efficiency of damping of oscillations in long hydraulic lines. Of the possible methods of oscillation damping, a method based on the use of optimal automatic regulators matching the output and the wave impedance is recommended. The technical feasibility of this method is verified by mathematical modeling of the hydraulic system. The adequacy of the mathematical model to the real system is confirmed by comparing the computer-calculated frequency characteristics with the experimental frequency characteristics.

To automatically match the output impedance with the wave impedance, two types of regulators are proposed. In one type, a flow controller of type "I" is used, in the other type an adaptive regulator is used.

To select the structure and parameters of the regulator, a multicriterion optimization method for a controlled system with distributed parameters is proposed. In the algorithm of the proposed adaptive regulator of wave oscillations, the special aspects of the nonlinear mechanics of the considered type of hydraulic system are taken into account [10, 11].

The result of the work is presented as a theoretically grounded solution of the problem of selecting the structure and parameters of the optimal regulator for a hydraulic system with distributed parameters and nonlinear boundary conditions.

References

[1] Popov D N, Sosnovskiy N G and Siukhin M V 2016 Wave Processes Control in Hydraulic Systems. *Science and Education: Scientific Publication of BMSTU* **10** pp 39–55

[2] Popov D N 1987 *Dynamics and regulation of hydraulic and pneumatic systems*(Moscow: Mashinostroenie Press) p 464

[3] Popov D Nand Astashev V K 2012 *Mechanical engineering: Encyclopedia* Vol. IV-2*: Electric drive. Hydraulic and vibratory drives.* Book 2*: Hydraulic and vibratory drives*(Moscow: Mashinostroenie Press) p 303

[4] Popov D N and Sosnovskiy N G 1995 Calculation of non-stationary processes and reduction of resonant pressure peaks during cleaning of hydraulic systems by a pulsating flow of liquid *Vestnik Mashinostroeniya* **2** pp 22–28

[5] Popov D N and Sosnovskiy N G 2015 Dynamic Characteristics of Communication Lines with Distributed Parameters to Control the Throttle-controlled Hydraulic Actuators *Science and Education: Scientific Publication of BMSTU* **9** pp 32–42

[6] Borovin G K and Popov D N 2007 *Multiobjective optimization of hydraulic systems* (Moscow: Bauman MSTU Press) p 94

[7] Sobol I M and Statnikov R B 2006*The choice of optimal parameters in problems with many criteria* (Moscow: Drofa Press) p 175

[8] Karpenko A P 2014 *Modern algorithms of search optimization*(Moscow: Bauman MSTU Press) p 446

[9] Borovin G K and Popov D N 2004 Computer Experiment in Problems of Optimal Designing of Controlled Systrms with Actuators Hydraulic*Mathematical Models and Computer Simulations* **6** pp 56–60

[10] Liu B 2005*Theory and practice of indefinite programming (*Moscow: BINOM, Laboratoriya Znaniy Press) p 416

[11] Trubetskov D I 2014 *Introduction to synergetics: Chaos and structures* (Moscow, Librokom Press) p 240

IOP Conf. Series: Materials Science and Engineering **468** (2018) 012015 doi:10.1088/1757-899X/468/1/012015

Joint inversion of geophysical (seismic and selenodetic) and geochemical data for internal structure and composition of the Moon

E. Kronrod [1], K. Matsumoto [2], O. Kuskov[1], V. Kronrod [1], R. Yamada [3] and S. Kamata[4]

[1]Vernadsky Institute of Geochemistry and Analytical Chemistry (GEOKHI RAS),
[2]RISE Project, National Astronomical Observatory of Japan,
[3]The University of Aizu, Research Center for Advanced Information,
[4]Creative Research Institution, Hokkaido University
E-mail: e.kronrod@gmail.com

Abstract. We investigate lunar internal structure models which are consistent with the seismic and the selenodetic data and thermodynamic constraints. The seismic travel time data constrain crustal and mantle structures and composition, while selenodetic data are expected to contribute to infer the structure below the deep moonquake region.

1. Introduction

In the recent paper [1] lunar interior models by complementing Apollo seismic travel time data with selenodetic data which have recently been improved by Gravity Recovery and Interior Laboratory (GRAIL) and Lunar Laser Ranging were explored. Important information on the thickness of the crust, Low-velocity/viscosity zone (LVZ), core structure and seismic velocities were obtained. But this problem statement retains lunar mantle composition to be uncertain. In [2, 3] mantle composition can be simulated based on thermodynamic approach and petrological evidence from seismic data, moment of inertia (MOI) and mass.

The goal of our investigation is to obtain geochemically and geophysically consistent information on the internal structure of the Moon. The first step was to use thermodynamic approach to calculate elastic properties (shear modulus μ, bulk modulus κ) and density ρ in the Moon's layers. But it lead to increasing number of model parameters, and it was impossible to obtain reasonable result. Here we modified the problem statement and included geochemical data (bulk Al and Fe composition) as observed data and used more constraints on the model (such as magma ocean condition) and fixed some model parameters.

2. Thermodynamic approach

Our general methodology is to combine the geophysical and geochemical constraints and thermodynamic approach, and to develop, on this joint basis, the self-consistent models of Moon, accounting for its chemical composition and internal structure. Thermodynamic modeling of phase relations and physical properties in the multicomponent mineral system (NaTiCFMAS) was used to develop a method for solving the inverse problem. The technique of Gibbs free energy minimization was used, and minerals equations of state and solid solutions were included in the database. While

Content from this work may be used under the terms of the Creative Commons Attribution 3.0 licence. Any further distribution of this work must maintain attribution to the author(s) and the title of the work, journal citation and DOI.
Published under licence by IOP Publishing Ltd

solving the inverse problem, we require a non-negative density gradient in the mantle and include additional geochemical and cosmochemical constraints. For the computation of elastic properties and density for a given chemical composition we have used the THERMOSEISM software [4].

3. Geochemical model of the Moon.

Current geochemical models for bulk FeO and Al_2O_3 concentrations in the Moon are shown in Fig. 1. Apparently there is a slight difference in FeO content and significant difference in refractory elements (Al_2O_3, CaO) content. In our study we consider only Al_2O_3 concentration presuming Ca/Al ratio to be chondritic. Models of the lunar composition can be conventionally divided into two groups: 1 – bulk Al_2O_3 content is close to Earth's value, 2 – bulk Al_2O_3 concentration is higher than 4,5 wt.%.
The analysis of majority of current Moon's composition models revealed the following $Al_2O_3$and FeO content with corresponding error bars for Group 1: $3,5 \leq Al_2O_3 \leq 4,5$ ("average" $Al_2O_3 = 4,05 \pm 0,35$ wt.%) and for Group 2: $4,5 \leq Al_2O_3 \leq 7,7$ ("average"$Al_2O_3 = 5,91 \pm 0,39$ wt.%). For both groups of models $10 \leq FeO \leq 14$ ("average" FeO = $12,25 \pm 1,33$ wt.%).

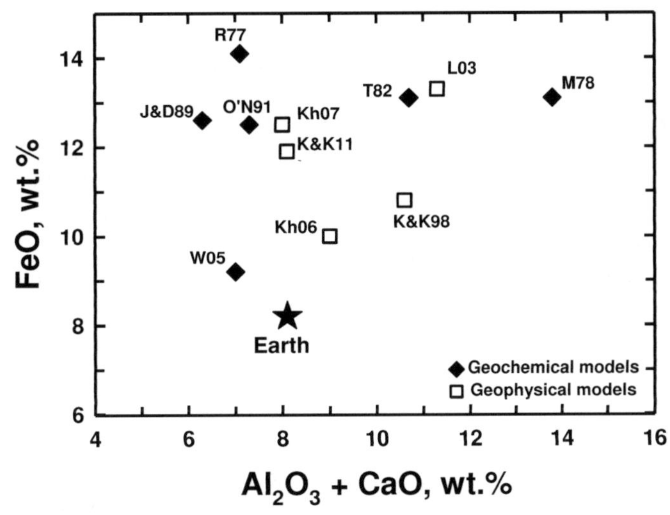

Figure 1. Geophysical and geochemical models of a bulk-Moon composition (crust + mantle) compared with that of the bulk silicate Earth [5].

The difference in concentrations of refractory elements can be interpreted by models of Moon's origin. Currently the hypothesis of giant impact is the prevalent theory of the Moon's origin.If we follow this theory, then the volatile loss occurred as a result of giant impact and further high-temperature processes but concentrations of refractory elements and there ratio must have remained equal to Earth's values, and it is assumed that isotope compositions of the Moon and colliding body were equal.
One of the possible arguments confirming giant impact hypothesis is possibility of modeling of Moon's composition with bulk Al_2O_3 content similar to models with Earth's bulk Al_2O_3 content.In the present study we consider geochemical models with smaller Al_2O_3 content (similar to Earth's value - Group 1).

4. Geophysical data

We employed the same data as [1]: six selenodetically observed data of the mean radius (R), mass (M), normalized mean solid moment of inertia (Is), and degree 2 potential tidal Love number k_2 reported by [6] who summarized recent results of selenodetic data analyses and monthly and annual quality factors ($Qm$and Qa) reported by [7]. These values are summarized in Table 1.The mean radius R constrains the size of the modeled Moon, and *M*, solid MOI, k_2, *Qm*, and *Qa*are used, together with the seismic travel time data, to form the likelihood function (equation (2))

Table 1. Selenodetic Data Used to Constrain Internal Structure Models of the Moon

Parameter	Value
Mean radius (*R*)	1737.151 km
Mass (*M*)	$(7.34630 \pm 0.00088) \times 10^{22}$ kg
Normalized mean solid moment of inertia (*Is*)	0.393112 ± 0.000012
Degree 2 potential tidal Love number (k_2)	0.02422 ± 0.00022
Monthly quality factor (*Qm*)	38 ± 4
Annual quality factor (*Qa*)	41 ± 9

We used the seismic travel time data selected by [8], i.e., 318 data (183 P-wave and 135 S-wave) from 59 sources (24 deep quakes, 8 shallow quakes, 19 meteoroid impacts, and 8 artificial impacts). The error for *n*th travel time $\sigma_{\tau n}$is given by

$$\sigma_{\tau n} = \sqrt{\sigma_{n,r}^2 + \sigma_{n,e}^2} \qquad (1)$$

where $\sigma_{n,r}$and $\sigma_{n,e}$are the read time error and the event time error of *n*th travel time data. In our model we considered $\sigma_{\tau n}$ increased by a factor of three, which resulted in better solution.

5. The model of the Moon:

We apply spherically symmetric viscoelastic hydrostatic model of the Moon (Fig.2). The Moon consists of nine layers: megaregolith, crust, four-layer mantle, low viscosity zone (LVZ), liquid outer core and fluid inner core. In each zone physical properties are assumed to be constant.

Crustal thickness was assumed to be 34 km [9] (including 1 km regolith layer). Composition of the crust was taken from [10].

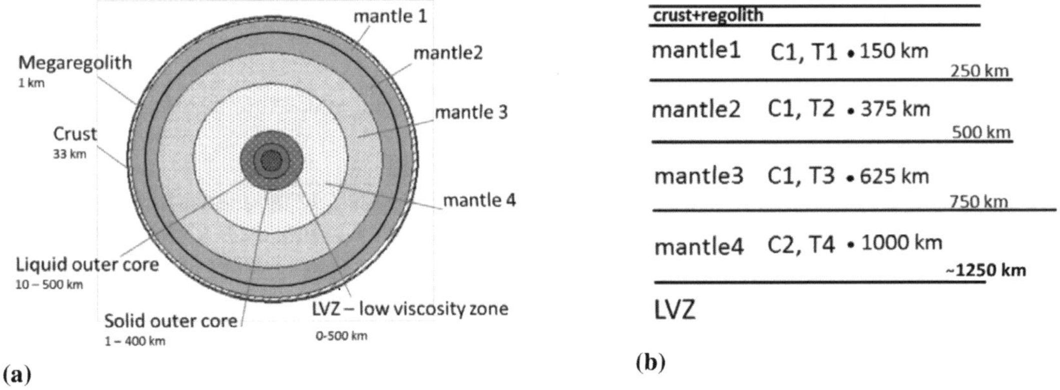

Figure 2 (a) The model of the Moon considered in the present study; **(b)**The mantle structure of the Moon. The depth of boundaries between mantle layers was fixed at the depths of 250, 500, 750 km. The depth of mantle-LVZ boundary was estimated during calculation. Temperature (T1-T4) and main oxides (Al_2O_3, FeO, MgO) concentrations (C1, C2) were specified at the depth of the middle of each mantle layer. C1 - equal for three upper layers, C2 in fourth layer (mantle 4) was calculated from balance equations.

In the present work, in the modeling of the chemical composition of the Moon, we considered four intervals of oxides concentrations in four mantle layers. Division mantle into layers was performed according to [11] model. Concentrations of main oxides were equal in 1-3 upper mantle layers and we applied the model of magma ocean to calculate oxide concentrations in fourth lower mantle layer (which implies that concentrations of main oxides in the lower mantle is equal to average concentrations in upper mantle and crust and equal to bulk concentrations). The models of the magma ocean in such a formulation were considered in our previous work [3].
Temperature gradient in the mantle was assumed to be approximate constant down to a depth of 1000 km [12] with the upper mantle temperature of 550 (±10) °C at the depth of 150 km and lower mantle temperature of 1150 (±10) °C at the depth of 1000 km.

6. Inversion:

A Bayesian inversion approach is an effective method to solve for a nonlinear problem such as planetary internal structure modeling [2, 13]. This study utilizes Markov chain Monte Carlo (MCMC) algorithm to infer the parameters of the lunar internal structure. The solutions of the parameters and their uncertainties are obtained from the posterior distribution which is sampled by the MCMC algorithm. Then, the likelihood function $L(m)$, which is a measure of misfit between the model predictions and the observations is written as (2) :

$$LHF = \exp\Bigg(-\frac{\left(d_{obs}^{mass}-d_{cal}^{mass}(m)\right)^2}{2\sigma_{mass}^2}-\frac{\left(d_{obs}^{MOI}-d_{cal}^{MOI}(m)\right)^2}{2\sigma_{MOI}^2}-\frac{\left(d_{obs}^{k2}-d_{cal}^{k2}(m)\right)^2}{2\sigma_{k2}^2}-\frac{\left(d_{obs}^{Qa}-d_{cal}^{Qa}(m)\right)^2}{2\sigma_{Qa}^2}-$$
$$-\frac{\left(d_{obs}^{Qm}-d_{cal}^{Qm}(m)\right)^2}{\sigma_{Qm}^2}-\sum_{n=1}^{318}\frac{\left(d_{obs}^{\tau n}-d_{cal}^{\tau n}(m)\right)^2}{\sigma_{\tau n}^2}-\frac{\left(d_{obs}^{Al\,bulk}-d_{cal}^{Al\,bulk}(m)\right)^2}{\sigma_{Al\,bulk}^2}-\frac{\left(d_{obs}^{Fe\,bulk}-d_{cal}^{Fe\,bulk}(m)\right)^2}{\sigma_{Fe\,bulk}^2}\Bigg) \quad (2)$$

where d_{obs}, $d_{cal}(m)$, σ, τn, denote observed data, data calculated from the model m, uncertainty of the observed data, and nth seismic travel time, respectively (similar to [1]. In the present model bulk Al_2O_3 content and bulk FeO content are included into LHF (Al_{bulk} and Fe_{bulk}).

7. Results:

The results of the inversion are shown in Fig. 3 and Fig. 4. Probable bulk silicate Moon oxide concentrations is Al_2O_{3bulk}~4,1 (±0,5) wt.% and FeO_{bulk} ~12,5 (±0,5) wt.%.
As a result of our investigation it turned out that the solution can be obtained only at short range of mantle temperatures. Even slight temperature variation as well as including additional variable parameters (such as crustal thickness and density) lead to geochemically unreasonable main oxides concentrations. The main reasons of this situation might be the following: 1. The model is overspecified. 2. Seismic and geochemical models possess narrow general solution region. 3. There might be errors in seismic model and travel time estimations.

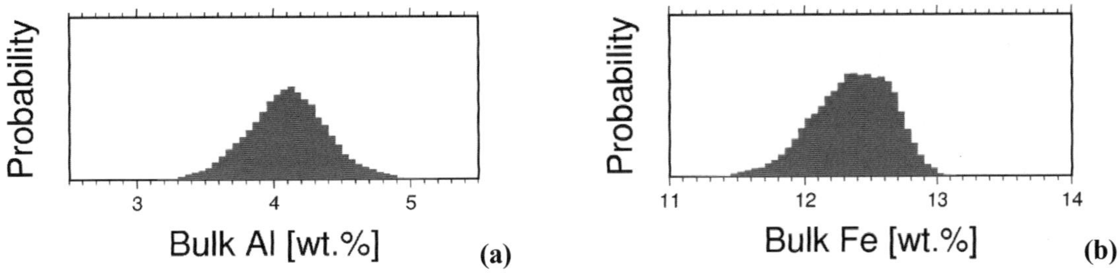

Figure 3 Posterior probability density functions for bulk concentrations of Al_2O_3 (a) and FeO (b).

IOP Conf. Series: Materials Science and Engineering **468** (2018) 012015 doi:10.1088/1757-899X/468/1/012015

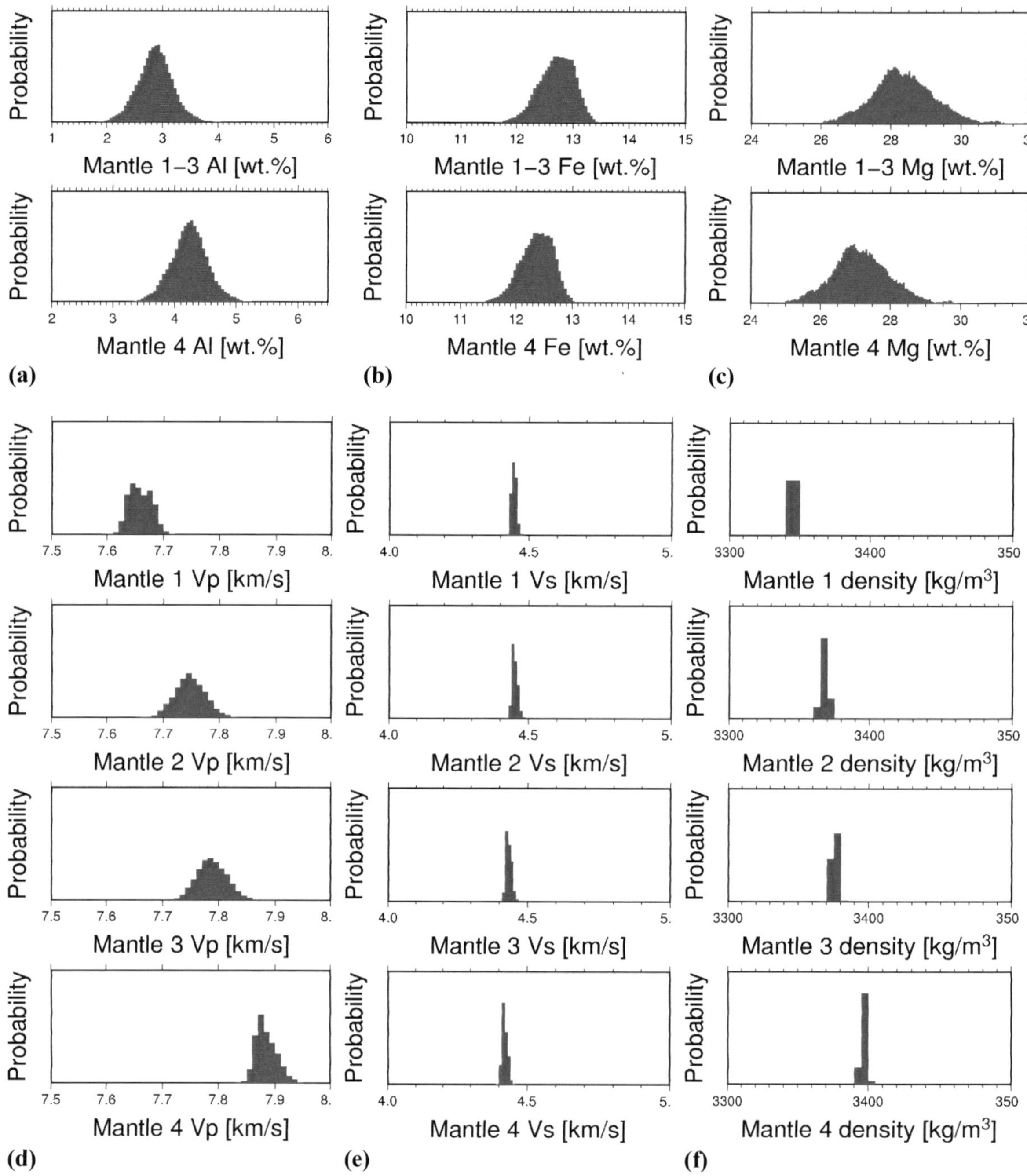

Figure 4 Posterior probability density functions in layers 1-4 of the lunar mantle for (a) Al_2O_3 concentration, (b) FeO concentration, (c) MgO concentration, (d) P-wave seismic velocity distribution, (e) S-wave seismic velocity distribution, (f) density distribution.

8. Conclusions:

We have studied lunar internal structure models which are consistent with the seismic data, the selenodetic (GRAIL and LLR) data and thermodynamic constraints. Our research revealed that the solution appears at short range of mantle temperatures and additional constraints on model parameters (constraints on bulk composition, magma ocean condition, fixed crustal parameters) are required. Probably this result substantially stems from paucity of observation data used or the problem is overspecified (too much model parameters), narrow solution space satisfying both the geophysical and geochemical constraints. In our future study we are going to investigate this problem.

References

[1] Matsumoto K, Yamada R, Kikuchi F, Kamata S, Ishihara Y, Iwata T, Hanada H, Sasaki S2015 Internal structure of the Moon inferred from Apollo seismic data and selenodetic data from GRAIL and LLR. *Geophysical Research Letters* **42** 18 pp 7351–7358

[2] Khan A., Connolly J, Maclennan J, Mosegaard K 2007 Joint inversion of seismic and gravity data for lunar composition and thermal state. *Geophys. J. Int.* **168** pp 243–258

[3] Kronrod V, Kuskov O2011 Inversion of seismic and gravity data for the composition and core sizes of the Moon*Izv. Phys. Solid Earth* **47** pp 711-730.

[4] Kuskov O2001 Core Sizes and Internal Structure of Earth's and Jupiter's Satellites'*Icarus*, **151**(2) pp 204–227

[5] Kuskov O, Kronrod V, Kronrod E2014 Thermo-chemical constraints on the interior structure and composition of the lunar mantle*Physics of the Earth and Planetary Interiors* **235** pp 84-95

[6] Williams J, Konopliv A, Boggs D, Park R, Yuan D, Lemoine F, Goossens S, Mazarico E, Nimmo F, Weber R, Asmar S, Melosh H, Neumann G, Phillips R, Smith D, Solomon S, Watkins M, Wieczorek M, Andrews-Hanna J, Head J, Kiefer W, Matsuyama I, McGovern P, Taylor G, Zuber M, 2014 Lunar interior properties from the GRAIL mission, *J. Geophys. Res. Planets* **119** pp 1546–1578

[7] Williams J, Boggs D 2015 Tides on the Moon: Theory and determination of dissipation. *J. Geophys. Res. Planets***120** pp 689–724

[8] Lognonné P, Gagnepain-Beyneix J, Chenet H. 2003 A new seismic model of the Moon: implications for structure, thermal evolution and formation of the Moon. *Earth and Planetary Science Letters* **211** pp 27-44

[9] Wieczorek M, Neumann G, Nimmo F, Kiefer W, Taylor G, Melosh H, Phillips R, Solomon S, Andrews-Hanna J, Asmar S, Konopliv A, Lemoine F, Smith D, Watkins M, Williams J, Zuber M 2012The crustoftheMoonas seenby GRAIL.*Science*,**339** (6120) pp 671–675

[10] Taylor S1982Planetary Science: A Lunar Perspective*LPI. Houston. TX*, **481**

[11] Gagnepain-Beyneix J, Lognonné P, Chenet H, Lombardi D, Spohn T2006 A seismic model of the lunar mantle and constraints on temperature and mineralogy *Phys. Earth and Planet Int.* **159** pp 140-166.

[12] Kronrod E, 2015Constraints on temperature regime and uranium concentrations in the Moon for the model of magma ocean with partial melting in the vicinity of the core. *Solar System Study: Some Milestones Proceedings Academician Mikhail Marov 80th anniversary session Forth Moscow Solar System Symposium, 4M-S3 IKI RAN, Moscow, October 14–18, 2013 Ed. A.V. Zakharov,* pp 89-102 *(Rus)*

[13] Khan A, Connolly J, Pommier A, Noir J2014 Geophysical evidence for melt in the deep lunar interior and implications for lunar evolution. *Journal of Geophysical Research E: Planets.* **119**(10) pp 2197 – 2221

Acknowledgments: This work was supported by JSPS KAKENHI grant 17K05643, the RFBR grant 15-05-01161 and Russian Academy of Sciences under Programs 7 and 22.

IOP Conf. Series: Materials Science and Engineering **468** (2018) 012016 doi:10.1088/1757-899X/468/1/012016

The fluctuation perturbations in the model of the Chandler wobble

V V Bondarenko, S S Krylov and V V Perepelkin

Moscow Aviation Institute (National Research University), Moscow, Russia
E-mail: vadimkin1@yandex.ru

Abstract. The combinative model of the fluctuations of the Earth Pole taking into account the stochastic components of the Pole tide is developed. The effects of the fluctuation-dissipative perturbations on the frequency that is near to the Chandler one are investigated on the basis of analytical approach. We note that allowance for the small fluctuation-dissipation perturbations due to the pole tide leads to variations in the Chandler component of the oscillations.

1. The Earth pole motion and pole tide gravity fluctuations

Creating a mathematical model, that is adequate to the observation data obtained by International Earth Rotation and Reference Systems Service (IERS) [1] and describes the real motion of the instantaneous axis of rotation in reference frame associated with Earth, is an interesting and complex problem in the fields of theoretical and celestial mechanics. It is significant for many practical problems, including the problems of improving the precision characteristics of the spacecraft coordinate-time and navigation support.

Apart from the regular properties of the Earth Pole oscillations studied in [2], accounting the stochastic small-scale high-frequency fluctuations of the pole tide on short time scales is of scientific and practical interest. The complex dynamics of the interaction of small and large-scale motions in the hydrosphere that includes a broad spectrum of turbulent fluctuations, given the dissipative properties of the Earth's viscoelastic mantle, is especially important in studying the damping of oscillations in geodynamical processes.

Differential equations of the oscillations of the Earth pole coordinates can be derived from the dynamic Euler–Liouville equations [3]:

$$\begin{gathered} \dot{x}_p + N_p y_p = \kappa_y r_0 + \mu_x, \\ \dot{y}_p - N_q x_p = -\kappa_x r_0 + \mu_y, \\ N = \sqrt{N_p N_q} \cong (0.84 \div 0.85)\omega_*, \quad \kappa_x = \frac{\delta J_{pr}}{A^*}, \quad \kappa_y = \frac{\delta J_{qr}}{B^*} \end{gathered} \tag{1}$$

Here, N is Chandler frequency; ω_* is the mean motion of the Earth in its orbit around the Sun; r_0 is the mean angular velocity of the Earth axial rotation; A^*, B^* are the effective central principal moments of inertia taking into account the deformations of the "frozen" Earth's shape; δJ_{pr}, δJ_{qr} are the small variations of the inertia tensor containing different harmonic components (zonal, tesseral,

Content from this work may be used under the terms of the Creative Commons Attribution 3.0 licence. Any further distribution of this work must maintain attribution to the author(s) and the title of the work, journal citation and DOI.
Published under licence by IOP Publishing Ltd

sectoral) due to the perturbing effect of gravitational tides. The terms $\mu_{x,y}$ define the perturbations at a frequency N_1 close to the Chandler one.

The tide caused by the perturbing centrifugal potential produces gravity variations. The gravity **g** at some point of the Earth's surface is the resultant of two forces, the gravitation due to the Newtonian attraction by the Earth's mass and the centrifugal force that arises as the deformable Earth rotates around its axis. The gravity g is given by the expression

$$g = \sqrt{\left(\frac{\partial W}{\partial r}\right)^2 + \frac{1}{r^2}\left(\frac{\partial W}{\partial \theta}\right)^2}, \quad (2)$$

$$W = \frac{1}{2}\left(\omega_0^2 r^2 \cos^2\theta - 2\omega_0^2 r^2 \cos\theta \sin\theta (x_p \cos\lambda + y_p \cos\lambda)\right) + V.$$

Here, r, θ and λ are the spherical coordinates, is the mean angular velocity of the Earth's axial rotation. The presence of an inertial force changing its direction in space leads to variations in gravity **g** whose radial component δg_ω is defined by an expression dependent on the Earth pole motion [1]:

$$\delta g_\omega = \omega_0^2 R_E \sin 2\theta \left[x_p \cos\lambda - y_p \sin\lambda\right]. \quad (3)$$

The variations in the vertical component of due to the Earth's additional potential are determined by the tesseral terms of the second zonal geopotential harmonic:

$$\delta g_{pt} = -\nabla_R \delta V_2. \quad (4)$$

$$\delta V_2 = \frac{Gm_E R_E^2}{R^3} \sum_{m=0}^{2} [\delta c_{2m} \cos m\lambda + \delta s_{2m} \sin m\lambda] P_{2m}(\cos\theta).$$

Here, $P_{2m}(\cos\theta)$ are the associated Legendre polynomials, and $Gm_E = 3{,}98600442 \times 10^{14}\,\text{м}^3\text{с}^{-2}$.

Present-day gravimetric measurements allow the changes in gravity to be determined with a high accuracy. For example, the required variations in **g** at a point of observation are attributable to the perturbation of the Earth's additional potential, the change in the height of the point of observation and the centrifugal acceleration.

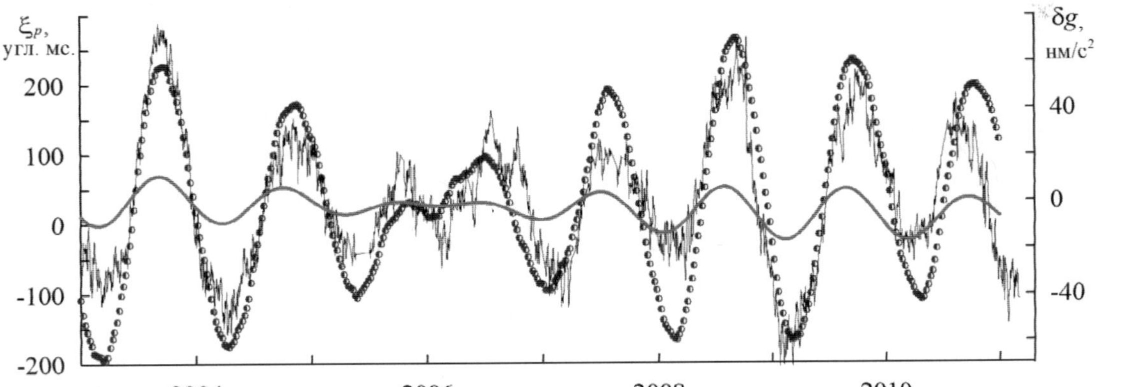

Figure 1. Gravity variations δg (thin black zigzag line) associated with the Earth pole motion in comparison with the variations δg_{pt} from the pole tide perturbation (smooth red line) and the Earth pole oscillations in coordinate ξ_p passing through the point of observation (dots).

To a first approximation, the term due to the change in the height of the point of observation will be

$$\delta g_h \approx -\frac{10\chi G m_E}{3 g R_E^3}\left(1+6\delta c_{20}P_{20}(\cos\theta)-\frac{3h_0}{R_E}\right)\delta V_{21}. \qquad (5)$$

h_0 is the constant (mean) height of the point of observation, the coefficient $\chi = 1$ for a homogeneous model Earth and 0.843 for the real Earth. Thus, the variation in the vertical component of **g** at the point of observation due to the pole tide perturbation is defined by the expression

$$\delta \hat{g}_{pt} = \left(1-\frac{10\chi G m_E}{g R_E^4}\left[1+6\delta c_{20}P_{20}(\cos\theta)-\frac{3h_0}{R_E}\right]\right)\delta g_{pt}. \qquad (6)$$

Figure 1 presents the observed gravity variations δg dependent on the Earth pole motion in comparison with the variations $\delta\hat{g}_{pt}$ from the pole tide perturbation and the Earth pole oscillation in projection onto the coordinate $\xi_p = x_p \cos\lambda - y_p \sin\lambda$ passing through the point of observation. The **g** variations were separated from the highly accurate measurements with the superconducting gravimeter (SG) of the Global Geodynamic Project (GGP) [4] located at Mt. Membach. The oscillations $\delta\hat{g}_{pt}$ were determined from the processed $\delta g - \delta g_\omega$ observations by the least-squares method.

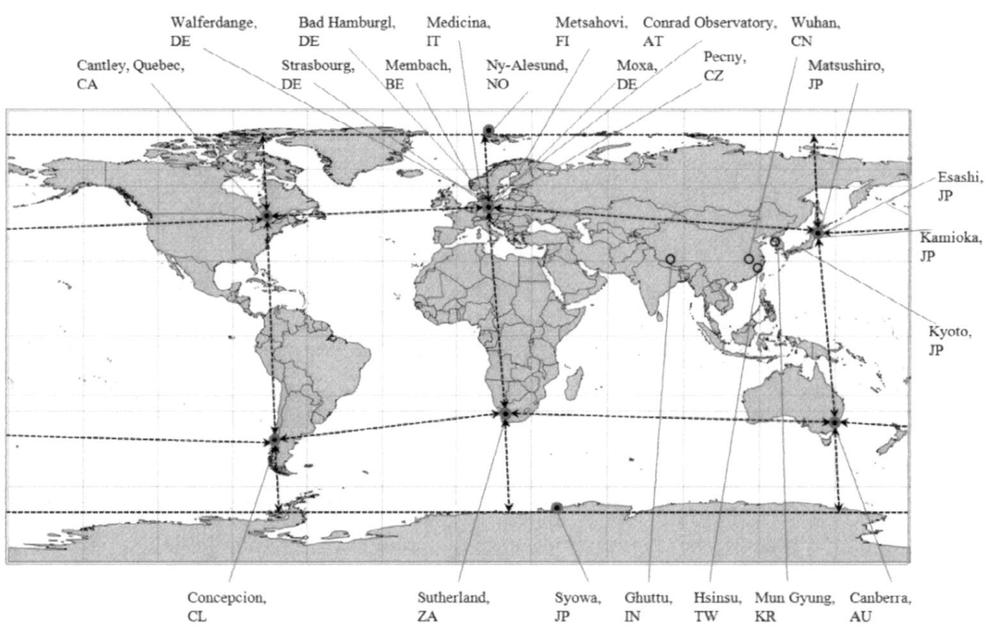

Figure 2. Map of the distribution of gravimetric stations. The dashed lines mark the regions of the map to interpolate the measurements at the grid points.

At present, a fairly large number of SGs located at various points of the world have been combined into the network of the developing GGP [4]. The density of the distribution of GGP stations is different (Fig. 2). Some of them are densely grouped (lie within one country) and some are far apart. The Earth's surface can be marked out in a region with a large area if the "mean" point with the mean values of the amplitude factor for the entire group of measuring instruments is considered for the densely grouped points. Such a marking is shown in Fig. 2. These "mean" points (for Europe and Japan) can be obtained from known papers (see, e.g., [5]) on the determination of pole tide parameters for various stations in which the results of processing the gravimetric measurements are analyzed. Within one region was obtained by interpolating the values at the grid points.

The quantities $\kappa_{x,y}$ are calculated based on the measurements of δg_{pt}. For example, κ_x is

$$\kappa_x = \frac{2m_E r_0 R_E^2}{\pi^2 \eta} \int_{-\frac{\pi}{2}}^{\frac{\pi}{2}} \int_0^{2\pi} \delta g_{pt}(\theta,\lambda) \sin 2\theta \cos\lambda d\lambda d\theta = \kappa_x^1 x_p + \kappa_x^2 y_p. \quad (7)$$

2. The fluctuation-dissipation perturbations in the model of Earth pole motion

Consider the Chandler Earth pole oscillations by taking into account the fluctuation terms of the pole tide. Because of the inhomogeneity of the Earth's mantle and the presence of a hydrosphere, the real tide can differ significantly due to a change in the centrifugal potential. Even a regular tidal potential gives rise to a random displacement field and random fluctuations in tidal processes due to the complex topography of the World ocean floor and the complex shape of the continental coastlines. Weak irregular perturbations of the components of the Earth's inertia tensor correspond to these perturbations. This can be integrally expressed in the corresponding fluctuation additions to the amplitude and phase of the pole tide for the centrifugal moments of inertia.

$$\begin{aligned} \dot{x}_p + N_p y_p &= (D_1^* + D_1' V_{1t}) x_p + (D_{12}^* + D_{12}' V_{2t}) y_p + V_{1t} + \mu_x, \\ \dot{y}_p - N_q x_p &= (D_{21}^* + D_{21}' V_{1t}) x_p + (D_2^* + D_2' V_{2t}) y_p + V_{2t} + \mu_y. \end{aligned} \quad (8)$$

Here $D_{i,ij}^*$ are the constant coefficients of the regular pole tide, while $D_{i,ij}'$ are the quantities defining its irregular component. The Gaussian broadband processes like white noise V_{it} have an intensity matrix $\nu = \|\nu_{ij}\|$. The terms $\mu_{x,y}$ define the perturbations at a frequency N_1 close to the Chandler one. As a result of adding random dissipative perturbations to the deterministic terrestrial pole model, Equations of the model are stochastic differential equations are considered in Stratonovich's sense [6, 7]. The equations for the mean pole coordinates are obtained after reducing the equations polar motion to stochastic differential equations in Ito's sense.

$$\begin{aligned} \dot{m}_x + \tilde{N}_{1p} m_y &= \tilde{D}_1 m_x + \frac{1}{2}(D_1' \nu_{11} + D_{12}' \nu_{22}) + \mu_x, \\ \dot{m}_y - \tilde{N}_{1q} m_x &= \tilde{D}_2 m_y + \frac{1}{2}(D_{21}' \nu_{11} + D_2' \nu_{22}) + \mu_y. \end{aligned} \quad (9)$$

We can estimate the amplitude A of the quasi-stationary Chandler pole oscillations at the external perturbation frequency N_1: where f_x and f_y are the amplitudes of the external perturbation N_1, at a frequency close to the Chandler frequency N. Hence, it follows that the total amplitude of the oscillations at the Chandler frequency is sensitive to the frequency difference $\sigma = N - N_1$ and the effective dissipation coefficients σ.

$$\begin{aligned} A &\approx \frac{1}{2}\sqrt{\frac{f_x^2 + f_y^2}{(\tilde{N}_1 - N_1)^2 + \tilde{D}}}, \quad \tilde{N}_1 = \sqrt{\tilde{N}_{1p}\tilde{N}_{1q}}, \\ \tilde{N}_{1p} &= N_p - D_{12}^* - \frac{1}{2}(D_{12}' D_1' \nu_{21} + D_2' D_{12}' \nu_{22}), \\ \tilde{N}_{1q} &= N_q - D_{21}^* - \frac{1}{2}(D_1' D_{21}' \nu_{11} + D_{21}' D_2' \nu_{12}), \end{aligned} \quad (10)$$

We discuss the fluctuation perturbations at a frequency close to the Chandler one in case taking into account the sea level observation data. We show that accounting the small irregular perturbations in tidal processes leads to variations in the Chandler component of the pole oscillations. We note that the amplitude of the Chandler pole oscillations is sensitive to the difference of the Chandler frequency and a frequency close to it and to the dissipation coefficients.

References

[1] International Earth Rotation and Reference Systems Service - IERS Annual Reports, http://www.iers.org.
[2] Markov Yu G, Perepiolkin V V, Rykhlova L V, Filippova A S 2015 *Astron. Rep.* **59** (4) pp 327 - 337
[3] Akulenko L D, Markov Yu G, Rykhlova L 2001 *Dokl. Phys.* **46** (4) pp 261-263
[4] http://isdc.gfz-potsdam.de
[5] Hu X-G, Liu L-T, Xu H J, Ducarme B, Sun H P 2007 *Geophys. J. Int.* **169** (3) pp 821-829
[6] Markov Yu G, Sinitsyn I N 2002 *Dokl. Phys.* **47** (7) pp 531-534
[7] Markov Yu G, Rykhlova L V, Sinitsyn I N 2010 *Astron. Rep.* **54** (9) pp 861-870

IOP Conf. Series: Materials Science and Engineering **468** (2018) 012017 doi:10.1088/1757-899X/468/1/012017

On safe orbits of a space station tethered to a precessing asteroid by two cables

A V Rodnikov

Bauman Moscow State Technical University, Moscow, Russia
E-mail: avrodnikov@yandex.ru

Abstract. We study the problem of existence of safe motion for a space station tethered to a small planet by two cables. We consider the case of the dynamically symmetric asteroid and of the cables which opposite ends are fixed to the poles of the small planet. We establish that the station equilibria with respect to axes of precession and of dynamical symmetry are the safe orbits if gravitation of the asteroid satisfy some restrictions. We illustrate these restrictions by geometrical diagrams. Also, we deduce general conditions for existence of the safe orbits of other types.

1. Introduction

One of the goals for the modern space missions is to master the Asteroid Belt. Each of the small planets might be interesting as an immediate base on the way into deep space, as a source of mineral raw materials or as a potentially dangerous object. Now we know that landing on the small planet surface is an elaborate procedure. Generally, the reasons are the asteroids complex shapes, and, consequently, theirs complex spin motion and theirs its intricate gravitational potentials.

Apparently, to place a space station close to an asteroid without direct landing might be more convenient decision. Taking into account that the asteroid gravitation is very small in comparison with the solar gravitation, one can place the space station on the asteroid heliocentric orbit. If the station will be sufficiently far from the small planet then the mission will be restricted to observations only. But if the station will be directly nearby the small planet then one can anchor the station to the asteroid surface, in other words, one might link the station and the asteroid by a cable. (For the first time, such linking have been suggested in [1] as non-gravitational method of orbits correction). Factually, in this case the tether is the simplest transport system between the station and the asteroid. Nevertheless, note that if the distance from the station to the asteroid surface is comparable-sized with the asteroid diameter then gravitation of the asteroid is several orders more than the other forces sum (of course, here we must take into account forces of inertia also). In this case the station becomes satellite of the asteroid and some problems of the station safety take place. Evidently, the tether eliminates running away of the station, but the station might fall down on the asteroid surface. Moreover, the tether might slack. In this case the tether might taut again after some time with impact that can be catastrophic for the station equipment. Moreover, the long-term missions is realizable only if the station motion is stable (may be in orbital sense only). One might suggest that the second anchor might stabilize the station orbit, but it is not guaranteed.

Note that the systems like ‘space elevators’ is not realizable for the problem under consideration by the reasons formulated above. Nevertheless, there are small planets that are close to bodies of rotation.

Content from this work may be used under the terms of the Creative Commons Attribution 3.0 licence. Any further distribution of this work must maintain attribution to the author(s) and the title of the work, journal citation and DOI.
Published under licence by IOP Publishing Ltd

One might assume that such asteroids are dynamically symmetric rigid bodies and theirs spin motions are close to regular precessions. In this case, one might fix the tether(s) end(s) at the asteroid poles, which arise as intersection of the asteroid surface and the axis of dynamical symmetry. (In practice, the 'anchors' may be some towers constructed on the poles). Evidently, the station doesn't fall down if there are two tethers and both of them are taut throughout the motion. Also, in this case, the station moves with respect to the asteroid surface, so the present construction is more universal then the traditional space elevator. Moreover, the station orbit can be stable [2, 3].

2. Formulation of the problem

Consider a dynamically symmetric rigid body modeling an asteroid. Let C be mass center of the body, Cz be the axis of dynamical symmetry, Cz_1 be the axis of precession with angular velocity ω and with angle of nutation ϑ (see figure 1). In our case $\omega = const$, $\vartheta = const$. Let us restrict to the general case of acute ϑ. Let S be a particle modeling a space station. Suppose that S tethered to poles P_1 and P_2 via two cables. As coordinate of P_1 on Cz is positive, this pole is called the "north pole", unlike P_2 that is called the "south pole".

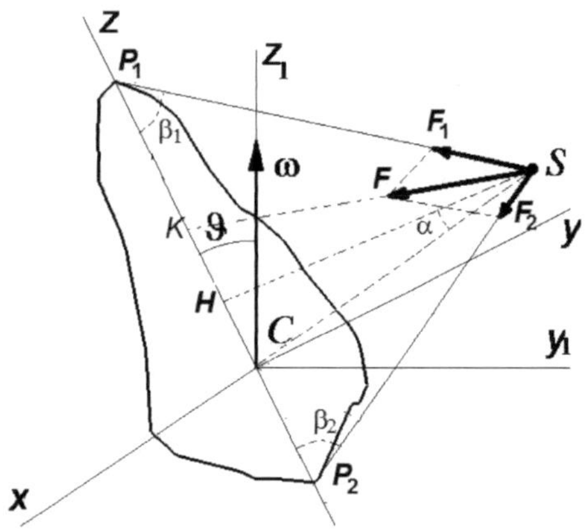

Figure 1. Designations and coordinate systems

The space station trajectory is called the safe orbit if the particle S motion along this trajectory is stable (maybe in orbital sense) and both tethers are taut throughout this motion. Suppose that the tethers are weightless, inextensible and absolutely flexible. Then the particle S motion is restricted by two unilateral constraints that are spheres with centers in the poles P_1 and P_2 and radii that are equal to lengths of the north and south tethers respectively. The spheres surfaces are called the constraint boundaries. Intersection of constraints boundaries is some circle with center on Cz (the point H in fig.1). This circle is orthogonal to the axis of dynamical symmetry. Therefore, the station S safe orbit is a point or an arc (maybe complete) of a circle with center on Cz in the plane that is orthogonal to Cz. Let us remark that we have two frame of references: axes Cz and Cz_1 and the asteroid. Evidently, if some safe orbit in the first frame of reference is point or arc then this orbit is the complete circle with respect to the asteroid, as the asteroid rotates about Cz. Motion along the safe orbit is called the constrained motion.
Let Cx be orthogonal to Cz and Cz_1 and $Cxyz$ and Cxy_1z_1 are right-handed coordinate systems. Let x, y, z be coordinates of S in $Cxyz$ and $x = \rho\cos\varphi$, $y = \rho\sin\varphi$. Moreover, let $\beta_1 = \angle SP_1C$, $\beta_2 = \angle SP_2C$, $\alpha = \angle HSC$. Note that if the particle S moves along the safe orbit then $\alpha, \beta_1, \beta_2, \rho, z$ don't change. Factually, they are parameters of the safe orbit . Without loss of generality one can assume that $z \geq 0$ (equivalently, $0 \leq \alpha < \pi/2$) and that β_2 is acute.
If the asteroid is close to body of rotation then one might assume that gravitational potential of such asteroid is invariant with respect to rotations about axis of dynamical symmetry. In this case, the asteroid gravitation force $\boldsymbol{F}$, acting on the station, throughout the constrained motion is directed on the same point on Cz (K in fig.1). Moreover, $|\boldsymbol{F}| = const$ in points of any safe orbit. One might present $\boldsymbol{F}$ as sum $\boldsymbol{F} = \boldsymbol{F}_1 + \boldsymbol{F}_2$, where each of $\boldsymbol{F}_i$ is directed to P_i.

3. Equations of motion and conditions for existence of safe orbits

If the station motion is constrained then reaction of each tether might be presented by product of Lagrange's multiplier and vector $\overrightarrow{P_iS}$ (here $i = 1$ for the north tether and $i = 2$ for the south one). In the case of constrained motion two of three equations of motion for the particle S can be written as

$$\varphi'^2 + 2(\cos\vartheta - \sin\varphi\sin\vartheta\,\mathrm{tg}\beta_1)\varphi' + 1 + \sin^2\vartheta\,(\mathrm{tg}\beta_1\mathrm{tg}\alpha - \sin^2\varphi) - \\ - \sin\vartheta\cos\vartheta\sin\varphi\,(\mathrm{tg}\beta_1 + \mathrm{tg}\alpha) - f_2\sin(\beta_1+\beta_2)/\cos\beta_1 = -\lambda_2, \quad (1)$$

$$\varphi'^2 + 2(\cos\vartheta + \sin\varphi\sin\vartheta\,\mathrm{tg}\beta_2)\varphi' + 1 - \sin^2\vartheta\,(\mathrm{tg}\beta_2\mathrm{tg}\alpha + \sin^2\varphi) - \\ - \sin\vartheta\cos\vartheta\sin\varphi\,(\mathrm{tg}\alpha - \mathrm{tg}\beta_2) - f_1\sin(\beta_1+\beta_2)/\cos\beta_2 = -\lambda_1, \quad (2)$$

where the prime $()'$ denotes derivative with respect to dimensionless time τ ($\tau = \omega t$), $f_i = |\boldsymbol{F}_i|/(m|SH|\omega^2)$ are dimensionless components of the gravitation force, m is the station mass, λ_i depend on Lagrange's multipliers. (Deriving (1-2) we took into account that $\rho' = \rho'' = z' = z'' = 0$ for constrained motion). Evidently, each of Lagrange's multipliers must be negative throughout constrained motion, and, consequently, for any safe orbit. Analyzing (1-2) one might see that Lagrange's multipliers will be negative for acute β_1 if $\lambda_1 < 0$, $\lambda_2 < 0$, for obtuse β_1 if $\lambda_1 < 0$, $\lambda_2 > 0$ and for right β_1 if

$$\lambda_1 < 0,\ 2\sin\varphi\left(\varphi' + \frac{\cos\vartheta}{2}\right) - \mathrm{tg}\alpha\sin\vartheta < -f_2\cos\beta_2\,/\sin\vartheta. \quad (3)$$

If the particle S motion is constrained then the last equation of motion is integrable and could be written as

$$\varphi'^2 + \sin\varphi\sin\vartheta\,(\sin\varphi\sin\vartheta + 2\mathrm{tg}\alpha\cos\vartheta) = 2h = const \quad (4)$$

after onetime integration.

4. Relative equilibria

Analyzing left part of (4) one might see that there are several equilibrium points of the particle S in rotating coordinate system $Cxyz$, i.e. in the frame of reference of axes of precession and dynamical symmetry. Such points set is intersection of planes Cy_1z_1 (coincides with Cyz and contains axes of precession and dynamical symmetry) and Cxy_1 (is orthogonal to axis of precession) with the circle of constrained motion. Evidently, this circle always has two points of intersection with Cy_1z_1. These points are the 'upper equilibrium' $\varphi = \pi/2$ and the 'lower equilibrium' $\varphi = -\pi/2$. The upper equilibrium is always unstable. Hence, this point is not a safe orbit. The lower equilibrium is stable only if $\alpha \geq \vartheta$. Note that if $\alpha > \vartheta$ then the circle of constrained motion doesn't intersect Cxy_1, and if $\alpha = \vartheta$ this circle is tangent to Cxy_1 in the point $\varphi = -\pi/2$. Thus, if $\alpha \geq \vartheta$ then there is only one stable equilibrium in $Cxyz$. If $\alpha < \vartheta$ then the circle of constrained motions has two intersection points with Cxy_1. These points are $\varphi = -\arcsin(\mathrm{tg}\alpha\ \mathrm{ctg}\alpha)$ and $\varphi = \pi + \arcsin(\mathrm{tg}\alpha\ \mathrm{ctg}\alpha)$. They are called the 'horizontal equilibria'. Both of them are stable.

Varying the tethers lengths one makes sure that any point of Cxy_1 could be the stable relative equilibrium (of course, excepting C). Moreover, any point inside acute angles between Cz and Cy_1 could be the stable relative equilibrium.

5. Diagrams of safety

Any of mentioned in the previous chapter stable equilibrium points could be the safe orbit, but only if both of Lagrange's multipliers will be negative in such point. These conditions for the lower equilibrium $\varphi = -\pi/2$ are written as

$$f_1\sin(\beta_1+\beta_2) < \cos(\vartheta+\beta_2)\cos(\alpha-\vartheta)/\cos\alpha, \quad (5)$$

$$f_2\sin(\beta_1+\beta_2) < \cos(\vartheta-\beta_1)\cos(\alpha-\vartheta)/\cos\alpha\,. \quad (6)$$

((5-6) are result of reducing (1-3)). Note that inequality (6) is possible only if $\beta_1 < \pi/2 + \vartheta$.
There is a simple geometrical interpretation of inequalities (5-6). Let H_1 be projection of S on Cz. Moreover, let $SQ_1H_1Q_2$ be parallelogram with opposite vertexes S and H_1 and with sides being parallel to tethers. ($SQ_1H_1Q_2$ is painted over in figure 2). Then the lower equilibrium point $\varphi = -\pi/2$ is the safe orbit if $\alpha \geq \varphi$ and if the head of vector $\boldsymbol{\Phi} = \boldsymbol{F}/(m\omega^2)$ with tail in S lays inside this parallelogram. Note that the lower equilibrium could be the safe orbit even if β_1 is obtuse (but less then sum of right angle and angle of nutation), i.e. even if the station moves higher the pole.

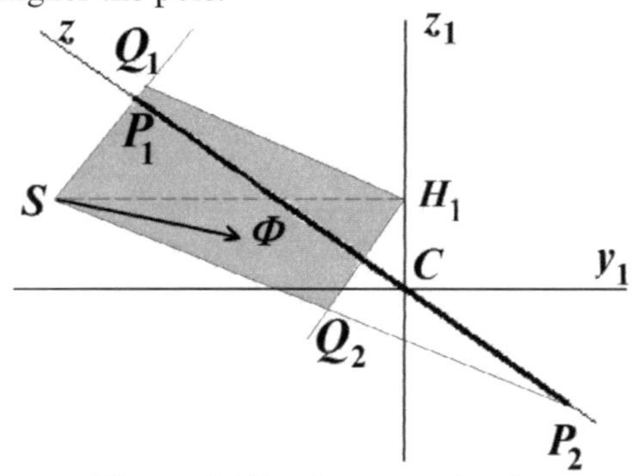

Figure 2. The diagram of safety for equilibrium point $\boldsymbol{\varphi = -\pi/2}$

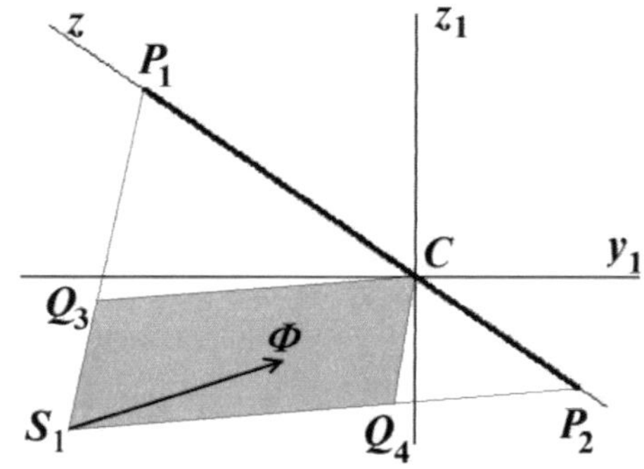

Figure 3. The diagram of safety for horizontal equilibrium points

For any equilibrium point in Cxy_1 Lagrange's multipliers is negative if

$$f_1 \sin(\beta_1 + \beta_2) < \cos(\alpha + \beta_2)/\cos\alpha, \tag{7}$$

$$f_2 \sin(\beta_1 + \beta_2) < \cos(\alpha - \beta_1)/\cos\alpha, \tag{8}$$

These inequalities are result of reducing of (1-3) also. They have a simple geometrical interpretation too. Let S_1 be the lower equilibrium that corresponds to studied horizontal equilibria. Moreover, let $SQ_3H_1Q_4$ be parallelogram with opposite vertexes S_1 and and with sides being parallel to tethers. .($SQ_3H_1Q_4$ is painted over in figure 3). Then horizontal equilibrium points are the safe orbits if the head of vector $\boldsymbol{\Phi} = \boldsymbol{F}/(m\omega^2)$ with tail in S_1 lays inside this parallelogram.

6. Safe orbits of other types

To check out orbits of the station one might analyze phase trajectories of Jacobi's integral (4). There are only three distinct types of phase portraits for such trajectories on the phase cylinder $(\varphi|\text{mod } 2\pi, \varphi')$ (see examples in figure 4-6). Note that some areas must be cut out from the mentioned cylinder. Such areas are called 'the constraint loss areas' and consist of points in which at least one of Lagrange's multipliers is nonnegative. The constraint loss areas have a variety of shapes. Some examples of them are shadowed over in figure 4-6. One might define boundaries of such areas substituting 0 for λ_i in (1-2) or replacing the sign '<' with '=' in (3).
Let phase trajectory be regular if it is not be unstable point of equilibrium, if it is not be separatrix and if it has no intersections with constraint loss areas. One might easily check out that anyone regular phase trajectory is stable in the orbital sense. Hence, each regular trajectory of (4) presents some safe orbit of the station. There are three types of such orbits.
The first type is librations near one of the stable equilibrium points (bold ovals in figure 4). In this case the station orbit in $Cxyz$ is the arc that has not more than one intersection with Cxy_1.
The second type is librations around both horizontal equilibria and the lower equilibria. (The last is unstable in this case). The bold curve in figure 5 is an example of such libration. In this case the station orbit in $Cxyz$ is the arc that has two intersections with Cxy_1.

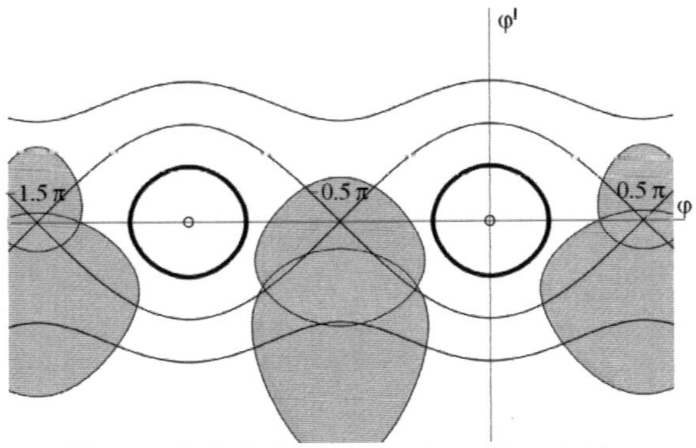

Figure 4. Safe librations about the stable equilibrium points.

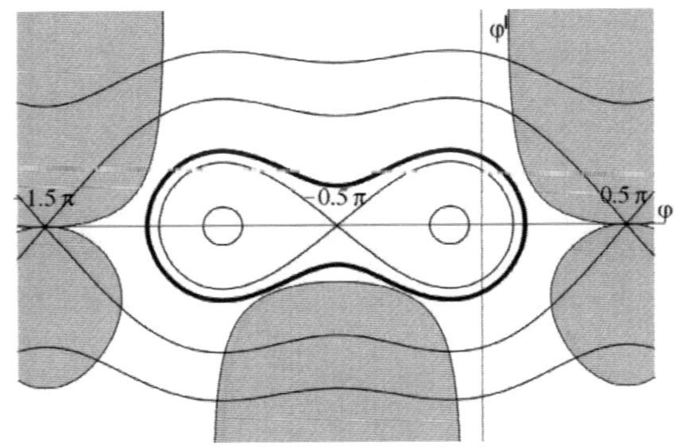

Figure 5. The safe libration around several equilibria

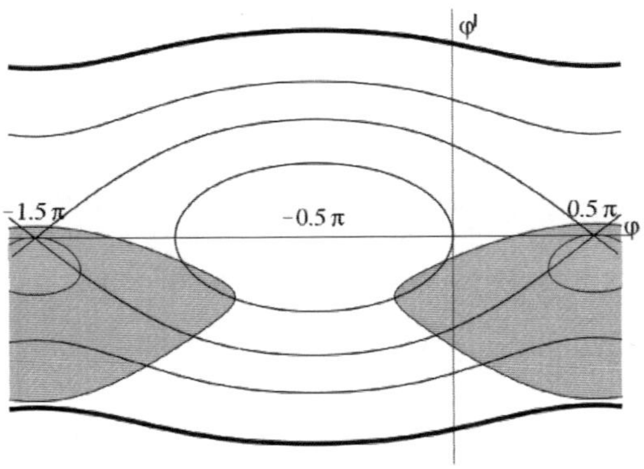

Figure 6. Safe rotations

The third type is rotations around the axis of dynamical symmetry (the bold curves in figure 6). In this case the station orbit in *Cxyz* is the full circle of constrained motion. Note that the safe rotations are impossible if at least one of angles β_1 and β_2 is not acute.

References

[1] Lanoix E L-M and Misra A K 2000 *Journal of Spacecraft and Rockets* **37** 475-480
[2] Rodnikov A V 2011 *Rus. J. Nonlin. Dyn* **7** pp 295-311
[3] Rodnikov A V 2012 *Rus. J. Nonlin. Dyn* **8** pp 309-322

V-shaped wings with the central body in supersonic flow

F A Maksimov[1,2], N A Ostapenko[1] and M A Zubin[1]

[1]Institute of Mechanics, Lomonosov Moscow State University, Moscow, Russia
[2]Institute of Computer Aided Design, Russian Academy of Sciences, Moscow, Russia

Abstract. The results of a theoretical investigation of the flow pattern of V-shaped wings at supersonic flow velocities are presented. Data are presented on the flow structure near V-shaped wings with a central body in the form of a part of a circular cone within the framework of the ideal gas model. The applicability of the criteria for the existence of nonviscid vortex structures has been determined, but with differences due to the shape of the body. Influence of the central body on the aerodynamic quality of its layout with a V-shaped wing was studied. A significant dependence of geometry of the optimal body on the lift coefficient was established.

1. Introduction

V-shaped wings belong to one of the promising forms of an aircraft with high aerodynamic quality at supersonic flight speeds [1, 2]. The study of the flow structure in shock layers near such wings in various flight modes has both an independent fundamental and practical interest. Learning characteristic types and properties of a real flow, containing substantial flow nonuniformity, and causes of their generation is necessary, in particular, when designing location of the engine air inlet of the aircraft, which is based on the supporting fuselage with the lower surface in the form of a V-shaped wing [3]. Calculations are performed to study their aerodynamic characteristics in the composition of arrangements with different shapes of the central body in parallel with studies of the disturbed flow properties near V-shaped wings.

2. The calculation method

The form of the Euler equations of motion in generalized coordinates for the conical flows (Figure 1) and solution method are described in [3].

Content from this work may be used under the terms of the Creative Commons Attribution 3.0 licence. Any further distribution of this work must maintain attribution to the author(s) and the title of the work, journal citation and DOI.
Published under licence by IOP Publishing Ltd

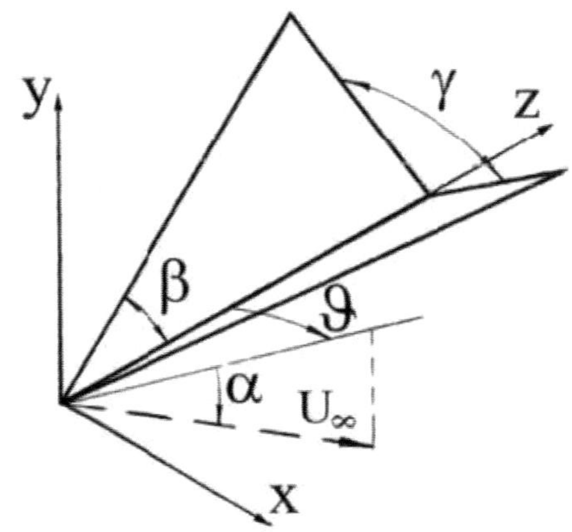

Figure 1. β, γ - angles at top of consoles and opening of wing. α, ϑ – angles of attack and slip.

In addition to the V-shaped wing with an opening angle less than 180°, investigated in [3], we consider a V-shaped wing with a central body in the form of a cone, a wedge and a V-shaped wing with an opening angle greater than 180° in this paper. Since modes with compression shocks, attached to the leading edges, or with centered rarefaction waves are considered, the turbulent flow property is effectively used in all cases, consisting in its conical character relative to the leading edges.
Figure 2a shows a grid for flow simulation near a V-shaped wing with a central body in the form of a cone. The grid is built at a single distance from the configuration top and has the form shown in Fig.2. The non-flow condition is stated on FG boundary corresponding to the surface of the wing and cone; conditions in the oncoming stream - on ABCD boundary; condition of flow conicity relative to corresponding E_1 and E_2 wing edges - at FA and GD lateral boundaries. The grid is built algebraically. In some area near the body, the size of which is selected for the flow mode, the grid consists of circles and rays, and some of the coordinate lines are directed to the corresponding wing edges (using the Bezier polynomial), starting from some distance from the cone.
Figure 2b shows a grid near the V-shaped wing with a central body in the form of a flat wedge. The condition of non-flow is imposed on FG boundary, corresponding to the surface of the wing and wedge; conditions in the oncoming stream - on AD boundary; condition of flow conicity relative to corresponding E_1 and E_2 wing edges - at FA and GD lateral boundaries. The grid with a given nodes distribution on the body surface is built by a parabolic generator from the surface of the body. Similarly, the grid is built for a variant of the V-shaped wing with an opening angle greater than 180°. When forming the distribution of nodes on the surface of a streamline body, it is important to take into account the salient points of the outline near which the grid nodes thicken.

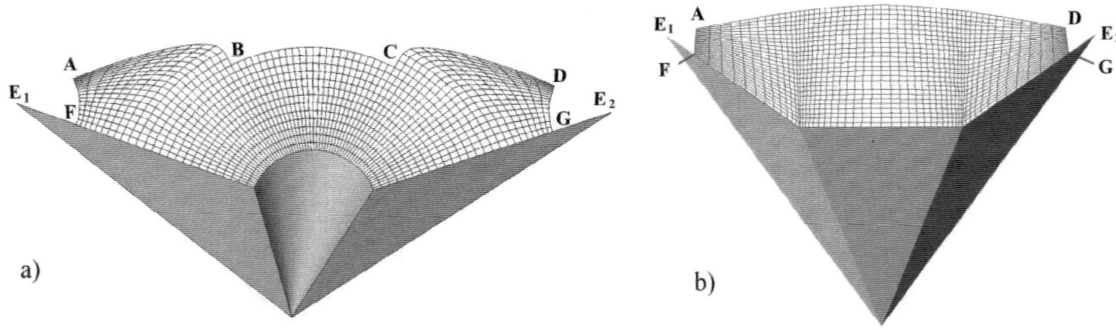

Figure 2. Grid near V-shaped wing with central body in form of cone and wedge.

To satisfy the conicity condition with respect to the edges, the requirement for the location of the grid nodes strictly on the straight lines, drawn from the centers of conicity, is provided at the appropriate boundaries in some neighborhood of these boundaries.

An algorithm is used to obtain a solution on a sequence of grids with a doubling of cells number in each direction to accelerate the establishment and improvement of the solution quality. A sequence of grids is built for each variant of the problem. Initially, the solution is obtained on a crude grid, which is then used by interpolation as the initial flow field for calculation on the next grid. In fact, the solution is resettled from a sufficiently good initial approximation on grids with a large number of nodes. A maximum of up to six grid embeddings is used, which ultimately allows getting the flow field on a sequence of grids with a different number of nodes. One can usually see the flow picture is established and basic physical features of flow are preserved regardless of the dimension of the grid size from the analysis of the results obtained.

3. Flow structure near V-shaped wings with a cone as the central body

Here we present the results of a numerical calculation of the flow structure near V-shaped wings with a central body in the form of a cone part within the ideal gas model with the Mach numbers 3 and 6 in the modes with a shock wave, attached to the leading edges. Data are given for a wing with an opening angle of 80° and angle at the top of consoles of 45° for different values of ν-cone semi-angle, such that the head shock wave remains attached to the leading edges at various angles of attack and slip.

Symmetrical flow near V-shaped wing of the selected geometry when the number of M=3 is characterized by appearance of Ferry vortex singularities in the shock layer only at angle of attack $\alpha=16°$. Figure 3a shows calculation results when $\alpha=15°$ with increasing angle ν of the central body. Position of the upper critical point of a saddle type (the upper curve) is plotted along the ordinate axis in the graph (fig.3b), which includes the current line, coming from the bridge-like compression shock of the Mach configuration of shock waves, position of the vortex singularity centers (the second curve from above), position of the lower critical point of the saddle type (a group of three lower points in the range of variation of ν to 2°), ceasing to exist above the surface of the cone in the plane of symmetry at $\nu\cong 2°$, when the node of streamlines disappears of the cone surface (the third curve from above). The fourth curve from above describes the distance of the vortex singularities from the plane of symmetry.

Similar data are shown in Figure 4 for a symmetric flow near the same V-shaped wing with Mach number M=3, but at angle of attack $\alpha=20°$. Vortex structures are implemented near the V-shaped wing and without a central body in this mode. Flow structure near the wing with different angle ν of the central cone with a symmetric flow with number M=6 at angle of attack $\alpha=20°$ is shown in Figure 5.

A common property of the presented flow structures (Figures 3-5) is disappearance of critical points of the nodal type on the cone surface in the plane of flow symmetry. Then two critical points of the saddle type remain in the symmetry plane. Another common property of flows is that, despite fulfillment of the criteria for existence of vortex structures [4, 5], related to values of intensity of the contact discontinuities, originating from the branch points of the head shock wave and the Mach number of the velocity component of the unperturbed flow, normal to the ray of the conical coordinate system, passing through that branching point (numerical data is not given), starting from the moment of their appearance in the shock layer, the moment of their degeneration comes and they disappear. As the analysis showed, this is due to the fact that when the displacement body is enlarged - the semiangle of the cone, two triple branch points of the head shock wave approach the leading edges and, consequently, the surface of the wing consoles so that contact discontinuities enter the corresponding vicinity of the salient points of the transverse outline of the body, where increased pressure is implemented which is close to its value at critical points. Slowing down of gas particles on both sides of contact discontinuities in the indicated vicinity leads to convergence of their total pressures on the sphere, and, consequently, to decrease in magnitude of the criteria relation, determining presence of the vortex structures in the shock layer.

IOP Conf. Series: Materials Science and Engineering **468** (2018) 012018 doi:10.1088/1757-899X/468/1/012018

When approaching flow modes with no vortex structures, first, the average of three critical points - the spreading point drops to the surface of the cone, and the streamline node disappears in the plane of symmetry of the shock layer. Then the saddle-type draining point, from which the contact discontinuities come to the Ferry vortex singularities to both sides from the plane of symmetry, approaches the surface of the cone and turns into a node of streamlines. At this point, the vortex structures cease to exist. With a further increase of angle ν, the streamlines node descends to the surface of the cone, after which a spreading line forms on the cone surface in the plane of symmetry, and there are two draining points on the periphery.

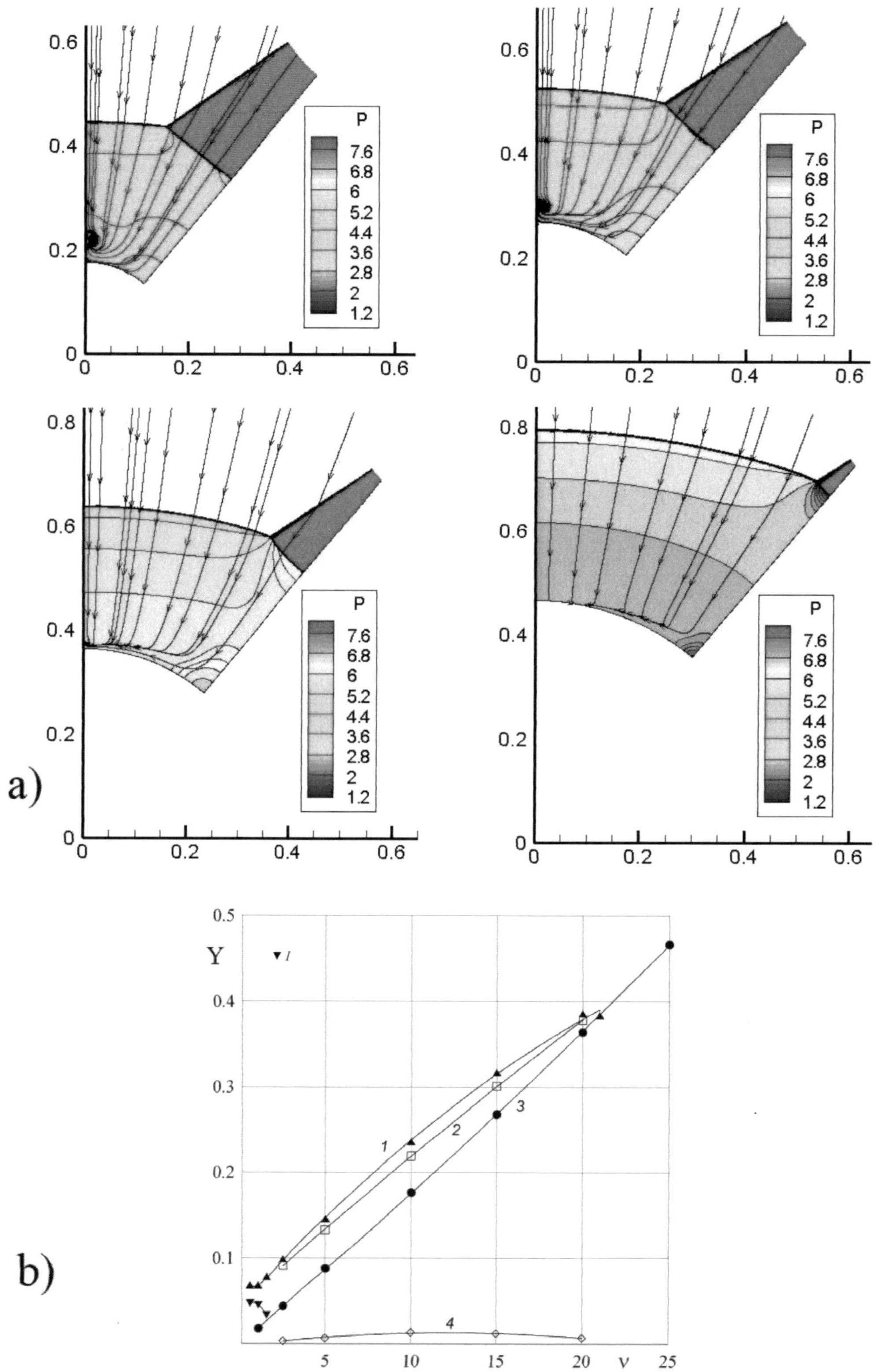

Figure 3. M=3, α=16° (ν=10, 15, 20 and 25°).

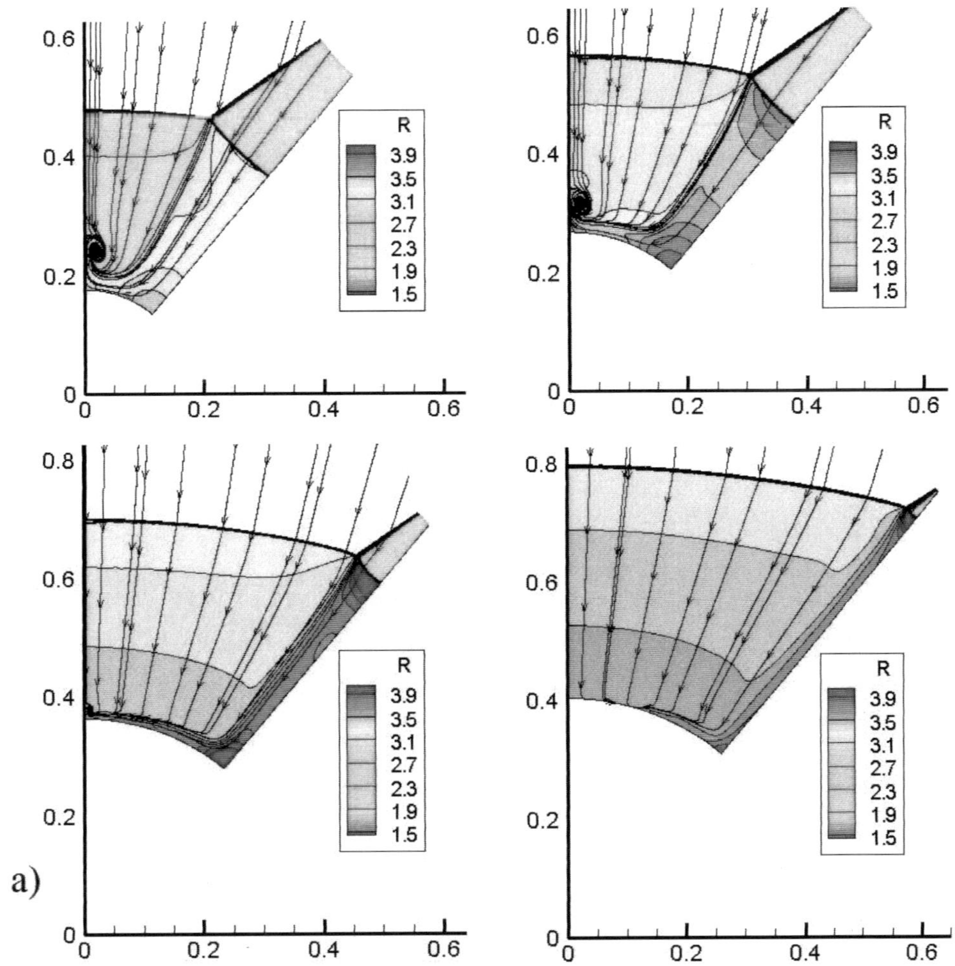

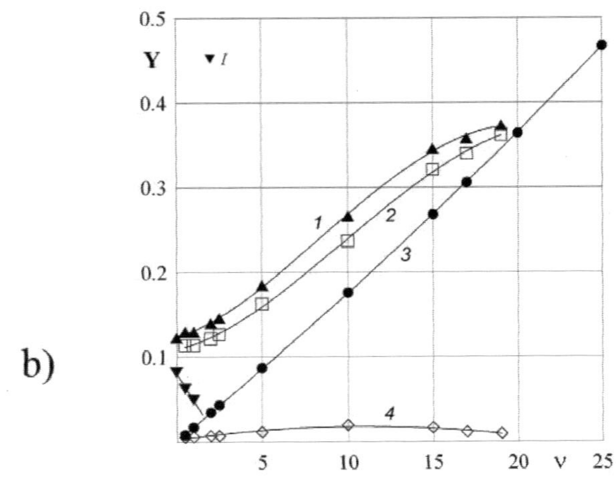

Figure 4. M=3, α=20° (ν=10, 15, 20 and 22°).

a)

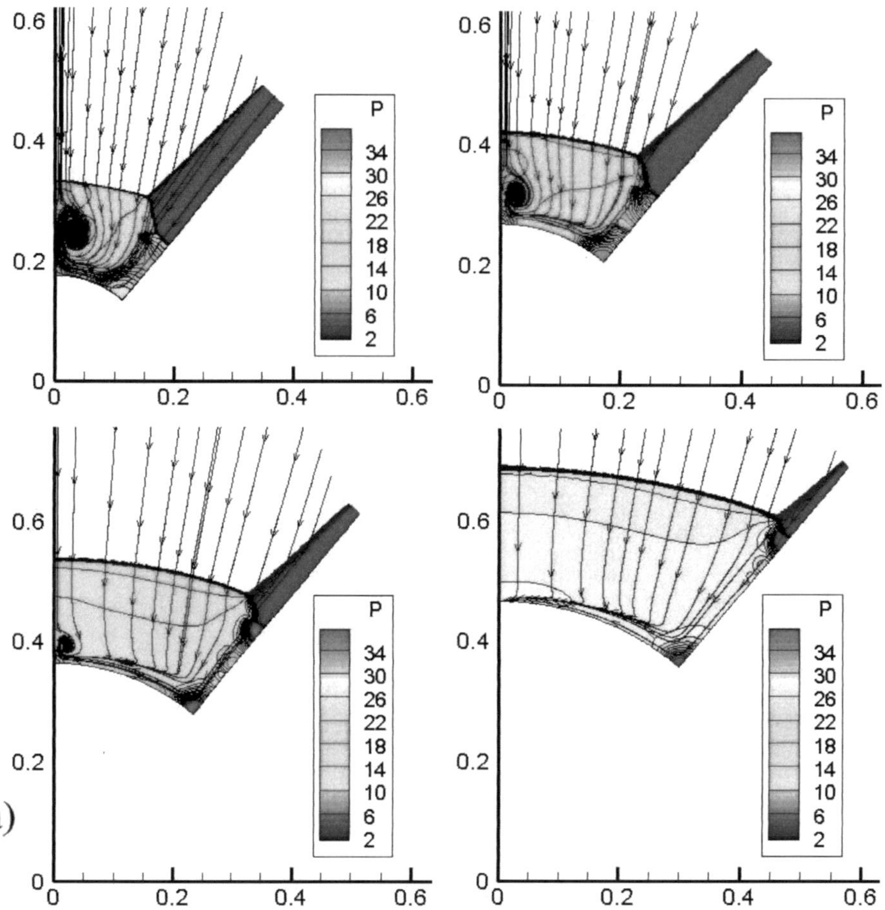

b)

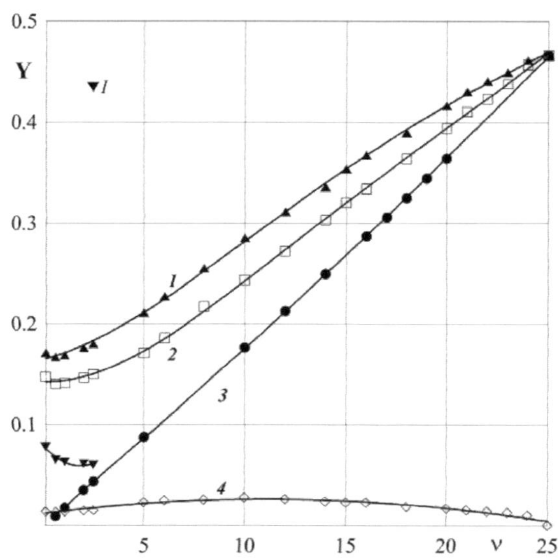

Figure 5. M=6, α=20° (ν=10, 15, 20 and 25°).

4. Aerodynamic quality of wings with a central body

Here are presented results of a numerical calculation of the aerodynamic quality and lifting force of the V-shaped wings with a central body in the form of a cone part. Figure 6 shows examples of body geometries and isobars a of the perturbed flow region. The surface on the windward side (this is the upper side in Fig. 6) is defined by a V-shaped wing with an opening angle γ between the blades from 90 to 210°. Angle β between the central chord of the wing and leading edges is $\beta \approx 31°$ ($tg\beta = 0.6$). A part of the circle with the center at the point of consoles intersection is placed between the consoles. The circle radius varies from 0.02 to the maximum possible one, depending on γ and given area of the midship section.

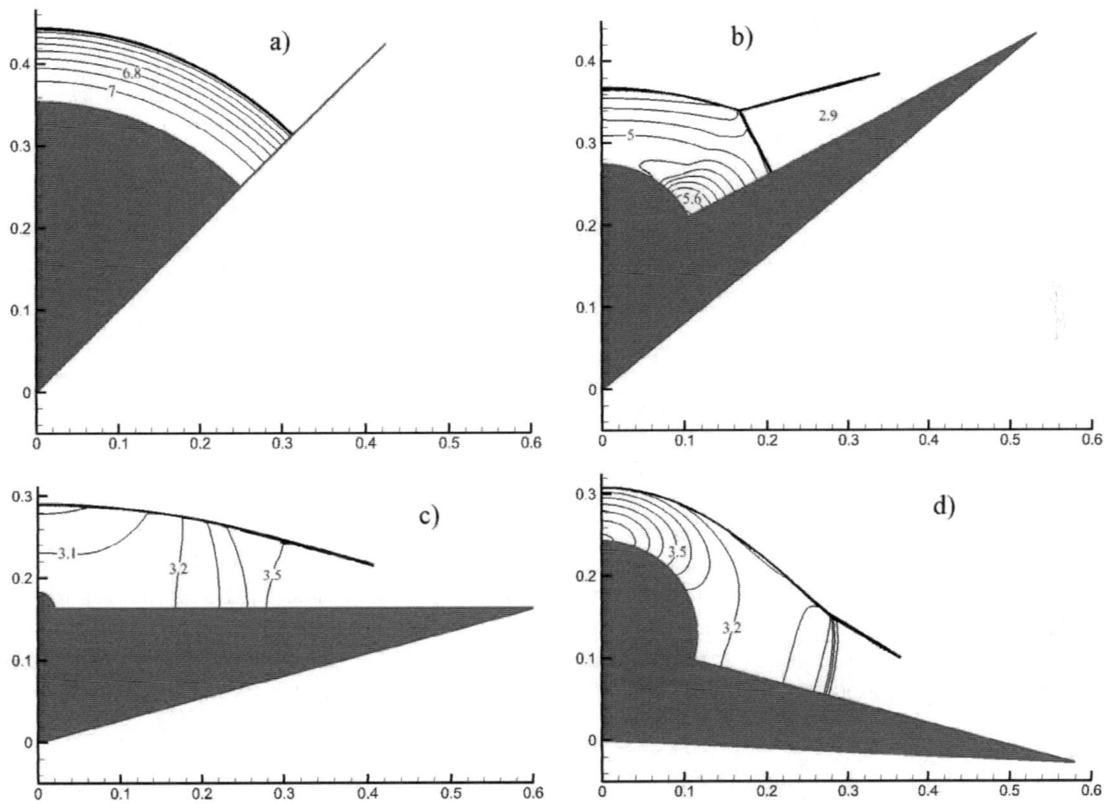

Figure 6. Examples of flows near V-shaped wing with a central body in the form of a cone part, pressure isolines.

The midship section area is fixed and equals $S_m = 0.5\pi R_m^2$ in the presented calculations where $R_m = 0.25$. Figure 6a shows a variant of the body when the entire volume of the body is in the cone, and consoles are of zero thickness. If radius of the cone R is small and the area of the cone is smaller than the given area of the midship section, the windward side is displaced so that the total area of the body's midship section corresponds to the given one. Flow at the angle of attack $\alpha = 0°$ is considered and the leeward (lower) side is located along the flow, pressure on it corresponds to that one in the oncoming stream. Depending on two parameters, radius of the cone R and angle of blades opening γ, a set of bodies is formed that can be calculated by the developed method and their aerodynamic characteristics can be determined. Calculations are performed for M=6. Aerodynamic coefficients of lifting force Cy and resistance Cx are related to the area of the windward side of the V-shaped wing blades, which is the same for all variants. When calculating the coefficient of resistance, pressure is assumed to be 0 at the bottom cut. (i.e. the maximum estimation of the sum of the wave and bottom resistance is obtained).

For a visual representation of the results, Figure 7 shows three-dimensional projections of aerodynamic quality values K=Cy/Cx and lifting force Cy depending on the parameters. Maximum value K≈4.2 is reached for the variant when R≈0 and γ≈180°. Figure 6в presents a variant of such a body; presence of a cone with a small radius practically does not affect the flow field.

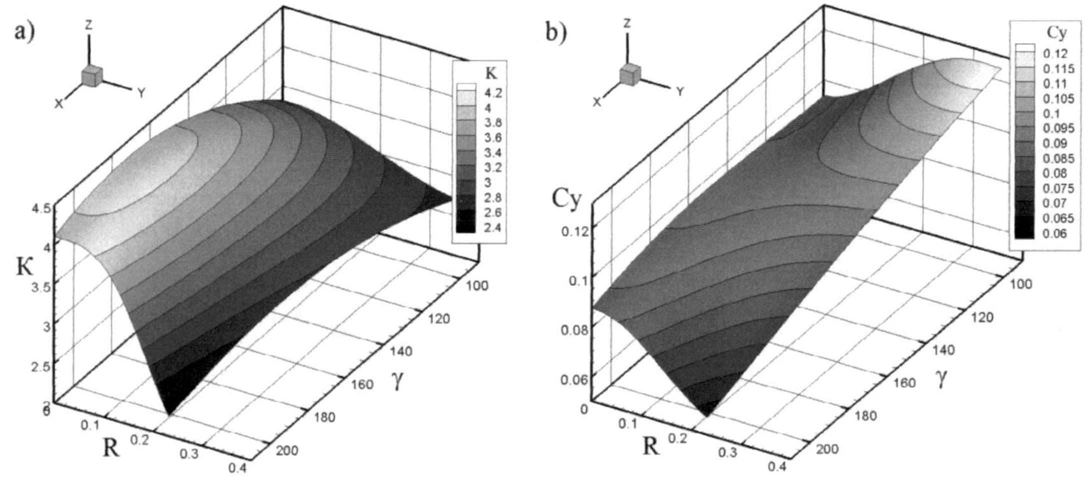

Figure 7. Aerodynamic quality K=Cy/Cx and lifting force Cy.

The maximum lifting force is implemented in case of volume placement in the cone and at the minimum considered value of γ≈90° (Fig. 6a); the quality is significantly reduced (up to K≈2.8) in this case with respect to the maximum possible one.

The aerodynamic quality in shades of gray and coefficient of the lifting force by isolines are presented in Figure 8a in the parametric plane. The data array for R=0 is supplemented by the calculation results of the V-shaped wing. For a given value of the lift coefficient, different scenarios are possible. If value Cy=0.08 or less is required, it is preferable to use geometry with a larger value of γ and a smaller R within the constraints imposed, which makes it possible to provide K≈3.8 (at Cy=0.08). In fact, there is an excess of the total area of the wing in this case. Figure 6d shows a variation of geometry: γ=210°, R=0.11.

If value Cy=0.09÷0.095 is required, then the choice of geometry should be γ≈180°, R=0. In this case, the entire windward side of the body is set at one angle to the incoming stream.

A part of the volume should be moved to the cone to further increase Cy≥0.1, which allows enhancing Cy, but leads to decrease of K. The relative positions of isolines Cy and K allow choosing the best option (in which K is the maximum possible). If value Cy=0.1 is required, then the choice of geometry should be γ≈125°, R≈0.12 (shown in Figure 6b).

The aerodynamic characteristics of the V-shaped wing configuration with a central body in the form of a flat wedge were calculated in a similar composition. Figure 8b show possible values of the aerodynamic quality K=K (Cy) with markers 2 (light circles), depending on the lift coefficient of the V-shaped wing with the cone; the calculation results of bodies, consisting of a wing and a wedge as the central body are given by markers 1 (dark triangles). Comparing the range of possible values of K, it can be seen that a V-shaped wing with a wedge in the central part makes it possible to obtain a higher aerodynamic quality with a given lifting force, or a higher lift for a given quality.

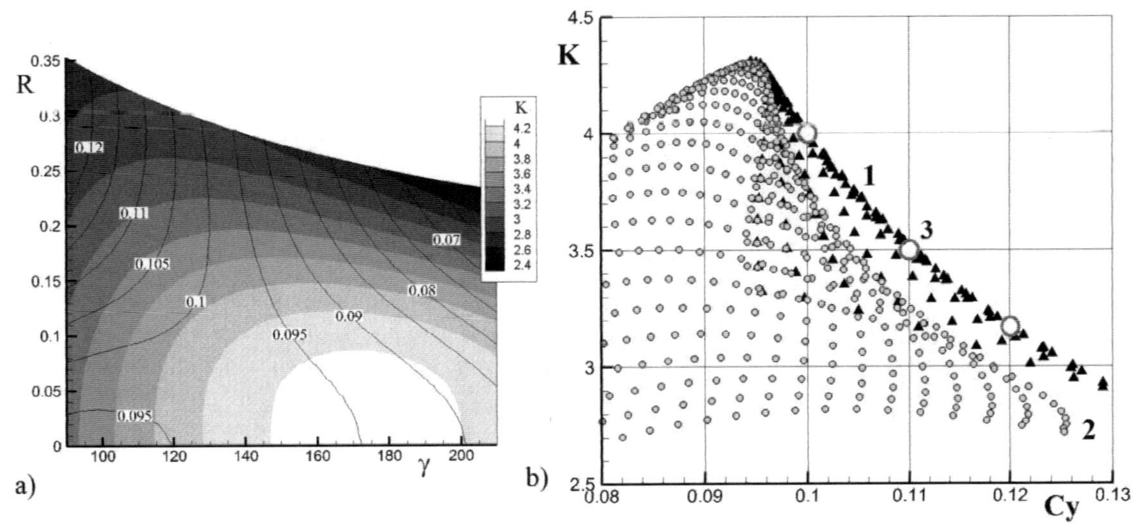

Figure 8

Figure 9 shows examples of flow calculation for three variants of geometries. The patterns of the disturbed flow are represented by pressure isolines. Figure 9a shows a variant of the body, providing quality K≈4 with lifting force Cy≈0.1: the wing parameters γ=140°, So≈0.2. Figure 9b - K≈3.5, Cy≈0.11: parameters γ=120°, S_o=0.36; figure 9c - K≈3.2, Cy≈0.12: parameters γ= 100°, S_o=0.49. These calculated points are identified by large markers 3 (circles) in Figure 8b.

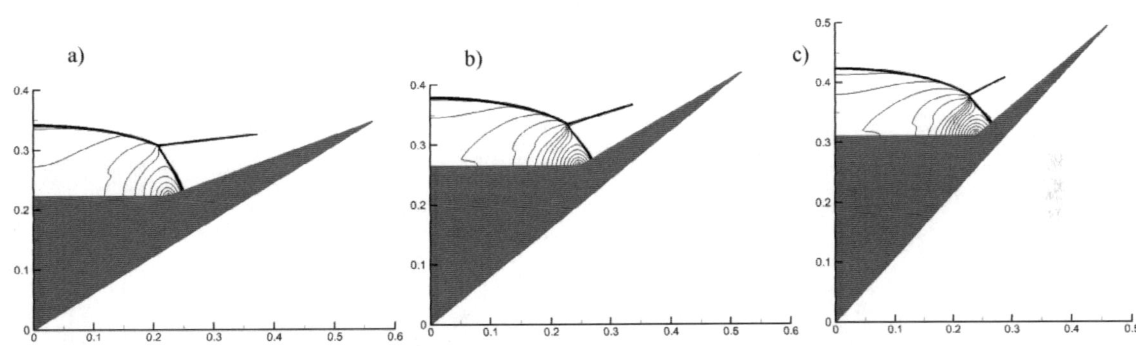

Figure 9 Examples of flows near V-shaped wing with wedge in central part, pressure isolines.

Calculations, carried out for the given parameters (the Mach number, sweep of the windward side of the V-shaped wing, bottom cut area), showed the use of a wedge as a central body is preferable than a cone. Such an arrangement may have a higher aerodynamic quality with a given lift.

References

[1] Ostapenko N A 2000 On Question of Optimal Waveriders Shape, Designed On Plane Shock Waves *Math., Russian Academy of Sciences, Fluid Dynamics* No. **6**, pp 46-63.

[2] Ostapenko N A, Semenov A V 2008 Lift-to-Drag Ratio Of Optimal Waveriders On Plane Shock Waves *'Problems Of Modern Mechanics' collection to the 85th anniversary of Academician G.G. Cherny, Moscow State University, Institute of Mechanics, edited by A. Barmin,* (Moscow University Publishing House, 'Omega-L' Publishing house), sec. No. **639**, pp 408-431.

[3] Maksimov F A, Ostapenko N A, Zubin M A 2015 Conical Flows Near V-shaped Wings With Shock Waves, Attached On Leading Edges *Progress in flight physics*, Vol. **7** / [Edited by D. Knight, I. Lipatov, and Ph. Reijasse], (Moscow: TORUS PRESS) pp 453-474.
[4] Zubin M A, Maksimov F A, Ostapenko N A 2014 Criteria Of Existence Of Nonviscous Vortical Structures In Shock Layers Of Supersonic Conical Gas Flows *Doklady Physics* Vol. 59, No. **1** pp 49–55.
[5] Zubin M A, Maksimov F A, Ostapenko N A 2015 Conditions Of Existence Of Vortex Ferry Singularities In Supersonic Conical Flows *Fluid Dynamics* Vol. 50, No. **1** pp 118–136.

Acknowledgments

The work was fulfilled with a partial financial support of the Russian Foundation for Basic Research (Project No. 18-01-00182). Calculations were carried out on MVS-100K at Interdepartmental supercomputer center of the Russian Academy of Sciences.

IOP Conf. Series: Materials Science and Engineering **468** (2018) 012019 doi:10.1088/1757-899X/468/1/012019

Preliminary evaluation of multiple atmospheric re-entries in Lunar return missions

D A Grishko, V V Leonov, M A Ayrapetyan and O S Shvirkina

Bauman Moscow State Technical University, Moscow, Russia
E-mail: dim.gr@mail.ru

Abstract. The paper focuses on the usage of multiple atmospheric re-entries while coming back to the Earth in Lunar-return mission. In order to deal with the flight duration and thermal loads the pericentre altitude should be variable. After each passing through the atmosphere an apocentre becomes significantly lower, so it is possible to achieve required target orbit in several revolutions, starting with the higher initial pericentre altitude.

1. Introduction

To date, the missions of "Apollo" program are the only ones that successfully realized the manned spaceflights [1] to the Moon with landing on its surface in the period of 1969-1972. During these flights the astronauts managed to collect the samples of Lunar soil, to install the laser reflectors (so the accuracy of measuring the distance between the Earth and the Moon significantly augmented), to study how the Moon environment acts on humans and hardware and to carry out some scientific research experiments. Unfortunately, "Apollo" program was closed and Soviet Moon program followed it, so the Moon investigations have been frozen for a long time.

During the last decade the interest to the Moon appeared again and was implemented in practice by Japanese mission "Kaguya", American missions LRO and LCROSS [2], space vehicles from Chinese "Chang'e" program and Indian "Chandrayaan" program [3]. We can observe the perspectives of manned spaceflights to the Moon [4, 5], that is why here it will be reasonable to study an opportunity of Earth atmosphere usage by the descending modules in return phase of the Lunar mission. This idea, to reduce the velocity by multiple atmospheric re-entries, is not exactly a new one. It was first formulated in 60-s years of XX century, but then was not studied properly because of the closure of Moon programs both in the USA and USSR; also the possibilities of computational hardware of that time were rather limited. This study is focused on preliminary evaluation of trajectory that actively uses the atmospheric drag by the example of "Apollo" descending module. The main advantage of such scheme is that the module sustains short-time thermal overloads while fast passing through the pericentre region, then it orbits around the Earth and finally starts to descend in the atmosphere like the vehicles of "Soyuz" family usually do. Speaking about the flight safety, the described scheme is better than the direct atmospheric entry with near-parabolic velocity.

2. Simplified mathematical model of motion

It is required to get the evaluation of trajectory that uses the multiple atmospheric re-entries, that is why the motion of the descending module can be considered in a very simplified form. It is supposed, that at the initial moment of time the module crosses the sphere of influence of the Moon and comes back to the Earth, being located at the Hohmann ellipse [6]. The apocentre altitude is 100 km higher than the average distance between the Earth and the Moon, the altitude of pericentre is variable. The

Content from this work may be used under the terms of the Creative Commons Attribution 3.0 licence. Any further distribution of this work must maintain attribution to the author(s) and the title of the work, journal citation and DOI.
Published under licence by IOP Publishing Ltd

following trajectory with respect to the Earth is considered to be plane, the atmosphere is taken into account when the module reaches the height 1500 km above the Earth. The simplified mathematical model of motion can be described in the inertial coordinate system, its centre is situated at the centre of the Earth:

$$\dot{V}_x = -\frac{\mu}{r^3}x - C_{xa}\cdot\frac{\rho v^2}{2}S_m\cdot\frac{V_x}{V}; \quad (1)$$

$$\dot{V}_y = -\frac{\mu}{r^3}y - C_{xa}\cdot\frac{\rho v^2}{2}S_m\cdot\frac{V_y}{V}; \quad (2)$$

$$\dot{x} = V_x; \quad (3)$$

$$\dot{y} = V_y \quad (4)$$

Here $\boldsymbol{r} = \sqrt{x^2 + y^2}$ is magnitude of module's position vector, $\boldsymbol{\mu}$ = 398600.33 km³/sec² – Earth gravitational constant, $\boldsymbol{v} = \sqrt{v_x^2 + v_y^2}$ is magnitude of module's velocity, $\boldsymbol{C_{xa}}$ is drag coefficient (it is supposed that the module's longitudinal axis always coincides with velocity vector), $\boldsymbol{\rho}$ is atmosphere density, calculated using Russian standards GOST P 25645.166–2004 (heights 120-1500 km) and GOST 4401-81 (heights 0-120 км), $\boldsymbol{S_m}$ is cross-section square of the module. The system of equations (1) - (4) was solved using 4th order Runge-Kutta numerical integration with the constant time step.

3. Insertion into the orbit with 200 km apocentre altitude

The descending operations from the International Space Station are the most approved for today. That is why in this study we consider the scheme when using the multiple atmospheric re-entries the module first should be transferred to an orbit with the apocentre altitude of 200 km and only after that propulsion system should provide the module with a breaking impulse with the following landing on the Earth.

It is known, that while orbiting the Earth at low heights, the apocentre altitude decreases significantly faster than the altitude of pericentre. That is why, it is possible to create an orbit with the required altitude of apocentre during the several revolutions and the altitude of pericentre will be approximately the same as initial value.

The drag intensity depends on pericentre altitude and this is a non-linear function of height above the Earth. There is such an altitude of pericentre that will make it possible to create an orbit with 200 km apocentre just after the first passing through the atmosphere (figure 1). In case of the "Apollo"-type module this critical pericentre altitude is equal to 66.5 km. Such trajectory corresponds to the minimal flight time, also the module will have to overcome maximal thermal overloads.

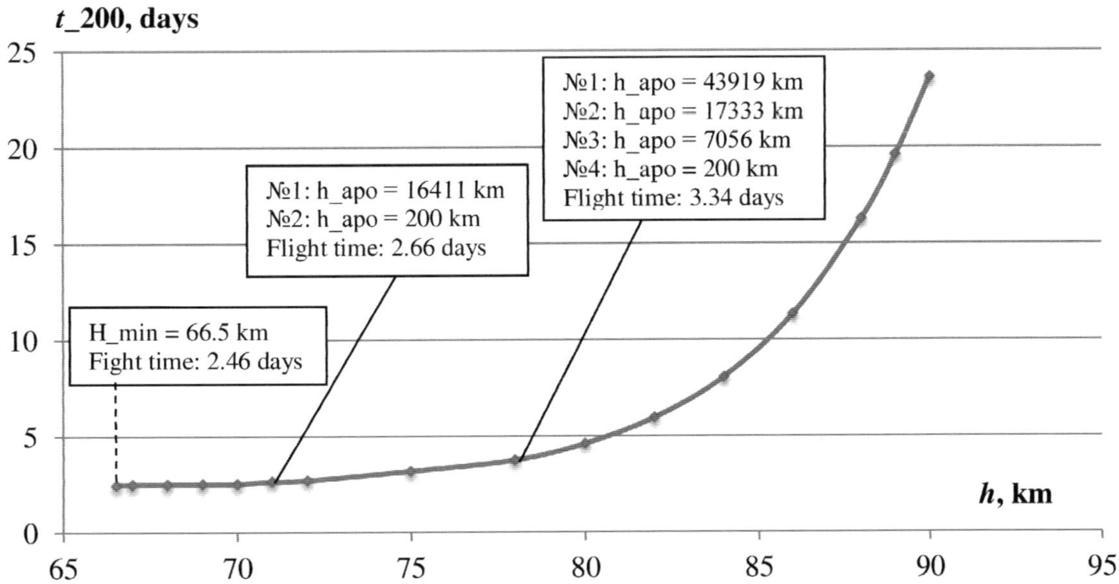

Figure 1. Dependence of flight time before reaching 200 km apocentre altitude on initial pericentre altitude

While the pericentre altitude increases, the augmentation of flight time (before 200 km apocentre altitude is attained) does not change uniformly. If the initial altitude of pericentre is changed from 66.5 to 71 km, the module will be able to pass atmosphere twice. If this pericentre altitude is increased for 5-7 km more, four re-entries are possible. If we continue to increase the altitude of pericentre till 90 km, the module will do multiple atmospheric re-entries. In that case the thermal loads will be reduced, but the flight duration significantly augments because during the first revolutions the module goes far away from the height of geostationary orbit, this will lead to the multiple passing through the radiation belts and it seems to be bad in terms of health of crew members.

4. Variable apocentre altitude after the first pass through the atmosphere

The altitude of 200 km for apocentre altitude should not be considered as a fixed value. While coming back from the Moon the real inclination to the Earth equator may differ from the planned, so the special maneuver will be required. Such type of maneuver should be realized in equatorial part of the orbit and as far as possible from the Earth.

As we mentioned in previous chapter, the attainment of the target orbit should be done as fast as possible. At the same time, if the target apocentre altitude is equal to 200 km, it will lead to the extremely low (66.5 km) pericentre altitude and to the significant thermal loads. There is a solution for this problem: after the first passing through the atmosphere we can get an orbit with much higher apocentre and at the following revolutions we can speed up the degradation of orbit using the propulsion system of the module. So we will not only decrease the thermal loads, but also will get a new level of accuracy to control the parameters of the obtained open-ended orbit with respect to the landing regions. The figure 2 represents the dependence between the pericentre altitude of initial orbit and the target apocentre altitude of the orbit that will appear after the first passing through the atmosphere.

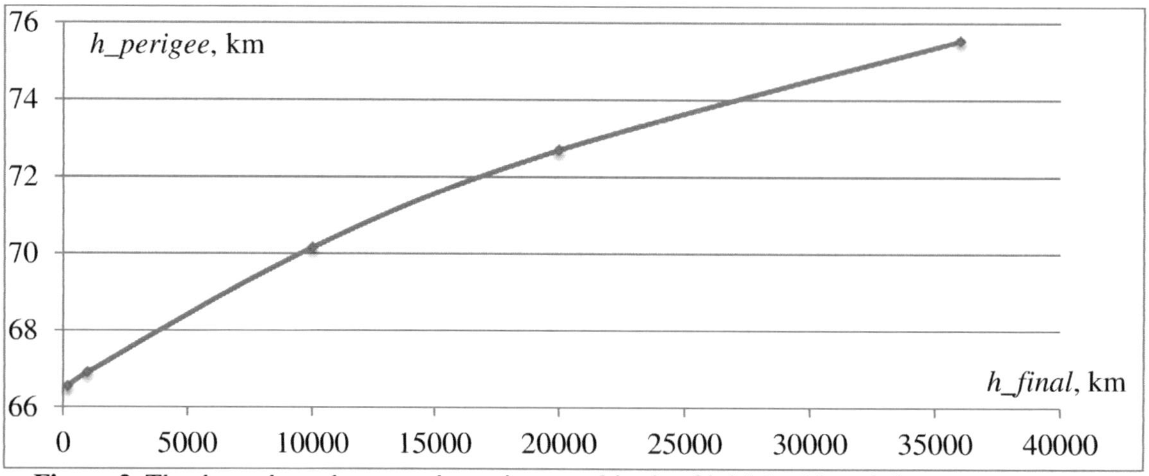

Figure 2. The dependence between the pericentre altitude of initial orbit and the target apocentre altitude of the orbit that will be obtained after the first passing through the atmosphere.

5. Attainment of the target apocentre altitude during several revolutions around the Earth

In chapter #3 and #4 we shown two ways how to increase the pericentre altitude of initial trajectory when approaching the Earth: increasing the number of re-entries in the atmosphere or increasing the target apocentre altitude. Both variants lead to the augmentation of flight duration, so let us compare them. The figure 3 contains the plot of flight duration as a function of pericentre altitude of initial trajectory. Each curve corresponds to the concrete target apocentre altitude, the points on each curve represent the number of atmospheric re-entries which are required to obtain this apocentre altitude.

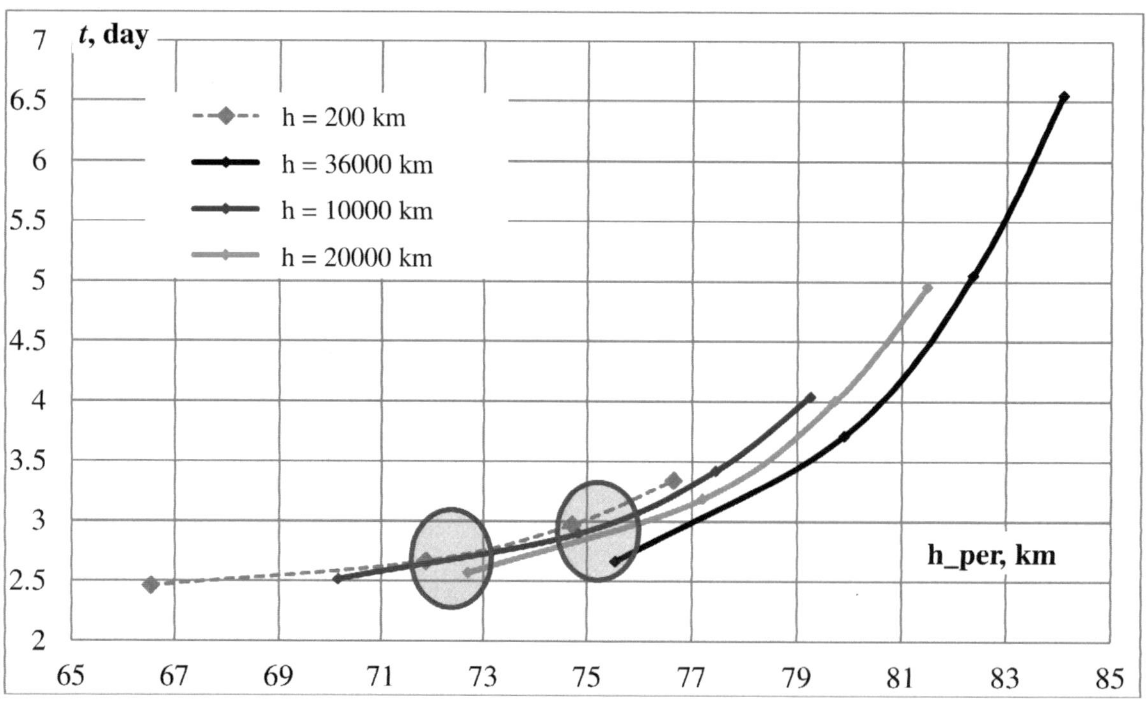

Figure 3. The dependence of flight duration on pericentre altitude of initial trajectory for different target apocentre altitudes and different quantity of atmospheric re-entries.

At figure 3 we can observe the red dotted curve of flight duration, that corresponds to the target orbit with the altitude of apocentre equal to 200 km. One can see that if the pericentre altitude of initial trajectory is 72 km we can either achieve 200km altitude of apocentre with two atmospheric re-entries or approximately 20 000 km altitude in case of single re-entry. The attainment of 200 km apocentre altitude after three re-entries takes the same time as to reach the geostationary heights after one pass of the atmosphere.

6. Conclusion

The usage of descending module's atmospheric braking makes it possible to achieve the near-Earth orbit with variable altitude of apocentre. If the initial pericentre altitude is low, then we obtain near-circular orbit and the flight duration is minimal, but at the same time the module will have to deal with significant thermal loads. If the initial pericentre altitude increases, it will lead to the elliptical orbit that degrades rather fast because its pericentre is situated in the upper atmosphere. The duration of each atmospheric re-entry remains approximately the same but the thermal overloads are lower due to the higher initial pericentre altitude. In order to decrease the flight duration it is possible to combine the primary atmospheric braking with the additional anti-velocity impulse after the current passing through the atmosphere.

7. References

[1] "Apollo Program Summary Report" 1975 *NASA Report JSC-09423 [Internet resource]: https://www.hq.nasa.gov/alsj/APSR-JSC-09423.pdf*
[2] Keller J W, Petro N E and Vondrak R R 2016 *Icarus* **273** pp 2
[3] Shuanggen J, Sundaram A and Hiroshi A 2013 *Advances in Space research* **2(52)** pp 285
[4] Murtazin R F 2014 *Acta Astronautica* **101** pp 151
[5] Grabois M R 2011 *Acta Astronautica* **68** pp 1353
[6] Hohman W 1925 *Die Erreichbarkeit der Himmelskörper* (Verlag Oldenbourg in München)

Acknowledgements

The research was supported by Russian Science Foundation (project No. 17-79-10450)

IOP Conf. Series: Materials Science and Engineering **468** (2018) 012020 doi:10.1088/1757-899X/468/1/012020

Motion planning of a space robot

G K Borovin[1] and V V Lapshin[2]

[1]Keldysh Institute of Applied Mathematics RAS, Moscow, Russia
[2]Bauman Moscow State Technical University, Moscow, Russia
E-mail: lapshin032@bmstu.ru

Abstract. The plane problem of motion control of a free-floating space robot is investigated. The robot is supposed to be in a state of passive flight. It is assumed that the vectors of momentum and kinetic moment of the robot equal to zero. The movement of the arm relative to the robot body changes the position of the center of masses of the body and leads to its rotation around the center of masses. There are constraints on the range of variation of the manipulator length and the angle of its rotation relative to the body. It is shown that while carrying out special movements of the manipulator, it is possible to ensure the movement of robot gripper from arbitrary initial to an arbitrary final position, if they are located inside the workspace, which is a ring with a centre in the centre of masses of the robot. Moreover, it is possible to get the desired value of the angle between the manipulator and the body in the final position.

1. Introduction

Space robots, consisting of a main body and provided with one or more manipulators, are promising for various type of works in outer space and orbiting Earth satellite to conduct of repair, maintenance and construction of various objects (space stations, orbital telescopes, etc.), and also for carrying out the work of removing space debris. Survey of work in this direction is given in [1, 2]. On Figure1 is shows the reusable spacecraft "Buran" (USSR) with the robotic arm in the cargo bay [3]. On Figure 2 is shows astronaut in open space, which is similar to anthropomorphic space robot. On Figure 3 is shows a project of Japanese free-flying of space robot [4].

In this work we consider planar motion of free-floating space robot flight. The robot consists of a body and manipulator arm. It does not have or does not use propulsions that control movement and orientation of the robot body.

The movement of the manipulator has a significant influence on the movement of the robot body due to the theorems about the change of quantity of motion and kinetic moment about the center of masses [1, 2, 5-9].

Content from this work may be used under the terms of the Creative Commons Attribution 3.0 licence. Any further distribution of this work must maintain attribution to the author(s) and the title of the work, journal citation and DOI.
Published under licence by IOP Publishing Ltd

Figure 1. The reusable spacecraft "Buran" with the manipulator in the cargo bay

Figure 2. Astronaut in open space (anthropomorphic space robot)

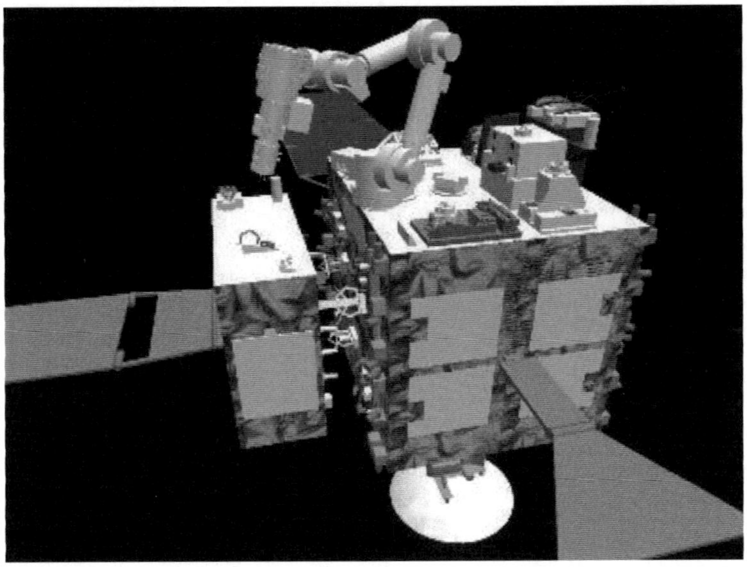

Figure 3. Project of the free-flying space robot ETS-VII (Japan)

In [1, 2, 5-7] considered the plane problem of the motion control free-floating robot consisting of the body and the manipulator arm. The manipulator arm has kinematic redundancy and consists of at least three hinged unit links. There are no constraints on the limits of variation of the deviation angles of the manipulator arm. It is assumed that the vector of quantity of motion of the robot and its kinetic moment relative the center of masses are zero.

In [1, 2] the motion control of the robot (the movement of the gripper of the robot from the given start position to the specified final position) is constructed locally from the current position determined by the small increments of the coordinates in the robot degrees of freedom with the aim to achieve the desired final position of the robot gripper. As a result, despite the efforts of the authors, failed to move a gripper in a significant portion of kinematically achievable final positions located in the robot's workspace.

In [5-7] this problem is solved by methods of the theory of motion control without considering the features of the robot dynamics associated with the existence of laws of conservation of the quantity of motion and kinetic moment. Control the movement of the gripper of the robot is constructed using feedback-based information on the position of the final goal point robot. The motion of robot body caused by the motion of the manipulator arm are treated as unknown perturbation. In the same way as in [1, 2] it is impossible to move the gripper of the manipulator from a given initial to an arbitrary final position inside its workspace.

In [8-9], it is shown that this result appears to be erroneous and caused by the methods used to control the motion of the robot based on the local principle of their formation. It takes transition to a global principle of constructing programmed motion of the robot, which would require the performance of special, not obvious movements in advance for the realization of this goal. On a simple model example, of planar motion of the robot with telescopic manipulator analytically shown that it is possible to provide the motion of the gripper of the robot from an arbitrary initial to an arbitrary kinematically achievable final position. The problem was solved under the assumption that there are constraints on the range of varying of the length of manipulator arm and there are no constraints on the range of varying of the angle of manipulator arm rotation relative to robot body.

In this work, which is the evolution [8-9], this problem is solved if there are constraints as on varying the length of the manipulator arm, and so on the limits of varying of the angle of manipulator arm rotation relative to robot body. It is shown that the workspace of the robot significantly more than that of a similar robot with a fixed body. By special motions of the manipulator arm can ensure a reversal of space robot and move the gripper of the robot from an arbitrary initial to an arbitrary final position, if they are located inside the ring with the center in the center of masses of the robot is bounded by circles with radii equal to the minimum and maximum distance from the gripper of the robot to its center of masses. Furthermore, it is shown that it is possible to provide the required (the most convenient for the execution of works) a value of the angle between the robot body and manipulator arm in the final position.

2. Statement of the problem

Consider planar motion of a free-floating space robot. In Figure 4 shows a schematic diagram of the robot, which consists of the body with mass m_1 and the telescopic manipulator arm attached to the body at its center of mass B. The moment of inertia of the body about its center of mass equal J_1. The manipulator arm consists of a cylinder that rotates relative to the point of his suspension to the body, and the rod is moved along the telescopic link of the arm. We denote m_2, m_3, J_2, J_3 respectively the mass of the cylinder and the rod of the manipulator arm and their moments of inertia about their centers of mass C_2 and C_3.

When the robot moves on it, only internal forces act in the degrees of freedom of the manipulator arm. There are a conservation laws of the robot's quantity of motion and its kinetic moment about the center of mass. Consider the problem of motion control of the robot at zero values of the vector of the of robot's quantity of motion and its kinetic moment about the center of mass. The center of mass of the robot C remains stationary and its position is taken as the beginning of inertial (orbital) coordinate system $C\eta\zeta$.

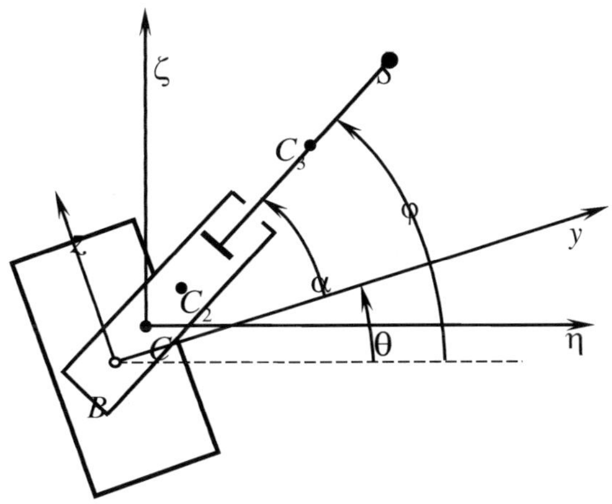

Figure 4. The constructive scheme of the space robot

The coordinate system Byz associated with the robot body, the axis Bz is constructive vertical and the axis By – the longitudinal axis of the body. The position of the robot in the absolute coordinate system $C\eta\zeta$ is defined by the coordinates of the body center of mass, the angle of body rotation θ, the angle, the angle between the axis By and manipulator arm α and the length of the arm $l = BS$. Point S – gripper of the manipulator. The angle φ specifies the orientation of the manipulator arm in absolute space:

$$\varphi = \theta + \alpha \quad . \tag{1}$$

The center of mass of the cylinder C_2 lies on the axis of the telescopic link at the distance $r_2 = BC_2$ from the point of its suspension to the body. The center of mass of the rod C_3 lies on the axis of the telescopic link at the distance $r_3 = SC_3$ from the gripper S. The center of mass of the robot C lies on the axis of telescopic link at the distance $\rho_B = BC$ from the point of suspension of the manipulator arm to robot body [7, 8]:

$$\rho_B = \frac{m_2 r_2 + m_3 (l - r_3)}{m_1 + m_2 + m_3} \quad ,$$

then length of the manipulator arm l and the distance from the center of mass of the robot to the gripper $\rho = l - \rho_B$ are related by

$$(m_1 + m_2 + m_3)\rho = (m_1 + m_2)l - m_2 r_2 + m_3 r_3 \quad . \tag{2}$$

The kinetic moment conservation law about the center of mass has the form [7]

$$k\dot{\varphi} + J_1\dot{\theta} = k\dot{\alpha} + (k + J_1)\dot{\theta} = 0 \quad , \tag{3}$$

where

$$k = k(l) = \mu_1(l-\lambda)^2 + \mu_2 \quad , \quad \mu_1 = \frac{(m_1+m_2)m_3}{m_1+m_2+m_3} \quad , \quad \lambda = \frac{m_2 r_2}{m_1+m_2} \quad ,$$
$$\mu_2 = J_2 + J_3 + \frac{m_3^2 r_3^2}{m_1+m_2+m_3} + \frac{m_1 m_2 r_1^2}{m_1+m_2} \quad . \tag{4}$$

Suppose that there are constraints on the limits of the length of the manipulator arm and the angle of its rotation of regarding the robot body:

$$l \in [l_{\min}, l_{\max}] \quad , \quad \alpha \in [\alpha_{\min}, \alpha_{\max}] \quad . \tag{5}$$

The position of the gripper of the robot is determined by the polar coordinates $\rho = CS$ and φ.

The purpose of control is to move the gripper of the robot from any arbitrary start position (ρ_0, φ_0) to arbitrary final position (ρ_D, φ_D) inside the workspace. The workspace is limited by two circles with center in the center of mass of the robot C and the radii $\rho_{\min} = \rho(l_{\min})$ and $\rho_{\max} = \rho(l_{\max})$. In addition, we require that the angle between the robot body and its arm in final position was a predetermined value α_D (for example, the most convenient to perform the work). From (1), (2) it follows that for this it is necessary and sufficient to translate the robot from initial state θ_0, α_0, l_0 to final state θ_D, α_D, l_D – where $\theta_D = \varphi_D - \alpha_D$, $l_0 = l(\rho_0)$, $l_D = l(\rho_D)$.

3. Planning of the programmed motion

The motion of the robot will be implemented in such a way that it was divided into time intervals (stages), during each of which operates only one actuator of manipulator arm. Either changes the length of the manipulator arm under a fixed angle α, in this case the angle θ remains constant due to (3), or change the angle of arm rotation relative to the robot body at a fixed length of the manipulator arm, then due to (3)

$$\dot{\theta} = -b(l)\dot{\alpha} \quad , \tag{6}$$

where

$$b(l) = \frac{k(l)}{k(l) + J_1} \quad .$$

Statement 1. The ratio $b(l)$ is an increasing function of l.

By (4) $k(l) > 0$ is a monotonically increasing function in the domain (5), as $l_{\min} > r_2 > \lambda$. Therefore, if $l_2 > l_1$, then

IOP Conf. Series: Materials Science and Engineering **468** (2018) 012020 doi:10.1088/1757-899X/468/1/012020

$$b(l_2)-b(l_1)=\frac{J_1\big(k(l_2)-k(l_1)\big)}{\big(k(l_1)+J_1\big)\big(k(l_2)+J_1\big)}>0 \quad .$$

Integrating (6) for $l = \text{const}$ from the initial position $\tilde{\alpha},\tilde{\theta}$ obtained

$$\theta = f(\tilde{\theta},l,\alpha,\tilde{\alpha}) = \tilde{\theta}-b(l)(\alpha-\tilde{\alpha}) \quad , \tag{7}$$

then a turn of the manipulator arm on the angle $\Delta\alpha=\alpha-\tilde{\alpha}$ relative of the robot body leads to its rotation at the angle $\Delta\theta=\theta-\tilde{\theta}$. On the plane of variables α and θ this dependence $\theta(\alpha)$ has the form of the family of straight lines, the slope of which depends at l. At the maximum length value of the manipulator arm $l = l_{max}$, its rotation produces a maximum effect on the motion of the robot body, so the straight line (7) in this case has the greatest slope. With a minimum length value of the manipulator arm $l = l_{min}$, its rotation has a minimal effect on the motion of the robot body, straight line (7) has the smallest slope. The rest of the lines passing through the point $\tilde{\alpha},\tilde{\theta}$ are implemented with intermediate values $l \in (l_{min}, l_{max})$.

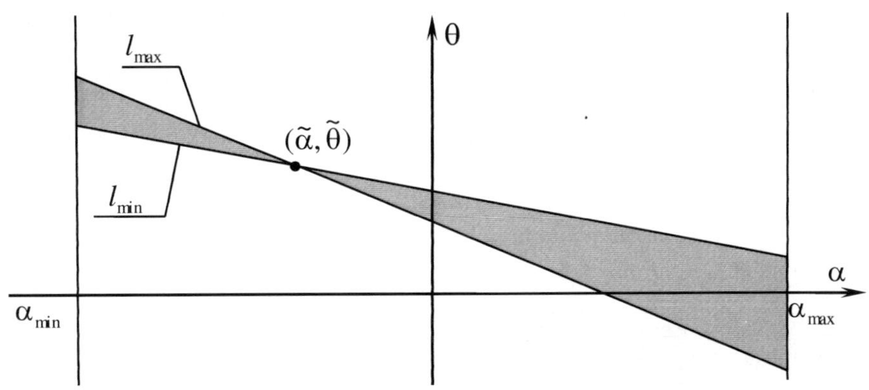

Figure 5. The attainability domain of the position $\tilde{\alpha},\tilde{\theta}$

Let's call the shaded domain and together with its boundaries in Figure 5 the attainability domain from the position $\tilde{\alpha},\tilde{\theta}$ at the fixed $l \in [l_{min}, l_{max}]$. Note that at each stage the robot motion its final position does not depend on the law of variation of the manipulator arm coordinates l and α, but determine only by the initial and final value of this coordinate.

Statement 2. The robot can move from start position α_0,θ_0,l_0 to terminal position α_D,θ_D,l_D, if they satisfy the constraints (5) and represents the point α_D,θ_D lies in the attainability domain of point α_0,θ_0.

In this case, from (7) it follows that

$$\theta_D = \theta_0 - b(l^*)(\alpha_D-\alpha_0) \quad , \tag{8}$$

where the value l^* determined from relations (4), (8)

$$l^* = \lambda + \sqrt{-\frac{1}{\mu_1}\left(\mu_2 + \frac{J_1(\theta_D - \theta_0)}{\theta_D - \theta_0 + \alpha_D - \alpha_0}\right)} \quad , \tag{9}$$

and in accordance with the statement 1 $l^* \in [l_{min}, l_{max}]$.

Remark. Here and further on instead l_{min} and l_{max} it is possible use other two values $\hat{l}_{min}$ and $\hat{l}_{max}$ that satisfy the conditions $l_{min} \le \hat{l}_{min} < \hat{l}_{max} \le l_{max}$.

Statement 3. The robot can move from any arbitrary start position α_0, θ_0, l_0 to arbitrary final position α_D, θ_D, l_D if they satisfy the constraints (5).

If the final position of the imaging point α_D, θ_D lies inside the attainability domain from its initial position, the programmed motion is constructed in accordance with the statement 2.

If the final point α_D, θ_D lies below the attainability domain from the initial position α_0, θ_0, there are two ways of planning a programmed motion. In Figure 6 shows the corresponding trajectory of the imaging point on the plane α, θ.

When using the first method (Figure 6,a) the gripper of the manipulator arm in the relative coordinate system *Byz* connected with the robot body moves along the trajectory shown in Figure 7. The initial position of the gripper is indicated S_0 , and the final – S_D. If the final position of the imaging point α_D, θ_D lies near the attainability domains from the initial position α_0, θ_0, the gripper of the manipulator arm moves along the path $S_0S_1S_2S_3S_4S_D$. Otherwise, the gripper of the manipulator arm moves along a trajectory $S_0S_1\_S_2S_3S_5S_6S_2\_S_3S_4S_D$, and allocated closed trajectory of the gripper $S_2S_3S_5S_6S_2$ can e repeated several times.

In this cyclic motion of the gripper of the manipulator arm in the sections S_2S_3 and S_5S_6 change only the length of the arm and the angle of rotation of the robot body in absolute space remains is unchanged. In the section S_3S_5 of the manipulator arm rotated relative to the body clockwise under $l = l_{min}$ and in accordance with (7), the robot body is rotated counterclockwise by the angle modulo equal to

$$|\Delta\theta_1| = b(l_{min})(\alpha_{max} - \alpha_{min}) \quad .$$

In the section S_6S_2 the manipulator arm rotated relative to the body counterclockwise under $l = l_{max}$ and in accordance with (7), the robot body is rotated clockwise by the angle is modulo equal to

$$|\Delta\theta_2| = b(l_{max})(\alpha_{max} - \alpha_{min}) \quad .$$

Because of statement 1 $b(l)$ is a monotonically increasing function of the manipulator arm length l, and, consequently,

 doi:10.1088/1757-899X/468/1/012020

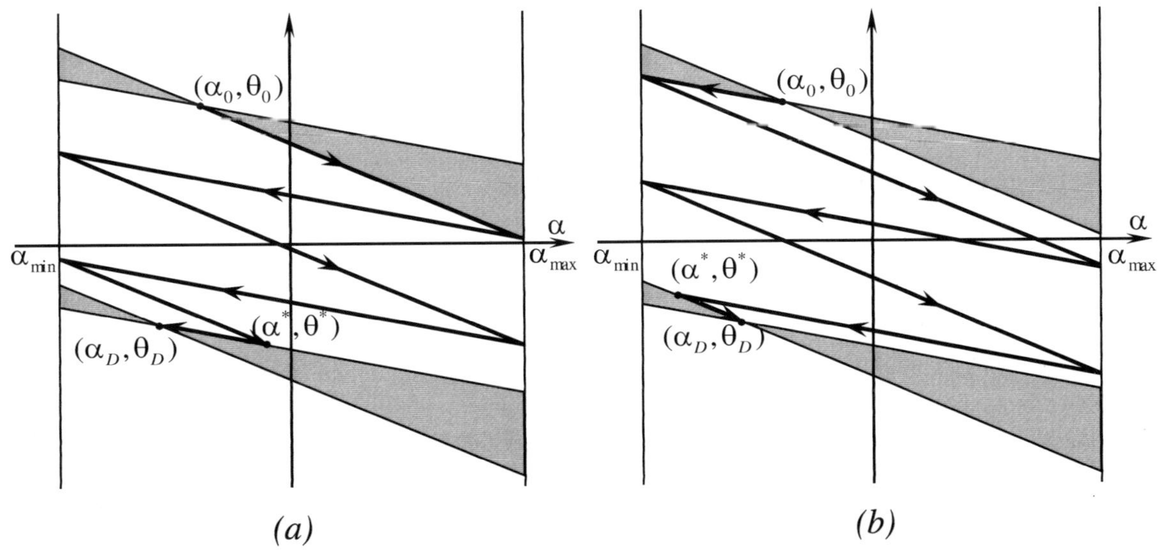

Figure 6. The trajectory of the representative point for the case when its final position lies below the attainability domain from the initial position

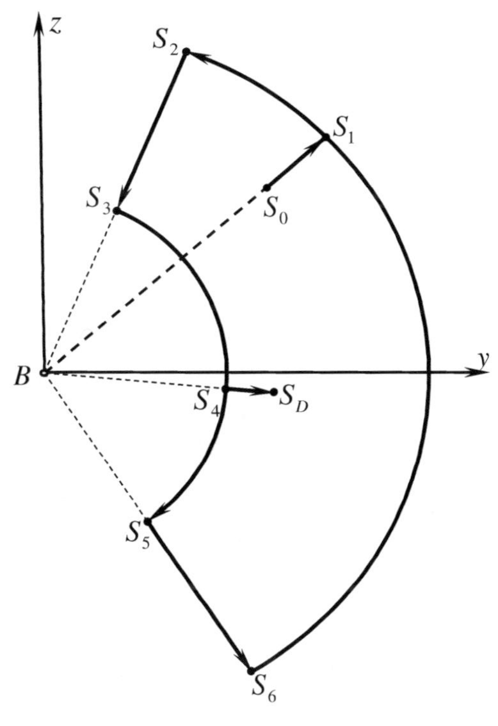

Figure 7. The trajectory of the manipulator arm gripper in relative coordinate system connected with the robot body.

$$|\Delta\theta_2| - |\Delta\theta_1| = (b(l_{max}) - b(l_{min}))(\alpha_{max} - \alpha_{min}) > 0 \quad .$$

As a result of each cycle of motion of the gripper of manipulator arm on closed section $S_2S_3S_5S_6S_2$ robot body turned clockwise at the angle $|\Delta\theta_2| - |\Delta\theta_1| > 0$.

If the final position of the imaging point α_D, θ_D lies above the attainability domain from the initial position α_0, θ_0, the programmed motion is planning analogous.

4. Conclusion

The planar motion of space robot in a state of passive flight under the assumption that the vector of momentum and kinetic moment of the robot equal to zero is investigated. Considered robot consist of a body and telescopic manipulator arm. It assumed that there are the constraints as to the limits of the length of the manipulator arm, and so at the its rotation angle above the body. It is shown that due to special cyclic motions of the manipulator arm relative to the body it is possible to turn the robot body to an arbitrary angle. In the result, the workspace of a free-floating space robot (the possible final position of the robot gripper) is more larger than the workspace of the robot with a fixed body. The workspace of the robot in absolute space is a ring bounded by two circles centered with the center of mass of the robot and the radii equal to the minimum and maximum distance from the center of mass of the robot to gripper. Moreover, in programmed motion planning can be provided not only the release of the gripper of the robot to a given final position, but also the required (convenient for work) the value of the angle between the robot body and manipulator arm in the end position.

References

[1] Dubovsky S and Papadopouls E 1993 *IEEE Trans. on Robotics and Automation* **9** No 5 pp 531-543
[2] Moosavian S, Ali A and Papadopouls E 2007 *J. Robotica* . **25** No 5 537-547
[3] http://www.buran.ru (accessed 24.06.2017).
[4] Inaba N and Oda M 2000 *Proc. of IEEE Int. Conf. on Robotics and Automation* pp 1169-1174
[5] Rutkovskiy V Yu, Sukhanov V M and Glumov V M 2010 *Automation and Remote Control* **71** No 1 pp 70-86
[6] Rutkovskiy V Yu, Sukhanov V M and Glumov V M 2010 *Automation and Remote Control* **71** No 11 pp 2345-2359
[7] Rutkovskiy V Yu, Sukhanov V M and Glumov V M 2014 *Proc. of XII All-Russian Meeting on the problems of control* 3853-3865 (in Russian)
[8] Lapshin V V 2017 *J. Computers and System Sciences International* **56** No 1 pp 157-163
[9] Borovin G K and Lapshin V V 2017 *Mathematica Montesnigri* **XXXIX** pp 67-78

Acknowledgements

The research was performed at the Bauman Moscow State Technical University with the financial support of the Ministry of Education and Science of the Russian Federation under the Federal Target Program "Research and development on priority directions of scientific and technological complex of Russia for 2014-2020". Agreement # 14.577.21.0247 (unique identifier RFMEFI57717X0247).

IOP Conf. Series: Materials Science and Engineering **468** (2018) 012021 doi:10.1088/1757-899X/468/1/012021

The stochastic theory of the turbulence

A V Dmitrenko[1,2]

[1]National Research Nuclear University «MEPhI», Moscow, Russia
[2]Russian University of Transport «MIIT», Moscow, Russia
E-mail: AVDmitrenko@mephi.ru

Abstract. The general results of new stochastic theory of turbulence for isothermal flows are presented. Stochastic equations for laws of conservation and for equivalence of measures between deterministic process and random process, the review of results of an analytical expressions are presented for the isothermal turbulent flow in the tube and for the flow on the flat plate depending on initial fluctuation in medium. An important result of a new theory is the fact that even if the fluctuation has an initial spectrum in the form of a delta-function, the modular fractal equation obtained in theory determines both the fractal equation for the diffusion process of energy transfer and the fractal generation equation of Landau type described the subsequent expansion of the spectrum. Derived formulas for critical Reynolds numbers and critical point, the velocity profile, the second-order correlation, and friction coefficients give the satisfactory agreement with the experimental data for both flows.

1. Introduction

Articles [1–10] were devoted to the search for equations and invariants that could determine the start of transition from a deterministic motion to a turbulent one. It is important to note an experimentally confirmed fact that random fluctuations are always present in a flow (initial turbulence). The impact of initial fluctuations on the result of the solution of equations is significant. This is especially important for numerical modeling and for developing computational codes. Numerical methods [11, 12, 13] do not remove the relevance of understanding of the nature of turbulence phenomena. So numerous numerical solutions do not allow to define the essence of the phenomenon of turbulence even together with experimental results. The new physical law for the studied phenomenon of turbulence was found along with new stochastic equations for a continuous medium [14-16]. The law was established theoretically [17] and was called an equivalence of measures between deterministic movement and random movements. As a result the systems of stochastic equations of energy, mass and momentum were defined for areas: 1) the beginning of the generation of turbulence, 2) the generation of turbulence, 3) the diffusion, 4) the dissipation of the turbulence.

2. Set of stochastic equations of conservation and equivalence of measures

Stochastic equations of conservation defined in [14-17] for the non-isothermal condition take the form:
the equation of mass (continuity)

$$\frac{d(\rho)_{colst}}{d\tau} = -\frac{(\rho)_{st}}{\tau_{cor}} - \frac{d(\rho)_{st}}{d\tau}, \qquad (1)$$

Content from this work may be used under the terms of the Creative Commons Attribution 3.0 licence. Any further distribution of this work must maintain attribution to the author(s) and the title of the work, journal citation and DOI.
Published under licence by IOP Publishing Ltd

the momentum equation

$$\frac{d\left(\rho u_i\right)_{colst}}{d\tau} = div(\tau_{i,j})_{colst} + div(\tau_{i,j})_{st} - \frac{(\rho U)_{st}}{\tau_{cor}} - \frac{d(\rho U)_{st}}{d\tau}, \quad (2)$$

and the energy equation

$$\frac{dE_{colst}}{d\tau} = div(u_i\tau_{i,j})_{colst} + div(u_i\tau_{i,j})_{st} - \left(\frac{E_{st}}{\tau_{cor}}\right) - \left(\frac{dE_{st}}{d\tau}\right). \quad (3)$$

Here, $\rho, \vec{U}, u_i, u_j, u_l, \mu, \tau, \tau_{i,j}$ are the density; the velocity vector; the velocity components in directions x_i, x_j, x_l (i, j, l = 1, 2, 3); the dynamic viscosity; the time; and stress tensor $\tau_{i,j} = P + \sigma_{i,j}$, $\delta_{ij}=1$ if i=j, $\delta_{ij}=0$ for $i \neq j$. P is the pressure of liquid or gas; $\sigma_{ij} = \mu\left(\frac{\partial u_i}{\partial x_j} + \frac{\partial u_j}{\partial x_i}\right) - \delta_{ij}\left(\xi - \frac{2}{3}\mu\right)\frac{\partial u_l}{\partial x_l}$.

In articles [14--16], the physical process is represented as a non-equilibrium thermodynamic system with i-subsets, which is characterized by the values of energy $\bigcup_{i=1}^{i=n}(E_i)_{g_{st}}$, momentum $\bigcup_{i=1}^{i=n}(M_iU_i)_{g_{st}}$, and mass $\bigcup_{i=1}^{i=n}(M_i)_{g_{st}}$. Here, U_i is the speed, $\bigcup_{i=1}^{i=n}(E_i)_{g_{st}}$ is the stochastic-field energy (index g_{st}); $\bigcup_{i=1}^{i=n}(E_i)_{col_{st}}$ is the fraction of the field energy, its deterministic component (index col $_{st}$) having the zero stochastic component of measure; and $\bigcup_{i=1}^{i=n}(E_i)_{st}$ is the fraction of the field energy, which is actually the stochastic field component (index st). Similarly, components of the momentum (ρ**U**) and the mass (ρ is the density) are determined. According to [15, 25], the correlator in space-time is

$$\lim_{m_i \to m_c; r_i \to r_c; \Delta\tau_i \to \tau_c}\left(D_{N,M}(m_i \to m_c; r_i \to r_c; \Delta\tau_i \to \tau_c)\right) = 0 , \quad (4)$$

$$D_{N,M}(m_c; r_c; \tau_c) = \sum_i \lim_{m_i \to m_c} \lim_{r_i \to r_c} \lim_{\Delta\tau_i \to \tau_c} \left\{ m\left(T^M Z^* \cap T^N Y^*\right) - R_{1_{T^M Z^* T^N Y^*}} m\left(T^M Z^*\right)\right\}. \quad (5)$$

Subscript i denotes parameters m_i (i = 3 means mass, momentum, and energy). For the case of the binary intersections, it was written that $X = Y + Z + W$. Here subscripts «*cr*» or «*c*» refer to critical point $r(x_{cr}, \tau_{cr})$ or r_c: the space-time point of the beginning of the interaction between the deterministic field and random field that leads to turbulence. In addition, subsets Y, Z, W are called extended in X, if measures $m(Y)$, $m(Z)$, and $m(W)$ have the properties:

$m(Y) = m(Y^*) = m(T^nY) + \bigcup_{k=0}^{k=n-1} m(T^k(G_1^{n-k}))$ and wandering subsets $\bigcup_{k=0}^{k=n-1} m(T^k(G_1^{n-k})) \subset Y$;

$m(Z) = m(Z^*) = m(T^nZ^*) + \bigcup_{k=0}^{k=n-1} m\left(T^k\left(G_2^{n-k}\right)\right)$ and wandering subsets $\bigcup_{k=0}^{k=n-1} m(T^k(G_2^{n-k})) \subset Z$;

$m(W) = m(W^*) = m(T^nW) + \bigcup_{k=0}^{k=n-1} m\left(T^k\left(G_3^{n-k}\right)\right)$ and wandering subset $\bigcup_{k=0}^{k=n-1} m(T^k(G_3^{n-k})) \subset W$.

The correlation function produces a set of equations that determine the equivalence of measures:

$$\left|m\left(T^M Z\right)\right| = \left(R_{T^M ZT^N Y}\right)_{n,m} \left|m\left(T^N Y\right)\right|, 0 \le \left(R_{T^M ZT^N Y}\right) \le 1.$$

Here, $R_{T^M ZT^N Y}$ is the fractal correlation function assumed to be equal to unity to obtain analytical solutions. Thus, $|m(Z)| = (R_{ZTY})_{n,m} |m(TY)|$ for the pair (*N*, *M*)=(1,0), and $|m(TZ)| = (R_{TZTY})_{n,m} |m(TY)|$ for (*N*, *M*) = (1,1). Here, T^n is the conservative transformation of *X* for all *n* such that there is $n > n_d$ when T^n is the dissipative transformation for $Y^* \subset X$ and $Z^* \subset X$.

In the critical point, the sets of stochastic equations of energy, momentum, and mass are defined for the next space-time areas: 1) the beginning of the generation (index 1,0 or 1), 2) generation (index 1,1), 3) diffusion (1,1,1), and 4) dissipation of the turbulent fields.

As a result in [14-16], for the transfer of substantial values F (the mass (the density-ρ), the momentum (ρ**U**), the energy (E)) of deterministic (laminar) motion into random (turbulent) motion (for the area 1) the beginning of the generation of turbulence, $D_{N,M}(r_c;m_{ci};\tau_c) = D_{1,0}(r_c;m_{ci};\tau_c)$, the pair (N, M) = (1,0)), equivalence of measures was written as $\left(d\Phi_{col,st}\right)_{1,0} = -R_{1,0}\left(\Phi_{st}\right)$. Additionally in [1], for «correlator» $D_{N,M}(r_c;m_{ci};\tau_c) = D_{1,1}(r_c;m_{ci};\tau_c)$, the pair (N,M)=(1,1), it was written that $\left(d\Phi_{col,st}\right)_{1,1} = -R_{1,1}\left(d\Phi_{st}\right)$, $\left(\frac{d(\Phi)_{col,st}}{d\tau}\right)_{1,1} = -R_{1,1}\left(\frac{d\Phi_{st}}{d\tau}\right)$. $R_{1,0}$, $R_{1,1}$ are fractal coefficients. Time-correlations [14-16] are: $\left(\tau^0_{cor}\right)_1 = \frac{L}{(E_{st}\backslash\rho)^{1\backslash 2}}$, $\left(\tau^0_{cor}\right)_2 = \frac{L^2}{\nu}$, $\left(\tau^0_{cor}\right)_3 = \frac{\nu}{(E_{st}\backslash\rho)}$.

3. The critical point and critical Reynolds number

Area 2) is the generation of turbulence. Using stochastic equations of equivalence of measures (4), (5) and stochastic equations of conservation for the area 1), the beginning of the generation of turbulence $r_{c0}(x_c + \Delta x_0, \tau_c + \Delta\tau_0) - r_c$, refering the pair (N, M) =(1,0), we have set (6) of equations of mass, momentum and energy:

$$\left(\frac{d(\rho)_{col,st}}{d\tau}\right)_{1,0} = -\frac{\rho_{st}}{\tau_{cor}}$$

$$\begin{cases}\left(\frac{d(\rho\vec{U})_{col,st}}{d\tau}\right)_{1,0} = -\left(\frac{(\rho\vec{U})_{st}}{\tau_{cor}}\right); \\ div(\tau_{i,j})_{col,st1} = \frac{(\rho\vec{U})_{st}}{\tau_{cor}}\end{cases} \qquad (6)$$

$$\begin{cases}\left(\frac{d(E)_{col,st}}{d\tau}\right)_{1,0} = -\left(\frac{(E)_{st}}{\tau_{cor}}\right)_{1,0}; \\ div(u_i\tau_{i,j})_{col,st1} = \left(\frac{(E)_{st}}{\tau_{cor}}\right)_{1,0}\end{cases}$$

For the flow in the round pipe and on the flat plate [14-16], the critical points are

$$\left(\frac{x_2}{R}\right) = \left[\frac{1}{4}\frac{(E)_{st}\backslash\rho}{U_0^2}\left(\frac{R}{L_U}\right)\right]^{\frac{1}{3}}. \qquad (7)$$

$$\left(\frac{x_2}{\delta}\right) = \left[0.4\frac{(E)_{st} \setminus \rho}{U_0^2}\left(\frac{\delta}{L_U}\right)\right] \tag{8}$$

The value of the first critical Reynolds number in the round pipe and on the flat plate [14-16] are defined as:

$$\mathrm{Re}_{d(U)} = \left\{6\left(\frac{1}{2}\right)^{1/3}\left(\frac{U_0}{\sqrt{E_{st}/\rho}}\right)^{5/3}\left(\frac{L_U}{R}\right)^{1/3}\right\} \tag{9}$$

$$(\mathrm{Re}_X)_1 = \left(\frac{U_0}{(E_{st} \mid \rho)^{1\backslash 2}}\right)^4\left[\frac{3K^2}{8}\right]\mathrm{Re}_{st} \tag{10}$$

Further L is the linear scale of the perturbation or the scale of turbulence. L_y is the scale along axis x_2=y and L_x is the scale along axis x_1=x, x_1 and x_2 coordinates along and normal to the wall. R, U_0, and u_1 are the pipe radius and velocities on the axis and along x_1. δ is the boundary layer thickness, Re_{st} = (u_1L/v), K~0,3. In accordance with (12), (13), and [14-16]. results for the first critical Reynolds number in the round pipe and on the flat plate are reasonably close to the experimental ones found for the first time by Reynolds: $\mathrm{Re}_d = 2300$, $\mathrm{Re}_{x1} = 3.2\ 10^5$.

4. Velocity profiles for turbulent flow in a pipe and on a flat plate

The area 2) $r_{c1}(x_c + \Delta x_0 + \Delta x_1, \tau_c + \Delta\tau_0 + \Delta\tau_1) - r_{c0}$ is the generation of turbulence. It is known that the experimental study of the averaged characteristics of fully developed turbulence showed that the velocity profiles have affine similarity. For example, for classical incompressible fluid or gas in the tube and in the, $\delta_{IIC} = \delta$ is the thickness of the boundary layer, $v=\mu/\rho$, and $U=U_0=U_m$ is the velocity on the boundary layer, we have equations, respectively: $(u_1/U) = (x_2/R)^{1/n}$, $(u_1/U) = (x_2/\delta_{IIC})^{1/n}$. Here, R is the tube radius tube axis and on the border of the boundary layer. Then using the correlator (4), (5) for the area 1) the generation of turbulence, refering the pair (N, M) =(1,1), we have a set of stochastic equations for mass, momentum and energy:

$$\left(\frac{d(\rho)_{col,st}}{d\tau}\right)_{1,1} = -\frac{d\rho_{st}}{d\tau}$$

$$\begin{cases} \left(\dfrac{d(\rho\vec{U})_{col,st}}{d\tau}\right)_{1,1} = -\left(\dfrac{d(\rho\vec{U})_{st}}{d\tau}\right); \\ div(\tau_{i,j})_{col,st2} = \dfrac{(\rho\vec{U})_{st}}{\tau_{cor}} \end{cases} \tag{11}$$

$$\begin{cases} \left(\dfrac{d(E)_{col,st}}{d\tau}\right)_{1,1} = -\left(\dfrac{d(E)_{st}}{d\tau}\right); \\ div(u_i\tau_{i,j})_{col,st2} = \left(\dfrac{d(E)_{st}}{d\tau}\right) \end{cases}$$

Therefore, in the case of the flow in the pipe and on the flat plate using critical point equation (7)-(10), we have the following expressions for «n» in (14) in accordance with [15-18]

$$\left|2\frac{1-n}{n^2}\sqrt{\left(\frac{x_2}{R}\right)^{(2/n)-4}}\right| = \left|\left(\mathrm{Re}_{st} - \frac{1}{\mathrm{Re}_{st}}\right)\right|^{0.5}, \tag{12}$$

.

$$\left|\frac{4(n-1)}{n^2}\left(\frac{x_2}{\delta}\right)^{(1/n)-1}\right| = \left|\left(\mathrm{Re}_{st} - \frac{1}{\mathrm{Re}_{st}}\right)\right|^{0.5} \quad . \tag{13}$$

For example, using and Nikuradze's and Klebanoff's experimental data [19-23] for $\mathrm{Re}_{st} = 10 \div 30$, the calculated values of exponents $n = 6 \div 10$ (12), (13) are in agreement with experimental results. Thus, velocity distributions as a function of initial turbulence for flows in a pipe and along a flat plate are found.

5. Spectral and correlation functions

The area 3) is the diffusion of the turbulence. So, for the area of 3) diffusion, in accordance with [14-16, 24], we have correlator $D_{N,M}(r_c;m_{ci};\tau_c) = D_{1,(1,0)}(r_c;m_{ci};\tau_c)$, (N=p,M=k,l)= (1,(1,0)),which creates the following set

$$\left(\frac{d(\rho)_{st}}{d\tau}\right)_{1,1,0} = -\left(\frac{\rho_{st}}{\tau_{cor}}\right)_{1,1,0};$$

$$\left(\frac{d(\rho\vec{U})_{st}}{d\tau}\right)_{1,1,0} = -\left(\frac{(\rho\vec{U})_{st}}{\tau_{cor}}\right)_{1,1,0}; \tag{14}$$

$$\left|\frac{d(E)_{st}}{d\tau}\right|_{1,1,0} = (R_{zTz})_{1,1,0}\left|\frac{(E)_{st}}{\delta\tau}\right|_{1,(1,0)} .$$

Then using the fractal equation

$$\left|\frac{d(E)_{st}}{d\tau}\right|_{1,1,0} = (R_{zTz})_{1,1,0}\left|\frac{(E)_{st}}{\delta\tau}\right|_{1,(1,0)} . \tag{15}$$

we have two fractal equations.

The first equation is written as

$$\frac{d(E_{st})_j}{d\tau} = -(R_{zTz})_{1;(1,0)}\frac{(E_{st})_j}{\delta\tau} \quad . \tag{16}$$

Then the solution may be written for $(R_{zTz})_{1;(1,0)} = 1$ as an expression for the correlation function in the statistic theory

$$B(\tau) = C\exp\left[-\alpha\tau^2\right] \quad . \tag{17}$$

For $\tau = 0$ we have the correlation function

$$C = (E_{st})_{\tau^1_{cor}} = (E_{st})_{\tau^0_{cor}}\left|\mathrm{Re}_{st} - \frac{1}{\mathrm{Re}_{st}}\right|_{\tau^0_{cor}} \tag{18}$$

So, the spectral function is equal to

$$E(\omega)_j = \frac{C}{\sqrt{\pi\alpha}}\exp\left[-\frac{\omega^2}{4\alpha}\right]. \tag{19}$$

Using (18) for constant C and in the case $\alpha = 0{,}5*\omega_1^2$, we have the following expression

$$E(\omega) = (E_{st})_{\tau^0_{cor}}\left|\mathrm{Re}_{st} - \frac{1}{\mathrm{Re}_{st}}\right| \cdot \left[\Phi(\mathrm{Re}_{st}) - \Phi(1/\mathrm{Re}_{st})\right]^{-1} \cdot \frac{1}{\sqrt{2}\omega_1} \cdot \exp\left(-\frac{\omega^2}{\omega_1^2}\right) \;;\; \left[\Phi(\mathrm{Re}_{st}) - \Phi(1/\mathrm{Re}_{st})\right]^{-1} \approx \sqrt{\frac{4}{\pi}} \tag{20}$$

Estimations of values of turbulent energy for frequency bands $\omega_1 \div \omega_2$ and $\omega_1 \div \omega_3$ are equal to

$$\left|\int_{\omega_1}^{\omega_2} E(w)dw\right| = \left[(E_{st})_{\tau^0_{cor}}\left|\left(\mathrm{Re}_{st}-\frac{1}{\mathrm{Re}_{st}}\right)\right|\right]_{\tau^0_{cor}} \cdot \sqrt{\frac{4}{\pi}}\int_{\omega_2/\omega_1}^{1} e^{-\frac{\omega^2}{2(\omega_1^2)}} d\left(\frac{\omega}{\sqrt{2}\omega_1}\right) = \left[(E_{st})_{\tau^0_{cor}}\left|\left(\mathrm{Re}_{st}-\frac{1}{\mathrm{Re}_{st}}\right)\right|\right]_{\tau^0_{cor}} \cdot \sqrt{\frac{4}{\pi}}(\Phi(1)-\Phi(\omega_2/\omega_1)),$$

$$\int_{\omega_1}^{\omega_3} E(\omega)d\omega = \sqrt{\frac{4}{\pi}}(E_{st})_{\tau^0_{cor}}\left|\mathrm{Re}_{st}-\frac{1}{\mathrm{Re}_{st}}\right| \cdot \int_{1}^{\omega_3/\omega_1} \exp\left(-\frac{\omega^2}{\omega_1^2}\right) \cdot d\left(\frac{\omega}{\sqrt{2}\omega_1}\right) = \sqrt{\frac{4}{\pi}}(E_{st})_{\tau^0_{cor}}\left|\mathrm{Re}_{st}-\frac{1}{\mathrm{Re}_{st}}\right| \cdot \left[\Phi\left(\frac{\omega_3}{\omega_1}\right)-\Phi(1)\right]. \quad (21)$$

Here, $\Phi(\omega_3/\omega_1) - \Phi(\omega_2/\omega_1) = \Phi(\mathrm{Re}_{st}) - \Phi(1/\mathrm{Re}_{st}) \to \sqrt{\pi}/2$.

Widths of the frequency bands equal $\omega_1 - \omega_2 = \omega_1(1+1/\mathrm{Re}_{st})$ and $\omega_3 - \omega_1 = \omega_1(1+\mathrm{Re}_{st})$.

Thus, knowing the parameters of the initial disturbance and using diffusion equation (18), which is derived from relations of equivalence of measures, in the case $(R_{zTz})_{1;(1,0)} = 1$, it is possible to write the value for the spectral function, which is well known in the statistical theory. The second fractal equation is written as

$$\frac{d(E_{st})_j}{d\tau} = +(R_{zTz})_{1;(1,0)}\frac{(E_{st})_j}{\delta\tau}. \quad (22)$$

This expression is the same as the equation for the square of the amplitude of oscillations $|A|^2$ in the theory of the turbulence by Landau:

$$\frac{d|A|^2}{d\tau} = 2\gamma|A|^2 - \delta|A|^4.$$

We expand the right side of equation (22) in a power series. Then we have in the first approximation

$$\frac{(R_{zTz})}{\delta\tau}(E_{st})_j = 2\gamma(E_{st})_j - \delta(E_{st})^2{}_j. \quad (23)$$

So, in the case of the equation

$$\frac{d(E_{st})_j}{d\tau} = 2\gamma(E_{st})_j, \quad (24)$$

the parameter in the right side of the equation may be written as $\gamma = \frac{(R_{zTz})_{1;(1,0)}}{\delta\tau}$. Thus, equation (24) includes Landau's equation as a special case. Solutions of equation (24) for the function

$$d\tau = \frac{d\tau}{\tau_{cor}} \cdot f(\tau);$$

$$f(\tau,\tau_{cor}) = \frac{1}{\tau_{cor}}\left(\frac{\tau}{\tau_{cor}}\right)^m, \quad m = 0,1,..... \quad (25)$$

in the case m = 0, m = 1 may be written as

$$(E_{st})_j = C_L \exp\left[(R_{zTz})_{1;(1,0)}\frac{|\tau|}{\tau^1_{cor}}\right], \quad (26)$$

$$(E_{st})_j = C_L \exp\left[\frac{1}{2}(R_{zTz})_{1;(1,0)}\left(\frac{|\tau|}{\tau^1_{cor}}\right)^2\right] \quad (27)$$

Thus, the spectrum of type delta function is converted into the continuous spectrum because of the sequential processes of the fractal diffusion, with the solutions derived above, and the fractal generation of turbulence with the solutions (26),(27). It should be noted that the functional dependence

$$f(\tau,\tau_{cor}) = \frac{1}{\tau_{cor}}\left(\frac{\tau}{\tau_{cor}}\right)^m, \quad m = 0,1,..... \quad (28)$$

for equation $\left|\frac{dE_{st}}{d\tau}\right| = (R_{zTz})_{1,1,0}\left|\frac{E_{st}}{\delta\tau}\right|$ allows us to find solutions satisfying the requirement of the continuous spectrum, but this dependence is not the exclusion, there is the following functional dependence

$$f(\tau,\tau_{cor}) = \frac{1}{\tau_{cor1} - \tau}. \quad (29)$$

This dependence allows us to define the solution, satisfying the requirement of the continuous spectrum for the hydrodynamic turbulence. Then, the integral of the fractal equation determines the solution

$$(E_{st})_j = C\exp\left[1 - \frac{|\tau|}{\tau^1_{cor}}\right] \quad ;|\tau| \le \tau^1_{cor} \quad . \quad (30)$$

This solution coincides with the expression for the correlation function

$$B(\tau) = C(1-\alpha|\tau|); \quad |\tau| \le \tau_{cor1} \quad .$$
$$B(\tau) = 0 \quad |\tau| > \tau_{cor1} \quad (31)$$

In addition, it is important to note that equation (26) admits the record as a particular case of the Bessel equation with the relevant decision containing the Bessel function of the second kind $K_\nu(\alpha\cdot\tau), \alpha = (\tau_{cor1})^{-1}$. Then we have the solution

$$(E_{st})_j(\tau) = C(\frac{\tau}{\tau_{cor1}})^\nu K_\nu(\frac{\tau}{\tau_{cor1}}) \quad . \quad (32)$$

This corresponds to the known expression for the correlation function, and taking into account expression for C (22), this solution may be written as

$$B(\tau) = C(\alpha\cdot\tau)^\nu K_\nu(\alpha\cdot\tau) \quad . \quad (33)$$

Thus, the physical regular pattern of the equivalence of measures and new stochastic equations for the region of the turbulence diffusion allow us to derive solutions for spectral and correlation functions depending on initial turbulence.

The fact is most important that even taking into account only the first term of the fractal equation of the Landau type, we can do the following conclusion. In the case even if fluctuation has the initial spectrum as the delta function, fractal equation (15) determines diffusion process (16), and the subsequent expansion of the spectrum with its further generation (19), due to the created gradient. Thus, fluctuation with the initial spectrum as the delta function is converted into a continuous spectrum in the implementation of successive processes of fractal diffusion according to equations (15-19) and generation according to equations (22-33).

6. Profiles of turbulent characteristics in the boundary layer

In addition, in the area 3), $r_{c2}(x_c + \Delta x_0 + \Delta x_1 + \Delta x_2, \tau_c + \Delta\tau_0 + \Delta\tau_1 + \Delta\tau_2) - r_{c1}$ or the diffusion of turbulence profiles of turbulent characteristics, may be defined as follows. The expression for the dynamic speed u^* equals

$$u^* = \sqrt{\left[\rho\nu\left(\frac{du_1}{dx_2}\right)\right]_w / \rho} = u_L \frac{x_2}{\delta_L} = \sqrt{\nu\frac{U_0}{\delta_L}\left(\frac{\delta_L}{\delta}\right)^{1/7}} = \sqrt{\nu\frac{U_0}{0{,}1\delta}\left(\frac{0{,}1\delta}{\delta}\right)^{1/7}} = \sqrt{\nu\frac{U_0}{\delta}(0{,}1)^{-6/7}} = 2{,}63\sqrt{\nu\frac{U_0}{\delta}} \qquad (34)$$

δ_L, u_L is thickness of the viscous sublayer and the velocity at the boundary of δ_L.

Then the value (q/u^*)**2 for the critical point is equal to

$$\frac{q}{(u^*)^2} = \frac{\left(\left[\sqrt{E_{st}/\rho}\right]_0\right)^2\left(\mathrm{Re}_{st,0} - \frac{1}{\mathrm{Re}_{st,0}}\right)}{\left[\rho\nu\left(\frac{du_1}{dx_2}\right)\right]_w/\rho} = \frac{\left(\left[\sqrt{E_{st}/\rho}\right]_0\right)^2\left(\mathrm{Re}_{st,0} - \frac{1}{\mathrm{Re}_{st,0}}\right)}{\left(2{,}63\sqrt{\nu\frac{U_0}{\delta}}\right)^2} = \frac{\left(\left[\sqrt{E_{st}/\rho}\right]_0\right)^2\left(\mathrm{Re}_{st,0} - \frac{1}{\mathrm{Re}_{st,0}}\right)}{6{,}92 * \nu\frac{U_0}{\delta}} \qquad (35)$$

Here q is the kinetic energy of the turbulence at the critical point. For the first critical number $\mathrm{Re}_x = \frac{x_1 U_0}{\nu} \approx 3*10^5$ and for the initial turbulent Reynolds number $\mathrm{Re}_{st,0}$, we determine the value $\mathrm{Re}_\delta = \frac{\delta U_0}{\nu}$. An index of 0 corresponds to the initial turbulence. Re_δ is Reynolds number calculated for the thickness of the boundary layer.

$$\mathrm{Re}_{\delta^*} = 950 \Rightarrow \rho U\delta^* = \int\rho(U-u)dy \Rightarrow \delta^* = \int_0^\delta \left(1 - \left(\frac{y}{\delta}\right)^{1/n}\right)dy = \delta - \frac{n}{n+1}\delta = \delta\frac{1}{n+1} \Rightarrow$$

$$\frac{\delta^*}{\delta} = \frac{1}{n+1} \Rightarrow \frac{\mathrm{Re}_{\delta^*}}{\mathrm{Re}_\delta} = \frac{1}{n+1} \Rightarrow n = 7 \div 8 \Rightarrow \mathrm{Re}_\delta = \mathrm{Re}_{\delta^*}(n+1) = 950 * (8 \div 9) = 7600 \div 8550$$

Then we have next values

$$\frac{q}{(u^*)^2} \approx 11{,}5 \qquad (36)$$

$$\frac{q_{xy}}{(u^*)^2} \approx 1{,}79 \qquad (37)$$

Here q_{xy} is turbulent stress. Values in (36)-(37) are in satisfactory agreement with the data [13, 14].

Values (q_{xy}/q) =0.08-0.2 are in agreement with experimental values [14-17].

7. The coefficient of friction along a smooth flat plate C_f and the coefficient of drag λ in a round smooth tube

According to [25], using expressions for the critical point and the critical Reynolds number and the dependences for the index of the velocity profile «n» as well as taking into account solutions for region 3, formulas for the friction of the isothermal flow in a pipe in accordance with the range of the experimental values of turbulence intensity Tu=1÷6 were derived. For a range $5\cdot 10^4 < \mathrm{Re}_d < 1\cdot 10^7$,

$$\lambda \approx 0{,}09 \cdot \left(\mathrm{Re}_d\right)^{-\frac{1}{7}}. \qquad (38)$$

Accordingly, for Tu ~ 4.5%, for the range $3\cdot 10^3 < \mathrm{Re}_d < 3\cdot 10^4$,

$$\lambda \approx 0{,}125 \cdot \left(\mathrm{Re}_d \right)^{-\frac{1}{7}} \quad (39)$$

Elementary calculations show that formulas (38) and (39) give satisfactory agreement with the data for a wider range of numbers Re [19-23]. Additionally, in accordance with [25], a value for the local drag coefficient in the case of the flow on a flat plate is

$$C_f^{'} = 0.0252 \cdot \left(\mathrm{Re}_{x_1} \right)^{-1/7} = 0.0252 \cdot \left(\mathrm{Re}_{x_1} \right)^{-0.14285} \quad (40)$$

Then we have next equations for the drag coefficient for a flat plate of finite length « »

$$C_f^{'} = 0.0302 \cdot \left(\mathrm{Re}_l \right)^{-1/7} = 0.0302 \cdot \left(\mathrm{Re}_l \right)^{-0.14285} \quad (41)$$

$$C_f^{'} = 0.0313 \cdot \left(\mathrm{Re}_l \right)^{-1/7} = 0.0313 \cdot \left(\mathrm{Re}_l \right)^{-0.14285} \quad (42)$$

Checking of formula (41) shows satisfactory agreement with the experimental data for the region $3 \cdot 10^6 < \mathrm{Re}_l < 1 \cdot 10^9$. For the range of turbulent flow $3 \cdot 10^5 < \mathrm{Re}_l < 3 \cdot 10^6$ that corresponds to experiments with a very narrow transition area, formula (42) yields an error less than ~6% for $\mathrm{Re}_l = 3 \cdot 10^5$ and less than 5.2% for $\mathrm{Re}_l = 1.10^6$ [25].

8. Conclusions

The main results of a new stochastic theory of turbulence for isothermal flows are presented. Based on sets of stochastic equations for laws of conservation and for equivalence of measures between deterministic process and random process, the review of results of analytical expressions are presented for the main parameters of the isothermal turbulent flow in a tube and for the flow on a flat plate depending on initial fluctuation in the medium. It is shown that derived formulas for critical Reynolds numbers and critical point, for the velocity profile, for the second-order correlation, and friction coefficients give the satisfactory agreement with experimental data for both flows. Correlation functions and spectral functions are derived by using equations, which include the equation of Landau in themselves as a special case, when fractal function equals unity. As a result, the fluctuation with the initial spectrum as the delta function is converted into a continuous spectrum in the implementation of successive processes of fractal diffusion according to equations (15-19) and generation according to equations (22-33). Such results were not predicated by another theories using the single law for the turbulence.

References

[1] Kolmogorov A N 1940, *Dokl. Akad. Nauk* **26** (1), pp 6
[2] Landau L D 1944 *Dokl. Akad. Nauk* **44** (8) pp 339
[3] Lorenz E N 1963 *J. Atmos. Sci.* **20** pp 130
[4] Ruelle D and Takens F 1971 *Communs. Math. Phys.* **20** pp 167
[5] Klimontovich Yu L 1989 *Usp. Fiz. Nauk* **158** B1, pp 59
[6] Arnol'd V I 1990 *Theory of Catastrophes* (Nauka: Moscow)
[7] Haller G 1999 *Chaos Near Resonance* (Springer: Berlin)
[8] Orzag S A and Kells L C 1980 J. *Fluid Mech.* **96** (1), pp 159
[9] Newton P K 2016 *J. Fluid Mech.* **786**, pp 1–4
[10] Barkley D *J. Fluid Mech.*, vol. **803**, 2016. DOI:10.1017/jfm.2016.465
[11] Dmitrenko A V 2008 *Fundamentals of heat and mass transfer and hydrodynamics of single phase and two -phase media. Criterial, integral, statistical methods and direct numerical simulation* (Moscow: Galleya print), http://search.rsl.ru/ru/catalog/record/6633402.
[12] Dmitrenko A V 2007 *Dokl. Phys.* **52** (7) pp 384

[13] Fursikov A V 1992 *Izv. Ross. Akad. Nauk*, **56**, No. 6, pp 1273
[14] Dmitrenko AV 2013 *Dokl. Phys.* **58** (6) pp 228
[15] Dmitrenko A V 2013 Equivalent measures and stochastic equations for determination of the turbulent velocity fields and correlation moments of the second order, *Abstr. Int. Conf. "Turbulence and Wave Processes,"Lomonosov Moscow State University, November 26−28, 2013, Moscow(2013)* pp. 39–40; http://www.dubrovinlab.msu.ru/turbulencemdm100ru.1576
[16] Dmitrenko A V 2013 *Theory of Equivalent Measures and Sets with Repeating Denumerable Fractal Elements. Stochastic Thermodynamics and Turbulence. Determinacy–Randomness Correlator* [in Russian], (Galleya-Print, Moscow), p 226, URL: http://search.rsl.ru/ru/catalog/record/6633402.
[17] Dmitrenko A V 2013 *Regular Coupling between Deterministic (Laminar) and Random (Turbulent) Motions —Equivalence of Measures*, Scientific Discovery Diploma No.458, registration No.583 of December 2,.
[18] Dmitrenko AV 2015 *Journal of Engineering Physics and Thermophysics* **88** (6) pp 1569
[19] Davidson P A 2004 *Turbulence* (Oxford Univ. Press) p 678
[20] Millionshchikov M D 1969 *Turbulent flow in boundary layers and in pipes* (Moscow: Nauka)
[21] Schlichting H 1968 *Boundary-Layer Theory* (6th Edition, McGraw-Hill)
[22] Monin A S and Yaglom A M 1971 *Statistical Fluid Mechanics* (M.I.T. Press.)
[23] Hinze J O 1975 *Turbulence* (McGraw-Hill.)
[24] Dmitrenko A V 2016 *Int. J. Fluid Mech. Res.* **43**(3), 82, doi:10.1615/InterJFluidMechRes.v43i3.
[25] Dmitrenko A V 2017 *Continuum Mech. Thermodyn.*, **29**(1),1 doi:10.1007/s00161-017-0566-x.

Acknowledgements This work was supported by the program of increasing the competitive ability of National Research Nuclear University MEPhI (agreement with the Ministry of Education and Science of the Russian Federation dated August 27, 2013, Project No.02.a03.21.0005).

Dedicated to the memory of Academician Yu. A. Ryzhov.

Computing experiment for unloading dumper truck at a sloping pad

Yu N Baryshnikov

Bauman Moscow State Technical University, Moscow, Russia
E-mail: mhts@list.ru

Abstract. The results of a computational experiment on unloading a dump truck are described. The experiment was carried out on the mathematical model of a car built on the basis of Lagrange's equations. Before the movement (unloading) of cargo, the lifting of the platform is considered as a quasi-static process. From the equations of the equilibrium of the car and geometric relationships, analytical expressions are derived for calculating the force interaction between the platform and the frame. When the cargo moves as a monolithic block, the mathematical model is represented in the form of a system of differential equations. After the adoption of a number of assumptions, its solution was substantially simplified. An algorithm for solving the problem is proposed that implements the method of incremental change of parameters. The results of calculation of forces in hydraulic cylinders during unloading of BelAZ mining dump truck are given.

1. Introduction

Determination of design loads is an important part of the strength analysis of any design. In many fields of technology, much attention has been paid to this issue. So, for example, for aircraft, ships, railway cars and subway cars, the main load regimes have been established and methods for calculating loads have been developed, which are approved by the relevant regulatory documents. In the automotive industry, there are no such standards yet. This situation can be explained both by the variety of loading regimes and by the complexity of calculating the operating loads.

One of these modes is hitting cars on uneven roads. The studies of loads in this case are devoted to the work of a number of authors [1] - [4]. Another calculation case, in particular for dump trucks, is the unloading of the cargo platform. Experience shows that within a day dump trucks can perform more than twenty loading and unloading operations, and over the entire period of operation - more than ten thousand. This can serve as one of the reasons for the destruction of their supporting structure [5]. That is why the calculation of the force interaction between the platform and the frame during unloading is of particular interest.

An attempt to numerically simulate the dump truck unloading process was done in [6]. However, it considers a simplified model obtained on the basis of the d'Alembert principle. In this connection, further study of the dump truck unloading process requires the creation of more sophisticated mathematical models.

Content from this work may be used under the terms of the Creative Commons Attribution 3.0 licence. Any further distribution of this work must maintain attribution to the author(s) and the title of the work, journal citation and DOI.
Published under licence by IOP Publishing Ltd

IOP Conf. Series: Materials Science and Engineering **468** (2018) 012022 doi:10.1088/1757-899X/468/1/012022

2. A mathematical model of the unloading of the damper truck

Usually the unloading process consists of two stages. At the first stage, the cargo remains stationary. In the second stage, the movement of the cargo begins when the platform is raised to the angle of the natural slope and ends with its full unloading.

Consider lifting the platform with a fixed cargo. The normal rise time of the platform is from 20 to 30 seconds, with the angular velocity of the platform turning small, constant and does not exceed 0.03-0.05 sec^{-1}. Therefore, lifting the platform with a fixed weight can be considered as a quasi-static process. In Figure 1 shows the calculation scheme for unloading a dump truck.

We denote by m_1 and m_2 - the mass of the platform 1 and the cargo 2; φ - platform inclination angle; α - angle of inclination of the hydraulic cylinder; s - movement of cargo; (X_K, Y_K) and (x_K, y_K) - the coordinates of the center of mass of the platform and the cargo in the stationary $OXYZ$ and in the mobile $Oxyz$ coordinate system, respectively.

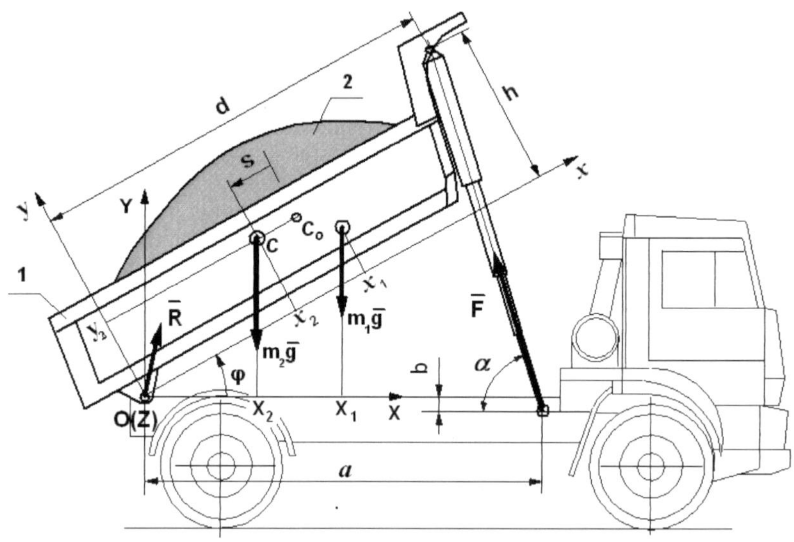

Figure 1. Calculation scheme for unloading the truck.

Before the movement of the cargo, as well as in the case of its adhesion (example), the force F of the hydraulic cylinder is found from the condition that the sum of the moments of all forces relative to the point O be zero.

$$F = \frac{m_1 g X_1 + m_2 g X_2}{a \sin\alpha - b\cos\alpha}. \tag{1}$$

Here the X_1 and X_2 coordinates centers of mass of the platform and cargo in a fixed coordinate system $OXYZ$.

Using the equations of coupling of a moving $Oxyz$ and a fixed $OXYZ$ coordinate system, we obtain:

$$X_K = x_K \cos\varphi - y_K \sin\varphi, \qquad (k = 1, 2) \tag{2}$$

In turn, the angle of inclination of the hydraulic cylinder is determined from the geometric relationships:

$$\alpha = arctg\left(\frac{b + d\sin\varphi + h\cos\varphi}{a - d\cos\varphi + h\sin\varphi}\right). \tag{3}$$

Thus, using equations (1) - (3), you can calculate the force F of the hydraulic cylinder before the movement of the cargo at any angle φ of inclination of the platform. The total reaction R in the hinged support O is also found from the equilibrium condition of the platform:

$$R_X = F\cos\alpha; \quad R_Y = m_1 g + m_2 g - F\sin\alpha; \quad R = \sqrt{R_X^2 + R_Y^2} \tag{4}$$

When the platform is raised to the angle of a natural slope ($\varphi \approx 30^{\circ}$), the cargo comes into motion. Let us consider the case when the cargo moves as a monolithic block symmetric with respect to the symmetry plane of the car (Figure 1). To derive the differential equations of motion of such a system, we use the Lagrange equations of the second kind

$$\frac{d}{dt}\left(\frac{\partial T}{\partial \dot{q}_i}\right) - \frac{\partial T}{\partial q_i} = Q_i, \tag{5}$$

The platform-cargo system has two degrees of freedom, which correspond to two generalized coordinates: the angle of the platform $q_1 = \varphi$ and the movement of the cargo $q_2 = s$ .The kinetic energy of the system is found as the sum of the kinetic energies of the platform and the cargo:

$$T = \frac{1}{2} J_{OZ}^{(1)} \dot{\varphi}^2 + \frac{1}{2} m_2 v_C^2 + \frac{1}{2} J_{CZ}^{(2)} \dot{\varphi}^2, \tag{6}$$

where $J_{OZ}^{(1)}$ - moment of inertia of the platform relative to the axis OZ ; $J_{CZ}^{(2)}$ - moment of inertia of the cargo relative to the main central axis CZ .

The speed v_C of the center of mass C of the cargo is found from the velocity vector diagram (Figure 2)

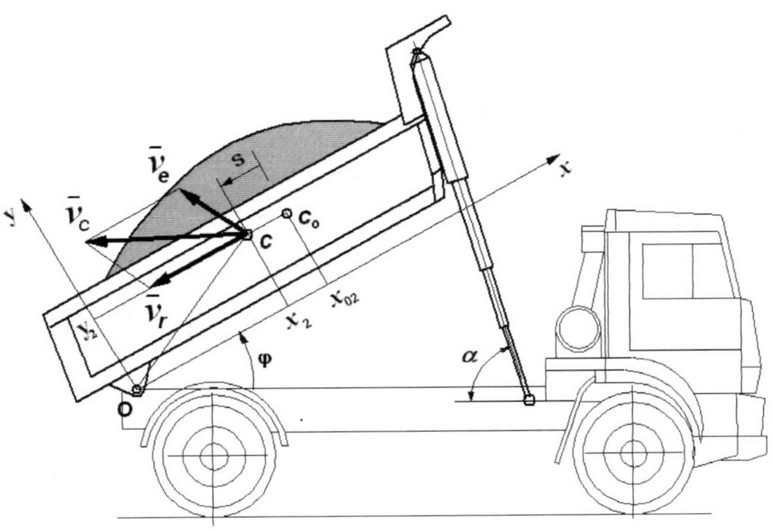

Figure 2. Vector diagram of cargo speeds.

$$v_C^2 = v_r^2 + v_e^2 + 2 v_r v_e \cos(\bar{v}_r, \hat{\ } \bar{v}_e) \tag{7}$$

Here $v_r = \dot{s}$ is the relative velocity of the center of mass of the cargo; $v_e = OC \cdot \dot{\varphi}$ - the portable speed of movement of the center of mass of the cargo. After the substitution v_r and v_e formula (7) takes the form

$$v_C^2 = \dot{s}^2 + \left(x_2^2 + y_2^2\right)\cdot\dot{\varphi}^2 + 2y_2\dot{s}\dot{\varphi}, \tag{8}$$

Finally, the kinetic energy of the platform-cargo system is

$$T = \frac{1}{2}\left[J_{oz}^{(1)} + m_2\left(x_2^2 + y_2^2\right) + J_{cz}^{(2)}\right]\cdot\dot{\varphi}^2 + m_2 y_2\dot{s}\dot{\varphi} + \frac{1}{2}m_2\dot{s}^2 \tag{9}$$

Generalized forces entering the right-hand sides of equations (5), we obtain in the form

$$Q_1 = -m_1 g\cdot X_1 - m_2 g\cdot X_2 + Fa\sin\alpha - Fb\cos\alpha; \tag{10}$$

$$Q_2 = m_2 g\sin\varphi - f(m_2 g\cos\varphi - 2m_2\dot{\varphi}\cdot\dot{s}), \tag{11}$$

where f is the sliding friction coefficient of the cargo along the platform.

Substituting these expressions into Lagrange's equations (5), we write the differential equations of motion of the platform-cargo system:

$$\left[J_{oz}^{(1)} + m_2\left(x_2^2 + y_2^2\right) + J_{cz}^{(2)}\right]\cdot\ddot{\varphi} + m_2 y_2\ddot{s} = -m_1 gX_1 - m_2 gX_2 + F(a\sin\alpha - b\cos\alpha); \tag{12}$$

$$m_2 y_2\ddot{\varphi} + m_2\ddot{s} + m_2 x_2\dot{\varphi}^2 = m_2 g(\sin\varphi - f\cos\varphi) - 2m_2 f\dot{\varphi}\cdot\dot{s}. \tag{13}$$

To solve the resulting system of differential equations, a numerical integration procedure is required, for example, based on the Runge-Kutta method.

However, these equations can be simplified. Indeed, since the angular velocity of rotation of the platform is small and constant, then $\ddot{\varphi}=0$ and $\dot{\varphi}^2 \approx 0$. In addition, the last term in Eq. (13) has a second order of smallness and can also be neglected. Then the differential equations (12) and (13) can be transformed to the form

$$m_2 y_2\ddot{s} = -m_1 gX_1 - m_2 gX_2 + F(a\sin\alpha - b\cos\alpha); \tag{14}$$

$$\ddot{s} = g(\sin\varphi - f\cos\varphi). \tag{15}$$

Eliminating the acceleration $\ddot{s}$ of the cargo from equation (14), we obtain the force of the hydraulic cylinder

$$F = \frac{m_1 gX_1 + m_2 gX_2 + m_2 g\cdot y_2(\sin\varphi - f\cos\varphi)}{a\sin\alpha - b\cos\alpha}. \tag{16}$$

To calculate the force of the hydraulic cylinder, we apply a step-by-step change of parameters. Starting with the angle of a natural slope, with a step $\Delta\varphi = 1^{\circ}$ from equation (15) we find the speed and the movement of the cargo. Then calculate the coordinate of the center of mass of the moving cargo $(x_2 = x_{02} - s)$ and from the communication equations (2) determine the coordinates X_1 and X_2 centers of mass of the platform and the cargo. From the equation (3) we find the angle of inclination of the hydraulic cylinder. We substitute all the obtained values into the formula (16). The process will be repeated until the angle of inclination of the platform reaches the maximum.

We find the total reactions in the hinged supports using the d'Alembert principle

$$R_X = F\cos\alpha - m_2\ddot{s}\cos\varphi; \qquad R_Y = m_1 g + m_2 g - F\sin\alpha - m_2\ddot{s}\sin\varphi$$

$$R = \sqrt{R_X^2 + R_Y^2} \tag{17}$$

IOP Conf. Series: Materials Science and Engineering **468** (2018) 012022 doi:10.1088/1757-899X/468/1/012022

3. The results of the calculation of the loads of dump trucks BelAZ

Based on the proposed mathematical model, the forces of the hydro cylinders of the tilting mechanism of the BelAZ dump truck platform are calculated. The cases of lifting the platform with a fixed cargo and unloading the cargo as a monolithic block were investigated. The results of the calculation are shown in Figure 3.

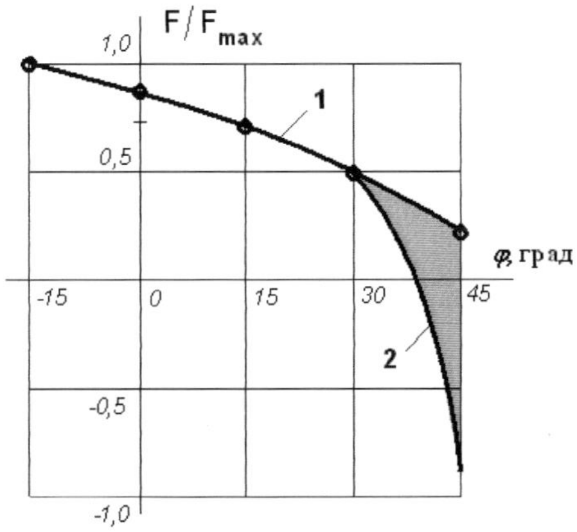

Figure 3. Efforts of the hydraulic cylinder during the lifting of the BelAZ platform with a fixed cargo (1) and with unloading of a monolithic block (2).

Graphs of effort F in hydraulic cylinders (curves 1 and 2) correspond to the two specified extreme cases that arise when unloading the platform. The analysis of the obtained results showed that in regular cases, when the cargo is bulk material, the force F change schedules are located in the area selected by hatching. Maximum effort F_{Max} in the hydraulic cylinders is observed at the beginning of the platform lift $(\varphi = -15^{\text{o}})$. When the platform is raised more than 40° an angle in the hydraulic cylinders, tensile forces may occur which can lead to the vehicle tipping over.

4. Unloading dumper truck at a sloping pad

Now we will examine the process of unloading the truck on an inclined platform. The calculation scheme of the car in this case is shown in Figure 4.

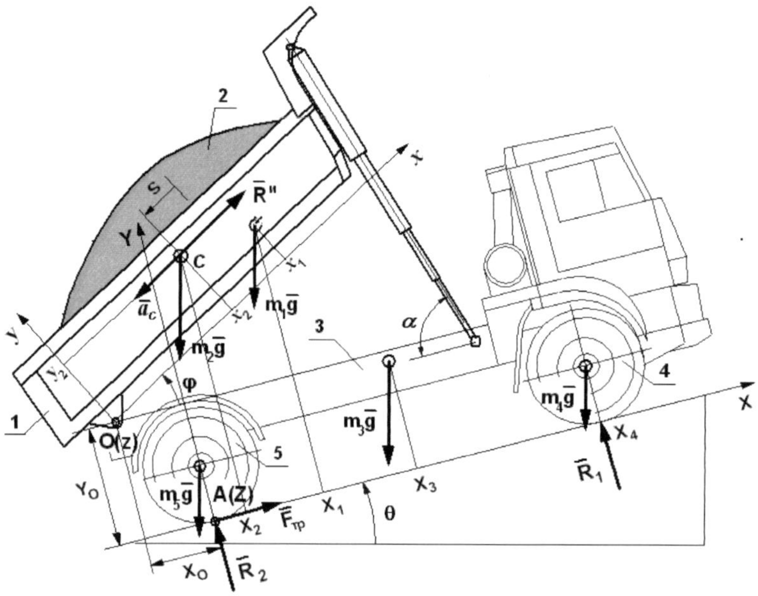

Figure 4. Calculation scheme for unloading the truck on an inclined platform.

Here is the m_1 - mass of the platform 1; m_2 - mass of cargo 2; m_3 - mass of other sprung aggregates 3; m_4 and m_5 - the mass of the front 4 and rear 5 bridges of the car; φ - platform elevation angle; θ - angle of inclination of the unloading site; s - movement of cargo; R_1 and R_2 - the total reaction on the wheels of the front and rear axles of the car.

To estimate the longitudinal stability of the truck, we use the d'Alembert principle:

$$\sum M_Z(m_\kappa \bar{g}) + \sum M_Z(\bar{R}_\kappa) + \sum M_Z(\bar{R}''_\kappa) = 0 \tag{18}$$

In our case, Eq. (18) takes the form:

$$-g\cos\theta\sum_{\kappa=1}^{4} m_\kappa X_\kappa + g\sin\theta\sum_{\kappa=1}^{5} m_\kappa Y_\kappa + R_1 X_4 - R''\cos\varphi\cdot Y_2 + R''\sin\varphi\cdot X_2 = 0, \tag{19}$$

where X_K and Y_K are the coordinates of the centers of mass of the platform and cargo $(k = 1, 2)$; $\bar{R}'' = -m_2\bar{a}_C$ - the main vector of inertia forces of the cargo $(a_C = \ddot{s})$.

Note that the main moment of forces of inertia of the load is equal to zero, since angular velocity of the platform is constant.

If the angle of inclination of the platform exceeds the critical one, the car will lose its stability and overturn. At the same time, the front wheels lose contact with the platform and the reaction on the wheels of the front axle R_1 turns to zero. Then we rewrite equation (19) in the following form:

$$g\cos\theta^*\sum_{\kappa=1}^{4} m_\kappa X_\kappa - g\sin\theta^*\sum_{\kappa=1}^{5} m_\kappa Y_\kappa + m_2\ddot{s}(Y_2\cos\varphi - X_2\sin\varphi) = 0. \tag{20}$$

To determine the critical angle θ^* of the slope of the site, we apply a step-by-step change of parameters. By incrementally increasing the angle θ, we simulate the rise of the platform at each step of changing the angle of inclination of the platform. When the sum of the angles $(\varphi+\theta)$ becomes equal to the angle of the natural slope, the movement of the cargo will begin. We calculate the acceleration of the center of mass of the cargo. To this end, we use equation (15), where the slope of the platform is replaced by $(\varphi+\theta)$:

$$\ddot{s} = g[\sin(\varphi+\theta) - f\cos(\varphi+\theta)]. \tag{21}$$

In turn, the coordinates of the centers of mass of the platform and the cargo at each step of the platform lifting can be found by means of the rotation matrix

$$X_{\kappa} = x_{\kappa}\cos\varphi - y_{\kappa}\sin\varphi - X_O; \tag{22}$$

$$Y_{\kappa} = x_{\kappa}\sin\varphi + y_{\kappa}\cos\varphi + Y_O, \tag{23}$$

where X_O and Y_O are the coordinates of the center of the hinged support (point O).

Substituting these values into equation (20), we check the condition for the detachment of the front wheels. If the sum of the moments entering into the specified equation will be less or equal to zero, the angle θ^{*} of an inclination of a platform has reached critical value and the dump truck will overturn. Otherwise, increase the angle θ of the platform by one step and repeat the platform unloading procedure.

5. Conclusion

As a result of the computational experiment, the force interaction of the platform and the frame of the dumper truck during the unloading of the monolithic block was investigated. The results of calculations of these loads during unloading of the BelAZ mining dump truck are presented. A mathematical model and an algorithm for calculating the maximum permissible inclination angle of the platform, excluding the overturning of the dump truck, is developed. The obtained results can be used in calculating the strength of the carrier system of dump trucks. The proposed approach allows carrying out multivariate calculations of the car with the purpose of choosing the optimal constructive solution.

References

[1] Erz K 1957 Uber die durch Unebenheiten der Fahrbauhn hervorgerufene Verdrehung von Strassenfahrzeugen *Z. Automobilltechnik* Bd 59 **4** pp 89-96
[2] Pavlovsky Ya 1977 *Automobile bodies*: trans. with Polish. (Moscow: Mashinostroenie) p 544
[3] Teser E 1979 *The body of heavy vehicles*: trans. with Polish. (Moscow: Mashinostroenie) p 232
[4] Baryshnikov Yu N 2014 Express analysis of loads when driving a car on uneven roads *Electronic J. Science and Education* (Moscow: BMSTU) **8** pp 224-236, DOI: 10.7463 /0814.0724703
[5] Belokurov V N, Gladkov O V, Zakharov A A and Melik-Sarkisyants A S 1987 *Dump trucks* ed. Melik-Sarkisyants (Moscow: Mashinostroenie) p 216
[6] Baryshnikov Yu N 2015 Numerical modeling of the process of unloading dump trucks *J.Natural and technical sciences* **11** pp 57-59

IOP Conf. Series: Materials Science and Engineering **468** (2018) 012023 doi:10.1088/1757-899X/468/1/012023

Determination and investigation of asteroid Apophis′ trajectories set potentially colliding with the Earth in 2036

Peng Guo[1], V V Ivashkin[1, 2], C A Stikhno[2] and P M Shkapov[2]

[1] Keldysh Institute of Applied Mathematics, RAS, Moscow, Russia
[2] Bauman Moscow State Technical University, Moscow, Russia
E-mail: ivashkin@keldysh.ru

Abstract. In the paper, we have computed a set of the asteroid Apophis' trajectories, which potentially result in the collisions with the Earth in 2036. Using the developed method and algorithms for determination of these collision trajectories and their impact points on the surface of the Earth, we have obtained its possible impact's path of risk on the world map. We have studied the characteristics in energy, impact time and geometry for these Apophis' trajectories and this impact area. Comparison with results of other authors on this issue shows that they coincide well enough.

1. Introduction

Asteroid (99942) Apophis is well known as one of the most hazardous near-Earth asteroids (NEA) in recent years. Since its first discovery by R. Tucker, D. Tholen and F. Bernardi on June 19, 2004, determination of its initial orbit with the use of optical and radar measurements, estimation of its Earth-impact probability in 2029 and also in the next several decades due to resonant return to the Earth have been studied in many works with different methods. It is shown that Apophis will pass close to the Earth at a distance about 38000 km from the Earth's center on April 13, 2029 and has a low possibility to pass through a "keyhole" about 600 m wide on the target plane of the close encounter, which will lead to a resonant return collision on April 13, 2036 [1–19].

In 2006, Schweickart et al. from the B612 foundation made estimates of the Apophis's possible impact path of risk (PoR) for 2036 [3]. It is shown that the impact PoR extends across many national borders, such as Russia, Nicaragua, Costa Rica, Colombia, Venezuela, etc. In our previous works [6–9] the characteristics of the possible prevention of collision with the Earth and correction of the Apophis' orbit were analyzed. Using the standard ORBFIT software package, Włodarczyk I. computed impact solutions of Apophis up to 2110 based on new observations in 2007 and 2013 and presented the possible impact path of risk for 2036, 2068, 2099 and 2102 [11–13]. Sokolov et al. [14–16] obtained potential impact trajectories of Apophis in the 21st century and investigated the characteristics of potential collisions (the minimum geocentric distance, the relative position of the semimajor axes of the keyholes at the initial time in 2006 and the dimensions of the keyholes).

The purpose of our study is to determine and investigate characteristics of collision trajectories of Apophis and its possible impact area on the Earth's surface in 2036 based on the initial orbit and its uncertainty estimated by IAA RAS in 2005 [1]. This work is an extension of our previous work [10].

Content from this work may be used under the terms of the Creative Commons Attribution 3.0 licence. Any further distribution of this work must maintain attribution to the author(s) and the title of the work, journal citation and DOI.
Published under licence by IOP Publishing Ltd

This task has been solved in the next two stages. Firstly, we have developed algorithms to search for all set of the Apophis' impact trajectories [6–10]. For this, we have found several sets of impact trajectories, for each of which the perigee distance of orbit is close to some fixed values from the range

$$r_\pi=\{2069\div2100;\ 2500\pm1;\ 3000\pm1;\ldots;\ 6360\pm1;\ 6375\pm1\}\ \text{km}. \quad (1)$$

Secondly, we have determined the intersection points of these trajectories with the Earth's surface. Based on this, we have built the impact path of risk on the world map. We have studied some geometric, temporal and energy characteristics of the Apophis' collision trajectories and the impact area on the Earth, in particular, its width and length, angle of entry into the atmosphere, the time and speed of collision, and also the direction of the asteroid's approach to the Earth.

2. Determination of the asteroid (99942) Apophis' impact trajectories set

2.1. The initial orbital parameters and dynamical model of Apophis

The orbit of the asteroid (99942) Apophis was computed by numerically integrating the differential equations of N-body motion model of Apophis in the heliocentric geoequatorial non-rotational coordinate system OXYZ at the J2000 epoch. In order to improve the accuracy of determining the trajectory of the asteroid Apophis, we have taken into account several perturbations, including the Newtonian attraction of the planets, the Moon and Pluto, as well as the solar radiation pressure, and the Earth's oblateness-J2. The celestial bodies' positions have been computed using the JPL planetary Ephemerides DE405. The influence of relativistic effects and Yarkovsky effect has been ignored in this work. The integration has been performed by the 8th order predictor-corrector method [20].

As the nominal initial parameters, we used the position vector $\mathbf{r}_n$ and velocity vector $\mathbf{v}_n$ of the asteroid and their 3σ uncertainties at epoch t_0: Jan. 30 2005=JD 2453400.5, which were previously estimated by IAA RAS based on optical and radar observations in 2005 [1]:

$$3\sigma_0(x,\ y,\ z)=3\ \text{km},\ 3\sigma_0(v_x,\ v_y,\ v_z)=2\ \text{mm/s}. \quad (2)$$

Here, in (2), $\sigma_0(x, y, z)$, $\sigma_0(v_x, v_y, v_z)$ are the standard deviations of the components of initial position vector and velocity of the asteroid.

2.2. Search for potential impact trajectories of Apophis

In order to find impact trajectories of Apophis as completely as possible in an acceptable time period and without going through the entire region of initial state due to uncertainties (2) by a very small step, we have created the fast and effective algorithms to search for several sets of trajectories with given values of the perigee distance r_π based on the Monte Carlo and gradient descent methods, which include the following six stages [6–10]:

1) To find a small number (~ 50) of the asteroid's trajectories with minimum geocentric distance less than about 5 million km for the 2036 encounter by Monte Carlo method in the uncertainty region of initial state;

2) By varying the initial states of these trajectories, we perform descent (decreasing the geocentric perigee distance r_π) to trajectories, which will lead to collision with the Earth in 2036, when r_π is less than the Earth's mean radius. Figure 1 shows this search process for one impact trajectory by the gradient descent method;

3) To compute the range of the heliocentric energy constant h_0 of the two-body system at the epoch t_0 for these found collision trajectories of Apophis. Energy depends on the solar gravitational parameter μ_S, the initial heliocentric distance r_0 and speed v_0:

$$h_0 = v_0^2 - 2\mu_S / r_0\ ; \quad (3)$$

4) To applicate the exhaustive method to search for impact trajectories from about 10 million ones, whose energy constants h_0 at t_0 are close to those of collision trajectories. After that, we will find about

20 000 additional impact trajectories, whose energy constants enter into the indicated range (Integration from epoch t_0 to the encounter in 2036 was performed only if the energy constant did not go far away from this range);

5) For each of these additional impact trajectories, we perform descent again, like in stage 2, to find more impact trajectories;

6) By varying the initial state of the asteroid's impact trajectories at t_0, we will obtain several sets of trajectories with given perigee distances r_π from the accepted range (1). The perigee distance r_π of close approach in 2036 is considered as a function of the initial state. Then, unlike in our previous works [6–9], in this stage, two directions are chosen – along the gradient of the energy constant and the opposite one:

$$x_1 = grad_{(r,v)}(h_0),\ x_2 = -grad_{(r,v)}(h_0)\,. \tag{4}$$

Along these directions, we find zero roots of the function $f_i = r_\pi(r_0, v_0) - c_i$, where c_i is the required magnitude of r_π.
Figure 2 shows a set of impact trajectories (~5000) in 2036 in the background of the dispersion region of the initial state (2) in the projection onto the (dr, dv) plane, where dr=r–r_n, dv=v–v_n.

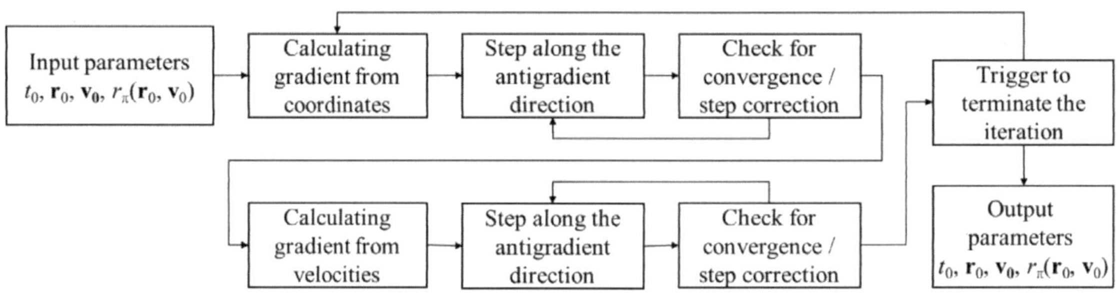

Figure 1. Flowchart to search for one impact trajectory by the gradient descent method

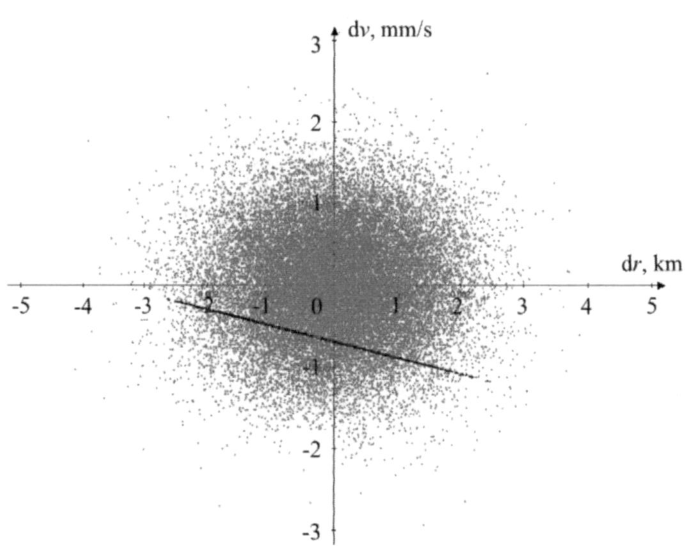

Figure 2. Dispersion region of the initial state and a set of impact trajectories (~ 5000) in 2036 on the (dr, dv) plane

According to computations [7, 9, 14], the minimum geocentric distance r_π for the 2036 encounter of the asteroid Apophis was estimated at about $r_{\pi\min}$~2070 km. These analogous results indicate the stability of the structure of potential collisions [14]. Therefore, to investigate further the characteristics of Apophis' impact trajectories, we have obtained several sets of impact trajectories with the geocentric distance r_π changing from the minimum distance $r_{\pi\min}$ to the Earth's radius. For each trajectory, we also have obtained position and velocity at the epoch when passing perigee in 2036:

$$t_\pi,\ \mathbf{r}_\pi(x_\pi,\ y_\pi,\ z_\pi),\ \mathbf{v}_\pi(v_{x\pi},\ v_{y\pi},\ v_{z\pi}). \qquad (5)$$

3. Analysis of the characteristics of the asteroid Apophis' collision trajectories and the impact path of risk

3.1. Determination of the impact positions for Apophis

Next, we determined geographical coordinates of the impact points on the Earth's surface. As the geometric model of the Earth, we applied the model of a rotational ellipsoid around its minor axis with equatorial radius R_e = 6378.137 km and ellipticity α=1/298.25. Note that the influence of atmosphere, Earth's nutation and precession on the impact positions of the asteroid has not been taken into account. An analysis shows that the structure of the impact path on Earth changes slightly, the impact points move approximately towards east 40–60 km due to nutation and precession.

By numerical integrating the equations of motion of the asteroid described in 2.1 with the initial condition (5), we obtained the radius vector **r**, the "absolute" velocity vector **v** of the asteroid at any epoch t, and then transformed them to the Greenwich geocentric geoequatorial coordinate system, got the geocentric latitude φ_g and longitude λ_g, altitude H of the asteroid. Finally, taking into account the current time t and the Earth's oblateness, we calculated the geographical latitude φ and longitude λ of the asteroid's position [21–23].

We consider that the asteroid impacts the Earth when H=0. Let's denote respectively the time of collision, the latitude and longitude of the impact point by t_c, φ_c and λ_c. When H=100 km, the asteroid passes through the space-atmosphere boundary. The time of entry into the atmosphere is denoted as t_{atm}. Taking into account the angular velocity of the Earth's rotation, we also determined the relative velocity of the asteroid $\mathbf{v}_r$ with respect to the Earth's surface. As H=0, we computed absolute and relative velocities $\mathbf{v}_C$ and $\mathbf{v}_{Cr}$ at the collision point, velocity $\mathbf{v}_{\eta\zeta}$ is the projection of the relative velocity $\mathbf{v}_{Cr}$ on the Earth's surface. As H=100 km, we determined angles of inclination for the absolute and relative velocities to the horizontal plane, i.e. angles of entry into the atmosphere θ_a and θ_r.

3.2. Asteroid Apophis' path of risk

Several sets of the collision trajectories A_n (n=0, 1, …, 9), B_n (n=1, 2, …, 10), for each of which the perigee distance varies within certain ranges: (A_0)={r_π: 2069÷2100 km}, (A_1, B_1)={r_π: 2500±1 km}, (A_2, B_2)={r_π: 3000±1 km}, (A_3, B_3)={r_π: 3500±1 km}, (A_4, B_4)={r_π: 4000±1 km}, (A_5, B_5)={r_π: 4500±1 km}, (A_6, B_6)={r_π: 5000±1 km}, (A_7, B_7)={r_π: 5500±1 km}, (A_8, B_8)={r_π: 6000±1 km}, (A_9, B_9)={r_π: 6360±1 km}, (B_{10})={r_π: 6375±1 km}. The difference between A_n and B_n will be explained below.

Figure 3 shows the impact area A_0 on the plane of geographical angles (φ, λ) for the set of impact orbits with minimum perigee distances r_π in the range [2069, 2100] km. A cross marks its center. This area is located in the Pacific, between the west coast of the United States and Hawaii. It can be seen that only part of the area A_0 is connected as 2069 km≤r_π≤2085 km. When r_π>2085 km, impact areas on the Earth's surface for other sets of orbits {r_π=const} are divided into two subsets: one corresponds to a decrease of the heliocentric energy h before the collision, the other – increase of the heliocentric energy h, comparing with the energy of orbit with minimum perigee distance $r_{\pi\min}$.

Next, we displayed impact positions of all sets of trajectories A_n (n=1, ..., 9) and B_n (n=1, 2, ..., 10) on the Earth. They generate the Apophis' impact path of risk in 2036 (figure 4). Note that due to the non-sphericity of the Earth and the inclination of the trajectories to the equator, the Earth's radius on the centers of areas A_9 and B_9 is different, about 6364 km for A_9 and 6376 km for B_9. Trajectories from A_9 almost touch the Earth's surface. However, trajectories from B_9 noticeably cross the Earth. So we added the set of orbits B_{10} with r_π=6375±1 km.

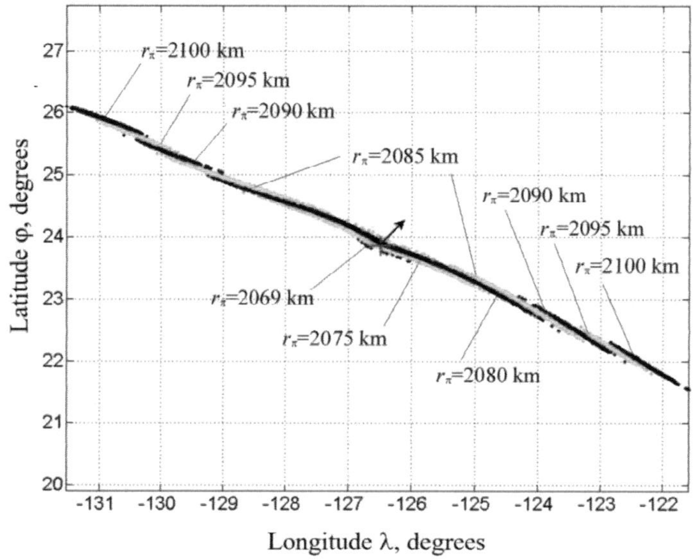

Figure 3. Impact area A_0 of the asteroid (99942) Apophis with minimum perigee distances $r_\pi \in [2069, 2100]$ km

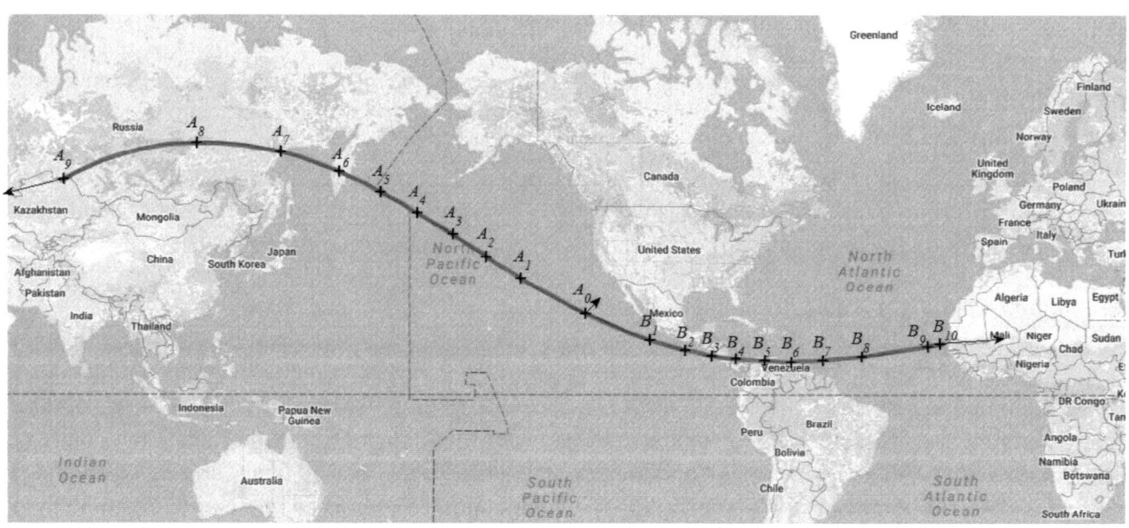

Figure 4. Asteroid (99942) Apophis' path of risk in 2036

The impact path of risk of Apophis lies in the Northern Hemisphere of the Earth. Its starting western point (area A_9) is located near Omsk, Tomsk, and Novosibirsk. Then it passes sequentially through the Krasnoyarsk region, Yakutia, the Far East, and Kamchatka of Russia, near the Aleutian Islands, across the North Pacific, Central America (Nicaragua, Costa Rica), South America (Colombia, Venezuela), ending in the Atlantic, just before reaching the west coast of Africa (near Dakar, area B_{10}), see figure 4. This path of risk is also given in [3, 4, 10, 12, 17]. Comparison of results by other authors with our ones, given here and in [10], shows that they almost completely coincide.

IOP Conf. Series: Materials Science and Engineering **468** (2018) 012023 doi:10.1088/1757-899X/468/1/012023

Numerical analysis has indicated that Apophis' path of risk lies approximately on a plane, and it is almost parallel to the ecliptic plane at collision epoch. We estimated the length and width of the PoR on the Earth's surface. In this case, we assumed approximately that the intersection of the PoR plane and the Earth is about a circle, its radius $R_P \approx 5370$ km. The central angle of PoR around the Earth $\phi \approx 4.534$ rad =259.8 degrees, then its length $L_P \approx \phi\, R_P = 24347$ km. The width of PoR was computed numerically according to the coordinates of the collision points in the normal direction to the longitudinal axis of PoR. The width of area A_0 is estimated at about 27 km. For A_i and B_i ($i \geq 1$), they are about 10–22 km wide. On average, the width of PoR is about 20 km. Note that in [17, 18], it was estimated at 30–50 km.

3.3. Characteristics of Apophis' collision trajectories

Figure 5 shows some temporal, energy, and geometric characteristics of the collision trajectories in the centers of all impact areas A_n and B_n. The impact area A_0 with the minimum perigee distance $r_{\pi\min}$ is located approximately at the center of this path of risk. For each set of trajectories (A_n, B_n) with 2085 km$<r_\pi<$6364 km, we obtained two areas, which are approximately symmetrical referring to the center of area A_0, areas A_n are at the west and B_n are at the east side of A_0, see figure 4. Areas A_n are generated by the orbits of the asteroid with higher energy before collision than in A_0, $h \in [-799.7689, -799.7554]$ km^2/s^2, here the semi-major axis of the Apophis orbit (7:6 exterior Earth resonance) $a \in [a_0, a_0+2784]$ km, $a_0 \approx 165938501$ km, see curve $h(\lambda)$, fig. 5. Areas B_n are generated by orbits with lower energy, $h \in [-799.7824, -799.7689]$ km^2/s^2, $a \in [a_0-2817, a_0]$ km.

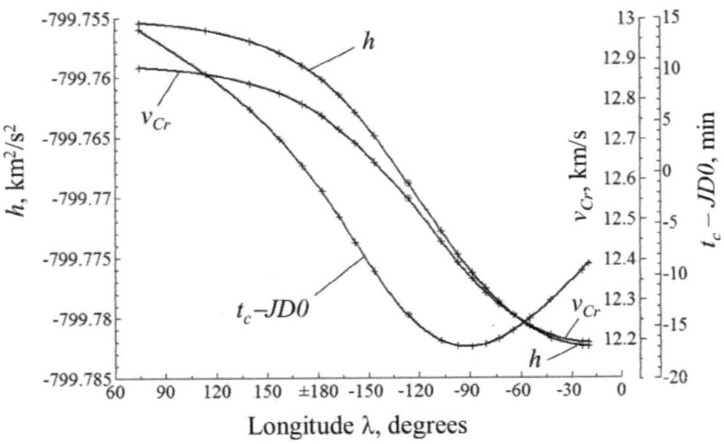

Figure 5. Dependence of relative impact speed v_{Cr}, energy before collision h and impact time t_c on the longitude of impact areas λ

Fig. 5 also shows the collision time t_c with respect to the reference time $JD0$= 9^h, April 13, 2036=JD 2464796.875, this is about the mean collision time. For the whole set of trajectories, time t_c is in the range from $JD0$–17 min to $JD0$+13.8 min, see fig. 5, curve $t_c - JD0$ (λ). The absolute speed of collision with the Earth for Apophis is $V_{ca} \approx 12.6$ km/s, also in [4–10]. The rotation of the Earth leads to an increase of the collision speed at the west side of path of risk to ~ 12.9 km/s and to a decrease at the east side to ~ 12.2 km/s, see fig. 5, curve v_{Cr}(λ). In figure 4, the relative velocity vectors $\mathbf{v}_{\eta\zeta}$ in A_9, A_0 and B_{10} are shown. They correspond to the directions of the asteroid approach to the Earth.

Fig. 6 shows the angle of entry into the atmosphere θ_r, $\Delta\theta = \theta_a - \theta_r$, as well as the mean value of the perigee distance r_π of the impact areas A_n and B_n depending on the longitude λ. The largest magnitude of θ_a and θ_r is about 59°, this is in the area A_0. With increasing r_π, it decreases to about 8° in A_9, B_{10}. The difference between θ_a and θ_r is less than 1.5°, see the curve $\Delta\theta$(λ) in fig. 6.

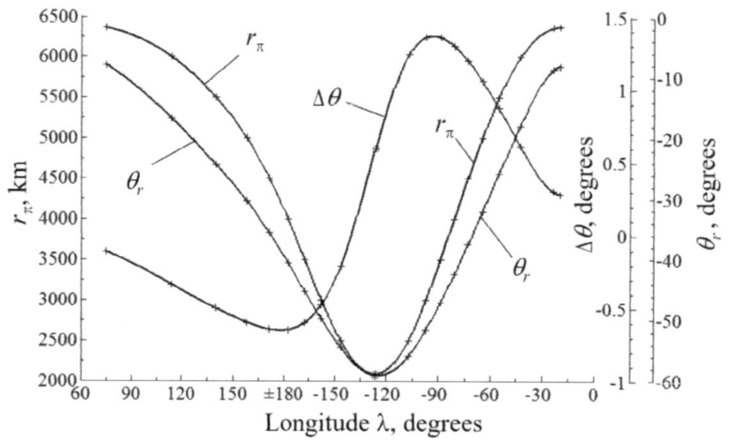

Figure 6. Dependence of the angle of entry into the atmosphere θ_r, $\Delta\theta=\theta_a-\theta_r$ and perigee distance r_π on the longitude of impact areas λ

3.4. Qualitative analysis of Apophis' close approach in 2036

The consideration of the target vector **b** in the B-plane and the velocity vector "at infinity" $\mathbf{v}_\infty$ of the hyperbolic geocentric orbits of Apophis helps to understand better the qualitative features of the set of Apophis' impact orbits. Numerical analysis has shown that for all collision orbits of Apophis, the endpoints of **b**-vectors, put aside from one starting point, form almost a line *L*, the direction vector of which is $\mathbf{S}=[-0.2443, 0.8776, 0.4124]^T$. Vectors $\mathbf{v}_\infty$ are almost constant and form an essentially plane passing through *L*, and then going to the plane of the PoR. The average value of vectors $\mathbf{v}_\infty$ is $\mathbf{v}_\infty=[5.5849, 0.7923, 1.6077]^T$ km/s, $|\mathbf{v}_\infty| = 5.87$ km/s, also in [2, 4]. For all orbits from A_0, the **b**-vector has an average magnitude $\mathbf{b}_0=[-1324.5, -3366.8, 6262.1]^T$ km, $|\mathbf{b}_0|= 7232$ km.

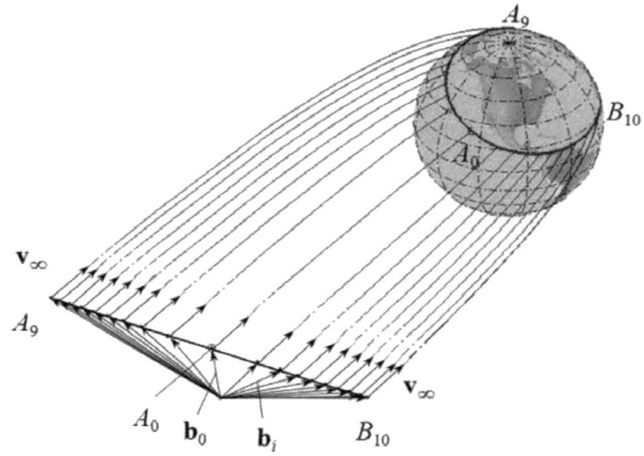

Figure 7. Possible impact trajectories of the asteroid Apophis in 2036

Figure 7 shows the geocentric motion of Apophis for all sets of impact trajectories in 2036. The velocity vectors "at infinity" and the **b**-vectors, as well as the possible impact path on the Earth's surface are shown in this figure. Isoline (contour line) $r_\pi=const$ will be approximately perpendicular to their corresponding target vectors **b**. Thus, as $r_\pi \approx r_{\pi\min} \approx 2069$ km, the isoline $r_\pi= const$ on the Earth's surface will be an arc along the longitudinal axis of PoR, which is shown in figure 3, and approximately perpendicular to vector $\mathbf{b}_0$. With increasing $r_{\pi i}$, the inclination of the isoline $r_{\pi i}=const$ to the longitudinal axis of the PoR increases monotonically, and these isolines, being almost perpendicular to the vector $\mathbf{b}_i$, become almost perpendicular to the longitudinal axis of the PoR on its edges, for A_9 and B_{10}, see figures 3, 4 and 7.

4. Conclusions

Algorithms for searching all probable impact trajectories, for determining the coordinates of the intersection points of the Apophis' orbits with the Earth's surface and for constructing the impact area on the Earth have been developed. We have obtained a set of collision orbits with the Earth in 2036 for the asteroid Apophis and several probable sets of asteroid's trajectories, corresponding to the magnitude of the perigee distance of the Apophis' orbit from minimum distance of about 2070 km to the Earth's radius. The possible impact areas, i.e. the Path of Risk of the asteroid Apophis in 2036 have been constructed on the Earth's surface. There have been studied the geometric, temporal and energy characteristics of the Apophis' collision trajectories and the Path of Risk, in particular, its width and length, the angle of the asteroid's entry into the atmosphere, time and the speed of collision, the direction of the approach of the asteroid to the Earth. Comparison of our results with other works on this issue shows that they coincide well.

References

[1] Yagudina E I and Shor V A 2005 *All-Russian Conf. "Asteroid-Comet Hazard-2005 (ACH-2005)"* (St. Petersburg, Oct. 3–7 2005) pp 355–358
[2] Chesley S R 2006 *Proc. 229th IAU Symp. 2005* (Cambridge Univ. Press) pp 215–228
[3] Russell Schweickart, C Chapman, D Durda, P Hut, B Bottke and D Nesvorny 2006 Threat Characterization: Trajectory Dynamics (White Paper 39) *B612 Foundation*
[4] D B Gennery 2007 Scenarios for dealing with Apophis *Plan. Def. Conf.* (G. Wash. Univ. USA, Mar. 5–8 2007)
[5] Giorgini J D *et al* 2008 *Icarus* **193** pp 1–19
[6] Ivashkin V V and Stikhno C A 2007 An analysis of the correction problem for the near-Earth asteroid (99942) Apophis-2004 MN4 *Plan. Def. Conf.* (G. Wash. Univ. USA, Mar. 5–8 2007)
[7] Ivashkin V V and Stikhno C A 2007 A problem of the orbit correction for the near-Earth asteroid Apophis *58th Inter. Astron. Cong.* (Hyderabad, India, Sep. 24–28 2007)
[8] Ivashkin V V and Stikhno CA 2008 *Dokl. Akad. Nauk* **419**(5) pp 624–627
[9] Ivashkin V V and Stikhno C A 2009 *Sol. Syst. Res.* **43**(6) pp 502–516
[10] Ivashkin V V, Stikhno C A and Guo P 2017 *Dokl. Phys.* **62**(8) pp 387–391
[11] Włodarczyk I 2008 *Contrib. Astron. Obs. Skalnate Pleso* **38** pp 21–32
[12] Włodarczyk I 2010 *Proc. Int. Conf. "Asteroid-Comet Hazard-2009"* (SPb.: Nauka) pp 299–301
[13] Włodarczyk I 2013 *Mon. Not. R. Astron. Soc.* **434** pp 3055–3060
[14] Sokolov L L *et al* 2012 *Sol. Syst. Res.* **46**(4) pp 291–300
[15] Sokolov L L, Borisova T P, Vasilev A A and Petrov N A 2013 *Sol. Syst. Res.* **47**(5) pp 408–413
[16] Sokolov L L, Kuteeva G A 2015 *Vestn. St.Petersb. Univ. Ser.* **2**(60) pp 148–156
[17] https://en.wikipedia.org/wiki/99942_Apophis
[18] http://galspace.spb.ru/index129.html
[19] https://en.wikipedia.org/wiki/B612Foundation
[20] Stepan'yants V A and L'vov D V 2000 *Math. Model.* **12**(6) pp 9–14
[21] Akim E L and Eneev T M 1963 *Cosm. Res.* **1**(1) pp 5–50
[22] Kaplan G H 2005 The IAU resolutions on astronomical reference systems, time scales, and Earth rotation models. Explanation and Implementation *US Naval Obser. Circular №179 Wash. DC* 20392 p 118
[23] Vallado David A 2007 Fundamentals of astrodynamics and applications 3rd ed. *Sp. Tech. Lib. Springer NY.* vol 24 p 1055

Fundamental problems of wave dynamics stratified medium modelling

V V Bulatov and Yu V Vladimirov

Institute for Problems in Mechanics RAS, Moscow, Russia
E-mail: internalwave@mail.ru

Abstract. In paper fundamental problems of wave dynamic stratified medium modeling (ocean, atmosphere) are considered. The basic mathematical models describing the processes of excitation, the propagation of internal gravity waves in vertical stratified horizontally inhomogeneous and non-stationary medium are presented. The uniform asymptotic forms of the internal gravity waves in horizontally inhomogeneous and non-stationary stratified medium are obtained. A modified spatio-temporal ray method is proposed, which belongs to the class of geometrical optics methods (WKBJ method). Analytical and numerical algorithms of internal gravity wave calculations for the real stratified medium parameters are presented. Some results of internal gravity waves measurements in real stratified medium (ocean, atmosphere) and it's interpretations are discussed. The universal nature of the developed asymptotic methods of modeling makes it possible to efficiently calculate the wave fields and qualitatively analyze the solutions obtained, which makes it possible to analyze wave patterns in general, to correctly set up mathematical models of wave dynamics, and to carry out express estimates in actual measurements of wave fields in natural stratified medium.

1. Introduction

The dynamics of wave motion in the natural stratified medium (ocean, Earth atmosphere) is currently of great interest because they are important geophysics and oceanology. As a rule, an analysis of these phenomena is based on asymptotic methods, because the study of unperturbed hydrodynamic equations leads to asymptotic expansions (Ansatzs, which is a German term for a type of solutions). These expansions permit solving the problems of perturbed equations, which can be used to describe the effects of nonlinearity, inhomogeneity, and non-stationary behavior of the real ocean. To obtain a detailed description of a wide range of physical phenomena related to wave dynamics of the stratified horizontally inhomogeneous unsteady ocean, it is necessary to start from sufficiently developed mathematical models, which are usually quite complicated, nonlinear, and multi-parametric. They can be investigated completely only using efficient numerical methods. However, there are several cases, in which a preliminary qualitative concept of the phenomena under study can be obtained on the basis of simper asymptotic models and analytic methods for studying these models. These models then enter a set of «blocks» used to construct the complete pattern of wave dynamics, which permits discovering the correlation between different wave phenomena and their relationship. Sometimes, despite the seeming simplicity of the model assumptions, a successive choice of the solution form allows one to obtain physically interesting results [1, 2].

The propagation of internal gravity waves (IGW) in the ocean is strongly affected by the horizontal inhomogeneity and unsteady behavior of the basic hydro-physical parameters. In this contribution, we generalize a method of geometrical optics, i.e., the space-time ray method, which permits solving the

Content from this work may be used under the terms of the Creative Commons Attribution 3.0 licence. Any further distribution of this work must maintain attribution to the author(s) and the title of the work, journal citation and DOI.
Published under licence by IOP Publishing Ltd

problem of mathematical modeling of IGW dynamics in the horizontally inhomogeneous and vertically stratified ocean. The ray representations agree well with the intuitive and empirical concepts of IGW propagation in the real ocean. This method is sufficiently universal, and in many cases, this is the only possible method for approximate calculations of wave fields in the ocean. The most typical horizontal inhomogeneities of the real ocean are the variations in the ocean floor shape, horizontal inhomogeneities of the density field, and unsteady ocean currents. An exact analytic solution can be obtained, for example, using the method of separation of variables only if the density distribution and the ocean floor shape can be described by sufficiently simple model functions. If the ocean floor shape and the ocean stratification are arbitrary, then one can construct only the asymptotic representations of the solution or solve the problem numerically. But the numerical solution does not permit obtaining and analyzing the qualitative characteristics of the wave field at large distances, which is necessary, for example, when solving the IGW detection problem by remote methods including, for example, the aerospace radiolocation.

The mathematical modeling of IGW wave dynamics in the horizontally inhomogeneous and vertically stratified medium (ocean, Earth atmosphere) is possible on the basis of a modified version of the space-time ray method (a method of geometrical optics, WKBJ method). The specific form of asymptotic representations can be determined by solving the problems, which describe the IGW dynamics in the vertically stratified, horizontally homogeneous, and steady-state ocean. As a rule, when studying the evolution of IGW packets in the ocean with slowly varying and unsteady parameters, it is assumed that this wave packet is locally harmonic. In contrast to the majority of works, in which this problem has been studied, the proposed modified method of geometrical optics allows one to describe the structure of wave packets near singular surfaces such as caustics and wave fronts. The term «geometrical optics» has different meanings in the scientific literature. The geometrical optics understood in the narrow (or ray) sense deals only with the methods for constructing images by using the rays, while the geometrical optics understood in the wide (or wave) sense is a method for obtaining approximate descriptions of wave fields. In the wave interpretation, which is used in this paper, the rays, as a rule, form only the geometric skeleton, on which the wave filed is «sewn on». According to the two previous interpretations of the geometrical optics, two periods in its development existed. The first ray period was ideologically completed by Hamilton's fundamental works, which significantly influenced the development of the classical mechanics. The construction of rays underlies the instrumental optics, which is mainly oriented to design various optical devices. The contemporary wave period originates from the Debye's works, which decisively influenced the formation of ray concepts in the wave theory. The asymptotic representation of solutions of the problem of wave packet propagation in the ocean with horizontally inhomogeneous density and numerical computations for typical oceanic parameters testify that the horizontal inhomogeneity significantly affects the real IGW dynamics in the ocean.

All results of wave dynamics modeling presented in this paper can be used for arbitrary density distributions and other parameters of the stratified ocean. It is necessary to consider them in the context of consistency with the available data of IGW full-scale measurements in the ocean. Such methods for analyzing the wave fields are important not only because they are illustrative, universal, and efficient in various problems, but also because they can serve as a certain semi-empirical basis for other approximate methods in the theory of wave packet propagation in the ocean. The waves in media with slowly varying parameters have been studied in the vast literature, while the amount of works dealing with the problem of studying IGW in the media with variable parameters is rather pure (mainly because of significant mathematical difficulties encountered in these problems). In this paper, we present the basics of the space-time ray method (a method of geometrical optics) with regard to IGW special characteristics, which permits studying the wave dynamics in the horizontally inhomogeneous and vertically stratified ocean [1,2].

IOP Conf. Series: Materials Science and Engineering **468** (2018) 012024 doi:10.1088/1757-899X/468/1/012024

2. Problem formulation

Our analysis starts from a linear system of hydrodynamic equations

$$\rho_0\frac{\partial u_1}{\partial t}=-\frac{\partial p}{\partial x},\qquad \rho_0\frac{\partial u_2}{\partial t}=-\frac{\partial p}{\partial y},\qquad \rho_0\frac{\partial w}{\partial t}=-\frac{\partial p}{\partial z}+g\rho, \tag{1}$$

$$\frac{\partial u_1}{\partial x}+\frac{\partial u_2}{\partial y}+\frac{\partial w}{\partial z}=0,\qquad \frac{\partial \rho}{\partial t}+u_1\frac{\partial \rho_0}{\partial x}+u_2\frac{\partial \rho_0}{\partial y}+w\frac{\partial \rho_0}{\partial z}=0.$$

Here (u_1, u_2, w) are components of the IGW velocity vector; p and ρ are perturbations of the pressure and density; g is the acceleration of gravity (the axis z is directed downwards). Using the Boussinesq approximation, which means that the unperturbed density $\rho_0(z,x,y)$ in the first three equations in system (1) is assumed to be constant, we reduce system (1.1) to the form

$$\frac{\partial^4 w}{\partial z^2\partial t^2}+\Delta\frac{\partial^2 w}{\partial t^2}+\frac{g}{\rho_0}\Delta(u_1\frac{\partial \rho_0}{\partial x}+u_2\frac{\partial \rho_0}{\partial y}+w\frac{\partial \rho_0}{\partial z})=0, \tag{2}$$

$$\frac{\partial}{\partial t}(\Delta u_1+\frac{\partial^2 w}{\partial z\partial x})=0,\qquad ,\quad \frac{\partial}{\partial t}(\Delta u_2+\frac{\partial^2 w}{\partial z\partial y})=0\qquad \Delta=\frac{\partial^2}{\partial x^2}+\frac{\partial^2}{\partial y^2}.$$

As the boundary conditions, we use the «rigid lid» condition at the surface: $W=0$, *(z=0,-H)*, where H is the ocean depth. We assume that, in the media with horizontally inhomogeneous density field, the steady-state flows due to this field can be neglected. Indeed, it follows from the hydrodynamic equations that if the unperturbed density is a function of horizontal coordinates, then the existence of the steady-state density distribution $\rho_0(z,x,y)$ implies the existence of steady-state flows. These flows are rather slow, and they can be neglected in the first approximation. Therefore, it is usually assumed that $\rho_0(z,x,y)$ is the background density field formed under the action of mass forces and non-adiabatic sources, and this field is given *a priori*, for example, by experimental data.

Now we consider harmonic waves (u_1, u_2, w) $=exp(i\omega t)(U_1,U_2,W)$. System (2) cannot be solved by the method of separation of variables, and therefore it is necessary to use asymptotic methods. The scales of horizontal variations in the ocean parameters can be greater than the scales of vertical variability. Further we introduce the dimensionless variables: $x^*=x/L,\quad y^*=y/L,\quad z^*=z/h,$ where L is the characteristic scale of horizontal variations in the density ρ_0 and h is the characteristic scale of vertical variations in ρ_0 (for example, the width of the thermocline). In the dimensionless coordinates, system (2) becomes (hereinafter, the asterisk in the indices is omitted)

$$-\omega^2(\frac{\partial^2 W}{\partial z^2}+\varepsilon^2\Delta W)+\varepsilon^2\frac{g_1}{\rho_0}(\varepsilon U_1\frac{\partial \rho_0}{\partial x}+\varepsilon U_2\frac{\partial \rho_0}{\partial y}+W\frac{\partial \rho_0}{\partial z})=0, \tag{3}$$

$$\varepsilon\Delta U_1+\frac{\partial^2 W}{\partial z\partial x}=0,\qquad \varepsilon\Delta U_2+\frac{\partial^2 W}{\partial z\partial y}=0,\qquad \varepsilon=\frac{h}{L}<<1,\qquad g_1=\frac{g}{h}.$$

3. Asymptotic forms of solution

We seek the asymptotic solution of (1.3) in the form typical of the method of geometrical optics

$$\mathbf{V}(z,x,y)=\sum_{m=0}^{\infty}(i\varepsilon)^m\mathbf{V}_m(z,x,y)exp(S(x,y)/i\varepsilon),\ \mathbf{V}(z,x,y)=(U_1(z,x,y),U_2(z,x,y),W(z,x,y)),$$

where function $S(x, y)$ and vector function $\mathbf{V}_m, m = 0, 1...$, are sought. As a rule, below, we determine only the leading term of this asymptotic expansion for the vertical velocity component $W_0(z, x, y)$. We obtain the following from the two last equations in (1.3)

$$U_{10} = -\frac{i\partial S/\partial x}{|\nabla S|^2}\frac{\partial W_0}{\partial z}, \qquad U_{20} = -\frac{i\partial S/\partial y}{|\nabla S|^2}\frac{\partial W_0}{\partial z}, \qquad |\nabla S| = \left(\frac{\partial S}{\partial x}\right)^2 + \left(\frac{\partial S}{\partial y}\right)^2.$$

Equating the terms of order O(1), we obtain the equation for function $W_0(z, x, y)$. This equation is written as

$$\frac{\partial^2 W_0(z,x,y)}{\partial z^2} + |\nabla S|^2\left(\frac{N^2(z,x,y)}{\omega^2} - 1\right)W_0(z,x,y) = 0, \quad W_0(0,x,y) = W_0(-H,x,y) = 0, \tag{4}$$

where $N^2(z,x,y) = \frac{g_1}{\rho_0}\frac{\partial \rho_0}{\partial z}$ is the Brunt–Väisälä frequency depending on the vertical and horizontal coordinates. It is well known that the basic boundary-value vertical spectral problem for internal waves (4) has countable many eigenfunctions W_{0n} and eigenvalues $K_n(x,y,\omega) \equiv |\nabla S_n|$. Functions $W_{0n}(z,x,y)$ and $K_n(x,y,\omega)$ are assumed to be known; the index n is omitted because we assume that all calculations are carried out for a separate wave mode. We use the eikonal equation $\left(\frac{\partial S}{\partial x}\right)^2 + \left(\frac{\partial S}{\partial y}\right)^2 = K^2(x,y)$ to determine function $S(x,y)$. In the plane case, the initial conditions for eikonal S are posed on line $L: x_0(\alpha), y_0(\alpha), \; S(x,y)|_L = S_0(\alpha)$. To solve the eikonal equation, we construct the rays, i.e., the characteristics of this equation, which have the following form

$$\frac{dx}{d\sigma} = \frac{p}{K(x,y)}, \quad \frac{dp}{d\sigma} = \frac{\partial K(x,y)}{\partial x}, \quad \frac{dy}{d\sigma} = \frac{q}{K(x,y)}, \quad \frac{dq}{d\sigma} = \frac{\partial K(x,y)}{\partial y}, \tag{5}$$

where $p = \partial S/\partial x, \quad q = \partial S/\partial y, \quad d\sigma$ is the ray length element. The initial conditions p_0 and q_0 for solution (5) are determined by solving the following system

$$p_0\frac{\partial x_0}{\partial \alpha} + q_0\frac{\partial y_0}{\partial \alpha} = \frac{\partial S_0}{\partial \alpha}, \quad p_0^2 + q_0^2 = K^2(x_0(\alpha), y_0(\alpha))$$

whose solution and the initial conditions $x_0(\alpha), y_0(\alpha), p_0(\alpha), q_0(\alpha)$ determine the ray $x = x(\sigma,\alpha), \; y = y(\sigma,\alpha)$. After the rays are constructed, eikonal S can be determined by integrating along the ray: $S = S_0(\alpha) + \int_0^\sigma K(x(\sigma,\alpha), y(\sigma,\alpha))d\sigma$. Eigenfunction $W_0(z,x,y)$ is calculated up to multiplication by arbitrary function $A_0(x,y)$: $W_0(z,x,y) = A_0(x,y)f_0(z,x,y)$, where $f_0(z,x,y)$ is the solution of the basic vertical spectral problem with normalization

$$\int_0^H (N^2(z,x,y)-\omega^2)f_0^2(z,x,y)dz=1.$$

Then, after rather cumbersome analytic calculations, we obtain the conservation law along the eikonal characteristics

$$\frac{d}{d\sigma}\left(ln\frac{A_0^2(x,y)I(x,y)}{K^2(x,y)}\right)=0,$$

where $I(x,y)$ is the geometric divergence of the rays (characteristics). We note that the wave energy flux is proportional to $A_0^2K^{-1}R$, where R is the width of an elementary ray tube; therefore, the quantity equal to the wave energy divided by the modulus of the wave vector is preserved in this case. The long-range IGW fields in the real ocean are, as a rule, non-harmonic wave packets. Indeed, at a far distance form perturbation sources, the complete wave field is a sum of separate wave modes whose asymptotics, depending on the stratification, depth, and other parameters of the ocean, can be expressed in terms of the Airy function or the Fresnel integrals. Therefore, to study the problem of wave packet evolution in a horizontally smoothly inhomogeneous and unsteady stratified medium, it is necessary to use another Ansatz.

We introduce slow variables $x^*=\varepsilon x,\ \ y^*=\varepsilon y,\ \ t^*=\varepsilon t$ (since z is not assumed to be a slow variable, we omit the asterisk in the index), where $\varepsilon=\lambda/L<<1$ is a small parameter characterizing the smoothness of the medium variations along the horizontal (λ is the characteristic wave length, and L is the scale of horizontal inhomogeneity). Then system (2) for determining the velocity components (U_1,U_2,W) in these slow variables becomes

$$\frac{\partial^4 W}{\partial z^2\partial t^2}+\varepsilon^2\frac{\partial^2 W}{\partial t^2}+\frac{g}{\rho_0}\Delta(\varepsilon U_1\frac{\partial\rho_0}{\partial x}+\varepsilon U_2\frac{\partial\rho_0}{\partial y}+W\frac{\partial\rho_0}{\partial z})=0, \quad (6)$$

$$\varepsilon\Delta U_1+\frac{\partial^2 W}{\partial z\partial x}=0\ \ ,\ \varepsilon\Delta U_2+\frac{\partial^2 W}{\partial z\partial y}=0.$$

Further we consider the superposition of harmonic waves (in slow variables x,y,t) $W=\int\omega\sum_{m=0}^{\infty}(i\varepsilon)^m W_m(\omega,z,x,y)exp\left(\frac{i}{\varepsilon}[\omega t-S_m(\omega,x,y)]\right)d\omega$, where functions $S_m(\omega,x,y)$ are assumed to be odd with respect to ω and $\min_{\omega}\frac{\partial S}{\partial\omega}$ is attained at $\omega=0$ (for all x and y). We substitute this representation into (6) and see that function $W_m(\omega,z,x,y)$ for $\omega=0$ has a pole of order m. Therefore, the model integrals, or phase functions $R_m(\sigma)$, for some terms of the asymptotic series are expressions $R_m(\sigma)=\frac{1}{2\pi}\int_{-\infty}^{\infty}(i/\omega)^{m-1}exp(i(\omega^3/3-\sigma\omega))d\omega$, where the contour of integration bypasses point $\omega=0$ from above, which ensures the exponential decay of functions $R_m(\sigma)$ for $\sigma>>1$. Functions $R_m(\sigma)$ have the following property $\frac{dR_m(\sigma)}{d\sigma}=R_{m-1}(\sigma)$, where $R_0(\sigma)=Ai'(\sigma)$,

$R_1(\sigma)=Ai(\sigma)$, $R_2(\sigma)=\int_{-\infty}^{\sigma} Ai(u)du$, etc. Obviously, starting from the corresponding properties of the Airy integrals, we can conclude that functions $R_m(\sigma)$ are related as $R_{1}'(\sigma)+\sigma R_1(\sigma)=0$, $R_{-3}(\sigma)+2R_0(\sigma)-\sigma^2 R_1(\sigma)=0$. For the model integrals $R_m(\sigma)$ describing the long-range IGW fields in the deep ocean, one can use the following expressions $R_0(\sigma)=Re\int_0^{\infty} exp\left(-it\sigma-it^2/2\right)dt \equiv Re\Phi(\sigma)$; in this case, functions $R_m(\sigma)$ satisfy the recurrence relations $R_{-3}(\sigma)-2iR_{-1}(\sigma)-i\sigma R_{-2}(\sigma)=0$ and $R_{-1}(\sigma)+i\sigma R_0(\sigma)=0$. It follows from the above and the structure of the first term of the uniform asymptotics (Airy or Fresnel wave) in a stratified and horizontally homogeneous medium that the solution of system (6) can be sought in the following form (index n is omitted for a separate wave mode)

$$W=\varepsilon^0 W_0(z,x,y,t)R_0(\sigma)+\varepsilon^a W_1(z,x,y,t)R_1(\sigma)+\varepsilon^{2a} W_2(z,x,y,t)R_2(\sigma)+\ldots,$$
$$\mathbf{U}=\varepsilon^{1-a}\mathbf{U}_0(z,x,y,t)R_1(\sigma)+\varepsilon\mathbf{U}_1(z,x,y,t)R_2(\sigma)+\varepsilon^{1+a}\mathbf{U}_2(z,x,y,t)R_3(\sigma)+\ldots,$$

where $\mathbf{U}$ is the vector of IGW horizontal velocity and the phase function argument $\sigma=\left(S(x,y,t)/a\right)^a\varepsilon^{-a}$ is assumed to be of the order of unity. This expansion agrees well with the general approach of the method of geometrical optics and the space-time ray method and is its generalization used to study the dynamics of IGW fields in the horizontally inhomogeneous stratified ocean. We also note that this structure of the solution implies that, in a horizontally inhomogeneous medium, the solution depends on both the «fast» (vertical coordinate) and «slow» (horizontal coordinates) variables. As a rule, the solution is sought in «slow» variables, and the structure elements depending on «fast» variables are obtained as integrals of some functions slowly varying along the space-time rays. This choice of the solution permits describing the uniform asymptotics of IGW fields propagating in the stratified ocean with slowly varying parameters, which is true both near and far from the wave fronts of a separate wave mode. If it is necessary to describe the behavior of the field only near the wave front, then one can use one of the methods of geometrical optics, i.e., the «traveling wave» method, and the weakly dispersion approximation in the form of the corresponding local asymptotics to seek the representation of the phase function argument σ in the form $\sigma=\alpha(t,x,y)(S(t,x,y)-\varepsilon t)\varepsilon^{-a}$; here function $S(t,x,y)$ describes the wave front position. It is found by solving the eikonal equation $\nabla^2 S=c^{-2}(x,y,t)$, where $c(t,x,y)$ is the maximal IGW group velocity of the corresponding wave mode, i.e., the first term in the expansion of the dispersion curve at zero. Function $\alpha(t,x,y)$ (the second term of the dispersion curve expansion) describes the space-time evolution of the pulse width of non-harmonic Airy or Fresnel waves and can then be found from some conservation laws along the eikonal equation characteristics whose specific form is determined by the physical conditions of the problems under study.

4. Comparison analytical and measurements results

Further we compare the analytic results with the results of the analysis of measurements of IGW variability in a real stratified medium with horizontally varying characteristics, namely, in the Northwest Pacific, according to the data recorded by moorings in the «Megapolygon» experiment in the Northwest Pacific. The measurements of the currents and the temperature recorded by the «Megapolygon» moorings allowed us to study the variability of tidal internal waves over the area of 460×520 km. The length of the tidal internal wave was calculated by integration of the basic IGW spectral equation with the real depth distribution of the Vaisala-Brunt frequency and with zero boundary conditions at the ocean surface and the ocean floor taking into account the Earth's rotation.

The wave length of the first mode in the «Megapolygon» area is equal approximately to 130 km, the wave length near the Emperor Ridge is greater (167 km), and it is equal to 156 km at a distance of 2000 km to the east. The wave propagation direction is also very stable and varies from 240 to 300 degrees, which corresponds to the actual wave propagation to the west and northwest from the Emperor Ridge. Some diffraction of tidal internal waves was observed in the «Megapolygon» study site, i.e., the direction of wave propagation varied from the northwest in the southeast of the site to the west in its northwest part. Let us consider the amplitude variations of the internal tide in the course of its propagation to the west and to the east from the Emperor Ridge. The IGW amplitudes were calculated from the deviations of the temperature values measured on moorings; then, the values were divided by the average vertical gradient of temperature. The calculations show that the IGW amplitude decreases approximately by 10% at the distance equal to the length of the tidal internal wave (130-150 km). We can also estimate the influence of different factors, including the horizontal inhomogeneity of density, on the IGW decay. In the framework of the theory discussed above, we consider the evolution of IGW frequency ω corresponding to the semidiurnal period $T = 12$ hours, which also admits slow variations in the stratification along the wave propagation path. The real geometry of the experiment allows us to assume that the problem under study is two-dimensional, which means that the stratification depends only on two variables: depth z and distance x along the wave propagation path [2].
Now we consider the case of constant depth H and stratification N linearly depending only on variable $x: N(x) = N_1 + (N_2 - N_1)x/L$, where L is the distance between the two observation points, $x = x_1 = 0$ is the initial point, $x = x_1 = L$ is the end point, and $N_{1,2} = N(x_{1,2})$. We consider only the first mode $\eta_1(z,x)$ of the amplitude of the vertical displacement of particles and omit its index. We seek the amplitude $\eta(z,x)$ in the form $\eta(z,x) = A(x)\, f(z,x)$, where $f(z,x)$ is the normalized eigenfunction of the standard boundary-value problem for the equation of internal waves with the normalization

$$\int_0^H (N^2(x) - \omega^2) f^2(z,x)dz = 1,$$

which has the form $f(z,x) = \sqrt{2}(H(N^2(x) - \omega^2))^{-1/2} sin(\frac{\pi z}{H})$. Amplitude $A(x)$ depending only on x is determined from the conservation law: $\frac{A^2(x_1)}{k^2(x_1)} da(x_1) = \frac{A^2(x_2)}{k^2(x_2)} da(x_2)$, where $k(x)$ is the absolute value of horizontal wave vector, and $da(x)$ is the width of an elementary wave tube. Since the problem is two-dimensional, the width of the ray tube does not vary along the ray and the conservation law is simpler: $A(x)/k(x) = const$. Since we consider small values of ω, the velocity of wave propagation is close to the maximum group velocity $c(x) = N(x)H/\pi$; hence, the wave number is equal to $k(x) = \omega\pi / N(x)H$ and the corresponding wave length is equal to $\lambda(x) = 2N(x)H/\omega$. Then, under the assumption that the observation points are at the same depth, it follows from the conservation law $(A_{1,2} = A(x_{1,2}))$ that $A_1N_1 = A_2N_2$ or $A_2 = A_1\lambda_1/\lambda_2$. Then the total amplitude attains the following values $W_{1,2} = A_{1,2}\sqrt{2}(H(N_{1,2}^2 - \omega^2))^{-1/2}$, which implies $W_2 = W_1N_1((N_1^2 - \omega^2)/(N_2^2 - \omega^2))^{1/2}/N_2$ or $W_2 = W_1\lambda_1^2\lambda_2^{-2}$, because $\omega << N$, i.e., the amplitude of the internal gravity wave is inversely proportional to the squared wave length. The wave travel time τ along the horizontal ray is determined from the equation of characteristics $\frac{dx}{dt} = c(x)$, where

$c(x)=(N_1+ax)H/\pi$ and $a=(N_2-N_1)/L$. Integrating this equation, we obtain the wave travel time $\tau=\pi ln(N_2/N_1)/aH=TL ln(\lambda_2/\lambda_1)/(\lambda_2-\lambda_1)$. The available data of full-scale tests give the following values of the basic parameters of the problem: $\lambda_1=167$ km, $\lambda_2=156$ km, $L=2000$ km. The wave attenuation coefficient without the wave length variations taken into account, which describes the amplitude decrease versus wave length, which is denoted by β, gives the value of $\beta:\beta=0.2^{167/2000}=0.874$ with regard to relation $W_2/W_1=0.2\equiv\beta^{t/T}=\beta^{L/\lambda}$ derived from the observation results. The attenuation with regard to the wave length variations along the ray, $W_2/W_1=\beta^{\tau/T}$, with the theoretically calculated time of the wave travel time τ gives the following value β = 0.878. Thus, the obtained estimates allow us to conclude that the influence of the density field inhomogeneities, which is taken into account in the above-described method for asymptotic representation of the wave fields, is one of the factors determining the scales of the space attenuation of IGW fields observed in field measurements.

5. Conclusions

Thus, in this paper, a general method for calculating IGW fields in the stratified medium outlined, namely:

– for an arbitrary distribution of the Brunt-Väisälä frequency, the basic vertical spectral IGW problem is solved and the corresponding normalized eigenfunctions and eigenvalues are determined;

– the characteristic systems with appropriate initial conditions are solved numerically;

– after the characteristics (rays) are calculated, the eikonal (phase value) of the phase functions is determined by numerical integration along these rays;

– the geometric divergence of the ray tubes is determined, for example, by numerical differentiation of closely located characteristics;

–the IGW amplitude is calculated from the equations of the corresponding conservation laws along the rays (characteristics), where the right parts of the relations are determined by using the locality principle, i.e., it is assumed that the ocean parameters remain horizontally unchanged over specific spatial intervals.

The universal character of the proposed asymptotic methods of modeling IGW fields in the stratified medium (ocean, Earth atmosphere) allows us to efficiently calculate the wave fields and, in addition, analyze qualitatively the obtained solutions. This opens wide opportunities for investigating the wave fields in general, which is also important for formulating correct statements of mathematical models of wave dynamics and for obtaining express evaluations in the field measurements of internal gravity waves [1, 2].

References

[1] Bulatov V V and Vladimirov Yu V 2007 *Internal gravity waves: theory and applications* (Moscow: Nauka Publishers).

[2] Bulatov V V and Vladimirov Yu V 2012 *Wave dynamics of stratified mediums* (Moscow: Nauka Publishers).

Effects of bolide parameters on the motion and destruction in the Earth's atmosphere

N G Syzranova and V A Andrushchenko

Institute for Computer Aided Design of the Russian Academy of Sciences, Moscow, Russia
E-mail: nina-syzranova@ya.ru

Abstract. The present work describes numerical studies of movement and destruction of four meteoroids in the Earth's atmosphere: the Sikhote-Alin (1947), the Beneshov (1991), the Kunya-Urgench (1998) and the Chelyabinsk (2013) ones, which differ considerably in their dimensions, material properties, and trajectory parameters. The meteoroid mass loss due to radiation and thermal effects, and fragmentation of bodies during the motion along their trajectories is considered. These meteorites were chosen for the verification of newly developed models and to illustrate the impact of their parameters on the processes of movement and destruction.
Keywords: meteoroid, motion, destruction, heat transfer, fragmentation

1. Introduction

The work is devoted to mathematical modeling of the dynamics and destruction of meteoroids in the Earth's atmosphere. The destruction of a meteor in the Earth's atmosphere is influenced by many factors. In most parts of the trajectory, there is an intensive loss of mass under the action of radiation and convective heat transfer. In addition, a single or multiple fragmentation into smaller bodies may occur. The processes of their interaction with the atmosphere are affected by the speed and the angle of entry into the atmosphere, as well as thermophysical and strength characteristics of the material. The earth's atmosphere completely "destroys" a relatively small body and is able to significantly impact large meteorites. When moving in the atmosphere meteoroids cause the phenomenon of a meteor or "shooting stars". Especially bright meteors are bolides, which can be pretty large. These will be considered in this work.

Each meteor body is quite individual. In this regard, the study of motion and destruction of each individual meteoroid is a separate task.

This paper analyses the movement and destruction in the atmosphere of four specific meteoroids: the Chelyabinsk (2013), the Sikhote-Alin (1947), the Beneshov (1991), and the Kunya-Urgench (1998) ones, under the action of heat and power loads. The subjects of the study vary in their size, material and trajectory parameters.

The first two bodies are relatively large, their size is of the order of several meters in diameter, the size of the other two is about 1 m. The parameters of these bodies are presented in Table.1.

These meteorites were selected for the verification of the developed models not only as objects with a sufficient amount of data available about them, but also as strongly distinguished objects. Observations data were employed for a better understanding of the phenomena occurring on the path of the bodies.

Content from this work may be used under the terms of the Creative Commons Attribution 3.0 licence. Any further distribution of this work must maintain attribution to the author(s) and the title of the work, journal citation and DOI.
Published under licence by IOP Publishing Ltd

Table 1. The parameters of meteoroids

Name of the meteorite	Entry speed, km/s	Entrance angle, degrees	Weight of the meteorite, t	Composition of the material; density, g/cm^3
Chelyabinsk (Chebarkul) (2013).	19.2	18	13000	stone chondrite; 3.3
Sikhote-Alin (1947, Primorsky Krai)	15	41	500	iron; 7.8
Beneshov (1991. Czech Republic)	21	45	3.5	stone; 3.7
Kunya-Urgench (1998, Turkmenistan)	13	30	3	stone chondrite; 3.3

2. Basic equations

A numerical study of the problem is based on the equations of the physical theory of meteors in the form of [1] considering the motion in an exponential atmosphere $\rho = \rho_0 \exp(-z/h)$ (z is the altitude position of the meteoroid over the Earth's surface, ρ_0 is the atmospheric density at $z = 0$, h is the characteristic height scale):

$$M\frac{dV}{dt} = Mg\sin\theta - C_D S_{mid}\frac{\rho V^2}{2} ,$$

$$V\frac{d\theta}{dt} = g\cos\theta - \frac{V^2\cos\theta}{R_з + z} ,$$

$$H_{eff}\frac{dM}{dt} = -C_H S_{mid}\frac{\rho V^3}{2} ,$$

$$\frac{dz}{dt} = -V\sin\theta ,$$

where V, M, θ are the velocity of the body, its mass and the inclination angle of the trajectory of the fall to the horizon; R_e is the radius of the Earth; C_D, C_H, H_{eff} are coefficients of resistance, heat transfer to the body surface and the effective enthalpy of vaporization of the meteoroid; S_{mid} – sectional area of the body.

Tremendous speed (~10-30 km/s) of meteoroids causes them to experience pressures and temperatures of the order of tens to thousands of atmospheres and thousands of degrees. Under these conditions there are two heat transfer mechanisms from the gas to the body surface: convective heat transfer and radiation heat transfer.

To determine the convective heat flux at the critical point of the surface of a spherical meteorite, the following formula is used [2]:

$$q_{c0} \approx 3.3 \cdot 10^{-5} \left(\frac{\rho_\infty}{R} \right)^{1/2} V_\infty^{3.2}, \ \text{W/m}^2$$

Here R is in m, ρ_∞ in kg/m^3, V_∞ in m/s. The index "∞" corresponds to the parameters of the incoming flow.

For the coefficient of radiant heat transfer at the critical point the ReVelle formula is applied, the parameters of which are presented in [1]:

$$C_{Hr} = f \cdot e^{A_1} \rho^{A_2 + A_3 V - 1} R^{A_4 + A_5 V + A_6 V^2} V^{A_7 + A_8 V + A_9 V^2 - 3}$$

Accordingly, the heat flux at the critical point will be presented as $q_{r0} = 0.5 \rho_\infty V_\infty^3 C_{Hr}$.

The distribution of heat transfer along a spherical surface for convective heat transfer is approximated by the formula [3]:

$$q_c = q_{c0}(0.55 + 0.45 \cos 2\beta),$$

where β is the angle of the meridian cross-section counted from the direction to the critical point; and the radiation flux [4] – $q_r = q_{r0} \cos^n \beta$, $n = 1/(0.051V - 0.43) + 1.811$.

The total heat transfer to the body surface is defined as the sum of convective and radiative heat flux.

When a large heat flux to the body surface is evaporating the surface of the body, it can lead to a significant weakening of the convective heat flux and even to full shielding of its effect [5]. Radiant heat flux can also be shielded by the injected gas, however, complete a shielding does not occur even with an intense injection [6]. The main quantitative determinant of the ratio of screened radiant energy is the spectral composition of the radiation, not the intensity of the injection, as in the case of evaporation under the action of convective heat transfer. Therefore, according to [7], the radiant heat transfer coefficient at the critical point of the evaporating surface can be written in the form of:

$$C_{Hkrw} = [1 - \psi_0(V)] C_{Hkr},$$

where $\psi_0(V)$ is the power function of the flight speed V: $\psi_0 = \alpha_0 V^{\beta_0}$: (where V is expressed in km/s, and $\alpha_0 = 0.0059$, $\beta_0 = 1.36$ for the air).

Statistics of meteorite falls show that a large part of them, including the ones considered in this work, hit the Earth's surface broken to pieces, so the calculation of the entrainment mass requires consideration of their fragmentation.

We consider the process of meteorite fragmentation within the model of sequential fragmentation with the influence of scale factor on the tensile strength of the object. The model of sequential fragmentation of the body is based the statistical theory of strength [8], when the fragmentation occurs at the defects and cracks that are inherent in such structurally heterogeneous bodies like meteorites. As a result, the fragmentation occurs as a process of successive elimination of defects with increased load followed by destruction of the body by these defects, thus the resulting fragments have greater strength than the original body. In this regard, the process of fragmentation ends when the velocity head starts to decrease. The detailed model of fragmentation is presented in the works [1, 9]. The problem of motion of disintegrating meteoroids is solved in three steps. On the first step we consider the motion of the monolithic body from the altitude of entry into the atmosphere before crushing, on the second – the movement of a swarm of fragments from the altitude of the beginning of fragmentation to the altitude of maximum velocity head. In the third step, since the fragments are considered to have the same size, the movement of a single fragment is tracked.

According to this model the strength of the fragment will take the following form:

$$\sigma^*_f = \sigma_0 \left(M_0 / M_f \right)^\alpha , \tag{1}$$

where σ_0, M_0 are the tensile strength and mass of the meteoroid before entering the atmosphere, σ^*_f, M_f are the same characteristics for a fragment; α is the indicator of heterogeneity of the material (larger α means greater heterogeneity).
The condition of the beginning of the destruction of the bolide in the atmosphere is:

$$\rho_* V_*^2 = \sigma^* , \tag{2}$$

where on the left is the magnitude of the velocity head, σ^* is one of the strength characteristics of the material of the meteoroid (compressive, tensile, shear).

The height of the beginning of fragmentation in the exponential atmosphere z_* is determined assuming that at this point, the body does not have time to decelerate and its speed is the initial speed of atmospheric entry V_0:

$$z_* = h \ln \left(\rho_0 V_0^2 / \sigma^* \right) \tag{3}$$

From this height, instead of a single body a swarm of fragments is falling, with increasing number N. Assuming that the resulting fragments of the sphere of equal mass M_f ($M_f = M / N$), from (1-3) is obtained, their number depend on the current values of dynamic pressure and the total mass of all fragments and the area of the Midsection (determined on the assumption that the fragments do not overlap):

$$N = \frac{M}{M_*} \left(\frac{\rho V^2}{\rho_* V_*^2} \right)^{1/\alpha} = \frac{M}{M_*} \left(\frac{\rho V^2}{\sigma^*} \right)^{1/\alpha} ; \quad S_{mid} = S_{mid*} \frac{M}{M_*} \left(\frac{\rho V^2}{\rho_* V_*^2} \right)^{1/3\alpha}$$

3. The results of the calculations

In [10] the following observations of the Chelyabinsk meteorite are provided: the body flew into the Earth's atmosphere at an angle ~ 18° to the horizon with a speed of ~ 19.2 km/s; the size of the meteorite ~ 19.8 ± 4.6 m; the final explosion occurred at an altitude of about 23 km. The collected fragments were ordinary chondrite structures, the effective enthalpy of evaporation of such material is assumed to be H_{eff} = 8kJ/g. Body collapsed in several stages: the destruction began at an altitude of ~ 45 km. The value of the strength parameter corresponding to the altitude of beginning of fragmentation, in this case is $\sigma_* = 10^6$ N/m^2. Assuming that the meteor is a ball of radius R ~ 9.8 m, and the characteristic chondrite density is 3.3 g/cm^3, its mass would be approximately 13,000 tons. These data are used in this paper in the simulation of flight and destruction of the Chelyabinsk meteorite in the atmosphere.
Sikhote-Alin meteorite is a typical iron meteorite. Effective enthalpy of vaporization of iron is significantly below this value for a chondrite and is about 2 kJ/g. According to the description of eyewitnesses, the fiery iron rain fell on the forest of the Sikhote-Alin in Primorsky Krai of Russia, on 12 February 1947. Witnesses said that the bolide was brighter than the sun. The earth's atmosphere was invaded by a space body several meters in diameter and weighing hundreds of tons. The fall of the meteorite occurred at an angle of 41°. When moving through the atmosphere it experienced repeated fragmentation. The first fragmentation occurred at an altitude of about 58 km [11]. According to these data, the critical value of the strength parameter, which begins the process of fragmentation is estimated to be 10^5 N/m^2.
Beneshov bolide was one of the most striking meteoroids, the speed at which the entry was made being over 21 km/s. The separation of fragments from the main body was observed at the height of 42 km. The final extinction of the bolide occurred at an altitude of 19 km at a speed of compact swarm of

debris - 5.2 km/s [12]. Twenty-three years scientists could not find the fragments of the meteorite, but thanks to modern techniques and more accurate calculations, was 4 pieces were discovered, with a total mass of approximately 12 kg. The estimated initial mass of the meteoroid Beneshov amounted to 2 - 4 t [12]. Meteorite Kunya-Urgench had the initial mass about the same as meteorite Beneshov, about 3 t. It flew into the atmosphere at a speed of 13 km/s at an angle of 30° to the horizon. According to the observations, the height of the beginning of the fragmentation was ~ 25 km, the explosion occurred at an altitude of 10-15 km, after which the meteorite began to fall almost vertically [13]. The largest part of the meteorite, weighing 820 kg, fell in a cotton field, forming a crater with a diameter of about 5 meters. Fig. 1 shows the heat transfer parameters for all of these meteorites depending on the flight altitude H, calculated in the framework of a single-body model (excluding fragmentation).

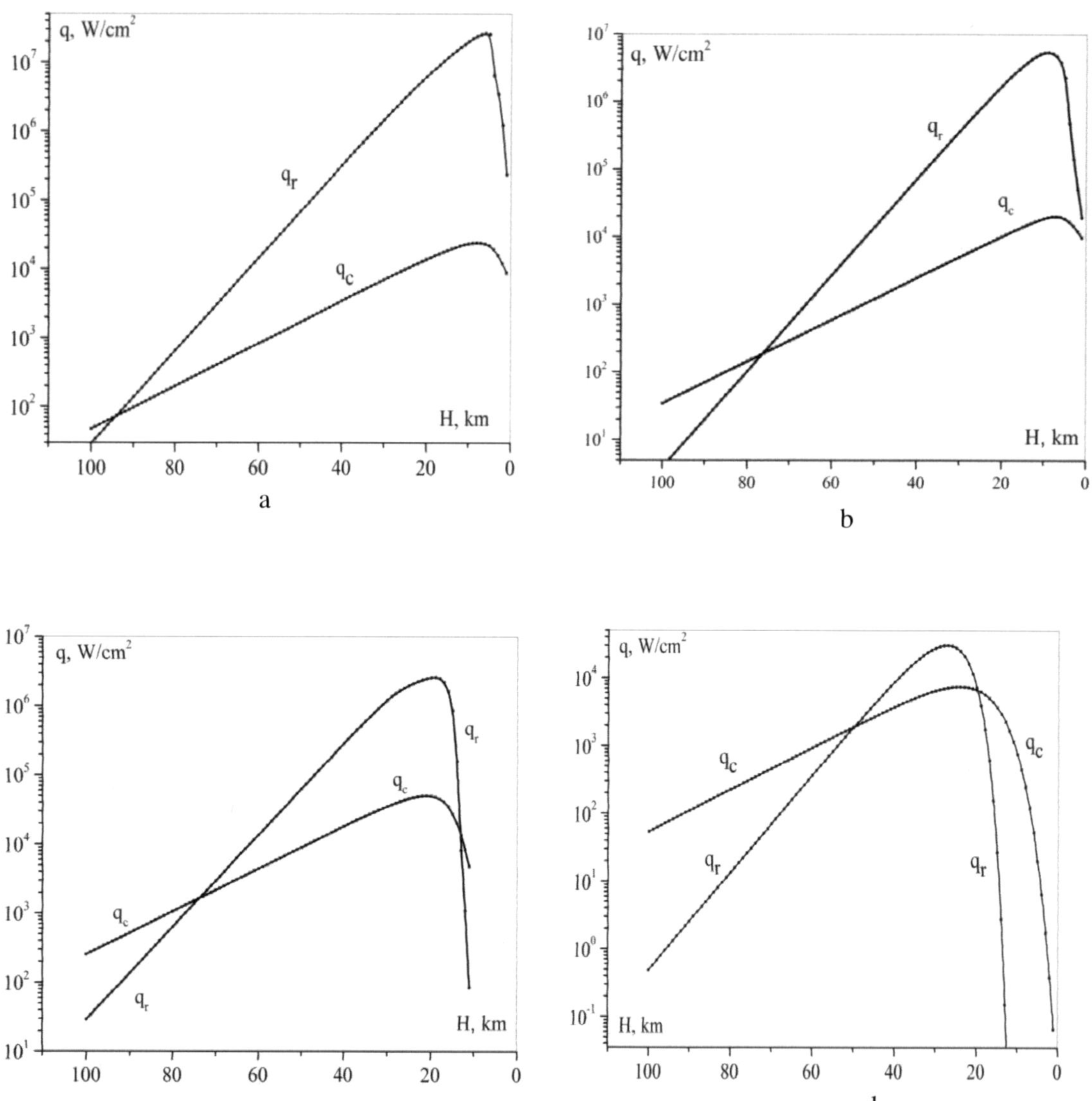

Figure 1. The values of convective and radiative transfer at the critical point depending on the height of the flight of meteorites: a - Chelyabinsk; b – the Sikhote-Alin, c - Beneshov, d - Kunya-Urgench.

The calculation of parameters of heat exchange conducted in the framework of the model show that for Chelyabinsk and the Sikhote-Alin meteorites the radiation heat flux is far superior to the convective one almost all the way down (high speed flight and large body size). While for a meteorite like Kunya-Urgench (small speed of flight and small body size) the values of radiative heat flux are several orders of magnitude lower, only at altitudes of 50-20km there is a slight predominance of the radiant flux over the convection. For the Beneshov meteorite with a size comparable to Kunya-Urgench, on the contrary, radiative heat transfer dominates the convective, mostly because of the high speed flight.

Change of the total mass of meteoroids depending on the altitude in the model of a single body and a disintegrating body is shown in Fig.2. It should be noted that ablation of mass at first increases dramatically due to the increase in surface area in the fragment stream. But this leads to increased suppression of the fragmentation of the body, leading to decreased velocities of the meteoroid and its fragments, which in turn reduces the heat flow to the surface and slows down the process of ablation. Therefore, the result of the entrainment of its mass compared to a non-fragmented bolide is unclear, as illustrated by the data in Fig.2.

The data in Fig.2 shows different values of the parameter α characterizing the degree of heterogeneity of the material. The number of the generated fragments strongly depends on the value of the parameter. For example, the Chelyabinsk meteorite with $\alpha = 1/2$ has the number of fragments $N \sim 233$, while $\alpha = 1/8$ corresponds to $N \sim 185900$; for the Sikhote-Alin fireball when $\alpha = 1/4$, the number of fragments is the order of $2 \cdot 10^5$. For bolide Beneshov the number of slices varies from 8 to 228 when the parameter α is 4/5 to 1/4, and for meteorite Kunya-Urgench from 2 to 6, with α changing from 1/2 to 1/8. The strength of meteoroids varies widely: 10^5-10^7 N/m^2. Usually little bodies are more uniform than large, the critical strength parameter is higher, so the number of generated fragments N is small.

It should be noted that according to Fig.2 the total mass of the fragments of the considered meteorites is much higher than the observed data. For instance, the weight of the found fragments of the Chelyabinsk meteorite does not exceed 1T, about 1T in the case of meteorite Kunya-Urgench, while for meteorite Beneshov it is negligible - not more than 12 kg. After the fall of the Sikhote-Alin bolide tens of thousands of fragments were collected with a total weight of about 27 tonnes.

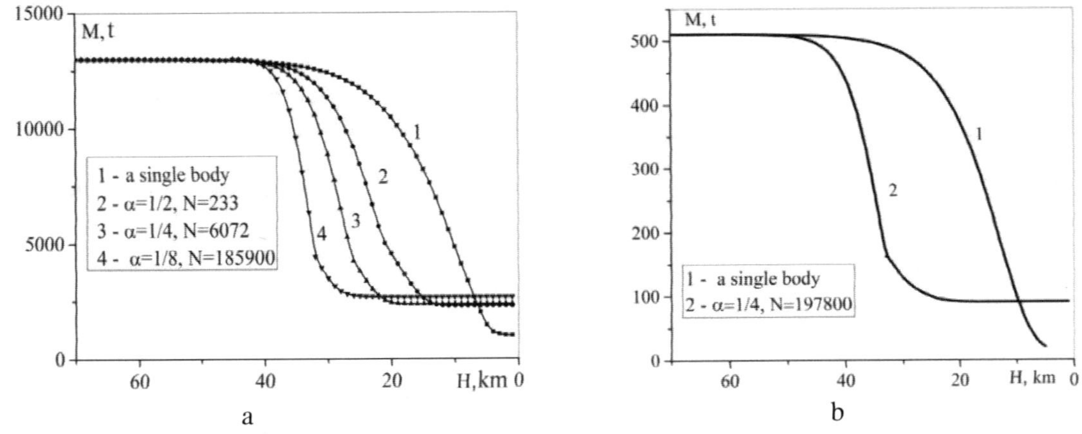

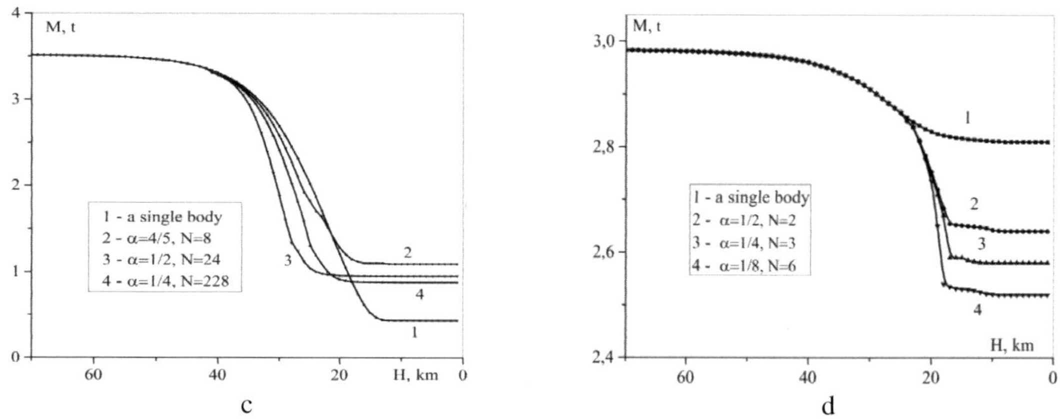

Figure 2. The change of the total mass of meteoroids depending on altitude for different values of parameter α. a – Chelyabinsk meteorite; b – Sikhote-Alin, c - Beneshov, d - Kunya-Urgench.

However, during the research it was found that at the final stage of the movement of meteoroid, destruction process can continue due to the thermal stress. Thermal stresses do not play a big role for large meteoroids, but if the size of the chunks reaches several centimeters, the resulting temperature gradients can break them into further small pieces to the size of coarse dust, which rapidly melts and evaporates in high temperature air. The complexity of the considered mechanism of fragmentation of meteorites is also demonstrated by evidence given in [10]. Lots of small (no larger than 2 cm) fragments of the Chelyabinsk meteorite were found over an extensive area. Among the collected cm of debris there were pieces with partial crust melting, which suggests that the split continued after the loss of meteor speed, and this process is caused by thermal stresses resulted in an additional mass loss. In [14] the evaluation showed that for bodies of 10 cm radius the time to reach the critical stress is ~ 4s. This means such a body during the fall through the atmosphere may be subjected to destruction due to thermal stress for several times. This mechanism may explain, for example, why the many tons of the Chelyabinsk meteorite were disintegrated into many small fragments.

Simultaneously with ablation, an other intensive meteorite destruction process can join in the so called Gertler vortexes. They arise in the boundary layer of the flow near the surface defects and present plasma microvortexes spinning at great velocities. The vortexes literally "bite" into the surface of the meteorite and "drill" her to deepen regmaglypt, which in turn stimulates the release of small fragments, which quickly decelerate in the atmosphere, fail to fully evaporate, and then fall as meteorites to the Earth along the flight path of the fireball.

It is confirmed that the possible fragmentation of medium-sized meteorites does not always influence its trajectory. For example, when observing and registering the meteoroid in Beneshov, the photos clearly showed fragmentation of meteoroids: it was a movement of 8 fragments before the final destruction. But according to observers [15] the mass of the separated fragments was much lower than initial body weight. In [16] it is shown that the trajectory of the fireballs is not determined by the model of progressive fragmentation, which involves the destruction of approximately equal fragments, and depends mainly on ablative processes. In this regard, for an approximate description of the trajectories and parameters of the Beneshov bolide the single-body model is more suitable, it gives the final mass of the meteorite closer to reality.

One of the important characteristics in meteor physics is the loss of kinetic energy of a body per unit length depending on the altitude. Suppression of meteorite fragmentation at the stage of falling leads to consequences such an explosion – an instantaneous release of energy at some point of the trajectory. In meteor physics an "explosion" of the meteoroid in flight is a rapid loss of kinetic energy of the body with its transition into the kinetic and internal energy of the surrounding gas.

Given the equations of the system (1) one can determine the energy released in the atmosphere, taking it equal to the loss of kinetic energy of the body E:

$$\frac{dE}{dz} = \frac{S_{mid}\rho V^2}{2\sin\theta}\left(C_D + C_H \frac{V^2}{2H_{eff}}\right)$$

Fig. 3 shows the calculated curves of the kinetic energy losses per unit length calculated for the considered bodies with and without fragmentation at various values of the parameter α. It is clear that the maximum energy loss, the so-called explosion, for the Chelyabinsk meteorite occurs at 12 km height when using the model of a unified body, and at 20-35 km when fragmentation model is used. The strength parameter α was changed form 1/2 to 1/8 (observations note the height of the main explosion, is equal to 23 km).

.

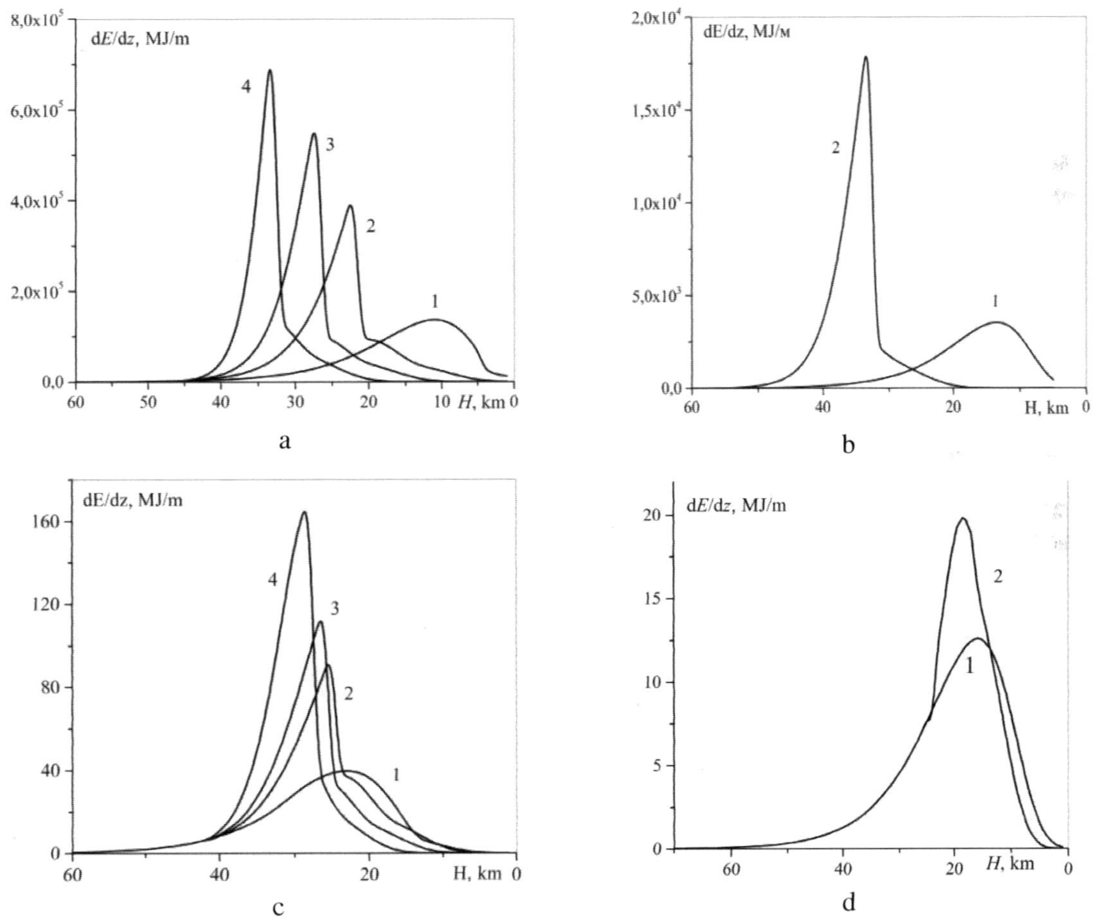

Figure 3. The loss of kinetic energy per unit length depending on flight height with different values of scale parameter for meteorites:
a – Chelyabinsk meteorite; 1 – a single body, 2 – disintegated meteorite, α=0.5, 3 – disintegrated meteorite, α=0.25, 4 – disintegrated meteorite, α=0.125;
b – Sikhote-Alin meteorite; 1 – a single body, 2 – disintegrated meteorite, α=0.25;
c – Beneshov meteorite; 1 – a single body, 2 – disintegrated meteorite, α=0.8, 3 – disintegrated meteorite, α=0.5, 4 – disintegrated meteorite, α=0.25;
d – Kunya-Urgench meteorite, 1 – single body, 2 – disintegrated meteorite, α=0.25.

For the Sikhote-Alin meteorite the calculated curve in Fig.4b, corresponds to fragmentation with the strength parameter α =0.25, has a maximum height of H=33km. In this case, the meteor body splits into many fragments (~ $2 \cdot 10^5$). For Beneshov meteorite the single-body model most correctly reflects the trajectory of the meteorite, gives the height of the explosion, equal to 23 km, which roughly corresponds to the observed height (24 km). The maximum energy loss for the meteorite Kunya-Urgench estimated to occur at a height of 15-18 km, which is also close to the observation data (10-15 km).

Thus, understanding the processes of destruction of meteoroids in the atmosphere appears to be extremely important: it can give more information about the structure of natural space bodies and contribute to the creation of new models for predicting the movement and destruction of meteoroids.

References

[1] Andrushchenko V A et al 2013 (In Russian) *Computer research and modeling* vol 5 no **6** pp 927-940
[2] Andrushchenko V A et al 2007 *J. Appl. Math. and Mech* .vol 71 no **5** pp 747-754
[3] Murzinov I N 1966 (In Russian) *Izv. Acad Nauk SSSR. Mekh. Zhidk. i Gasa* no **2** pp 184-188
[4] Apstein E Z et al 1986 (in Russian) *Izv. Acad Nauk SSSR. Mekh. Zhidk. i Gasa* no **1** pp 92-97
[5] Shevelev Yu 1986 (In Russian) *Spatial problems of computational aero-and hydrodynamics* (Moscow: Nauka)
[6] Rumynsky A N and Churkin V P 1974 (In Russian) *Zh. Vychisl. Mat. i Mat. Fiz.* vol 14 no **6** pp 1553
[7] Stulov V P et al 1995 (In Russian) *The aerodynamics of the meteorite* (Moscow: Nauka)
[8] Weibull W A 1939 *Proc. Roy Swedish Inst. Eng. Res.* no **151** pp 1 - 45
[9] Tirskiy G A and Khanukaeva D Y 2008 *Cosmic Res.* vol 46 no **2** pp 120-132
[10] Emelyanenko V V et al 2013 *Solar System Res.* 2013 vol 47 no **4** pp 240-254
[11] Divari N B. 1959 *Phenomena accompanying the fall of a meteor shower, and its atmospheric trajectory / Sikhote-Alin iron meteoritic rain* vol 1 Moscow USSR Academy of Sciences pp 26-48
[12] Borovicka J et al 1998 *Astronom. and Astrophys.* vol **334** pp 713-728
[13] Bronshten V A 1999 *Astronom. Let.* vol 25 no **2** pp 119-121
[14] Andrushchenko V A et al 2016 *Mathem. Models and Comput. Simulation* vol 8 no **5** pp 506–512
[15] Borovicka J and Spurny P 1996 *Icarus* vol **121** pp 484-510
[16] Barry N G and Stulov V P 2003 *Solar System Res.* vol 37 no **4** pp 332-335

IOP Conf. Series: Materials Science and Engineering **468** (2018) 012026 doi:10.1088/1757-899X/468/1/012026

Determination of the satellite attitude motion according to the onboard measurements data

V A Pankratov and M Yu Barkin

Bauman Moscow State Technical University, Moscow, Russia
E-mail: v.a.pankratov@bmstu.ru

Abstract. The methods for reconstruction of the rotational motion of the Earth scientific satellites from measurements of onboard sensors are presented in this article. The goal of this reconstruction is to gain the data for the analysis of the residual microaccelerations that were affecting during space experiments. It is sufficient to know the quasi-static (low frequency) component of microacceleration, to analyze the results of many experiments with gravitationally sensitive systems and processes. The quasistatic component of the microacceleration has frequencies less than 0.01 Hz and is determined from information of the satellite motion by calculation.

1. Introduction

The methods based on the complete (dynamic and kinematic) equations of the satellite's rotational motion were applied in these articles [1–7]. These methods were used only one type of measurements: the Earth's Magnetic Field (EMF) [1, 2, 6, 7], or the satellite's angular velocity [3], or the satellite's microacceleration [5]. The motion was reconstructed by the least squares method using the solutions to the equations of satellite rotational motion. The time intervals in which these equations make reconstruction possible were from one to five orbital revolutions. To get an idea on microaccelerations and satellite motion during an entire flight, the motion was reconstructed in several tens of such intervals [1, 2, 6, 7]. And in the paper [7] were proposed a method for motion reconstruction suitable for an interval of arbitrary length. The method is based on the Kalman filter.

The complete equations contain explicit expressions for the moments of external forces applied to the satellite, and are intended to describe uncontrolled motion. Such expressions may not be accurate enough. And results of methods, which are based on complete equations, may be checked by the presented methods. Presented methods are based only on the kinematic equations and are used the measurements data of both types – the EMF and the satellite's angular velocity. The possibility of constructing a technique for determining rotational motion on the basis of only kinematic equations has been known for a long time. In the article [8] was submitted, that the onboard measurements of the Earth magnetic field and the angular velocity, provide the opportunity for the real-time satellite's motion determining. In this article are presented two views to the determining of the satellite's rotational motion: the integral-statistic method, which based on the least squares method, and the method, based on the Kalman Filter. The rotational motion of the satellites Foton-12 [9], Foton M-2 [10] and Foton M-3 [11] were constructed, using the integral-statistic method. An updated version of this method is represented in this paper.

The monitoring method, based on kinematic equations, will become the main one for processing experimental data, received from perspective satellites, which will be flying in the oriented mode. In this article is presented method to solve the problem, which was mentioned above, of constructing

Content from this work may be used under the terms of the Creative Commons Attribution 3.0 licence. Any further distribution of this work must maintain attribution to the author(s) and the title of the work, journal citation and DOI.
Published under licence by IOP Publishing Ltd

rotational motion during an entire flight. This method is based on the Kalman Filter and the kinematic equations.
The methods of reconstructing uncontrolled rotational motion of a low-orbit satellite are described using the example of Foton M-3.

2. The angular rate and the Earth's magnetic field strength measurement data

On board the satellite were four three-component magnetometers included in the *DIMAC* equipment [6, 7, 11]. The equipment was designed to measure microaccelerations on board the satellite. Its main sensors were accelerometers.

Let $Ox_1x_2x_3$ is the instrumental coordinate system rigidly fixed to the satellite. The EMF and the satellite's angular velocity onboard measurement data are interpreted in this system. Point O is the satellite's center of mass, axis Ox_1 is parallel to the satellite's length wise axis and is directed from the descending vehicle toward the instrument module.

The *DIMAC* system carried out 9 sessions of angular velocity measurements that covered 84-minute lengths of time. The measurement data were transmitted to the Earth by the telemetric channel, after the session.

The measurement data obtained during one session is a collection of numbers

$$t_n^{\Omega}, \Omega_1^{(n)}, \Omega_2^{(n)}, \Omega_3^{(n)} \quad (n = 0, 1, \ldots, N_{\Omega}),$$

where $\Omega_i^{(n)}$ $(i = 1,2,3)$ – the values of the satellite angular velocity components at the time t_n^{Ω}, $t_{n+1}^{\Omega} - t_n^{\Omega} = 12$ s in the instrumental coordinate system. These data were approximated by discrete Fourier series, independently for each vector component, for a posteriori approximation of satellite motion [9–11].

Magnetic measurements were carried out during the flight. Measurements of different magnetometers were digitized for combined time instants, the intervals between which were varied from 1 to 12 s; on average, they were about 5 s. The data were represented as a set of numbers:

$$t_n^{H}, h_1^{(n)}, h_2^{(n)}, h_3^{(n)} \quad (n = 0, 1, \ldots, N_H),$$

where $h_i^{(n)}$ $(i = 1,2,3)$ are the measured values of the vector components of the magnetic field strength at instant t_n^H, $t_0^H < t_1^H < \ldots < t_{N_H}^H$. We believe that these components with an accuracy up to constant shifts, as well as small errors of measurements and coordinate transformations, coincide with the components of the EMF strength in the $Ox_1x_2x_3$ coordinate system.

3. Kinematic equations of the satellite motion

We assume satellite is a rigid body. The motion of the center of mass of the satellite is the Kepler ellipse. The elements of this motion are determined from the data of trajectory measurementsents [9–11].We will use the instrumental coordinate system $Ox_1x_2x_3$ and the Greenwich coordinate system $CY_1Y_2Y_3$ for writing the equations of motion of the satellite relative to the center of mass and the formulas for the measurement data processing. Point C is the center of mass of the Earth, plane CY_1Y_2 contains the equator, the positive semiaxis CY_1 intersects the Greenwich meridian, and the CY_3 axis is directed toward the North Pole. The position of the coordinate system $Ox_1x_2x_3$ relative to the Greenwich coordinate system will be given by the unit quaternion $\mathbf{Q}$, when the points C and O are connected.

The kinematic equations for the components of the quaternion $\mathbf{Q}$ takes form

$$2\dot{Q}_0 = -\sum_{i=1}^{3} Q_i\omega_i, \quad 2\dot{Q}_i = Q_0\omega_i + \sum_{j,k=1}^{3} e_{ijk}Q_j\omega_k, \quad \omega_i = \Delta_{\Omega i} + \chi_i(t-\tau) - \omega_e g_{3i} \quad (i=1,2,3). \quad (1)$$

Here e_{ijk} is the Levi-Civita symbol, ω_i is the components of the absolute angular velocity vector of the satellite in the system $Ox_1x_2x_3$, $\chi_i(t)$ is the trigonometric polynomials, approximating the angular velocity measurements, $\Delta_{\Omega i}$ are the constants of the bias in these measurements, τ is the offset of the time scale of the *DIMAC* instrument with respect to the time scale of the satellite motion control system, ω_e is the angular velocity of the rotation of the Earth, $\| g_{ij} \|_{i,j=1}^{3}$ is the transition matrix from the system $Ox_1x_2x_3$ to the system $CY_1Y_2Y_3$.

4. Reconstruction of the satellite motion by the least squares method, using kinematic equations

Let $I_\Omega(\tau) = \left[t_0^\Omega + \tau, t_{N_\Omega}^\Omega + \tau\right]$. The equations (1) are considered for values τ и t, satisfying conditions $I_\Omega(\tau) \subset \left[t_0^H, t_{N_H}^H\right]$, $t \in I_\Omega(\tau)$. As an approximation to the actual motion of the satellite on the $I_\Omega(\tau)$ was taken the solution to the equations (1), giving a minimum to the functional

$$\Phi = \sum_{i=1}^{3} \Big\{ \sum_{n \in U(\tau)} [h_i^{(n)} - \hat{h}_i(t_n^H)]^2 - N_\tau \tilde{\Delta}_{Hi}^2 \Big\},$$

$$\tilde{\Delta}_{Hi} = \frac{1}{N_\tau} \sum_{n \in U(\tau)} [h_i^{(n)} - \hat{h}_i(t_n^H)], \quad U(\tau) = \{n : t_n^H \in I_\Omega(\tau)\}.$$

Here, $\tilde{\Delta}_{Hi}$. are estimates of constant shifts in the EMF measurements, $\hat{h}_i(t) = \sum_{j=1}^{3} H_j(t) g_{ji}$, $H_j(t)$ is the calculated values of the components of the EMF strength, and N_τ is the number of elements of the set $U(\tau)$. Φ is minimized according to the initial conditions of the solution $\mathbf{Q}(t_0^\Omega + \tau)$ and shifts $\tilde{\Delta}_{Hi}$ $(i = 1,2,3)$, τ taking into account the normalization condition $\| \mathbf{Q}(t_0^\Omega + \tau) \| = 1$.

The motion of the Foton M-3 satellite was restored in 9 time intervals, corresponding to the angular velocity measurements, using a method, based on the least squares method. An example of approximating the actual motion of the satellite Foton M-3 is shown in the figure 1. The time in this figure is measured in minutes from the moment $t_0^\Omega + \tau$ to $t_{N_\Omega}^\Omega + \tau$. The graphs of expressions $\chi_i(t)$, approximating the angular velocity measurements, and broken lines, passing through the points $(t_n^\Omega, \Omega_i^{(n)})$, were placed in the left part of the figure. The graphs of functions $\hat{h}_i(t) + \tilde{\Delta}_{Hi}$ and broken lines, passing through the points $(t_n^H, h_i^{(n)})$ $n \in U(\tau)$, were placed in the right part of the figure. Each broken line and the plot approximating its function are shown in the universal coordinate system. The broken lines and their approximating graphs are almost identical, with the chosen scale. The results of the constructed reconstructions are consistent with the results obtained in the articles [9–11].The difference between the above method and the technique used [9–11] is the refusal of EMF measurement data compression. All EMF measurements are included in the processing, integration of the motion equations is carried out with an optimal step in the presented method. The calculated analogs of the measurements are calculated in the measurement points, using polynomials that interpolate the computed solution within the integration step [7]. Such polynomials for Runge-Kutta methods have entered the wide practice of computations in the last quarter of the 20th century [12]. Interpolation polynomials, enlargement of personal computers speed and memory capacity have made it possible to modify the software of integral-statistic methods.

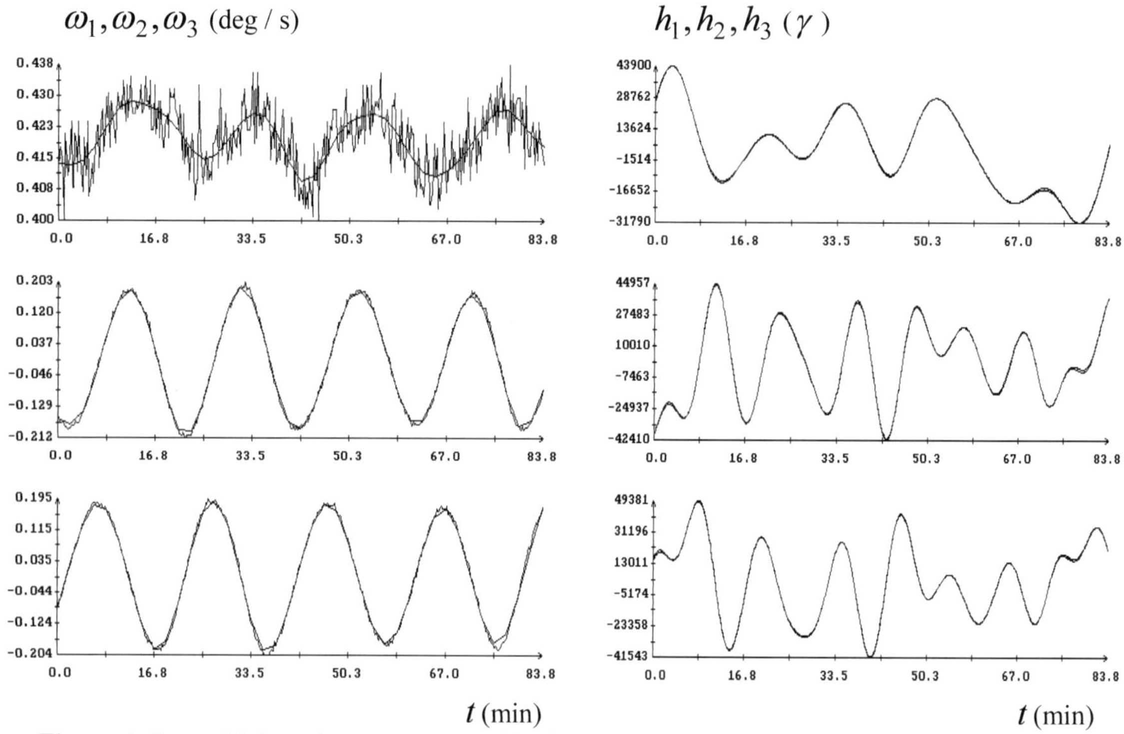

Figure 1. Foton M-3 motion reconstructed by least squares method. Time instants t = 0 in plots correspond to 12:27:24 DMT on September 22, 2007, $N_\tau = 1109$, $\sigma = 567\gamma$.

5. Reconstruction of the satellite motion, using the Kalman filter

The satellite motion mathematical model and the measurement process in the nonlinear Kalman filter, is represented in the form [13]

$$x_n = F_n(x_{n-1}) + \eta_n, \quad y_n = G_n(x_n) + \xi_n \quad (n = 1, 2, \ldots). \tag{2}$$

The state vector components x_n are independent parameters, by which the variables of the satellite motion equations (1) values are expressed in the nodes of the time grid $\{t_n\}$, $t_n < t_{n+1}$ $(n = 0, 1, 2, \ldots)$. Shifts in measurements of the angular velocity and the magnetic field, which are constants on the $t_{n-1} < t \le t_n$, are in the set of the vector x_n components, thus $\dim x_n = 10$.

All measurements of the EMF, falling within the $t_{n-1} < t \le t_n$, are included into the measurement vector y_n. The dimension of the vectors y_n depends on the n. These vectors may be missing, for some values of n, but typically $\dim y_n = 6$. The functions $F_n(x)$ and $G_n(x)$ are calculated by integrating the equations of the satellite rotational motion in the interval $t_{n-1} < t \le t_n$. These functions are smooth, their Jacobi matrices have full rank, under the assumptions were made. And, in the (2), ξ_n is the vector of random measurement errors, η_n is the vector of random disturbances of the system. Let the vectors ξ_m and η_n are uncorrelated for all values of m and n. The vectors ξ_m and ξ_n, as well as η_m and η_n, are uncorrelated for all $m \ne n$. The expected values and covariance matrices of these vectors are determined by the formulas

$$\mathrm{M}\xi_n = 0, \quad \mathrm{M}\eta_n = 0, \quad \mathrm{M}\xi_n\xi_n^T = K_n, \quad \mathrm{M}\eta_n\eta_n^T = L_n.$$

Here, the K_n and L_n are positive definite matrices; the order of the matrix K_n is equal the dimension of the vector y_n, the order of the matrix L_n is 9.

An estimate of the vector x_n by the measurements y_k $(k=1,2,\dots,n)$ is given by the expected values $\hat{x}_n$ and the covariance matrix P_n. An estimate of the vector x_n by the measurements y_k $(k=1,2,\dots,n)$ is given by the expected values $\hat{x}_n$ and the covariance matrix P_n. Recurrent formulas, for calculating these values, can be obtained by the least squares method. It is more comfortable to work with inverse values of the covariance matrices, if variable dimensionality of measurement vectors is a large. Let $\hat{x}_{n-1}$ и P_{n-1}^{-1} are known. The vector's x_n prediction has a expected value and a covariance matrix $P_n^{'}$, where

$$P_n^{'-1} = Z_n - Z_n(L_n^{-1}+Z_n)^{-1}Z_n, \quad Z_n = (A_n^{-1})^T P_{n-1}^{-1} A_n^{-1}, \quad A_n = \frac{\partial F_n(\hat{x}_{n-1})}{\partial x}.$$

The refined value of the vector, x_n by the new set of measurements y_n, is found from the condition of the minimum of the function

$$f(x_n) = (x_n - x_n')^T P_n^{'-1}(x_n - x_n') + [y_n - G_n(x_n)]^T K_n^{-1}[y_n - G_n(x_n)],$$

following the least squares method, i. e. $\hat{x}_n = \operatorname{argmin} f(x_n)$. The refined covariance matrix satisfies equation

$$P_n^{-1} = P_n^{'-1} + B_n^T K_n^{-1} B_n, \quad B_n = \frac{\partial G_n(\hat{x}_n)}{\partial x}.$$

The nodes of the time grid $\{t_n\}$ are the midpoints of the $[t_{n-1}^{\Omega}, t_n^{\Omega}]$. The function is minimized by the Gauss-Newton [14] method, the initial point is the $\hat{x}_{n-1}$. The normal equations of the least squares method are

$$\left(P_n^{'-1} + B_n^T K_n^{-1} B_n\right)\Delta x_n = B_n^T K_n^{-1}[y_n - G_n(x_n^*)] - P_n^{'-1}(x_n^* - x_n'),$$

$$\Delta x_n = x_n - x_n^*, \quad B_n = \frac{\partial G_n(x_n^*)}{\partial x}.$$

Here x_n^* is the current approximate value of $\hat{x}_n$, which after the solution of the normal system is assigned the value $x_n^* + \Delta x_n$. The refinement of x_n^* stops, when the norm $\|\Delta x_n\|$ becomes an acceptable small. Finally, $\hat{x}_n = x_n^*$ is accepted.

The prediction is carried out by integrating the equations of rotational motion and the corresponding equations in variations from the point t_{n-1} to the point t_n. The iteration refinement of the function $f(x_n)$ minimum point was carried out by repeatedly integrating these equations from the point t_{n-1} to the point t_n.

The equations of rotational motion are integrated by the difference scheme

$$\mathbf{Q}(t_n) = \mathbf{Q}(t_{n-1})^{\circ}\boldsymbol{\kappa}_n,$$

$$\boldsymbol{\kappa}_n = \left(\frac{16 - \tau_n^2 |\boldsymbol{\omega}(t_n^{\Omega})|^2}{16 + \tau_n^2 |\boldsymbol{\omega}(t_n^{\Omega})|^2}, \frac{8\tau_n \boldsymbol{\omega}(t_n^{\Omega})}{16 + \tau_n^2 |\boldsymbol{\omega}(t_n^{\Omega})|^2}\right), \ \omega_i(t_n^{\Omega}) = \Omega_i^{(n)} + \Delta_i(t_n), \ \tau_n = t_n - t_{n-1}.$$

This scheme saves the quaternion norm, its local truncation error is on the order of $O(\tau_n^{\ 3})$. The equations in variations are integrated by a similar scheme with order local truncation error is on the order of $O(\tau_n^{\ 3})$.

An example of approximating the motion of the satellite by the Kalman Filter is demonstrated in the figure 2. This approximation is constructed in the same time segment as the reconstruction in the figure 1. The graphs of broken lines, passing through the points $(t_n^\Omega, \Omega_i^{(n)} + \Delta_i(t_n))$, were placed in the left part of the figure. These broken lines demonstrate the angular velocity used to construct the rotational motion by the Kalman filter. The graphs of broken lines, passing through the points $(t_k^H, h_i^{(k)})$ and $(t_k^H, \hat{h}_i(t_k^H) + \Delta_{Hi}(t_n))$, were placed in the right part of the figure 2. The broken lines are almost identical, with the chosen scale. The results of the reconstruction by the Kalman filter are also completely consistent with the results obtained by integral-statistic method.

This work was supported by the Russian Foundation of Basic Research (projects 17-01-00143 and 17-07-00653).

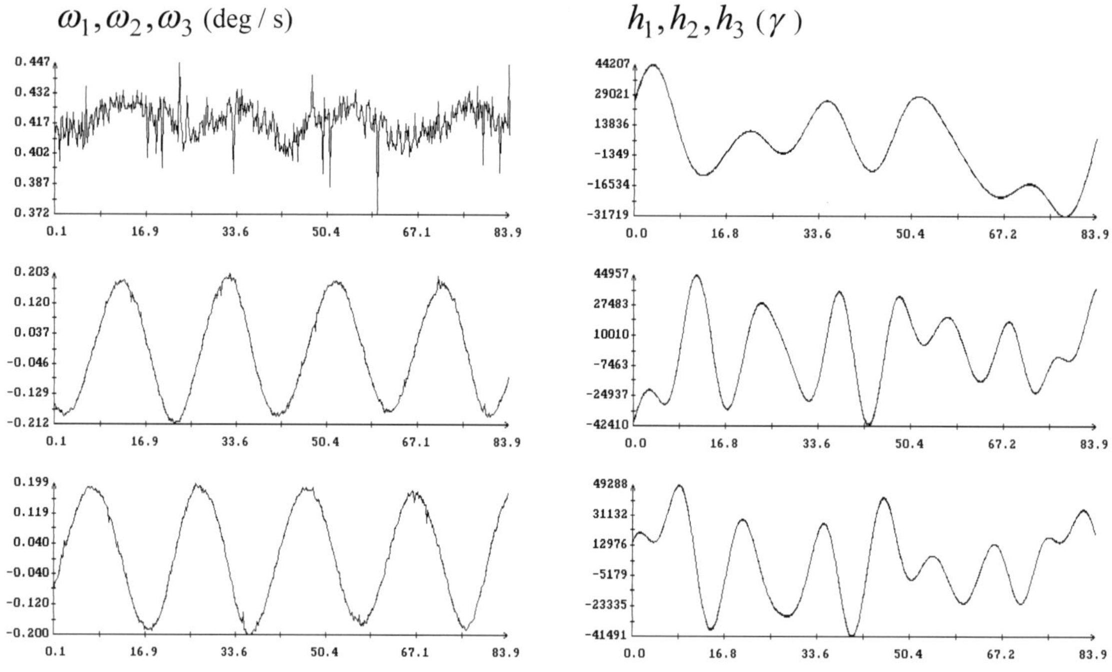

Figure 2. Foton M-3 motion reconstructed with Kalman filter. Time instants t = 0 in plots correspond to 12:27:18 DMT on September 22, 2007.

References

[1] Abrashkin V I, Balakin V L, Belokonov I V et al. 2003 Uncontrolled attitude motion of the Foton-12 satellite and quasi-steady microaccelerations onboard it *Cos. Res.* vol. **41**, no. 1, pp 39–50.

[2] Abrashkin V I, Bogoyavlenskii N L, Voronov K E et al. 2007 Uncontrolled motion of the Foton M-2 satellite and quasistatic microaccelerations on its board *Cos. Res.* vol. **45**, no. 5, pp 424–444.

[3] Abrashkin V I, Kazakova A E, Sazonov V V and Chebukov S Yu 2008 Determination of attitude motion of the Foton M-2 satellite using the data of onboard measurements of its angular velocity *Cos. Res.* vol. **46**, no. 2, pp 146–165.

[4] Beuselinck T, Van Bavinchove C, Sazonov V V and Chebukov S Yu 2008 An analysis of low-frequency component in microacceleration measurements made onboard the Foton M-2 satellite *Cos. Res.* vol. **46**, no. 5, pp 436–455.

[5] Beuselinck T, Van Bavinchove C, Sazonov V V and Chebukov S Yu 2009 Determination of

Foton M-2 satellite attitude motion by the data of microacceleration measurements *Cos. Res.* vol. **47**, no. 6, pp 500–512.
[6] Beuselinck T, Van Bavinchove C, Abrashkin V I et al. 2010 Determination of attitude motion of the Foton M-3 satellite according to the data of onboard measurements of the Earth's magnetic field *Cos. Res.* vol. **48**, no. 3, pp 246–259.
[7] Pankratov V A, Sazonov V V 2016 Reconstruction of spacecraft rotational motion using a Kalman filter *Cos. Res.* vol. **54**, no. 3, pp 251–267.
[8] Pivovarov M L 1985 Determination of Satellite Orientation Using Measurements of Angular Velocities *Kosm. Issled.* vol. **23**, no.3, pp 331–334.
[9] Abrashkin V I, Volkov M V, Voronov K E, Egorov A V, Kazakova A E, Pankratov V A, Sazonov V V, Semkin N D 2005 Determination of the attitude motion of a satellite based on kinematic model of motion and using data of measurements of its angular velocity and the Earth's magnetic field strength, *Cosmic Res.*, vol. **43**, no. 4, p. 280.
[10] Abrashkin V I, Bogoyavlensky N L, Voronov K E, Kazakova A E, Pankratov V A, Sazonov V V, Semkin N D, Stratilatov N R 2006 Determining the attitude motion of the Foton M-2 satellite on measurements of its angular rate and the strength of the Earth magnetic field using the kinematical model of the motion *Keldysh Institute preprints*, **060**, p 28
[11] Beuselinck T, Van Bavinchove C, Abrashkin V I, Kazakova A E, Pankratov V A, Sazonov V V 2009 Determining the Attitude Motion of the Spacecraft Foton M-3 on Measurements of its Angular Rate and the Strength of the Earth Magnetic Field, *Keldysh Institute preprints*, **069**, p 19
[12] Hairer E, Nørset S P and Wanner G 1993 *Solving Ordinary Differential Equations. I. Nonstiff Problems* (Berlin–Heidelberg Springer)
[13] Bryson A and Yu-Chi Ho 1969 *Applied Optimal Control: Optimization, Estimation and Control* (Waltham, Mass Blaisdell)
[14] Gill Ph, Murray W and Wright M 1982 *Practical Optimization* (London Emerald Group)

IOP Conf. Series: Materials Science and Engineering **468** (2018) 012027 doi:10.1088/1757-899X/468/1/012027

The mathematical model of short-time forecast of the Earth axial rotation irregularity

M Yu Barkin[1], **V V Bondarenko**[2] **and V V Perepelkin**[2]

[1]Bauman Moscow State Technical University, Moscow, Russia
[2]Moscow Aviation Institute (National Research University), Moscow, Russia
E-mail: vadimkin1@yandex.ru

Abstract. The precision of the universal time UT1 forecast for the short time intervals (up to two months) is researched in this paper based on the celestial mechanical model of the Earth axial rotation irregularity. It is shown that use of the approximate low-parametric model for the problem of the Earth axial rotation velocity variation forecast on the short time intervals is justified. The approximate model is obtained by the averaging of the variable parameters that is exposed to small variations in consequence of perturbation factors nonstationarity. The method of weight coefficients distribution in approximation algorithm that takes into account the irregular oscillations of the Earth axial rotation velocity is suggested. The precision of the forecast that is identical to published by IERS one is achieved with the help of weighted least squares method.

1. The mathematical model of the Earth axial rotation irregularity

The creation of verified model of the Earth rotation irregularity forecast is important for increasing of precise characteristics of the navigation systems coordinate-temporal support. The forecast of the Earth axial rotation irregularity on short time intervals (up to two months) has essential significance for applications.

The variations of the Earth axial rotation velocity lead to the oscillations of the day time *l.o.d.* and the scale of universal time *UT1* [1]. The fundamental low-parametrical model of within-the-year variations of *l.o.d.* and *UT1* based on Euler-Liouville classic dynamical equations with variable tensor of inertia is quoted in paper [2].

Euler-Liouville classic dynamical equations with variable tensor of inertia are used in this paper for investigation of the deformable Earth axial rotation [2-4]:

$$\begin{gathered}\frac{d}{dt}J\boldsymbol{\omega}+\boldsymbol{\omega}\times J\boldsymbol{\omega}=\mathbf{M},\quad \boldsymbol{\omega}=(p,q,r)^T,\quad J=J^*+\delta J,\quad J^*=\text{const},\\ J^*=\text{diag}(A^*,B^*,C^*),\quad \delta J=\delta J(t),\quad \|\delta J\|<<\|J^*\|,\\ \mathbf{M}=\mathbf{M}^S+\mathbf{M}^L-\dot{\mathbf{h}}-\boldsymbol{\omega}\times\mathbf{h}.\end{gathered}\tag{1}$$

Here $\boldsymbol{\omega}$ is the angular velocity vector in fixed-in-the-Earth coordinate system [1], which is coordinated with ITRF qualitatively and quantitatively. The axes of the chosen coordinate system coincide approximately with the main central axes of inertia J^* of the Earth "frozen" figure with consideration of "equatorial bulge". It is postulated that small variations of tensor of inertia δJ include different harmonic components which are caused by regular perturbed action of gravitational

Content from this work may be used under the terms of the Creative Commons Attribution 3.0 licence. Any further distribution of this work must maintain attribution to the author(s) and the title of the work, journal citation and DOI.
Published under licence by IOP Publishing Ltd

diurnal tides from the Sun and the Moon and possibly other ones, for example, a year's, month's, etc. $\mathbf{M}^S$ and $\mathbf{M}^L$ are perturbed moment from the Sun and the Moon correspondingly, and $\mathbf{h}$ is the vector of angular momentum of atmosphere.
It is known [1-3] that data of measurements of the Earth axial rotation velocity include harmonics with different periods. The main components (with the largest amplitudes) have periods one year, half year, 13.66, 27.56 days and also 9.1 days and a third year which are less significant.
Differential equations of the model of tidal alteration of the Earth axial rotation velocity may be obtained from the third equation of system (1) which with the following designation $dUT1(\varphi,t) = UT1 - TAI$ assumes the form:

$$\frac{d}{dt}\left[(C^* + \delta C)d\dot{U}T1(\varphi,t)\right] = \mu(\varphi,t), \tag{2}$$

$$l.o.d.(\varphi,t) = -D_0\, d\dot{U}T1(\varphi,t),$$

here C^* and δC are axial angular momentum subject to deformations of the Earth "frozen" figure and its time variations correspondingly; φ is the angle of own rotation; $D_0 = 86400$ sec. is the duration of standard day. The value μ has the following structure:

$$\mu = r_0^{-1}(M_r^S + M_r^L - \dot{h}_3), \tag{3}$$

here M_r^{SL} are lunar-solar gravitational-tidal moments of forces with the main frequencies [2], h_3 is the moment of atmosphere zonal circulation momentum [4], $r_0 = 7.292115\times10^{-5}$ radian/sec.
The system of equations (2)-(3) ignoring the summands with components of moment of atmosphere momentum describe tidal variations of the Earth axial rotation velocity which were studied in papers [2, 3]. The approximate expressions of tidal regular variations of duration of the day and the correction *dUT1* for the universal time on within-the-year interval assume the form after own rotation averaging:

$$l.o.d.(\tau) = a_0 + \sum_{i=1}^{4} a_i \sin(2\pi\nu_i\tau + \alpha_i),$$

$$\frac{1}{365.25} dUT1(\tau) = c - a_0\tau - \frac{1}{2\pi}\sum_{i=1}^{4}\frac{a_i}{\nu_i}\sin(2\pi\nu_i\tau + \alpha_i), \tag{4}$$

Here $\nu_1 = 1$, $\nu_2 = 2$, $\nu_3 = 13.25$, $\nu_4 = 26.73$ are the frequencies of annual, a half year's, a month's and two week's oscillations caused by lunar-solar perturbations; α_i are phases of corresponding oscillations; the unknowns a_i, c are the values that must be calculated with the help of the least squares method from IERS measurements. The parameter τ in (4) and here is counted in standard years.

2. The moment of atmosphere momentum and fluctuations of the day duration

It is necessary to account the geophysical perturbations that is occurred as result of irregular and quasi-stationary motion of masses in atmosphere and hydrosphere [5] and that influences on variations of axial moment of inertia and on redistribution of total angular momentum for refinement of main celestial-mechanics model. But let's get the objective to account these perturbations in some "average" sense remaining in the frames of considered mathematical model of the Earth rotation. Incidentally in practice objectives let's try to save the of low-parameters property of the model.
It is known [4] that the most significant quasi-regular oscillations of axial component of atmosphere angular momentum are the annual and half year oscillations. The main lunar harmonics is observed in data of the measurements also. It may be referred to the corresponding summands of the model (4) on the short time interval (up to two months) if we introduce quasi-constant amplitudes for these components.

It follows from analysis of the main components of spectral density of atmosphere angular momentum axial component that fluctuations of the day duration caused by atmosphere perturbations ignoring annual, half year's and lunar harmonics may be approximated by the set of quasi-periodic oscillations with the average frequencies $\vartheta_1 = 4.56$, $\vartheta_2 = 9.13$, $\vartheta_3 = 18.26$, $\vartheta_4 = 0.42$ cycles in a year on the short time intervals.
The availability of oscillations with these frequencies influences on the result of determination of constant coefficients of the main harmonics of the model (4). The reduction of its influence reduces to improvement of precision of the model local coefficients. In particular this may be achieved by introduction in special way the weights in the algorithm of the least squares method and its relative increasing to the end of interpolation interval.

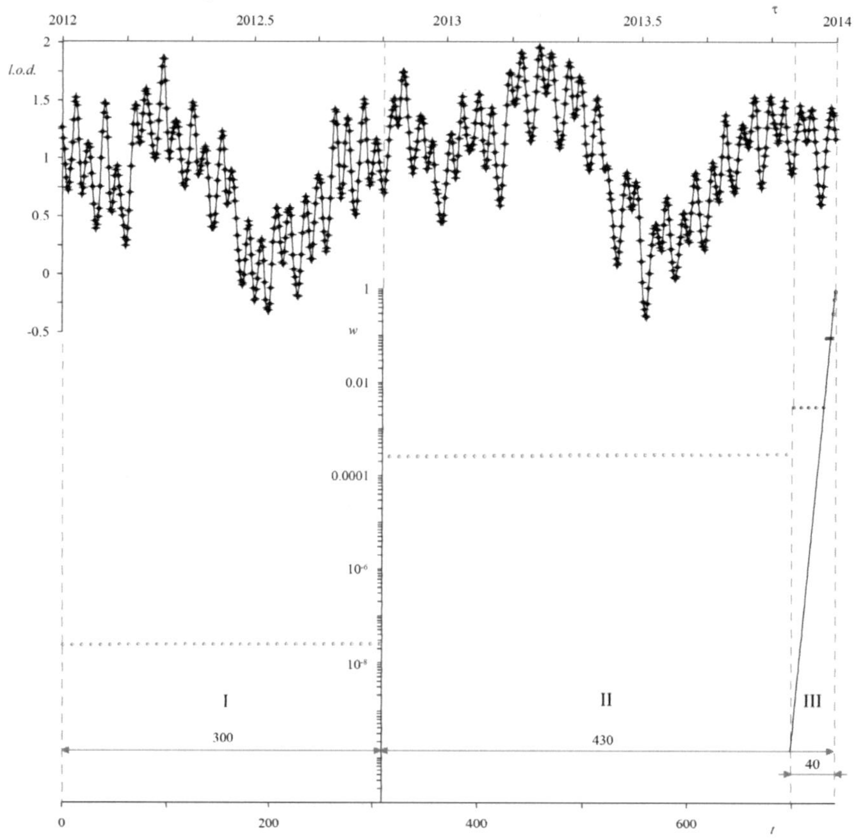

Figure 1. The variations of the day duration on the time interval from 2012 till 2014 according to data of IERS observations and measurements (upper chart); the distribution of weights w on two years interval (lower chart); the straight line is the exponent with parameters of increasing of the weight coefficients on the end of interval

It may be shown with the help of numerical calculations that the precision of determination of annual component coefficients is better on interpolation intervals being multiple of 2 years due to presence of quasi-two-years oscillation with approximate 2.4 years period, and determination of lunar harmonics coefficients is better on intervals that equals the half of quasi-two-years cycles that is 1.2 years or 440 days. In ones turn presence of harmonics with the frequencies ϑ_1, ϑ_2, ϑ_3 causes the increasing of weight coefficients on 40-days interval being abut directly with the forecast interval. The precision of coefficients determination of month's and two-week's harmonics for building of short-time forecast is higher on time intervals being multiple of such duration. Also the relative increasing of weight

IOP Conf. Series: Materials Science and Engineering **468** (2018) 012027 doi:10.1088/1757-899X/468/1/012027

coefficients on ten-day's interval being abut directly with the forecast interval yields higher precision of two-weeks harmonic forecast. In the end of interpolation interval the desired function is attracted to the last measurements with the help of weights which have exponential character of distribution. The choice of weight coefficients is illustrated on Fig. 1 for two-year's interpolation interval. The final parameters of weight functions were chosen during numerical calculations on test interval. The changing of interval durations for introducing of weight coefficients during the verification of the model on long-time interval in average carries to loss the precision of short forecast.

3. The numerical simulation: the estimation of short-time forecast precision

The approach of model of the Earth axial rotation irregularity to the main four-frequency model with constant coefficients by the averaging of variable parameters on the chosen intervals turns out be justified for simulation on short-time intervals.

As it was drawn attention above it is possible to increase precision of the forecast on short-time interval that being abut with the end of interpolation interval by introduction of weight coefficients in the algorithm of the least squares method and by its relative increasing to the end of the interval. The parameters that adjusts the weight factors and optimizes the algorithm depending on previous results of interpolation and forecast may be introduced in the filtration algorithm. The used adjustment of the model allows to dispense without additional long-periodic perturbed factors.

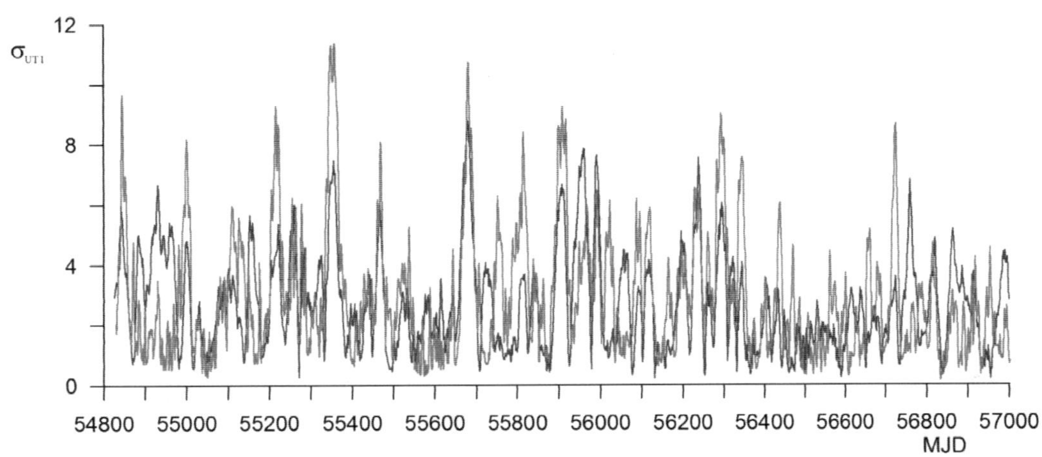

Figure 2. The average-square deviations of 45-day's forecasts of the elaborated model (red curve) and IERS forecasts (blue curve)

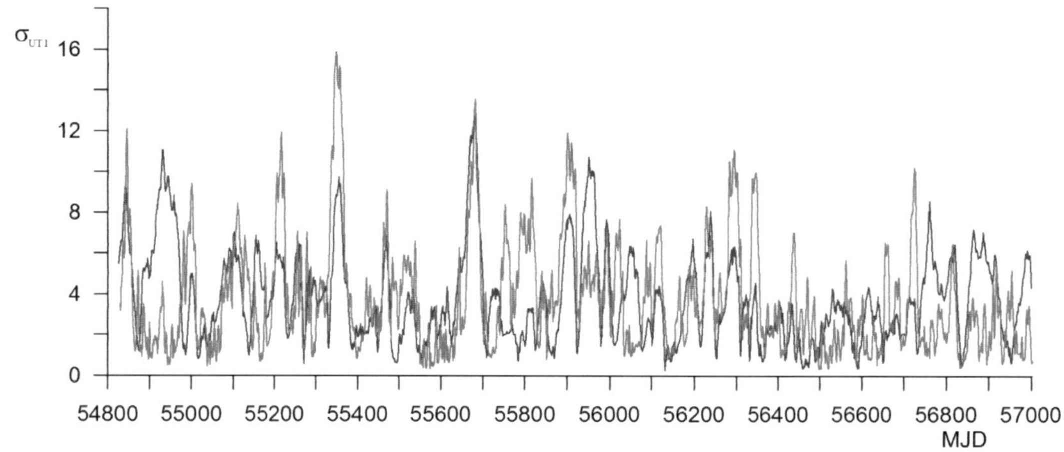

Figure 3. The average-square deviations of 60-day's forecasts of the elaborated model (red curve) and IERS forecasts (blue curve)

The numerical calculations were carried based on using of bearing functions of model (4) with the main frequencies of lunar-solar perturbations observed by IERS [1]. The time interval from 54826 MJD till 57060 MJD was chosen for estimation of precision of presented forecast model *dUT1*. The algorithm of determination of the model parameters bases on the weighted least-square method. The IERS forecasts published in "Bulletin A" were taken and the average-square deviations for 45 day and 60 days were calculated for comparison with obtained data.

Table 1.

Interval, days	σ_{UT1}, msec.	σ_{UT1}^{IERS}, msec.
45	2,917	2,685
60	3,755	3,805

The comparisons of the average-square deviations of 45-days and 60-day's forecasts of the elaborated model (red curve) and IERS model (blue curve) are presented on Fig. 2 and Fig. 3 accordingly. The average-square deviations are shown along ordinate axis on the graphs in milliseconds. The average deviations values of 45-day's and 60-day's forecasts of the model $dUT1$ (the second column) and IERS published forecasts (the third column) for whole forecast interval are presented in the Table.
The considered method of forecast based on low-parametric model in precision of process approximation gets a good agreement with observation data and IERS forecast.

References

[1] International Earth Rotation and Reference Systems Service - IERS Annual Reports, http://www.iers.org.
[2] Akulenko L D, Markov Yu G, Perepelkin V V 2009 *Cosmic Res.* **47**(5) pp 452-459
[3] Perepelkin V V 2011 *Mechanics of Solids* **46**(6) pp 960-968
[4] Akulenko L D, Markov Yu G, Perepelkin V V 2011 *Doklady Physics* **56**(1) pp 16 - 21
[5] Sidorenkov N S 2009 *The interaction between Earth's rotation and geophysical processes* (WILEY-VCH Verlag GmbH and Co. KGaA) p 317

IOP Conf. Series: Materials Science and Engineering **468** (2018) 012028 doi:10.1088/1757-899X/468/1/012028

Smart parallel robots for massage

S N Sajapin[1,2]

[1] Bauman Moscow State Technical University, Moscow, Russia
[2] Department of Vibration Biomechanics, Institute for Machine Science named after Blagonravov of Russian Academy of Sciences, Moscow, Russia
E-mail: S.Sayapin@rambler.ru

Abstract. A new concept smart robots for massage with plane and spatial parallel mechanisms is shown. Plane and spatial parallel robots (PPR and SPR) are based on triangular and octahedral structures. Every one of sides of triangle and ribs of octahedron is linear drive. Ends of the adjacent linear drive are connected with by cylindrical (for triangle) and spherical (for octahedron) joints. As result PPR and SPR have three and twelve degrees of freedom, and they can adapt to body of patient and. Possibility of full automation some kinds massage by autonomous portable smart PPR and SPR is shown. The use of these robots will allow the masseur remotely via the Internet to serve several patients at the same time in real time. PPR and SPR are portable multifunctional smart robots for various rehabilitation applications. Both kinds parallel robots will be discussed in detail.

1. Introduction

It is known that a massage is now considered as one of the most effective non-drug methods of effective meaning of disease prevention, treatment, rehabilitation, removal of weariness. Among the body parts most widely massaged are the back, the chest, and the upper and lower limbs, including the knee and elbow joints, as well as the neck. In Figure 1 some applying massage to the back (a), and the upper (b, and c) and lower (d) extremities, including the elbow joint (c), and also to patients' neck (e) are shown [1]. The masseur uses the following main approaches: vacuum cup massage (VCM), stroking and rubbing, lengthwise and transverse kneading of the upper and lower extremities, mobilization and manipulation at the elbow or knee joints, manipulation with vibration of limbs, manipulation by forced movement of the free part of the limb relative to the fixed part, etc. Because of the need for forceful actions on these body parts, the masseur undertakes prolonged physical loading during the working day, leading to fatigue and decreased productivity. Furthermore, only one patient and by two hands can be massaged at a time, decreasing the number of patients treated. Thus, automation of the process of massaging the massage of the upper and lower extremities (MULE), including the elbow and knee joints, as well as the neck, is a relevant current task in rehabilitation medicine. The many current massage robotic technical devices (RTD) are intended for application of massage only to certain body parts, for example massage chairs, RTD for massage of the elbow, knee, and ankle joints and the neck, and RTD with anthropomorphic manipulators, etc. [2]. However, none of these is a universal RTD able to automate the massage procedure of all the body parts listed above. Currently, there are some types anthropomorphic robots, which can be used as prototypes for massage robots [3-7], for example, such as anthropomorphic one-armed robot with hand prosthesis and three active fingers (Fig. 2, a) [3] or two-armed robot with three active fingers (Fig. 2, b) [4], and also two-armed robot with four active fingers (Fig. 2, c) [5]. Using an RTD with one or two anthropomorphic manipulators, their movement trajectories can be specified by a control program not requiring the

Content from this work may be used under the terms of the Creative Commons Attribution 3.0 licence. Any further distribution of this work must maintain attribution to the author(s) and the title of the work, journal citation and DOI.
Published under licence by IOP Publishing Ltd

direct involvement of the masseur. However, during the procedure, in contrast to the situation with the adaptive movements of a masseur, the patient's body parts being massaged must be fixed relative to the basal system of coordinates of the anthropomorphic manipulator. In addition, the lack of a remote connection between the RTD and the masseur outside the massage room can lead not only to decreases in the efficacy of massage, but also to emergency situations threatening the patient's health. Because of their size and weight, such RTD are not mobile or suitable for use in domestic or field conditions. Therefore, it is needed of modern portable smart massage robots, being able to work autonomously without the presence of the masseur [8].

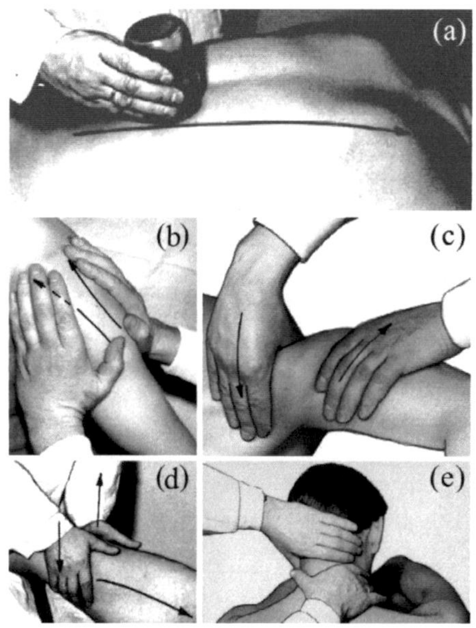

Figure 1. Some examples of massage manipulations for the back (a), and the upper (b, and c) and lower (d) extremities, the elbow joint (b), and patients' neck (e).

To overcome these drawbacks, I propose new concepts of adaptive self-moving robotize masseurs based on active triangle with 3 DOF called plane parallel robot (PPR) and active octahedron with 12 DOF called spatial parallel robot (SPR). The description of design principles of PPR and SPR as future smart base blocks for various rehabilitation and other robotic systems that can self-movement and self-reconfigure, and also its benefits with respect to anthropomorphic one/two-armed robots are presented below.

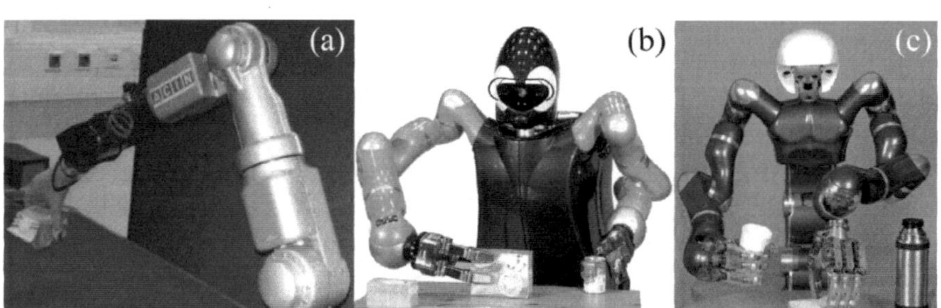

Figure 2. The anthropomorphic one-armed robot with hand prosthesis and three active fingers (a), two-armed robot with three active fingers (b), and two-armed robot with four active fingers (c).

2. Description of PPR

PPR is based on the 2D triangular parallel mechanism with high stiffness and light structure. In Figure 3 a structural scheme of PPR and some examples it's using are shown. PPR is smart portable robotized system, which can be use for the following applications:

- VCM to several patients simultaneously by several cups in the sitting or standing, for example by three cups on chest and three cups on back of patient at the same time.
- VCM of back and chest by sliding and stretching of skin at the same time the three cups.
- Controlled Self-VCM of inaccessible parts of back.
- Controlled Self-VCM at the home.
- The Triangle released extra time, which can be used to conduct other types of massage and, as a consequence, there is a possibility of increasing the number of patients served per shift.

A summary of the results we have the reduction of physical fatigue of the masseur, thereby increasing his productivity, and improving health at end his work shift.

PPR (Figure 3) has a massage device based on triangular parallel structure with cup 1 in each of a vertex. The vertices (A, B, and C) connect by joins ends of adjacent similar linear drives (LD) 2. Each of the LD 2 has a sensor of force (SF) 3, a sensor of displacement (SD) 4, and a sensor of velocity (SV) 5.

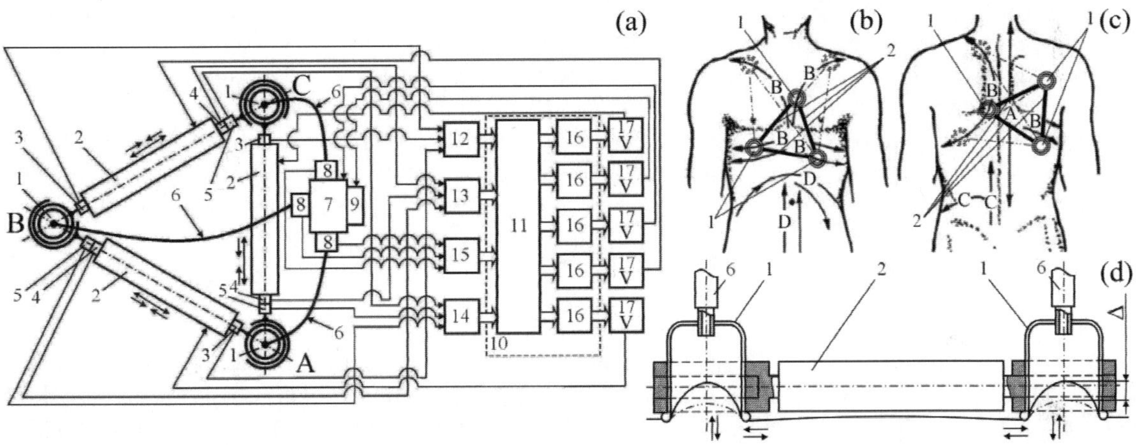

Figure 3. Scheme of PPR (a). A installation of PPR on the chest (b) and back (c) of patient and scheme of VCM; local VCM by PPR (d).

Each of cups 1 is connected by a tubing 6 with a system of air allotment (SAA) 7 and a sensor of pressure (SP) 8. Also SAA 7 and vacuum system (VS) 9 are connected with a system of control (SC) 10, which includes a computer 11. The inputs of computer 11 are connected with analog–digital converters (ADC) 12, 13, 14 and 15 of sensors 3, 4, 5, and 8. The outputs of computer 11 are connected digital-analog converters (DAC) 16 and amplifiers of power (AP) 17 with LD 2, SAA 7, and VS 9. SC 10 works in real-time.

VCM by PPR includes following operations. Massage device ABC installations on the chest (Figure 3, b) or the back (Figure 3, c), and cups 1 degas by VS 9 [1]. VS 9 and SAA 7 make two immovable cups 1 and one mobile cup 1 by turns.

In Figure 3 (b, and c) schemes of the VCM by PPR in various diseases are shown, for examples, spinal osteochondrosis (A), lumbago (A), pneumonia (B), bronchitis (B), myositis (C), sciatica (C), colitis (D), and hypertension (D) [1].

The movement of the cups1 on the surface of the patient's body is carried out by appropriately changing the lengths of LD 2, controlled SD 4. Thus the required speed of the massage movements of the cups 1 on the patient's body is provided by the speed control of axial movements of the linear drives 2 according to signals from the respective relative-velocity sensors 5. Geometric resistance of the massage device ABC (Figure 3, b, c) allows you to define the coordinates of its vertices A, B, and C by SD 4 of the lengths of all LD 2 and control their movements similarly to the organization of spatial movements of the *l*-coordinate manipulator [9]. SV 5 allow to control the speed of movement of the sliding the cups 1 at this stage of massage in accordance with the valid values, defined for this type of massage movements and introduced to SC 10. It should be noted that all of the cups 1 in the course of their rearrangements through sliding motions on the massaged surface of the patient body, alternately become floating and unmoveable. Here in, the process of their permutation. In the massage device ABC one of the cup 1 closest to his move and the other two are roaming and unmoveable, respectively. While the maximum allowable air pressure sets in one roaming sliding cup 1, and in two fixed cups 1 – minimum allowable air pressure. After that, SC 10 is commanded to switch on the LD 2, is connected to the sliding cup 1 and produced a consistent change in their lengths at a given speed controlled by SV 5.

Using of PPR it's also able to produce a local vacuum massage and the massage by sliding and stretching of the muscle tissue [1]. In Figure 3 (c) the scheme of their conduct is shown. Local vacuum massage is performed at a fixed on the patient's body cups 1 by the vacuum in the air up to a pressure whose value is in the range from the maximum pressure to the minimum pressure. For its implementation by commands from the control system in cups 1 with established frequency and amplitude of the pressure change from the minimum permissible to the maximum and Vice versa. Management of alternate higher and lower pressure is carried out using SAA 7 and VS 9 according to the commands from SC 10, formed as a result of processing the signals from SP 8. The distance Δ corresponds to the maximum value of the amplitude in the conduct to local vibration by varying the pressure in the cups 1 (Figure 3).

Massage by sliding and stretching the muscle tissues is produced, while still fixed on the patient's body cups 1 by discharging air to the minimum allowable pressure. It is possible to carry out this type of massage as alternate movement of the linear actuators, as the simultaneous movement of two or more linear actuators. After fixing the cups 1 on the massaged area of the body of the patient (Figure 3) at the position at which the length of LD 2 correspond to their average values, SC 10 is commanded to LD 2 on change its length at a predetermined value, after reaching which LD 2 is stopped and is switched on for reverse, increasing (decreasing) its length to a set value. The process is repeated at a given rate, the required number of cycles. The maximum change in length of LD 2 corresponds to a given vibration amplitude of the massage movements of compression-tension. Changing the length of LD 2 and the amplitude are controlled by SD 4 and the speed – with the help of SV 5. The efforts of the mechanical effects of cups 1 on muscle tissue of patient performing all types of massage are controlled by SF 3.

A summary of the results novel concept of portable smart mobile PPR for autonomously robotized VCM was created. In Figure 4 the pneumatic prototype of PPR (a) and VCM of back by pneumatic prototype of PPR (b) are shown.

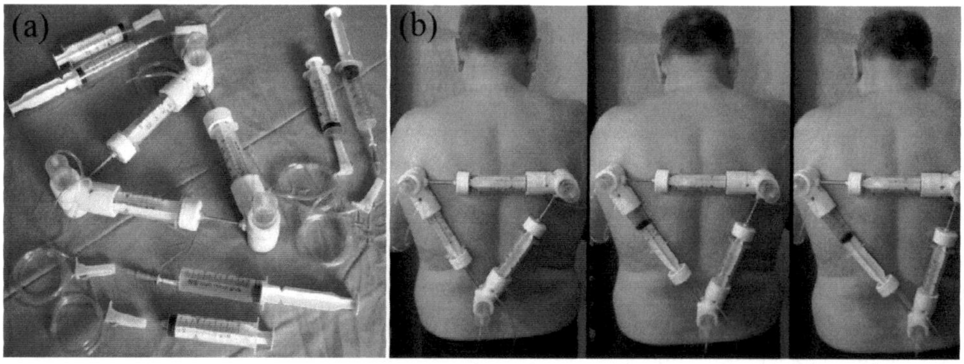

Figure 4. Pneumatic prototype of PPR (a) and VCM of back by pneumatic prototype of PPR (b).

3. Description of SPR

SPR is based on spatial parallel mechanism with 12 DOF in the form of octahedron with variable geometry (OVG) 1, which has a high rigidity and low weight [9, 10]. Such robot versus hexapod has the larger workspace. Each sides of OVG 1 can compress a limb, and move yourself along it. Reference [11, 12] shows new functional capabilities of OVG 1. They were used us for creating smart adaptive SPR for automation process of massage of the upper and the lower limbs. The kinematic (a) and structural (b) schemes PPR are shown in Figure 5.

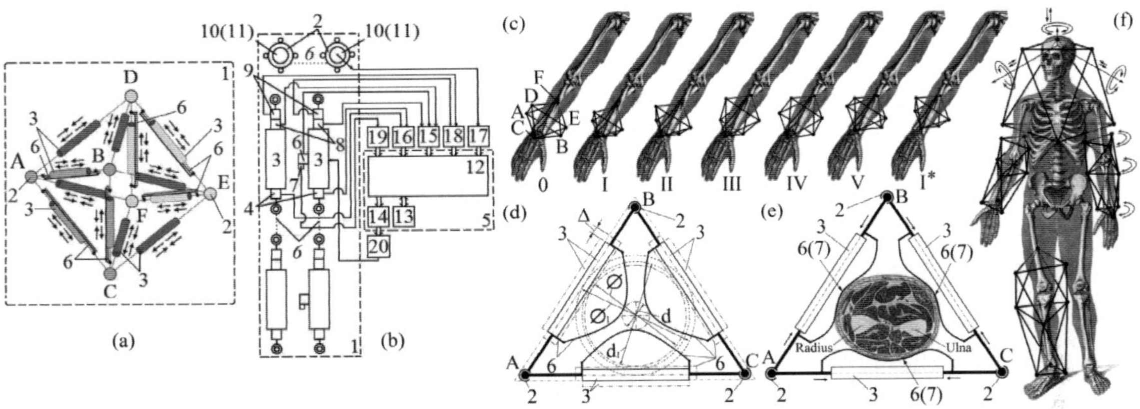

Figure 5. Kinematic (a) and structural (b) schemes PPR; movement of OVG along arm (c), a geometry of grasping face (d) and pressing of arm (e), exoskeletons (f).

All ribs of OVG 1are executed as the rods with the linear drives (LD) 2 each of which has sensor of force (SF) 3, medial sensor of force (MSF) 4, sensor of displacement (SD) 5, and sensor of velocity (SV) 6. The ends of adjacent LD 2 are connected by the spherical joints in the tops 7 of OVG 1. The tops 7 contain the radial stops and the middles of the rods contain the clamps (on Figure 5 were not shown) each of which have sensor of temperature (ST) 8. OVG 1 is a spatial farm as soon as all LD 2 are turned off. All tops 7 have the sensors of spatial position (SSP) 9 which are integrated with the three-axial blocks of accelerometers (TABA) 10. The control system (CS) 11 includes: the neurocomputer (NC) 12, the software 13 and the digital-analogue integrated converters (DAIC) 14. The entrances of CS 11 are connected to exits of the analogue-digital integrated converter (ADIC) 15 for the force sensors 3 and 4, ADIC 16 for the relative linear movement sensors 5, ADIC 17 for SSP 9 with the three-axial blocks of the accelerometers 10, ADIC 18 for the relative velocity sensors 6, and ADIC 19 for the temperature sensors 8. Exits of CS 11 are connected to entrances of the software 13 and DAIC 14. The exits DAIC 14 are connected to the power amplifier 20 which is connected to each of LD 2.

OVG 1 may be used as a base element not only at the single-modular. OVG 1, but also at the multi-modular ones. The radial stops and the clamps (on Figure 5 were not shown) provide the transmission of the efforts from linear drives toward the internal and external contact surfaces. SF 3, MSF 4 and ST 8 provide the operative control of these efforts and temperature in the contact places. SSP 9 with TABA 10 provide the operative control of the spatial position of tops 7 and of vibration along each of axes of rods with LD 2. SD 5 and SV 6 (observers of conditions) of LD 2 register their relative movements and velocities. Before using it we will have to place OVG 1 in inside or outside of the closed surface and then carry out the necessary movements depending on tasks. LD 2 and CS 11 fulfill the coordinated changes of the lengths of the ribs of OVG 1 herewith. As a result the tops 7 have got spatial movement concerning a base system of coordinates. A geometrical invariability of OVG 1 allows to define the spatial coordinates of all tops 7 as a result of the measurement of the lengths of all rods and to control their spatial movements similarly to the organization of spatial movements of the *l*-coordinate manipulator [9]. SSP 9 allow elevating a precision of these measurements herewith. The

NC 12 and the software 13 provide the control of real time.
SPR is based on the use of OVG 1 without mounting of SF 3 (Figure 5), ST 8 and the radial stops of the tops 7, and the claps of lateral sides of OVG 1. SPR may be used to do the massage of the upper extremities and the lower ones. The self-propelled of OVG 1 on the arm is shown in Figure 5 (c) and it includes the following stages:

- Initial Stage or Position 0, I: It is the placement of SPM on the arm (Figure 5, c). The rods AB, BC, AC, DE, EF, and DF are shortened by LD 2. These drives will be stopped by CS 11 from signals of SF 4 (Figure 5).
- Stage 1 or Position I: The rods DE, EF, and DF are lengthened by LD 2. These drives will be stopped by CS 11 from signals of SD 5.
- Stage 2 or Position II: The rods AD, AF, BD, BE, CE, and CF are lengthened by the linear drives 2. These drives will be stopped by CS from signals of SD 5.
- Stage 3, 4 or Position III: The rods DE, EF, and DF are shortened by LD 2. These drives will be stopped by CS 11 from signals of SF 4. The rods AB, BC, and AC are lengthened by the LD 2. These drives will be stopped by CS 11 from signals of SD 5.
- Stage 5 or position IV: The rods AD, AF, BD, BE, CE, and CF are shortened by LD 2. These drives will be stopped by CS 11 from signals of SD 5.
- Initial Stage or Position I*: The rods AB, BC, AC, DE, EF, and DF are shortened by LD 2. These drives will be stopped by CS 11 from signals of SF 4. Position I*: It is the initiation of next a cycle. The rods DE, EF, and DF are lengthened by LD 2. These drives will be stopped by CS 11 from signals of SD 5. The rods AD, AF, BD, BE, CE, and CF are lengthened by LD 2. These drives will be stopped by CS 11 from signals of SD 5.

SPR has 12 DOF and it may therefore be used for massage operations such as stroking, friction, malaxation, and vibration on all surfaces of upper extremities and the lower ones. The interest of this concept is that it can be fully automated. ST 8 permit to measure the temperature of body. TABA 10 may be used not only for remote monitoring of the vibration influences, but also for the pulse.
SPR may also be used for a limitation of an articular mobility and for a decrease of compressive (extending) load. The contact places may be changed to decrease stagnant effects.
OGV 1 (Figure 5, a) can connect together, forming some new mobile structures for various applications. For example two octahedral modules can be connected together, forming exoskeleton of the forearm with the elbow joint and that of the lower leg with the knee joint. These exoskeletons are shown in Figure 5 (f). They are based on the use of the octahedral modules without mounting of ST 8 and the radial stops of the tops 7, and the claps of lateral sides of OGV 1(Figure 5).
This concept possesses the following functional capabilities and the diagnostic ones:

- Adaptation: These exoskeletons can adapt to the arm (leg) with different sizes. The common side of two octahedral modules 1 can change sizes and it increases the workspace for the elbow (knee) joint.
- Forced Movement: These exoskeletons can be used for the forced movements of the forearm with the elbow joint and that of the lower leg with the knee joint in full compliance with the program of rehabilitation.
- Force-Moment Measurements: The force sensors 3, 4 may be used for the force-moment measurements of arm (leg).
- Remote Monitoring: It may be used with SD 5, and SV 6, SSP 9, TABA 10 for the monitoring of arm (leg) movements and vibration influences.

4. Conclusion

Any suitable self-propelled robots for autonomous VCM and for MULE, including massage of the elbow and knee joints, and neck haven't found.
We presented the novel approach to the creation of smart self-propelled autonomous portable massage robot for VCM for back and chest, and for MULE, including massage of the elbow and knee joints, and neck based on original planar and spatial parallel mechanisms. The proposed novel concepts of

PPR and SPR can be used in physiotherapy offices, hospitals, rehabilitation, sports and Wellness centres, as well as at home to conduct autonomous self-massage hard to reach areas of the body.
The use these robots will allow you to automate the processes of VCM and MULE, including massage of the elbow and knee joints, and neck to improve its efficiency and reduce physical weariness and fatigue of therapist. Thus it's possible additional installation of sensors at the vertices of PPR and SPR, e.g. ultrasound, which will allow you to extend the functionality of the massager moving.
PPR and SPR could be used as future intelligent base blocks for various rehabilitation and other robotic systems that can self-propelled and self-reconfigure.
Current objective is to design software and equip of PPR and SPR with reliable and miniature electronics, linear drives and communication hardware.
Another future task is in applying of PPR and SPR towards on-line diagnostics of patient's health during the rehabilitation procedure.

References

[1] Dubrovsky V I and Dubrovskaya A V 2004 *Therapeutic massage* (Moscow: GEOTAR-Media)
[2] Bischoff R, Kurth J, Schreiber G, Koeppe R, Albu-Schäffer A, Beyer A, Eiberger O, Haddadin S, Stemmer A, Grunwald G and Hirzinger G 2011 *Proc. of the Robotics, 41st Int. Symp. on and 6th German Conf. on robotics* (Berlin) (Berlin: VDE Verlag GmbH) pp 741-748
[3] Merlet J-P 2006 *Parallel Robots* (Dordrecht: Springer)
[4] Golovin V, Arkhipov M and Zhuravlev V 2014 *New Trends in Medical and Service Robots* (Springer International Publishing) pp 95-107
[5] Huang Y, Li J, Huang Q and Liu C 2013 *Proc. of the 2013 IEEE/RSJ Int. Conf. on Intelligent Robots and Systems (Tokyo) (*IEEE) pp. 1960-1965
[6] http://www.innovationmagazine.com/volumes/v15n1/in-brief2.html
[7] Sajapin S N and Sajapina M S 2015 *Russian Patent 2 551 939*
[8] Sajapin S N 2015 *Russian Patent 2 564 754*
[9] Koliskor A S 1982 *Russia. Stanki i Instrument* **12** pp 21-24
[10] Belisle R, Yu C H, Nagpal R 2010 *Proc. of the 2010 IEEE Int. Conf. on Robotics and Automation, (Anchorage, USA)* (IEEE) pp 17-23.
[11] Sajapin S N and Sinev A V 2011 *Russian Patent 2 564 754*
[12] Sayapin S N 2012 *Journal of Machinery Manufacture and Reliability* **41** pp 457–466

IOP Conf. Series: Materials Science and Engineering **468** (2018) 012029 doi:10.1088/1757-899X/468/1/012029

Nambu mechanics and its applications

A O Shishanin

Bauman Moscow State Technical University, Moscow, Russia
E-mail: guaicura@gmail.com

Abstract. Here mechanical systems with the Nambu bracket is discussed. We consider some examples of the Nambu mechanics. The main idea that the superintegrability of mechanical system leads to the Nambu bracket. Some applications for the field theory are given.

1. Introduction

Usual textbook mechanics [1], [2], [3] is based on the fact that the coordinates and momenta form a certain manifold M. Some function of the coordinates and momenta the Hamiltonian H is defined dynamics on M. This manifold M has an algebra of function that is defined by the Poisson bracket. We can say that M is the manifold with the Poisson bivector. An example of the Poisson manifolds are the symplectic manifolds. The equations of motion (Hamiltonian's equations) can be written in terms of the Poisson bracket. Y. Nambu in 1973 published an article in which described another situation. Let us assume that an algebra of function on same manifold N is trinary. Then the Nambu bracket can be define by a trivector on N. The dimension of the manifold N has the form 3k, where k is an integer number. Then dynamics on N is defined by two Hamiltonians (the integrals of motion) H and F. Important example of Nambu mechanics is the asymmetric Euler top where the Hamiltonians are energy and magnitude of the angular momentum. It is simple generalize for the Nambu mechanics with n-linear bracket and n-1 Hamiltonians. We will discuss some applications of Nambu mechanics for the theory of integrable systems and the field theory.

For the phase space $\mathbb{R}^{2n}$ (space of momenta and coordinates) we can define the Poisson bracket

$$\{\mathrm{f},\mathrm{g}\} = \sum_{i=1}^{n}\left(\frac{\partial f}{\partial p_i}\frac{\partial g}{q_i} - \frac{\partial g}{\partial p_i}\frac{\partial f}{\partial q_i}\right) = \varepsilon_{ij}\partial_i f \partial_j g.$$

For one-dimensional case the Poisson bracket is conveniently written in terms of the Jacobian

$$\{\mathrm{f},\mathrm{g}\} = \frac{\partial(f,g)}{\partial(p,q)}.$$

For Poisson's manifold with the bivector

Content from this work may be used under the terms of the Creative Commons Attribution 3.0 licence. Any further distribution of this work must maintain attribution to the author(s) and the title of the work, journal citation and DOI.
Published under licence by IOP Publishing Ltd

$$\pi = \frac{1}{2}\pi^{ij}\partial_i \wedge \partial_j$$

the Poisson bracket for smooth functions f and g has form

$$\{\mathrm{f}, \mathrm{g}\} = \pi^{ij}\partial_i f \partial_j g = \pi(df, dg).$$

The Poisson bracket has some usual properties of antisymmetry, linearity, the Leibniz rule and the Jacobi identity. Put some function of momenta and coordinates H which is Hamiltonian of system. Then the evolution equation for some function f is defined Hamiltonian H

$$\frac{df}{dt} = \{\mathrm{H}, \mathrm{f}\}.$$

Nambu mechanics is constructed for some kn-dimensional smooth manifold (k is some natural number) with n-vector π for functions $f_1, f_2, ..., f_n$

$$\{\mathrm{f}_1, \mathrm{f}_2, ..., \mathrm{f}_n\} = \pi^{i_1...i_n}\partial_{i_1} f_1 ... \partial_{i_n} f_n = \pi(df_1, ..., df_n).$$

The Nambu bracket has similar properties to the Poisson bracket.

It is important to note that the Nambu bracket on a flat space $\mathbb{R}^n$ has form

$$\{\mathrm{f}_1, \mathrm{f}_2, ..., \mathrm{f}_n\} = \frac{\partial(f_1, ... f_n)}{\partial(x_1, ... x_n)} = \varepsilon^{i_1...i_n}\partial_{i_1}\mathrm{f}_1 ... \partial_{i_n}\mathrm{f}_n,$$

where $x_1, ..., x_n$ are Cartesian coordinates. For evolution some function f one needs n-1 Hamiltonians $\mathrm{H}_1, ..., \mathrm{H}_{n-1}$ (integrals of motion)

$$\frac{df}{dt} = \{\mathrm{H}_1, ..., \mathrm{H}_{n-1}, \mathrm{f}\}$$

2. Exampes of Nambu systems

Let us consider some examples of Nambu mechanics. Most popular one is the Euler asymmetric top. Put I_1, I_2, I_3 are principal axes of inertia and $\mathrm{L}_1, \mathrm{L}_2, \mathrm{L}_3$ are components of the angular momenta. The Euler equations are

$$\frac{dL_1}{dt} = \frac{I_3 - I_2}{I_2 I_3} L_2 L_3, \quad \frac{dL_2}{dt} = \frac{I_1 - I_3}{I_1 I_3} L_1 L_3, \quad \frac{dL_3}{dt} = \frac{I_2 - I_1}{I_1 I_2} L_1 L_2.$$

There are Hamiltonians the energy H_1 and the full angular momentum H_2

$$H_1 = \frac{L_1^2}{2I_1} + \frac{L_2^2}{2I_2} + \frac{L_3^2}{2I_3}, \quad H_2 = \frac{1}{2}\left(L_1^2 + L_2^2 + L_3^2\right).$$

One can obtain the Euler equations by substitution

$$\frac{dL_i}{dt} = \varepsilon^{kji}\partial_k H_1 \partial_j H_2.$$

The Nahm system (the model for static SU(2)-monopoles) has the following equations of motion

$$\frac{dx_1}{dt} = x_2 x_3, \quad \frac{dx_2}{dt} = x_1 x_3, \quad \frac{dx_3}{dt} = x_1 x_2.$$

This system can be written in Nambu's form

$$\frac{dx_i}{dt} = \{\mathrm{H}_1, \mathrm{H}_2, x_i\}$$

with $H_1 = x_1^2 - x_2^2, \quad H_2 = x_1^2 - x_3^2$.

Another evident example is the n-dimensional harmonic oscillator [8] with the Hamiltonian

$$H = \sum_{i=1}^{n} \left(p_i^2 + x_i^2 \right).$$

This Hamiltonian is invariant under SU(n) group.

In addition to the angular momenta

$$L_{ij} = p_i x_j - x_j p_i,$$

there are additional integrals of motion

$$A_{ij} = p_i p_j + x_i x_j,$$

where $i, j = 1, \dots, n$.

For example, when n=2 one can choose following set of integrals [8]

$$\begin{aligned} H_1 &= p_1^2 + x_1^2 = C_1, \\ H_2 &= p_2^2 + x_2^2 = C_2, \\ H_3 &= x_1 p_2 - x_2 p_1 = C_3, \\ H_4 &= p_1 p_2 + x_1 x_2 = C_4. \end{aligned}$$

Here there are many possibilities to choose the Nambu bracket. For example, one case is

$$\{\mathrm{H}_1, \mathrm{H}_2, \mathrm{H}_3, \mathrm{f}\} = -\frac{1}{2C_4} \frac{\partial(H_1, H_2, H_3, f)}{\partial(p_1, p_2, \mathrm{x}_1, \mathrm{x}_2)},$$

and another way is following

$$\{\mathrm{H}_1, \mathrm{H}_2, \mathrm{H}_4, \mathrm{f}\} = -\frac{1}{2C_3} \frac{\partial(H_1, H_2, H_4, f)}{\partial(p_1, p_2, \mathrm{x}_1, \mathrm{x}_2)}.$$

It is simple to see that following system has the same integrals as the harmonic oscillator

$$H_1 = F(\mathrm{z}), \quad \mathrm{z} = \mathrm{p}_1^2 + \mathrm{p}_2^2 + \mathrm{x}_1^2 + \mathrm{x}_2^2,$$

IOP Conf. Series: Materials Science and Engineering **468** (2018) 012029 doi:10.1088/1757-899X/468/1/012029

where $F(z)$ is any function. We conjecture that it is also Nambu's system.

Let us consider new example of Nambu's structure with the Hamiltonian

$$H_1 = p_x p_y - xp_x - yp_y + xy.$$

It is clear that that this system has following integrals of motion

$$H_2 = xp_x - yp_y, \quad H_3 = p_x - y.$$

The Nambu bracket here is

$$\{\mathrm{H}_1, \mathrm{H}_2, \mathrm{H}_3, \mathrm{f}\} = \frac{1}{C}\frac{\partial(H_1, H_2, H_3, f)}{\partial(p_x, p_y, x, y)},$$

where $C = H_3$.

The Kepler problem has Nambu's structure. The Kepler Hamiltonian has form

$$H = \frac{p^2}{2} - \frac{\alpha}{r}.$$

The angular momentum $\vec{L}$ is an integral of motion. It is well-known that also there is an integral which called the Laplace-Runge-Lenz vector

$$\vec{A} = \vec{p} \times \vec{L} - \frac{\alpha \vec{r}}{r}.$$

If give any five integrals between components of $\vec{L}$ and $\vec{A}$ then it is possible to obtain Nambu's structure in the Kepler system [8].

Superintegrable system is an integrable system with additional integrals of motion. If usual integrable system has n integrals of motion, then superintegrable system owns (2n-1) Hamiltonians. It can be observed that for the existence of Nambu mechanics, a superintegrability of the mechanical model is necessary. Examples in [10], [11] and another superintegrable systems have Nambu's structure.

As an example of superintegrable systems, let's consider the following construction.We believe that we can construct a Nambu bracket in an arbitrary algebra with the Casimir operator. Put some Lie algebra g with representation of its generators by differential operators $F(\mathrm{x}_i, \partial_i)$ in $\mathbb{R}^n$. Then we replace $\partial_i \to p_i$ and put H is the Casimir operator of algebra g. Here functions $F(\mathrm{x}_i, \mathrm{p}_i)$ are the remaining integrals of motion. For example, consider 4-dimensional Lie algebra with following nonzero commutators

$$[\mathrm{e}_1, \mathrm{e}_4] = e_1, \ [\mathrm{e}_2, \mathrm{e}_4] = e_1 + e_2, \ [\mathrm{e}_3, \mathrm{e}_4] = e_2 + e_3.$$

Then we can present here the Poisson brackets $\{\mathrm{Q}_i, \mathrm{Q}_j\} = \mathrm{f}_{ij}^k \mathrm{Q}_k$, where

$$Q_1 = p_1, \ Q_2 = p_1 x_2, \ Q_3 = \frac{1}{2} p_1 x_2^2, \ Q_4 = p_1 x_1 + p_2.$$

Let us pick up following Hamiltonians:

the Casimir operator $H_1 = p_1 e^{x_2}$, $H_2 = p_1 = Q_1$, $H_3 = p_1 x_1 + p_2 = Q_4$.

Then for this example the Nambu bracket has form

$$\{H_1, H_2, H_4, f\} = \frac{1}{Q_1}\frac{\partial(H_1, H_2, H_4, f)}{\partial(p_1, p_2, x_1, x_2)}.$$

3. Some applications to field theory

L. Takhtajan had found remarkable action for n-dimensional objects [7]. Useful to recall that in the Poisson mechanics [1] is considered the following 1-form

$$\omega^1 = p_i dq_i - Hdt.$$

The Poincare-Cartan invariant is an integral along a closed way from this form. Minimizing the integral along a smooth open curve from this form leads to the Hamilton equations. The generalized Poincare-Cartan integral invariant form [7] is the following 2-form

$$\omega^2 = x_1 dx_2 \wedge dx_3 - H_1 dH_2 \wedge dt.$$

In order to write here the variational principle, we assume that the surface over which the integral is taken is parametrized t, σ and $0 \le t \le 1$, $0 \le \sigma \le 1$. Here x_1, x_2, x_3 are functions of t, σ. Let us consider like sigma-model action

$$S = \frac{1}{3}\int dt d\sigma \varepsilon^{ijk} x_i \partial_t x_j \partial_\sigma x_k - \int dt d\sigma H_1 \partial_\sigma H_2.$$

Here the equation of motion for string-like object is

$$\partial_t x_i + \frac{1}{2}\varepsilon^{ijk}\frac{\partial(H_1, H_2)}{\partial(x_j, x_k)} = \alpha(t,\sigma)\partial_\sigma x_i,$$

where $\alpha(\tau,\sigma)$ is any function. One can generalize this action to an n-dimensional brane (see for example [12]).

Most famous application of the Nambu bracket in field theory is BLG model [13], [14] which describes low-energy action for stack of M2-branes. There is some kind of sigma-models which called the topological Poisson sigma-models

$$S = \int_\Sigma (A_i dX^i - \frac{1}{2}\pi^{ij} A_i \wedge A_j),$$

where Σ is a 2-dimensional world sheet, the fields X are embedding functions $X : \Sigma \to M$ and the fields A are 1-forms on Σ with values in $T_X^* M$. There exists Nambu's analog of this construction, which is called the Nambu sigma-model [15], [16]. There is opportunity to add Wess-Zumino term to the action of the Nambu sigma-model [17] using the Courant-algedroid sigma model introduced by D. Roytenberg [18].

Conveniently to use the Nambu bracket to describe extended objects like strings and membranes. One most important problem associated with the Nambu bracket is its quantization. We don't know an analog of the Kontsevich formula for Nambu manifolds.

References

[1] Arnol'd V 1989 *Mathematical Methods of Classical Mechanics* (Springer-Verlag, New- York-Heidelberg-Berlin) p 520
[2] Goldstein H, Poole Ch, Safko J 2002 *Classical Mechanics* (Addison Wesley. San Francisco) p. 680
[3] Landau L D, Lifshitz E M 1982 *Mechanics* (Butterworth-Heinemann) p 224
[4] Nambu Y 1973Generalized Hamiltonian mechanics. *Phys. Rev.D*, vol. **7**, pp 2405-2412
[5] Bayen V, Flato M 1976Remarks concerning Nambu's generalized mechanics. *Phys. Rev. D*, 1975, vol. **11**, pp 3049-3053
[6] Mukunda N., Sudarshan E. 1976 Relation between Nambu and Hamiltonian mechanics *Phys.*

Rev.D vol. **13**, pp. 2846-2850

[7] Takhtajan L 1994 On Foundation of the generalized Nambu mechanics . *Comm.Math.Phys.* vol. **160**, pp. 295-316

[8] Chatterjee R 1996 Dynamical symmetries and Nambu Mechanics. *Lett.Math Phys* vol. **36**, pp. 117-126

[9] Chatterjee R, Takhtajan L 1996 Aspects of classical and quantum Nambu mechanics. *Lett.Math.Phys.* vol. **37**, pp 475-482

[10] Evans N W 1990 Superintegrability in classical mechanics, *Phys. Rev. A*, vol. **41**, pp. 5666-5766

[11] Bonatsos D, Daskaloyannis C, Kokkotas K 1994 *Phys. Rev. A* vol. **50**, pp 3700-3709

[12] Ho P-M, Matsuo Y 2015 Nambu bracket and M-theory *Prog. Theor. Exp. Phys.* vol.**6** pp 1-25

[13] Bagger J, Lambert N 2007 Modeling Multiple M2's. *Phys. Rev. D*, vol. 75, p.045020

[14] Gustavson A 2009 Algebraic structures on parallel M2-branes. *Nucl. Phys. B*, vol. **811**, pp. 66-76

[15] Jurco B, Schupp P 2012 Nambu-Sigma model and effective actions. *Phys.Lett. B* vol. **713** pp 313-316

[16] Schupp P, Jurco B 2011 Nambu Sigma Models and Branes *proceedingPoS CORFU2011* pp 045-053, arxiv:1205.2595

[17] Jurco B, Schupp P, Vysoki J 2013 p-Brane action and Higher Roytenberg Brackets *JHEP*, vol.**1302**, pp. 042-063

[18] D. Roytenberg. 2002 A Note on quasi Lie bialgebroids and twisted Poisson manifolds, *Lett.Math.Phys.*, vol. **61**, pp. 123–137, arxiv: math/0112152

Microrotation Waves Propagating in a Cylindrical Waveguide

E V Murashkin[1,2], Y N Radayev[2] and V A Kovalev[3]

[1]Bauman Moscow State Technical University, Moscow, Russia
[2]Ishlinsky Institute for Problems in Mechanics of the Russian Academy of Sciences, Moscow, Russia
[3]Moscow City Government University of Management, Moscow, Russia
E-mail: evmurashkin@gmail.com, radayev@ipmnet.ru

Abstract. The present study is devoted to analysis of coupled harmonic waves of translational displacements and microrotations propagating along the axis of a long cylindrical waveguide with circular cross-section. Microrotation waves modelling is realized within the frameworks of linear micropolar elasticity by introducing microrotations as independent degrees of freedom of the elastic continuum. The coupled system of vector differential equations of the micropolar elasticity is given. The translational displacements and microrotations vectors in the coupled wave are decomposed into potential and vortex parts. The coupled differential equations are uncoupled for some distinguished cases. The Helmholtz equations solutions for the translational and microrotation waves are obtained for a high-frequency waves in a cylindrical domain.

1. Introduction

By now the micropolar continuum models are intensively used in applied mechanics in order to resolve various problems arising while employing the classical theory of elasticity. Asymmetric stress and strain tensors are intrinsic to the micropolar theory of elasticity (see discussion in [1] for the subject). Mechanics of metamaterials are often requires the micropolar elastic models. The aim of the present study is to give a method which provides determination of the translational and microrotation waves propagating along an infinite waveguide with circular cross-section. Microrotation waves in an elastic solid are modelled within the frameworks of linear micropolar elasticity by introducing in the basic equations of the continuum mechanics microrotations as independent degrees of freedom. Such a modelling can be regularly carried out in terms of the classical field theory [2], starting from the action integral and the least action variational principle.

2. Boundary Value Problem

The vector partial differential equations determining the coupled translational and microrotation waves in a micropolar elastic continuum are formulated as (see [1]):

$$\begin{cases}(\lambda + 2\mu)\boldsymbol{\nabla}\boldsymbol{\nabla}\cdot\mathbf{u} - (\mu+\eta)\boldsymbol{\nabla}\times(\boldsymbol{\nabla}\times\mathbf{u}) + 2\eta\boldsymbol{\nabla}\times\boldsymbol{\phi} = \rho\ddot{\mathbf{u}},\\ (\beta + 2\gamma)\boldsymbol{\nabla}\boldsymbol{\nabla}\cdot\boldsymbol{\phi} - (\gamma+\varepsilon)\boldsymbol{\nabla}\times(\boldsymbol{\nabla}\times\boldsymbol{\phi}) + 2\eta\boldsymbol{\nabla}\times\mathbf{u} - 4\eta\boldsymbol{\phi} = \Im\ddot{\boldsymbol{\phi}},\end{cases} \tag{1}$$

where λ, μ, γ, β, ε, η are the constitutive constants of micropolar elastic continuum; $\boldsymbol{\nabla}$ is the three-dimensional Hamilton operator (Hamilton nabla); $\mathbf{u}$ is the translational displacement vector; $\boldsymbol{\phi}$

Content from this work may be used under the terms of the Creative Commons Attribution 3.0 licence. Any further distribution of this work must maintain attribution to the author(s) and the title of the work, journal citation and DOI.
Published under licence by IOP Publishing Ltd

IOP Conf. Series: Materials Science and Engineering **468** (2018) 012030
doi:10.1088/1757-899X/468/1/012030

is the microrotation vector; ρ is the mass density; $\Im$ is the microelement inertia constant; superimposed dot denotes partial differentiation with respect to time.
The constitute equations for the force stress tensor $\boldsymbol{\sigma}$ and the moment stress tensor $\boldsymbol{\mu}$ can be represented as follows

$$\boldsymbol{\sigma} = (\mu + \eta)\boldsymbol{\gamma} + (\mu - \eta)\boldsymbol{\gamma}^{T} + \lambda \mathbf{I}\,\mathrm{tr}\boldsymbol{\gamma},$$
$$\boldsymbol{\mu} = (\gamma + \varepsilon)\boldsymbol{\kappa} + (\gamma - \varepsilon)\boldsymbol{\kappa}^{T} + \beta \mathbf{I}\,\mathrm{tr}\,\boldsymbol{\kappa}. \tag{2}$$

In the given equations $\mathbf{I}$ denotes the three-dimensional unit tensor, $\boldsymbol{\gamma}$ is the strain tensor, $\boldsymbol{\kappa}$ is the bending-torsion tensor. The latter tensors are expressed via translational displacement vector and the microrotation vector according to

$$\boldsymbol{\gamma} = \boldsymbol{\nabla} \otimes \mathbf{u} - \boldsymbol{\phi} \cdot \boldsymbol{\varepsilon},$$
$$\boldsymbol{\kappa} = \boldsymbol{\nabla} \otimes \boldsymbol{\phi}. \tag{3}$$

We employ the notation $\boldsymbol{\varepsilon}$ for the three-dimensional permutation tensor.
In view of equations (2) and (3) after chain of transformations the following equations for the traction vector $\mathbf{t}$ and the moment vector $\mathbf{m}$ can be obtained:

$$\mathbf{t} = \lambda(\boldsymbol{\nabla} \cdot \mathbf{u})\mathbf{n} + 2\mu(\mathbf{n} \cdot \boldsymbol{\nabla})\mathbf{u} + (\mu - \eta)(\mathbf{n} \times \boldsymbol{\nabla}) \times \mathbf{u} + 2\eta\mathbf{n} \times \boldsymbol{\phi},$$
$$\mathbf{m} = \beta(\boldsymbol{\nabla} \cdot \boldsymbol{\phi})\mathbf{n} + 2\gamma(\mathbf{n} \cdot \boldsymbol{\nabla})\boldsymbol{\phi} + (\gamma - \varepsilon)(\mathbf{n} \times \boldsymbol{\nabla}) \times \boldsymbol{\phi}, \tag{4}$$

where $\mathbf{n}$ is the unit normal vector determining the spatial orientation of the plane element.
These equations are used in order to formulate boundary conditions on the free from tractions and moments sidewall of the waveguide:

$$\mathbf{t} = \mathbf{0}, \ \mathbf{m} = \mathbf{0}. \tag{5}$$

The system of vector partial differential equations (1) is coupled and can be uncoupled following a scheme described below. First, introduce dynamic potentials of the translational displacements and microrotations. They are decomposed into potential and vortex parts according to the Helmholtz representations:

$$\mathbf{u} = \boldsymbol{\nabla}\Phi + \boldsymbol{\nabla} \times \boldsymbol{\Psi},$$
$$\boldsymbol{\phi} = \boldsymbol{\nabla}\Sigma + \boldsymbol{\nabla} \times \boldsymbol{H}, \tag{6}$$

where Φ, Σ are the scalar potentials, and $\boldsymbol{\Psi}$, $\boldsymbol{H}$ are the vector potentials of translational displacements and microrotations respectively.
Second, in the present study we shall impose the usual calibration conditions for the dynamic potentials

$$\boldsymbol{\nabla} \cdot \boldsymbol{\Psi} = \mathbf{0},$$
$$\boldsymbol{\nabla} \cdot \boldsymbol{H} = \mathbf{0}. \tag{7}$$

Then, after substituting equations (6) in (1) and taking into consideration the calibrating equations (7), it is seen that the system (1) is satisfied if the scalar and the vector potentials are to be solutions of the uncoupled scalar equations

$$\Delta\Phi - \frac{1}{c_{\parallel}^{2}}\ddot{\Phi} = 0,$$
$$\Delta\Sigma - \frac{1}{{}_{\mu}c_{\parallel}^{2}}\ddot{\Sigma} - \frac{\Omega^{2}}{{}_{\mu}c_{\parallel}^{2}}\Sigma = 0, \tag{8}$$

and the following coupled system of vector partial differential equations:

$$\begin{cases} \Delta\boldsymbol{\Psi} - \dfrac{1}{{}^{\prime\prime}c_{\perp}^{2}}\ddot{\boldsymbol{\Psi}} + 2d_{\perp}^{2}\boldsymbol{\nabla}\times\boldsymbol{H} = \mathbf{0}, \\ \Delta\boldsymbol{H} - \dfrac{1}{{}_{\mu}c_{\perp}^{2}}\ddot{\boldsymbol{H}} - \dfrac{\Omega^{2}}{{}_{\mu}c_{\perp}^{2}}\boldsymbol{H} + \dfrac{\Omega^{2}}{2\,{}_{\mu}c_{\perp}^{2}}\boldsymbol{\nabla}\times\boldsymbol{\Psi} = \mathbf{0}, \end{cases} \tag{9}$$

wherein we have introduced the notations for a set of constitutive constants

$$\Omega^{2} = \frac{4\eta}{\Im},\ c_{\parallel}^{2} = \frac{\lambda+2\mu}{\rho},\ {}_{\mu}c_{\parallel}^{2} = \frac{\beta+2\gamma}{\Im},\ {}_{\mu}c_{\perp}^{2} = \frac{\mu+\eta}{\Im},$$

$${}^{\prime\prime}c_{\perp}^{2} = \frac{\mu+\eta}{\rho},\ {}^{\prime}c_{\perp}^{2} = \frac{\eta}{\rho},\ d_{\perp}^{2} = \frac{{}^{\prime}c_{\perp}^{2}}{{}^{\prime\prime}c_{\perp}^{2}}. \tag{10}$$

In the present study only coupled high-frequency waves of thranslational displacements and microrotaions are considered (i.e. $\omega > \Omega$). In such a case the following differential equations for the dynamic potentials can be obtained

$$\begin{cases} (\Delta + \alpha_{\parallel}^{2})\Phi = 0, \\ (\Delta + \beta_{\parallel}^{2})\Sigma = 0; \\ (\Delta + \alpha_{\perp}^{2})\boldsymbol{\Psi} + 2d_{\perp}^{2}\boldsymbol{\nabla}\times\boldsymbol{H} = \mathbf{0}, \\ (\Delta + \beta_{\perp}^{2})\boldsymbol{H} + \dfrac{\Omega^{2}}{2\,{}_{\mu}c_{\perp}^{2}}\boldsymbol{\nabla}\times\boldsymbol{\Psi} = \mathbf{0}. \end{cases} \tag{11}$$

In the above equations we have employed the constants defined as

$$\alpha_{\parallel}^{2} = \frac{\omega^{2}}{c_{\parallel}^{2}},\ \beta_{\parallel}^{2} = \frac{\omega^{2}-\Omega^{2}}{{}_{\mu}c_{\parallel}^{2}};$$

$$\alpha_{\perp}^{2} = \frac{\omega^{2}}{{}^{\prime\prime}c_{\perp}^{2}},\ \beta_{\perp}^{2} = \frac{\omega^{2}-\Omega^{2}}{{}_{\mu}c_{\perp}^{2}}. \tag{12}$$

The equations (11) for the vortex potentials $\boldsymbol{\Psi}$ and $\boldsymbol{H}$ are still coupled. Uncoupled equations can be derived, however this requires differential operators of the fourth order. At last, for the vortex potentials $\boldsymbol{\Psi}$ and $\boldsymbol{H}$ the separate equations are furnished by

$$\begin{aligned} (\Delta + K_{1}^{2})(\Delta + K_{2}^{2})\boldsymbol{\Psi} = \mathbf{0}, \\ (\Delta + K_{1}^{2})(\Delta + K_{2}^{2})\boldsymbol{H} = \mathbf{0}, \end{aligned} \tag{13}$$

where

$$K_{1,2}^{2} = -\Delta_{1,2}\ ,$$

$$\Delta_{1,2} = \frac{-(\alpha_{\perp}^{2} + \beta_{\perp}^{2} + \sigma_{\perp}^{2}) \pm \sqrt{(\alpha_{\perp}^{2} - \beta_{\perp}^{2} + \sigma_{\perp}^{2})^{2} + 4\beta_{\perp}^{2}\sigma_{\perp}^{2}}}{2}, \tag{14}$$

$$\sigma_{\perp}^{2} = \frac{d_{\perp}^{2}\Omega^{2}}{{}_{\mu}c_{\perp}^{2}}.$$

The coupled translational and microrotation wave fields in a cylindrical domain are determined in the cylindrical coordinates r, φ, z by the separation of variables technique (see details in [2]). This technique permits investigation of waves of an arbitrary azimuthal number n propagating along the waveguide.

For the scalar potentials Φ , Σ the following representations are obtained:

$$\Phi = C_1 I_n(p_1 r)\begin{Bmatrix}\cos n\varphi \\ -\sin n\varphi\end{Bmatrix} e^{\pm ikz},$$
$$\Sigma = -C_2 I_n(p_2 r)\begin{Bmatrix}\sin n\varphi \\ \cos n\varphi\end{Bmatrix} e^{\pm ikz}, \tag{15}$$

wherein the harmonic exponents have been omitted, k denotes the wavenumber; C_1 and C_2 are arbitrary constants; $I_n(\cdot)$ is the Bessel function of the first kind of an imaginary argument;

$$p_1^{\ 2} = k^2 - \alpha_{\|}^2,\ p_2^{\ 2} = k^2 - \beta_{\|}^2.$$

The vortex potentials of translational displacements and microrotations are given by the following formulae for their physical components in the cylindrical coordinate net:

$$\begin{cases} \Psi_r = \left[C_3' I_{n-1}(q_1 r) + C_4' I_{n+1}(q_1 r) + C_3'' I_{n-1}(q_2 r) + C_4'' I_{n+1}(q_2 r)\right]\begin{Bmatrix}\sin n\varphi \\ \cos n\varphi\end{Bmatrix} e^{\pm ikz}, \\ \Psi_\varphi = \left[C_3' I_{n-1}(q_1 r) - C_4' I_{n+1}(q_1 r) + C_3'' I_{n-1}(q_2 r) - C_4'' I_{n+1}(q_2 r)\right]\begin{Bmatrix}\cos n\varphi \\ -\sin n\varphi\end{Bmatrix} e^{\pm ikz}, \\ \Psi_z = \left[C_5' I_n(q_1 r) + C_5'' I_n(q_2 r)\right]\begin{Bmatrix}\sin n\varphi \\ \cos n\varphi\end{Bmatrix} e^{\pm ikz}; \end{cases} \tag{16}$$

$$\begin{cases} H_r = \left[L_3' I_{n-1}(q_1 r) + L_4' I_{n+1}(q_1 r) + L_3'' I_{n-1}(q_2 r) + L_4'' I_{n+1}(q_2 r)\right]\begin{Bmatrix}-\cos n\varphi \\ \sin n\varphi\end{Bmatrix} e^{\pm ikz}, \\ H_\varphi = \left[L_3' I_{n-1}(q_1 r) - L_4' I_{n+1}(q_1 r) + L_3'' I_{n-1}(q_2 r) - L_4'' I_{n+1}(q_2 r)\right]\begin{Bmatrix}\sin n\varphi \\ \cos n\varphi\end{Bmatrix} e^{\pm ikz}, \\ H_z = \left[L_5' I_n(q_1 r) + L_5'' I_n(q_2 r)\right]\begin{Bmatrix}-\cos n\varphi \\ \sin n\varphi\end{Bmatrix} e^{\pm ikz}, \end{cases} \tag{17}$$

wherein the harmonic exponents have been omitted as previously, $C_3' - C_5'$, $C_3'' - C_5''$, $L_3' - L_5'$ and $L_3'' - L_5''$ are arbitrary constants and

$$q_1^2 = k^2 - K_1^2,\ q_2^2 = k^2 - K_2^2.$$

The coordinate representations (16) and (17) of the vortex potentials are based on a solution

$$\begin{cases} \Gamma_r = \left[C_1 I_{n-1}(q_* r) + C_2 I_{n+1}(q_* r)\right]\begin{Bmatrix}\sin n\varphi \\ \cos n\varphi\end{Bmatrix} e^{\pm ikz}, \\ \Gamma_\varphi = \left[C_1 I_{n-1}(q_* r) - C_2 I_{n+1}(q_* r)\right]\begin{Bmatrix}\cos n\varphi \\ -\sin n\varphi\end{Bmatrix} e^{\pm ikz}, \\ \Gamma_z = C_3 I_n(q_* r)\begin{Bmatrix}\sin n\varphi \\ \cos n\varphi\end{Bmatrix} e^{\pm ikz}. \end{cases}$$

of the vector Helmholtz equation

$$(\Delta + k_*^2)\mathbf{\Gamma} = \mathbf{0}$$

which can be obtained by separating of variables in the cylindrical coordinates r, φ, z; we should note that

$$q_*^2 = k^2 - k_*^2.$$

In view of equations (15), (16) and (17) the physical components of displacement and microrotation vectors in the propagating coupled wave can be obtained. We give these components for any azimuthal number and omitting as usually the harmonic exponents:

$$u_r = \begin{bmatrix} C_1\left(p_1 I_{n+1}(p_1 r) + \frac{n}{r} I_n(p_1 r)\right) + \frac{n}{r}\left(C_5' I_n(q_1 r) + C_5'' I_n(q_2 r)\right) + \\ +(\mp ik)\left(C_3' I_{n-1}(q_1 r) - C_4' I_{n+1}(q_1 r) + C_3'' I_{n-1}(q_2 r) - C_4'' I_{n+1}(q_2 r)\right) \end{bmatrix} \times \tag{18}$$
$$\times \begin{Bmatrix} \cos n\varphi \\ -\sin n\varphi \end{Bmatrix} e^{\pm ikz},$$

$$u_\varphi = \begin{bmatrix} -\frac{n}{r} C_1 I_n(p_1 r) - C_5'\left(q_1 I_{n+1}(q_1 r) + \frac{n}{r} I_n(q_1 r)\right) + \\ +(\pm ik)\left(C_3' I_{n-1}(q_1 r) + C_4' I_{n+1}(q_1 r) + C_3'' I_{n-1}(q_2 r) + C_4'' I_{n+1}(q_2 r)\right) - \\ -C_5''\left(q_2 I_{n+1}(q_2 r) + \frac{n}{r} I_n(q_2 r)\right) \end{bmatrix} \times \tag{19}$$
$$\times \begin{Bmatrix} \sin n\varphi \\ \cos n\varphi \end{Bmatrix} e^{\pm ikz},$$

$$u_z = \left(C_1(\pm ik) I_n(p_1 r) + (C_3' - C_4') q_1 I_n(q_1 r) + (C_3'' - C_4'') q_2 I_n(q_2 r)\right) \times$$
$$\times \begin{Bmatrix} \cos n\varphi \\ -\sin n\varphi \end{Bmatrix} e^{\pm ikz}; \tag{20}$$

$$\phi_r = \begin{bmatrix} -C_2\left(p_2 I_{n+1}(p_2 r) + \frac{n}{r} I_n(p_2 r)\right) + \frac{n}{r}\left(L_5' I_n(q_1 r) + L_5'' I_n(q_2 r)\right) + \\ +(\mp ik)\left(L_3' I_{n-1}(q_1 r) - L_4' I_{n+1}(q_1 r) + L_3'' I_{n-1}(q_2 r) - L_4'' I_{n+1}(q_2 r)\right) \end{bmatrix} \times \tag{21}$$
$$\times \begin{Bmatrix} \sin n\varphi \\ \cos n\varphi \end{Bmatrix} e^{\pm ikz},$$

$$\phi_\varphi = \begin{bmatrix} -\frac{n}{r} C_2 I_n(p_2 r) - L_5'\left(q_1 I_{n+1}(q_1 r) + \frac{n}{r} I_n(q_1 r)\right) + \\ +(\pm ik)\left(L_3' I_{n-1}(q_1 r) + L_4' I_{n+1}(q_1 r) + L_3'' I_{n-1}(q_2 r) + L_4'' I_{n+1}(q_2 r)\right) - \\ -L_5''\left(q_2 I_{n+1}(q_2 r) + \frac{n}{r} I_n(q_2 r)\right) \end{bmatrix} \times \tag{22}$$
$$\times \begin{Bmatrix} -\cos n\varphi \\ \sin n\varphi \end{Bmatrix} e^{\pm ikz},$$

$$\phi_z = \left(-C_2(\pm ik)I_n(p_2 r) + (L_3' - L_4')q_1 I_n(q_1 r) + (L_3'' - L_4'')q_2 I_n(q_2 r)\right)\times \\ \times\begin{Bmatrix}\sin n\varphi \\ \cos n\varphi\end{Bmatrix} e^{\pm ikz}. \quad (23)$$

Arbitrary constants in the formulae (18)-(23) are related by the linear equations

$$(C_3' + C_4')q_1 + (\pm ik)C_5' = 0,$$
$$(C_3'' + C_4'')q_2 + (\pm ik)C_5'' = 0;$$
$$(L_3' + L_4')q_1 + (\pm ik)L_5' = 0,$$
$$(L_3'' + L_4'')q_2 + (\pm ik)L_5'' = 0.$$

which follows from the calibration conditions (7). Additional linear algebraic equations for them are derived from the boundary conditions (5) on sidewall of the waveguide. This conditions are to be expanded in virtue of formulae (4), representing the traction and moment vectors via the displacement and microrotation vectors, and the Helmholtz decompositions (6).

References

[1] Nowacki W 1986 *Theory of Asymmetric Elasticity.* (Pergamon Press, Oxford)
[2] Kovalev V A and Radayev Y N 2010 *Wave Problems of the Field Theory and Thermomechanics.* (Saratov University Press)

Acknowledgements

The present study is financially supported by the Ministry of Science and Higher Education of the Russian Federation (State Registration Number AAAA-A17-117021310381-8) and by the Russian Foundation for Basic Research (project No. 18-01-00844).

Experimental and theoretical studies of oscillations of stratified fluid

Win Ko Ko and A N Temnov

Bauman Moscow State Technical University, Moscow, Russia
E-mail: antt45@mail.ru

Abstract. This article examines the problem of small movements of liquids and the motion of a solid with a circular cylindrical cavity that is completely filled three of incompressible ideal fluids. Assuming irrotational fluid motion is formulated boundary value problem and solutions are obtained for potentials of the displacements of the particles of liquids. Using the Lagrange equations of the 2nd kind the equations of motion of a hydromechanical system. Illustrates the difference between motion of a rigid body having a cavity with fluids from the case considered by N. E. Zhukovsky. Examines the problem of natural vibrations of liquids in a stationary tank and the motion of a rigid body fluids. Given formula, determine the natural frequencies and mode shapes. The article presents the results of an experimental study of motion of rigid body and liquids.
Keywords: rigid body, liquid, natural frequencies, mode shapes, incompressible ideal fluids.

1. Introduction.

Currently, the problem of the motion of a rigid body with a cavity filled with fluid is a classical problem in mechanics. The first such task was considered N. E. Zhukovsky and published in [1]. Recent advances in the aerospace industry, cryogenic industry and gas industry require research of a more general problem of the motion of bodies containing a complex liquid medium. One of such problems is the problem of the motion of a rigid body with a cavity filled with layered liquids. Fluctuations of liquids partially filling the cavity of the stationary and movable rigid body devoted to a fairly large number of works, see for example [2]-[15].

2. Statement of the problem

Let a circular cylinder of radius r_0 , completely filled with three ideal incompressible fluid having a density ρ_0, ρ_1, ρ_2, $(\rho_0 \langle \rho_1 \langle \rho_2)$ (see figure 1). The cylinder is mounted on a movable platform with a total mass M, capable of moving on a smooth horizontal base. Assuming that the fluid in the cylinder, making a small displacement will define the main characteristics of the movement of liquids to a circular cylinder.

Content from this work may be used under the terms of the Creative Commons Attribution 3.0 licence. Any further distribution of this work must maintain attribution to the author(s) and the title of the work, journal citation and DOI.
Published under licence by IOP Publishing Ltd

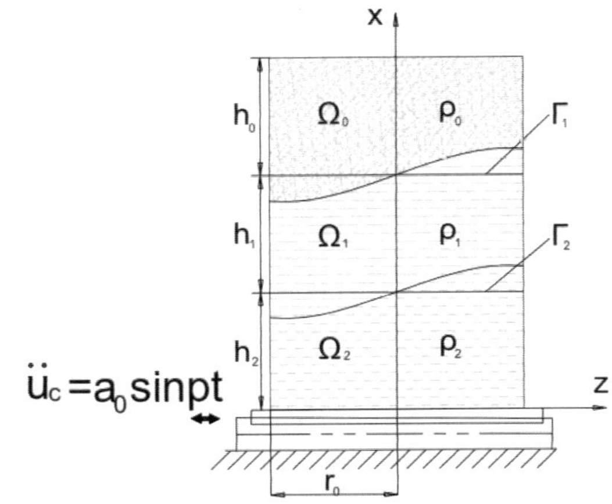

Figure 1. A cylindrical vessel layered liquid

To compile the equations of motion of the considered mechanical system will use the equations of motion the Lagrange 2-nd kind.
The kinetic energy of the system can be written in the form

$$T_i = \frac{1}{2} M \dot{u}_c^2 + \frac{1}{2} \sum_{i=0}^{2} \rho_i \iiint_{V_i} (\nabla \Phi_i)^2 dV_i , \tag{1}$$

where V_i, Φ_i $(i = 0,1,2)$ -the volumes and capacities of the absolute velocities of the liquids. At low fluid motion potentials velocities Φ_i can be expressed the potentials of displacement of the particles of liquids.

$$\Phi_i = \frac{\partial \chi_i (x_i, y_i, z_i, t)}{\partial t}, (i = 0,1,2). \tag{2}$$

The potentials of displacement χ_i are solutions of boundary value problems that satisfy the Laplace equation and the boundary conditions:

$$\Delta \chi_i = 0, (i = 0,1,2), \tag{3}$$

(a) the conditions of no motion on wetted surfaces

$$\left. \frac{\partial \chi_i}{\partial v_i} \right|_{S_i} = u_c \cdot \sin(\eta); \left. \frac{\partial \chi_0}{\partial x_0} \right|_{x_0 = h_0} = 0, \left. \frac{\partial \chi_2}{\partial x_2} \right|_{x_2 = 0} = 0, \tag{4}$$

(b) kinematic conditions on the surfaces of the sections

$$\frac{\partial \chi_0}{\partial v_0} = \frac{\partial \chi_1}{\partial v_1}; \text{ at } \Gamma_1 \text{ и } \frac{\partial \chi_1}{\partial v_1} = \frac{\partial \chi_2}{\partial v_2}; \text{ at } \Gamma_2, \tag{5}$$

(c) dynamic conditions on the surfaces of the sections

$$\left(\rho_0 \cdot \frac{\partial^2 \chi_0}{\partial t^2} - \rho_1 \cdot \frac{\partial^2 \chi_1}{\partial t^2} \right) = (\rho_1 - \rho_0) \cdot g \cdot \frac{\partial \chi_0}{\partial v_0}; \text{ at } \Gamma_1, \tag{6}$$

$$\left(\rho_1 \cdot \frac{\partial^2 \chi_1}{\partial t^2} - \rho_2 \cdot \frac{\partial^2 \chi_2}{\partial t^2} \right) = (\rho_2 - \rho_1) \cdot g \cdot \frac{\partial \chi_1}{\partial v_1}; \text{ at } \Gamma_2, \tag{7}$$

The solution of the boundary value problems (2)-(7) the displacement potentials can be written as,

$$\chi_0 = 2r_0 \sin\eta \sum_{n=1}^{\infty} \frac{J_1(k_n r)}{J_1(\xi_n)(\xi_n^{\,2}-1)} \left\{ u(t) - thk_n \mathrm{h}_1 \cdot \frac{chk_n(x_0 - h_0)}{shk_n \mathrm{h}_0} \sigma_{1n}(t) \right\}, \tag{8}$$

$$\chi_1 = 2r_0 \sin\eta \sum_{n=1}^{\infty} \frac{J_1(k_n r)}{J_1(\xi_n)(\xi_n^{\,2}-1)} \left\{ u(t) + thk_n \mathrm{h}_1 \cdot \frac{chk_n x_1}{shk_n \mathrm{h}_1} \sigma_{1n}(t) - \frac{chk_n(x_1 - h_1)}{chk_n \mathrm{h}_1} \sigma_{2n}(t) \right\}, \tag{9}$$

$$\chi_2 = 2r_0 \sin\eta \sum_{n=1}^{\infty} \frac{J_1(k_n r)}{J_1(\xi_n)(\xi_n^{\,2}-1)} \left\{ u(t) + thk_n \mathrm{h}_1 \cdot \frac{chk_n x_2}{shk_n \mathrm{h}_2} \sigma_{2n}(t) \right\}. \tag{10}$$

The potential energy for the considered case is given by,

$$\Pi_\Sigma = \frac{1}{2} g(\rho_1 - \rho_0) \sum_{n=1}^{\infty} \int_{\Gamma_1} \left(\frac{\partial \chi_1}{\partial x_1} \right)^2 r \sin\eta \, dr + \frac{1}{2} g(\rho_2 - \rho_1) \sum_{n=1}^{\infty} \int_{\Gamma_1} \left(\frac{\partial \chi_2}{\partial x_2} \right)^2 r \sin\eta \, dr. \tag{11}$$

Putting the solution of (8) to (10) for potentials of displacements in formulas (1), (11), we obtain final expressions for the kinetic and potential energy,

$$\begin{gathered} T_\Sigma = \frac{1}{2} M^* \dot{u}_c^{\,2} + \frac{1}{2} \sum_{n=1}^{\infty} \left\{ (m'_{1n} - m'_{0n}) \dot{u}_c \dot{\sigma}_{1n} \right\} + \frac{1}{2} \sum_{n=1}^{\infty} \left\{ (m'_{2n} - m'_{1n}) \dot{u}_c \dot{\sigma}_{2n} \right\} + \\ + \frac{1}{2} \sum_{n=1}^{\infty} \left\{ (m'_{1n} + m'_{0n} \overline{f_{0n}}) \dot{\sigma}_{1n}^{\,2} \right\} + \frac{1}{2} \sum_{n=1}^{\infty} \left\{ (m'_{2n} \overline{f_{1n}} + m'_{1n}) \dot{\sigma}_{2n}^{\,2} \right\} \end{gathered}, \tag{12}$$

$$\Pi_\Sigma = \frac{1}{2} \sum_{n=1}^{\infty} (m'_{1n} - m'_{0n}) \omega_n^2 \sigma_{1n}^2 + \frac{1}{2} \sum_{n=1}^{\infty} (m'_{2n} - m'_{1n}) \omega_n^2 \sigma_{2n}^2. \tag{13}$$

In formulas (12), (13) adopted notation: σ_{1n}, σ_{2n} -generalized coordinates of wave motion of surfaces of sections of the fluids.

$m_0 = \pi r_0^2 \rho_0 h_0$, $m_1 = \pi r_0^2 \rho_1 h_1$, $m_2 = \pi r_0^2 \rho_2 h_2$ - the mass of each liquid.

$m'_{0n} = \rho_0 V_n, m'_{1n} = \rho_1 V_n, m'_{2n} = \rho_2 V_n$ - mass of oscillating fluids.

$$V_n = \frac{2\pi r_0^3}{\xi_n(\xi_n^{\,2}-1)} thk_n h_1,\ \overline{f_{0n}} = thk_n h_1 cthk_n h_0,\ \overline{f_{1n}} = thk_n h_1 cthk_n h_2,\ \omega_n^2 = gk_n thk_n h_1.$$

$M^* = m_0 + m_1 + m_2 + M$ - the total mass of the whole hydro-mechanical system.

3. Derivation of equations of motion

Used the Lagrange equations of the 2-nd kind, we get the system of equations of motion in the form

$$M^* \ddot{u}_c + \sum_{n=1}^{\infty} (m'_{1n} - m'_{0n}) \ddot{\sigma}_{1n} + \sum_{n=1}^{\infty} (m'_{2n} - m'_{1n}) \ddot{\sigma}_{2n} = 0, \tag{14}$$

$$\ddot{\sigma}_{1n} + \frac{(m'_{1n} - m'_{0n})}{(m'_{1n} + m'_{0n} \overline{f_{0n}})} \omega_n^2 \sigma_{1n} - \frac{m'_{1n} \ddot{\sigma}_{2n}}{(m'_{1n} + m'_{0n} \overline{f_{0n}}) chk_n h_1} + \frac{(m'_{1n} - m'_{0n})}{(m'_{1n} + m'_{0n} \overline{f_{0n}})} \ddot{u}_c = 0, (\mathrm{n} = 1,2,3...), \tag{15}$$

$$\ddot{\sigma}_{2n} + \frac{(m'_{2n} - m'_{1n})}{(m'_{2n} \overline{f_{1n}} + m'_{1n})} \omega_n^2 \sigma_{2n} - \frac{m'_{1n} \ddot{\sigma}_{1n}}{(m'_{2n} \overline{f_{1n}} + m'_{1n}) chk_n h_1} + \frac{(m'_{2n} - m'_{1n})}{(m'_{2n} \overline{f_{1n}} + m'_{1n})} \ddot{u}_c = 0, (\mathrm{n} = 1,2,3...). \tag{16}$$

For the convenience of further research, we introduce new notations: $m_{1n}^{*}=(m_{1n}'-m_{0n}')$, $m_{2n}^{*}=(m_{2n}'-m_{1n}')$, $\alpha_{1n}=(m_{1n}'+m_{0n}'\overline{f_{0n}})$, $\alpha_{2n}=(m_{2n}'\overline{f_{1n}}+m_{1n}')$, $a_{12n}=m_{1n}'/chk_nh_1\cdot\alpha_{1n}$, $a_{21n}=m_{1n}'/chk_nh_1\cdot\alpha_{2n}$, $b_{1n}=\dfrac{m_{1n}^{*}}{\alpha_{1n}}$, $b_{2n}=\dfrac{m_{2n}^{*}}{\alpha_{2n}}$, $\beta_{1n}^2=b_{1n}\omega_n^2$, $\beta_{2n}^2=b_{2n}\omega_n^2$.

Let the tank completely filled with liquid, and mounted on the movable plate can only move progressively in the horizontal plane. We assume that the plate acts in the horizontal plane of the external force $F_u=F_0\sin pt$, the line of action which pass through the plane of symmetry of the tank. Considering only the first tone of the vibrations of the fluid ($n=1$), we define the amplitude of the external force F_0, if the movable plate moves with acceleration $\ddot{u}_c=a_0\sin pt$.

The equations of motion of a plate with liquid in ($n=1$) will be form

$$M^{*}\ddot{u}_c+m_1^{*}\ddot{\sigma}_1+m_2^{*}\ddot{\sigma}_2=F_0\sin pt, \tag{17}$$

$$\ddot{\sigma}_1+\beta_1^2\sigma_1-a_{12}\ddot{\sigma}_2+b_1\ddot{u}_c=0, \tag{18}$$

$$\ddot{\sigma}_2+\beta_2^2\sigma_2-a_{21}\ddot{\sigma}_1+b_2\ddot{u}_c=0. \tag{19}$$

When considering forced oscillations of the liquid, we assume that the generalized coordinates $\sigma_1(t)$, $\sigma_2(t)$ change the law $\sigma_1=\sigma_1^0\sin pt$, $\sigma_2=\sigma_2^0\sin pt$ and $u_c=u_c^0\sin pt$, and rewrite equations (17)-(19),

$$-p^2(M^{*}u_c^0+m_1^{*}\sigma_1^0+m_2^{*}\sigma_2^0)=F_0, \tag{20}$$

$$\left(-p^2+\beta_1^2\right)\sigma_1^0-a_{12}p^2\sigma_2^0=b_1p^2u_c^0, \tag{21}$$

$$-a_{21}p^2\sigma_1^0+\left(-p^2+\beta_2^2\right)\sigma_2^0=b_2p^2u_c^0, \tag{22}$$

From equations (21) and (22) define values σ_1^0 and σ_2^0,

$$\sigma_1^0=\frac{b_1\left(-p^2+\beta_2^2\right)+b_2p^2a_{12}}{(\Omega_1^2-p^2)(\Omega_2^2-p^2)}p^2u_c^0,\ \sigma_2^0=\frac{b_2\left(-p^2+\beta_1^2\right)+b_1p^2a_{21}}{(\Omega_1^2-p^2)(\Omega_2^2-p^2)}p^2u_c^0, \tag{23}$$

where Ω_1 и Ω_2 -natural frequencies of the three liquids in a stationary tank,

$$\Omega_{1,2}{}^2=\frac{-b\pm\sqrt{b^2-4ac}}{2a}; \tag{24}$$

$$a=\left(1-a_{21}a_{21}\right),\ b=-\left(\beta_1^2+\beta_2^2\right),\ c=\left(\beta_1^2\cdot\beta_2^2\right). \tag{25}$$

Substitute the obtained expression (23) into equation (20). The result of the equation of motion of a rigid body transformed to the

$$F_0=a_0M_{np}, \tag{26}$$

where M_{np} - mass converted rigid body,

$$M_{np}=M^{*}\left[1+\frac{m_1^{*}}{M^{*}}\frac{\left[b_1\left(-p^2+\beta_2^2\right)+b_2p^2a_{12}\right]p^2}{(\Omega_1^2-p^2)(\Omega_2^2-p^2)}+\frac{m_2^{*}}{M^{*}}\frac{\left[b_2\left(-p^2+\beta_1^2\right)+b_1p^2a_{21}\right]p^2}{(\Omega_1^2-p^2)(\Omega_2^2-p^2)}\right]. \tag{27}$$

One of the results of studies of the motion of a rigid body with a cavity completely filled with a homogeneous liquid drop is that mass converted rigid body is equal to the sum of the masses of rigid

body and solidified liquid [1]. In this case, as shown by the formula (27) in the motion of a rigid body with a layered liquid mass converted rigid body depends on the frequency of the oscillations.
Graph of the ratio M_{np} / M^* shown in figure 2. The graph was the following notation,

$$f(\bar{p}) = M_{np} / M^*, \bar{p}^2 = \frac{p^2}{\Omega_1^2}, \bar{\Omega}_2 = \frac{\Omega_2^2}{\Omega_1^2}, \bar{\beta}_1^2 = \frac{\beta_1^2}{\Omega_1^2}, \bar{\beta}_2^2 = \frac{\beta_2^2}{\Omega_1^2}, \quad (28)$$

$$\Delta_2 = (\rho_2 - \rho_1), \Delta_1 = (\rho_1 - \rho_0). \quad (29)$$

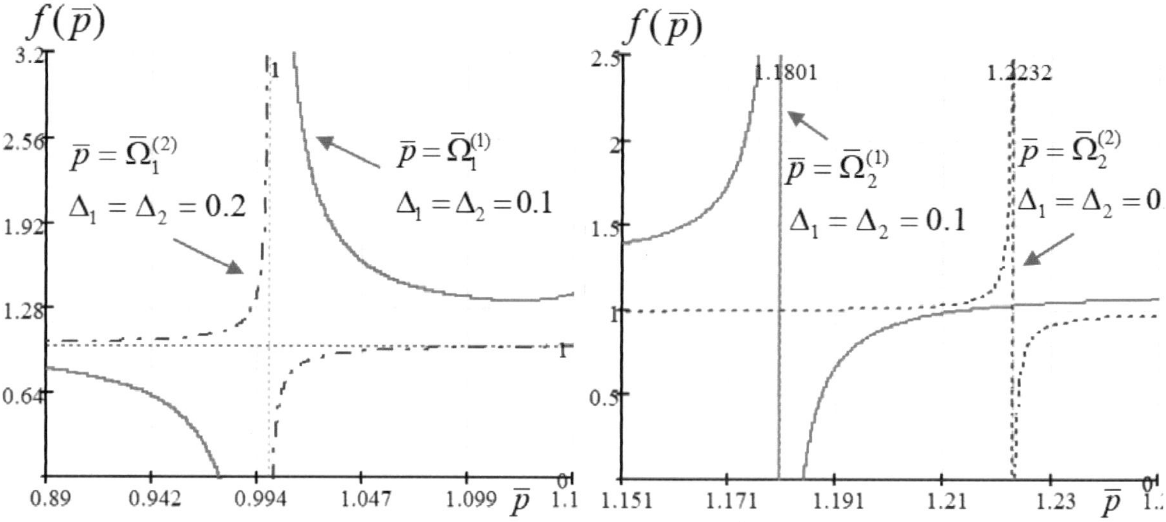

Figure 2. The dependence of the dimensionless mass converted rigid body $f(\bar{p}) = M_{np} / M^*$ from the dimensionless frequency of oscillation, ($\bar{\Omega}_1^{(1)}, \bar{\Omega}_1^{(2)}, \bar{\Omega}_2^{(1)}, \bar{\Omega}_2^{(2)}$ - the first and second main frequency for the cases $\Delta_1 = \Delta_2 = 0.1$ and $\Delta_1 = \Delta_2 = 0.2$).

4. Experimental study of fluctuations of the stratified liquid

Subjects liquids selected were water ($\rho_2 = 1000 kg / m^3$), sunflower oil density ($\rho_1 = 920 kg / m^3$) and formic alcohol ($\rho_0 = 830 kg / m^3$), with coefficients of surface tension of water - sunflower oil ($\sigma_1 = 0.013 N / m$), and sunflower oil-formic alcohol ($\sigma_2 = 0.02 N / m$), the value of the kinematic viscosity of water ($\nu_2 = 1,006.10^{-6} m^2 / s$), sunflower oil ($\nu_1 = 39,6.10^{-6} m^2 / s$) and formic alcohol ($\nu_0 = 1,54.10^{-6} m^2 / s$).

4.1. Description of the experimental setup

Experimental setup figure 3 consisted of the base -1, the movable plate -2, -3 transparent tank, electro-mechanical exciter of vibrations -4 and measuring complex. The base was a stationary plate of the textolite, which was fixed electromechanical exciter and metal guides ensure smooth motion without tank vibration of the movable plate.

Control and measurement system consisted of a strain test station VI6-6 T -5, pressure sensors and displacement -6 and -7, USB digital oscilloscope Hantek 6022BE -8 and -9 laptop and power supply HY5003-2 -10, which were placed on the laboratory table, close to the fixed base. Power supply HY5003-2 is designed to power the strain test station and motor. Control and data display oscilloscope

Hantek 6022BE implemented personal computer/laptop/netbook/tablet running Windows OS via USB interface.
General view of the experimental setup is shown in figure 3.

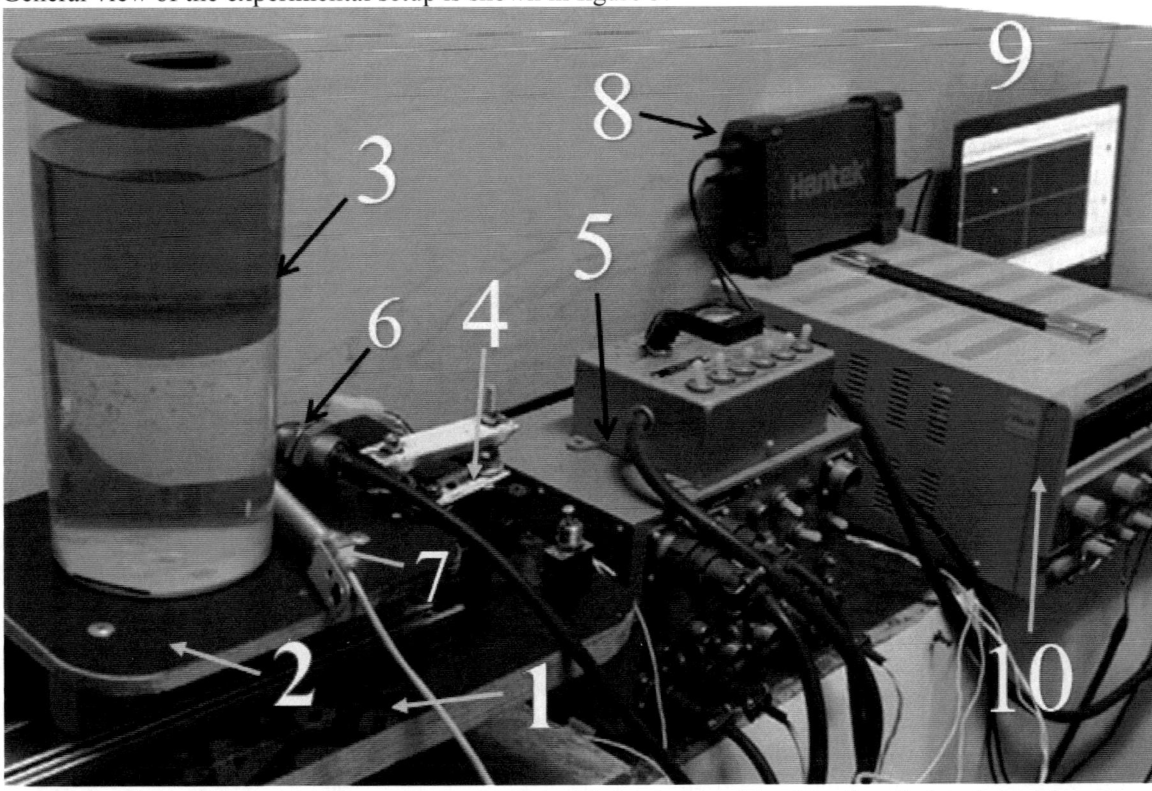

Figure 3. General view of the experimental setup: 1-base, 2-movable plate, 3-cylindrical container, 4 - electromechanical exciter, 5 - strain station VI6-6 T, 6 - pressure transducer, 7 - the displacement transducer, 8 - USB digital oscilloscope Hantek 6022BE, the 9 - ASUS laptop 10 power HY5003-2.

4.2. The kinematic diagram of the setup

Electromechanical oscillation exciter consisted of a DC motor and a crank mechanism that reciprocates the movable plate.
As the vibration measuring instrument used inductive apparatus VI6-6 T-compact six-channel, with solid-state power supply. Installation VI6-6 T is designed to measure vibrations, accelerations, pressures and displacements. On the basis of the instrument lies principle amplitude modulation frequency inductive sensors on differential-transformer scheme.

4.3. Experimental determination of the main vibration characteristics of layered fluid

To find the oscillation frequencies of a layered fluid was used the method of free oscillations. To create the initial conditions of the free oscillations of the fluids moving platform with a tank was set vibrations through the reciprocating motion close to a resonant frequency of oscillations of the fluids. After some time joint fluctuations of the tank with liquids platform instantly stopped, and the liquid made free oscillations in a stationary tank. In the experiment on a laptop recorded the forced and free oscillations of fluids and the movement of the tank itself.
In the experiment it is possible to allocate following stages:
(1) before the experiment was determined the theoretical value of the oscillation frequency of liquids in a stationary tank by the formulas (24), (25),

(2) carried out calibration of displacement sensors and pressure calibration of pressure sensors was carried out by additions of liquid (water) at a certain depth,
(3) the test cylinder was freed from some part of the water and then filled with layers of liquids of different thickness for the experiment,
(4) the experiment was repeated again with a different average depth of the liquid,
(5) the results of the records of the experiments the laptop was determined the natural frequencies of liquids.
The values of the frequencies defined by the formulas (24), (25) and obtained from the experiment shown in figure 4.

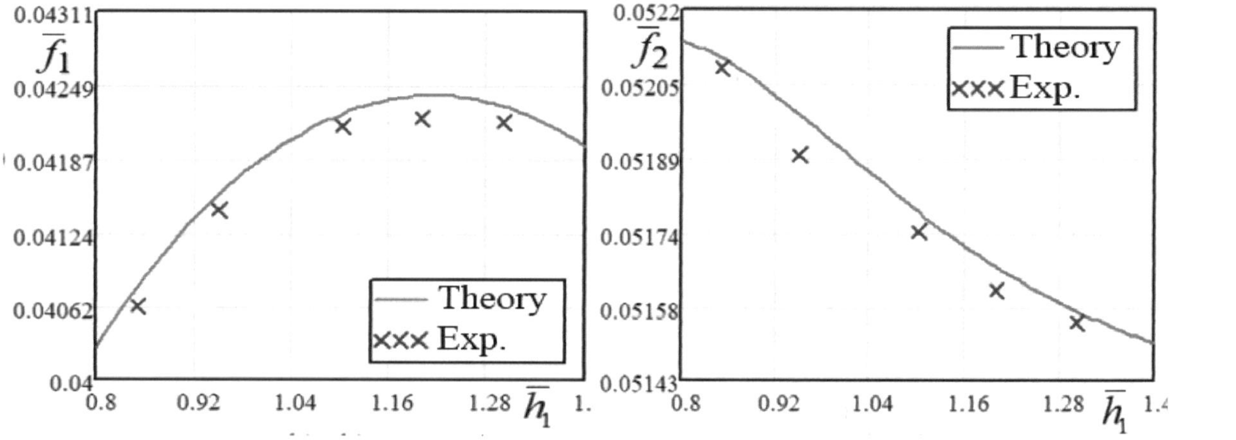

(a) (b)

Figure 4. Dependence to the first main frequency (a) and the second main frequency (b) on the average thickness of the liquid defined by the formula (24), (25), and defined in the experiment ($\bar{h}_0 = 0.6, \bar{h}_2 = 2.1 - \bar{h}_1$)

In figures 5 and 6 shows typical graphs of the processes recorded during the experiments

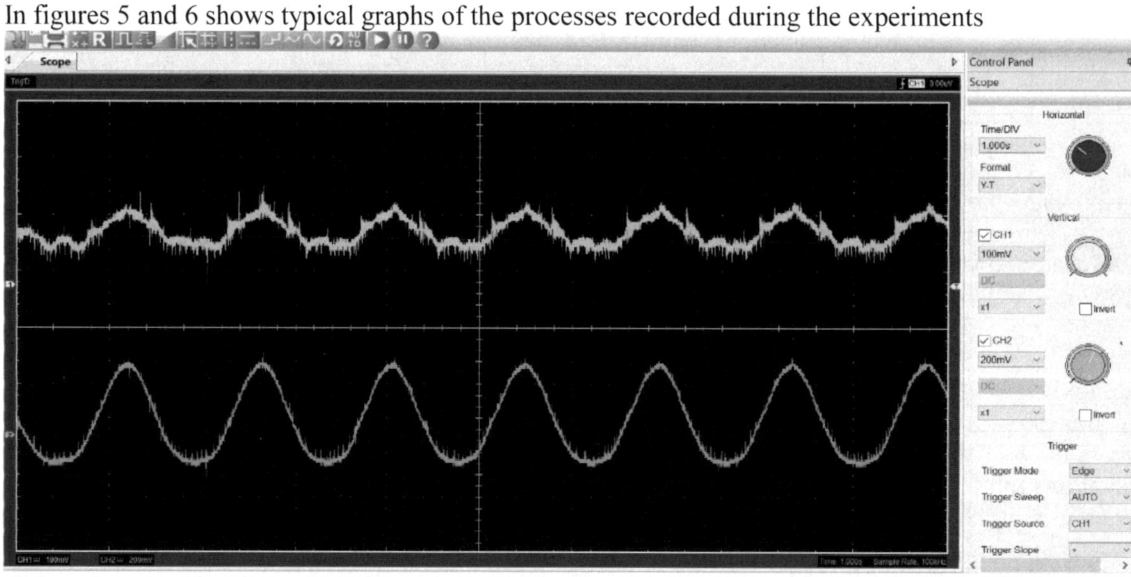

Figure 5. Example of a record of the oscillations of the moving platform and forced vibration of fluids

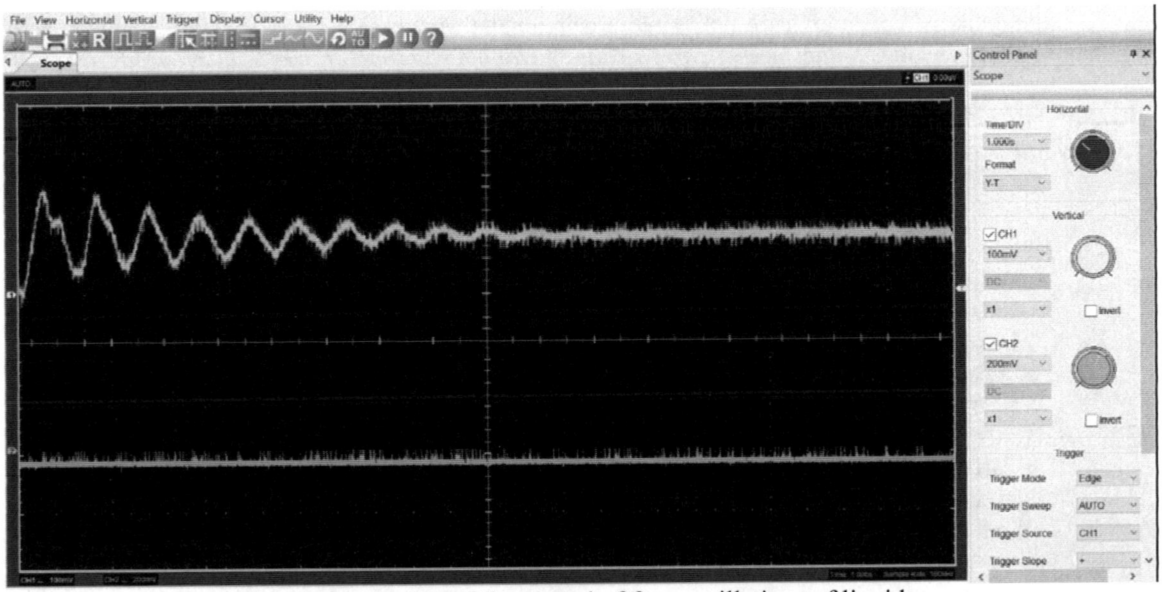

Figure 6. Example of the record of free oscillations of liquids

Conclusion. In work the equations of motion of a rigid body performs translational motion and having a hollow cylindrical shape, is completely filled with three ideal homogeneous immiscible liquids. The resulting equations of motion, formulas for the added masses and the results of numerical calculations. Given the results of the pilot study.

The results can be used in the design of transport systems with cryogenic liquids or liquefied gases.

References:

[1] Zhukovsky N E 1955 On the motion of a rigid body having cavities filled with a homogeneous drop of liquid *FAV. CIT.*, Vol. **I**.

[2] Chashechkin Yu D 2014 Fluid mechanics: consistent analytical, numerical and laboratory models of stratified flows *Vestnik MGTU im. Bauman. Ser. Natural science,* No.**6[57]** pp 53.

[3] Babsky V G, Zhukov M Yu, Kopachevskii N D, Myshkis A D, Slobozhanin L A, Tyuptsov A D 1992 Methods of solving problems of hydromechanics of weightlessness *K.: Naukova Dumka,* pp 592

[4] Ganichev A I, Kachur V P, Temnov A N 1974 Small oscillations of two immiscible liquids movable in a cylindrical vessel *In the book: Oscillations of elastic constructions with liquid, Novosibirsk, NETI*, pp 82-88.

[5] Goncharov D A 2013 Dynamics of a two-layer liquid divided by an elastic partition with the consideration of surface tension forces *Science and education of the BMSTU. N. E. Bauman. Electron. zhurn.,* No. **11**.

[6] Gaziev E L 2013 Modeling of natural vibrations of the system "capillary ideal fluid-barotropic gas" in a cylindrical container *Book of Abstracts of Crimean International Mathematics Conference (CIMC-2003). Simferopol: Kola science center of the national Academy of Sciences*, Vol. **3** pp 51-52.

[7] Kolesnikov K S, Pozhalostin A A, P M Shkapov 2012 Problems of the dynamics of hydromechanical systems in the work of the department of theoretical mechanics named after Professor N E Zhukovsky №.**7(7)** *URL: http://engjournal.ru/catalog/eng/teormech/285.html (reference date: 21.09.2015)*

[8] Mikhayluk A V, Timokha A. N. 1995 Variational formulations for a spectral problem with parameter on the interface of two mediums *Dopov. Nats. Akad. Ukr. Mat. Prirodozn. Tekh. Nauki*. Issue 45, No. **10**, pp 38-40.
[9] Popov D N, Panaioti S S, Ryabinin V M 2014 *Fluid Mechanics*. (Moscow:Tutorial), p 320.
[10] Martinson L K, Malov Y I 2002 *Differential equations of mathematical physics*. (Moscow: publisher BMSTU im. N. E. Bauman) p 185.
[11] Kolesnikov K S 2003 *Dinamika Raket*. (Moscow: Mashinostroenie), p 520
[12] Kalinichenko V A, Aung Naing 2013 Wave Faraday in the movable vessel and their mechanical analog *Engineering journal: science and innovation*. **12(24)** URL:http://engjournal.ru/cata...g/teormech/1138/html (date accessed: 20.05.2015)
[13] Win Ko Ko, Temnov A N 2016 Oscillations of a discrete stratified fluid in a cylindrical vessel and their mechanical counterparts *Vestnik BMSTU im. Bauman. Ser. "Natural science"*. No. **3** pp 57-69.
[14] Nesterov C B 1983 Natural frequencies of internal waves in a fluid with an arbitrary frequency of Brent-Vaisala *Reports of as USSR*. T. 271. No. **3**. pp 570-573.
[15] Akulenko L D, Nesterov S V 1994 Parametric excitation of oscillations of a body with a cavity filled with a stratified fluid *Izv. Russian Academy of Sciences. MIG*. No. **3**. Pp 53-60.

IOP Conf. Series: Materials Science and Engineering **468** (2018) 012032 doi:10.1088/1757-899X/468/1/012032

Multifrequency model of Earth tidal deformations within framework of three-body problem

I V Skorobogatykh

Moscow Aviation Institute, Moscow, Russia
E-mail: igorsko@rambler.ru

Abstract. The Earth tidal deformation is studied. Earth is considered as a visco-elastic axially symmetrical body with a solid core. For the description of the visco-elastic Earth part the linear theory of visco-elasticity is used. It is assumed that the Earth - Moon center of mass moves on the constant elliptic orbit around the Sun, that is considered as attractive center. From the d'Alembert – Lagrange principle the system of motion equations is obtained. Then the displacement vector is expanded in series of Earth's oscillation modes. The system of ordinary differential equations for modal variables is obtained. This system describes process of oscillations, but in real situation we can consider these deformations as quasi-stationary ones. Hence, we can discard inertial members in equations. We can expand right parts of equations in series by orbit's eccentricities of the Earth – Moon center of mass and of the Earth (Moon). Then we can look, that the right parts of the equations are the series with coefficients depending of angles combinations, hence modal variables can be represented also as the similar series. From these conditions we can find approximately of Sun – Moon's tidal frequencies. We find well known frequencies, corresponding periods as year, half year, month, half month, day, half day, frequencies combinations, near above mentioned and large quantity of other combinations.

1. Introduction

In this paper we consider the problem of computation the deformations of the Earth under the influence of the gravitational fields of the Sun and the Moon. The Earth is considered as a visco-elastic body having a solid core. The orbit of the barycentre of the Earth-Moon system is assumed to be slowly evolving, so we can treat it as constant. On the basis of approximate differential equations, it is possible to obtain an expression for the displacement vector. Analysis of the displacement vector expression allows us to determine the set of frequencies of tidal deformations of the Earth. In the literature, in most cases, there are the models of numerical computer simulation based on the observations of the ICRM [1]. This article is an attempt to construct a numerical-analytic model of a deformed Earth. For example, this model can be used in problems related to the dynamics of the Earth's rotation [2,3].

2. Formulation of the problem

To simplify the formulation of the problem, we consider the Moon and the Sun as material points. The Earth is considered as a body consisting of an axisymmetric solid core and a visco-elastic axisymmetric (in an undeformed state) shell, obeying the Kelvin-Voigt model. On the inner boundary of the shell, there are no displacements, and the outer boundary is free. It is assumed that the process

Content from this work may be used under the terms of the Creative Commons Attribution 3.0 licence. Any further distribution of this work must maintain attribution to the author(s) and the title of the work, journal citation and DOI.
Published under licence by IOP Publishing Ltd

of deformations of the Earth can be considered as quasistationary. The origin of the inertial coordinate system $O\xi_1\xi_2\xi_3$ is placed in the attracting center (the Sun), with the barycenter C we connect the König axes $O\xi_1'\xi_2'\xi_3'$. The Earth's core is rigidly connected with axes $C_1x_1x_2x_3$ directed along the main central axes of inertia of the planet in the undeformed state, and it is also assumed that the point C_1 coincides with the center of mass of the Earth in the undeformed state. Further, let $\boldsymbol{R}_C = \boldsymbol{R}_C^0 R_C$, $R_C = p(1+e\cos\vartheta)^{-1}$ - radius-vector of barycenter, p and e – focal parameter and eccentricity of its orbit, ϑ - true anomaly. Similarly we represent a vector $\boldsymbol{R}_{21} = \boldsymbol{R}_{21}^0 R_{21}$ - from the Moon to the Earth, where $R_{21} = p_1(1+e_1\cos\vartheta_1)^{-1}$, $w_1 = \vartheta_1 + \omega_1$,

$$\boldsymbol{R}_{21}^0 = \begin{pmatrix} \cos w_1 \cos\Omega_1 - \sin w_1 \sin\Omega_1 \cos i \\ \cos w_1 \sin\Omega_1 + \sin w_1 \cos\Omega_1 \cos i \\ \sin w_1 \sin i_1 \end{pmatrix}.$$

We write the potential energy of the Earth in the form

$$\Pi = \Pi_S + \Pi_M,$$

$$\Pi_S = -\int_\Omega \frac{fM\rho}{[(O^{-1}\boldsymbol{R}_C + O^{-1}\boldsymbol{R}_1 + \boldsymbol{r} + \boldsymbol{u})^2]^{1/2}} dx, \qquad \Pi_M = -\int_\Omega \frac{fm_2\rho}{[(O^{-1}\boldsymbol{R}_{21} + \boldsymbol{r} + \boldsymbol{u})^2]^{1/2}} dx, \tag{1}$$

where Π_S is Earth's potential energy in gravitation field of Sun, Π_M is potential energy in gravitation field of Moon, Ω - the area occupied by the undeformed Earth in the axes $C_1x_1x_2x_3$, M – mass of Sun, m_1 - mass of Earth, m_2 - mass of Moon, $O^{-1}(t)$ - the matrix of the transition from the inertial coordinate system with the origin at the Earth's center of the mass, to the coordinate system $C_1x_1x_2x_3$ rigidly bound to the solid core. The matrix $O^{-1}(t)$ is represented in the form

$$O(t) = \Gamma_1\Gamma_2\Gamma_3,$$

where

$$\Gamma_1 = \begin{pmatrix} \cos\psi & -\sin\psi & 0 \\ \sin\psi & \cos\psi & 0 \\ 0 & 0 & 1 \end{pmatrix}, \quad \Gamma_2 = \begin{pmatrix} 1 & 0 & 0 \\ 0 & \cos\theta & -\sin\theta \\ 0 & \sin\theta & \cos\theta \end{pmatrix}, \quad \Gamma_3 = \begin{pmatrix} \cos\varphi & -\sin\varphi & 0 \\ \sin\varphi & \cos\varphi & 0 \\ 0 & 0 & 1 \end{pmatrix},$$

and φ,ψ,θ –Euler's angles. Hence

$$O^{-1}(t) = \Gamma_3^{-1}\Gamma_2^{-1}\Gamma_1^{-1}.$$

The equations defining deformations follow from the variational principle of d'Alembert-Lagrange:

$$\int_\Omega (O^{-1}\ddot{\boldsymbol{R}}_C + O^{-1}\ddot{\boldsymbol{R}}_1 + \ddot{\boldsymbol{u}} + \boldsymbol{\varepsilon}\times(\boldsymbol{r}+\boldsymbol{u}) + \boldsymbol{\omega}\times(\boldsymbol{\omega}\times(\boldsymbol{r}+\boldsymbol{u})) + 2\boldsymbol{\omega}\times\dot{\boldsymbol{u}})\rho\delta\boldsymbol{u}dx + \tag{2}$$

$$+(\nabla E[\boldsymbol{u}],\delta\boldsymbol{u}) + (\nabla D[\dot{\boldsymbol{u}}],\delta\boldsymbol{u}) + (\nabla\Pi[\boldsymbol{u}],\delta\boldsymbol{u}) = 0$$

Here $\boldsymbol{\omega}$ and ε are the angular velocity and angular acceleration vectors, respectively; $E[\boldsymbol{u}]$ - functional of elastic deformations energy, $D[\dot{\boldsymbol{u}}] = \chi b E[\dot{\boldsymbol{u}}]$ - dissipative functional.

3. Modal approach

Using the modal approach, we expand the vector $\boldsymbol{u}$ in a series on its own forms (modes) of oscillations similarly to [4, 5]:

$$\boldsymbol{u}(\boldsymbol{r},t) = \sum_{k,m=0}^{\infty} [q_{km}(t)\boldsymbol{V}_{km} + p_{km}(t)\boldsymbol{W}_{km}],$$

and choosing $\delta \boldsymbol{u} = \boldsymbol{V}_{ij}$ or $\delta \boldsymbol{u} = \boldsymbol{W}_{ij}$, we obtain an infinite system of ordinary differential equations for the modal variables q_{km} and p_{km}. There are equalities

$$(\nabla E, \boldsymbol{V}_{km}) = v_{km}^2 q_{km}, \qquad (\nabla E, \boldsymbol{W}_{km}) = v_{km}^2 p_{km}, \tag{3}$$

and also

$$(\nabla D, \boldsymbol{V}_{km}) = \chi b v_{km}^2 \dot{q}_{km}, \qquad (\nabla D, \boldsymbol{W}_{km}) = \chi b v_{km}^2 \dot{p}_{km}. \tag{4}$$

In our case, in order to obtain a qualitative effect, we restrict ourselves to only two modal variables q_{20} and p_{20}, which describe the oscillations on the forms $\boldsymbol{V}_{20}$ and $\boldsymbol{W}_{20}$ which model the tidal humps of the Earth quite well. In addition, we omit the inertial terms, taking into account the quasistationary nature of the deformations. When calculating the gradient of the gravitational potential, we assume that the equivalence relations $\frac{r}{R_{21}} \sim \frac{r}{R} \sim \varepsilon_1$, $\frac{u}{r} \sim \varepsilon_1^2$, $\boldsymbol{R} = \boldsymbol{R}_C + \boldsymbol{R}_1$, ε_1 - small parameter, $\boldsymbol{R}_1$ - vector, directed from barycenter to Earth's mass center. Then, restricting ourselves to terms of order not higher ε_1^2, and taking into account the relations $\int_\Omega \boldsymbol{r}\boldsymbol{V}_{20}\rho dx = \int_\Omega \boldsymbol{r}\boldsymbol{W}_{20}\rho dx = 0$,

$$\int_\Omega (O^{-1}\boldsymbol{R}_{21}^0)(O^{-1}\boldsymbol{R}_{21}^0 \cdot \boldsymbol{r})\boldsymbol{V}_{20}\rho dx = \gamma_1\gamma_2(b_{2021} + b_{2012}),$$

$$\int_\Omega (O^{-1}\boldsymbol{R}_{21}^0)(O^{-1}\boldsymbol{R}_{21}^0 \cdot \boldsymbol{r})\boldsymbol{W}_{20}\rho dx = c_{2011}\gamma_1^2 + c_{2022}\gamma_2^2,$$

$$b_{2021} = \int_\Omega V_{202}x_1\rho dx, \quad b_{2012} = \int_\Omega V_{201}x_2\rho dx, \quad c_{2011} = \int_\Omega W_{201}x_1\rho dx, \quad c_{2022} = \int_\Omega W_{202}x_2\rho dx,$$

where V_{20i} и W_{20i} - projections of vectors $\boldsymbol{V}_{20}$ and $\boldsymbol{W}_{20}$ on the axis of the Cartesian coordinate system $x_1x_2x_3$, we obtain

$$(\nabla\Pi_M, \boldsymbol{V}_{20}) = -\frac{3fm_2}{R_{21}^3}(\gamma_1\gamma_2)(b_{2021} + b_{2012}), \qquad (\nabla\Pi_M, \boldsymbol{W}_{20}) = -\frac{3fm_2}{R_{21}^3}(c_{2011}\gamma_1^2 + c_{2022}\gamma_2^2), \tag{5}$$

and similarly

$$(\nabla\Pi_S, \boldsymbol{V}_{20}) = -\frac{3fM}{R^3}(\kappa_1\kappa_2)(b_{2021} + b_{2012}), \qquad (\nabla\Pi_S, \boldsymbol{W}_{20}) = -\frac{3fM}{R^3}(c_{2011}\kappa_1^2 + c_{2022}\kappa_2^2). \tag{6}$$

Here is denoted

$$O^{-1}\boldsymbol{R}_{21}^0 = (\gamma_1, \gamma_2, \gamma_3)^T, \quad O^{-1}\boldsymbol{R}^0 = (\kappa_1, \kappa_2, \kappa_3)^T.$$

Taking into account the terms with centrifugal acceleration due to polar compression in the Earth figure, we omit the same in equations, and rewrite the equations for the modal variables in the form

$$v_{20}^2 q_{20} + \chi b v_{20}^2 \dot{q}_{20} + 2\kappa_{20}\omega_3 \dot{p}_{20} = \frac{3fm_2}{R_{21}^3}(\gamma_1\gamma_2)(b_{2021} + b_{2012}) + \frac{3fM}{R^3}(\kappa_1\kappa_2)(b_{2021} + b_{2012}), \tag{7}$$

$$v_{20}^2 p_{20} + \chi b v_{20}^2 \dot{p}_{20} - 2\kappa_{20}\omega_3 \dot{q}_{20} = \frac{3fm_2}{R_{21}^3}(c_{2011}\gamma_1^2 + c_{2022}\gamma_2^2) + \frac{3fM}{R^3}(c_{2011}\kappa_1^2 + c_{2022}\kappa_2^2).$$

Omitting the small terms arising from the Coriolis acceleration, we rewrite equations (7) in vector form

$$v_{20}^2 \boldsymbol{p} + \chi b v_{20}^2 \dot{\boldsymbol{p}} = \boldsymbol{F}. \tag{8}$$

Here is denoted

$$\boldsymbol{p} = (q_{20}, p_{20})^T,$$

$$\boldsymbol{F}=\begin{pmatrix}\frac{3fm_2}{R_{21}^3}(\gamma_1\gamma_2)(b_{2021}+b_{2012})+\frac{3fM}{R^3}(\kappa_1\kappa_2)(b_{2021}+b_{2012})\\ \frac{3fm_2}{R_{21}^3}(c_{2011}\gamma_1^2+c_{2022}\gamma_2^2)+\frac{3fM}{R^3}(c_{2011}\kappa_1^2+c_{2022}\kappa_2^2)\end{pmatrix}$$

The solution of equation (8) we can found in the form

$$\boldsymbol{p}=\sum_{n=0}^{\infty}(-\chi)^n\frac{\partial^n\boldsymbol{p}_0}{\partial t^n}, \tag{9}$$

where $\boldsymbol{p}_0$ - is solution of equation $v_{20}^2\boldsymbol{p}_0=\boldsymbol{F}$, in other words $\boldsymbol{p}_0=v_{20}^{-2}\boldsymbol{F}$.
Restricting ourselves to the first approximation in (9), we obtain

$$\boldsymbol{p}=\boldsymbol{p}_0-\chi\dot{\boldsymbol{p}}_0 \tag{10}$$

4.Tidal frequencies

Next, calculate the coordinates of the vectors $O^{-1}\boldsymbol{R}_{21}^0$ and $O^{-1}\boldsymbol{R}^0$, and substitute in (10). To simplify the calculations, we can assume $i\approx const,\psi\approx const,\Omega_1\approx const$, i.e. assume that the orbit of the barycenter has a constant inclination and longitude of the ascending node, and neglect the precession of the Earth's axis. We expand both R_{21} and R in series by power of eccentricity and restrict ourselves to terms of not higher than the first order. Then to the right in (10) we obtain series of harmonic functions of the form

$$\sum_i(A_i\cos\alpha_i+B_i\sin\alpha_i), \tag{11}$$

with arguments α_i, which are a combination of angles φ,w,w_1,θ, $w=\vartheta+\omega$. These combinations will determine the frequencies of tidal deformations. In particular, there are following combinations

$$2\varphi,\quad 2w_1,\quad \theta,$$
$$w_1\pm w\pm\theta,\qquad 2w\pm\theta,\qquad 2w\pm\theta,$$

defining periods of half a day, half of the month, the period of nutation of the Earth's axis (Chandler's period), two periods close to a month, two close to the half-year. These periods are consistent with known periods of oceanic and solid Earth tides. In addition, there are also a large number of more complex combinations corresponding to periods close to semidiurnal ones, for example

$$2\varphi\pm 2w_1,\quad 2\varphi\pm\theta,\quad 2\varphi\pm 2\theta,\quad 2\varphi\pm 2w_1\pm\theta,\quad 2\varphi\pm 2w_1\pm 2\theta,$$
$$2\varphi\pm w\pm w_1,\quad 2\varphi\pm w\pm w_1\pm\theta,\quad 2\varphi\pm w\pm w_1\pm 2\theta,$$
$$2\varphi\pm 2w,\quad 2\varphi\pm\theta,\quad 2\varphi\pm 2\theta,\quad 2\varphi\pm 2w\pm\theta,\quad 2\varphi\pm 2w\pm 2\theta,$$
$$2\varphi+w_1-\omega_1,\quad 2\varphi-w_1+\omega_1\quad 2\varphi+3w_1-\omega_1,\quad 2\varphi-3w_1+w_1,$$
$$2\varphi\pm w_1\pm\omega_1\pm\theta,\quad 2\varphi+3w_1-\omega_1\pm\theta,\quad 2\varphi-3w_1+\omega_1\pm\theta,$$
$$2\varphi\pm w_1\pm\omega_1\pm 2\theta,\quad 2\varphi+3w_1-\omega_1\pm 2\theta,\quad 2\varphi-3w_1+\omega_1\pm 2\theta,$$
$$2\varphi\pm\omega\pm w_1,\quad 2\varphi+2w-\omega\pm w_1,\quad 2\varphi-2w+\omega\pm w_1,$$
$$2\varphi\pm 2w+w_1-\omega_1\pm\theta,\quad 2\varphi\pm 2w-w_1+\omega_1+\theta,\quad 2\varphi\pm 2w+w_1-\omega_1-\theta,\ldots$$

half-monthly

$$2w_1\pm\theta,\quad 2w_1+w-\omega,\quad 2w_1-w+\omega,\quad 2w_1+w-\omega+\theta,\quad 2w_1-w+\omega+\theta,$$
$$2w_1+w-\omega-\theta,\quad 2w_1-w+\omega-\theta,$$

monthly

$$w_1+2w-\omega_1\pm\theta,\quad w_1+2w+\omega_1\pm\theta,$$

and Chandler $\theta+w_1-\omega_1,\quad \theta-w_1+\omega_1,\quad \theta+w_1-\omega_1,\quad \theta-w_1+\omega_1$.
There are also combinations with other periods, for example:

$$3w_1 \pm \omega_1, \quad 3w_1 \pm \omega_1 \pm \theta,$$
$$\theta + w - \omega, \quad \theta - w + \omega\,.$$

Although equations (10) also allow us to write formulas for the tidal amplitudes corresponding to different periods, these coefficients require knowledge of many parameters - the Earth's modes of oscillation, the exact shape of the Earth, the Lame coefficients, the dissipative coefficient, and therefore, more simply, they can determined from the results of observations. As is known, the figure of the Earth as a result of tidal deformations (oceanic and solid tides) is changing in time and tends to find the form of a real surface, called a geoid. The gravitational potential of such an Earth can be found from (1) by replacing the displacement vector by an expression of the form (11), where the angles can be represented with a high degree of accuracy as known functions of time. We also note that increasing the order of the expansions in series by powers of the eccentricity in (10) will add new terms in (11) and, correspondingly, new frequencies. Also results can be improved by using all modes in series for displacement vector $\boldsymbol{u}$.

References

[1] IERS Annual Reports 2000/2002. Frankfurt am Main:KG 2001/2003
[2] Akulenko L D, Markov Yu G, Perepelkin V V, Rykhlova L V 2008 *Astr. J.* vol **85**,№7, pp 657–664
[3] Barkin M Yu, Perepelkin V V, Skorobogatykh I V 2012 *Kosmicheskie issledovaniya.* vol **50**, №3 pp 271–280
[4] Markov Yu G, Minyaev I S. 1994 *Kosmicheskie issledovaniya.* vol 32, №6 pp 89–98
[5] Skorobogatykh I V, Timoshin D S, Filippova A V 2012 *Kosmonavtika i raketostroenie.* Vol **4**(69) pp 121-127

Flight trajectory optimization without decomposition into separate stages

S A Kumakshev and A M Shmatkov

Institute for Problems in Mechanics of the Russian Academy of Sciences, Moscow, Russia
E:mail: kumak@ipmnet.ru

Abstract. Optimization problem is solved for the trajectory of a typical medium range aircraft flying between two airdromes. The optimization is considered with respect to the fuel consumption and is performed for the entire trajectory, without decomposition into separate stages. The optimal trajectory is calculated on the basis of the complete model of the trajectory motion of an aircraft in a vertical plane. To account for all constraints imposed by both the engineering features of the aircraft, as well as other requirements, including passenger safety and comfort, Bellman's dynamic programming method is used.

1. Introduction

Airliner flight parameters depend upon different technical and legal factors. There are many restrictions for wide-body jets and that is why the corresponding trajectories cannot be optimized entirely to a great extent. In contrast to them single aisle aircrafts have much more possibilities to select the optimal trajectory. But usually the investigation follows the same way – the entire flight is divided into parts and then each part is optimized separately. In this paper we will find the whole trajectory with minimal fuel consumption.

The common method to solve such a problem is to use Pontryagin's maximum principle [1]. But this principle leads to extremely complicated equations in the presence of numerous phase restrictions. For example, the adjoint variables may be discontinuous time functions about the points where the constraints become active and it is impossible to guess where such points can arise.

Bellman's dynamic programming method [2] requires a lot of calculations, but the idea of it is very simple and makes it possible to take into account many complicated constraints. In addition, the more restrictions are imposed, the less number of variants should be tested. On the other hand, most of recent papers on trajectory optimization for non-manoeuvrable aircrafts have been founded on substantially simplified models [3-12]. So, an investigation including a momentous number of innate factors of real-world would be of interest.

2. The problem description

We will claim that the flight trajectory is always located in the vertical plane. The centre of mass coordinates x, y are assigned in a fixed coordinate system with the vertical Y axis and the horizontal X axis. The well-known equations of motion of the airliner are:

$$\dot{V} = g(n_x - \sin\theta), \quad \dot{\theta} = \frac{g}{V}(n_y - \cos\theta), \quad \dot{y} = V\sin\theta, \quad \dot{x} = V\cos\theta, \quad \dot{m} = -Q_t(M, y, P) \qquad (1)$$

Content from this work may be used under the terms of the Creative Commons Attribution 3.0 licence. Any further distribution of this work must maintain attribution to the author(s) and the title of the work, journal citation and DOI.
Published under licence by IOP Publishing Ltd

Here, V is the aircraft velocity magnitude, g is the acceleration of free fall, θ is the pitch angle, m is the mass of the aircraft, n_x is the tangential load, n_y is the corresponding normal load factor, Q_t equals the fuel mass which the aircraft uses per second. This value depends on the M (Mach number), y (the aircraft altitude), and P (the thrust).

We will disregard the normal component of the thrust. The tangential load n_x and the normal load factor n_y can be written in the following form:

$$n_y = \frac{C_y q S}{mg}, \quad n_x = \frac{P}{mg} - \frac{qSC_x}{mg} \tag{2}$$

where S is the area of the wing and equals to 168.3 square meters, $\rho(y)$ is the atmosphere density and depends up on the altitude, q is the dynamic pressure, $C_x(C_y, M)$ is the drag coefficient. It depends on the value C_y (lift coefficient)and the value M (the Mach number).

We will eliminate the time variable from the relations (1.1) using the fourth equation:

$$\frac{dV}{dx} = \frac{g}{V}\left(\frac{n_x}{\cos\theta} - \theta\right), \quad \frac{d\theta}{dx} = \frac{g}{V^2}\left(\frac{n_y}{\cos\theta} - 1\right), \quad \frac{dy}{dx} = \theta, \quad \frac{dm}{dx} = -\frac{Q_t(M, y, P)}{V\cos\theta} \tag{3}$$

The lift coefficient C_y and the thrust P are the controlling parameters. We will substitute the expressions for n_y and n_x from(1.2) into the first two equations in (1.1) and obtain:

$$C_y = \frac{2m(V^2 d\theta + g dx)\cos\theta}{\rho(y)V^2 S dx}, \quad P = \frac{1}{2}\rho(y)V^2 S C_x + \frac{mVdV\cos\theta}{dx} + mg\sin\theta \tag{4}$$

From system (1) it is easy to see that the airliner motion can be described using the three functions $V(x)$, $\theta(x)$ and $y(x)$. These functions determine the controls C_y and P according to formula (4). The airliner mass can be found from last equation of the system (3) and the flight time T from the fourth equation (1).

The airliner initial and final positions are as follows: x(0)=0, y(0)=100 m, V(0)=100 m/s, θ(0)=0, m(0)=100000 kg; x(T)=L=500 km, y(T)=100 m, V(T)=100 m/s, θ(T)=0.

We need to obtain the controlling parameters to make the airliner travel from the initial position to the final position with minimum fuel consumption. The form of the optimization criterion is

$$\int_0^T Q_t dt \to \min$$

The fuel consumption function was approximated by a polynomial of first order in the value M and of second order in the values y and P. The lift coefficient C_y varies from -0.1 to 1.2, so

$$C_y^{\min} \le C_y \le C_y^{\max}, \quad P_{\min}(y, V) \le P \le P_{\max}(y, V) \tag{5}$$

We approximated the functions $P_{min}(y, V)$ and $P_{max}(y, V)$ by polynomials of second-order in the values M and y. The airliner velocity complies with the restrictions shown in the figure 1 with dashed line.

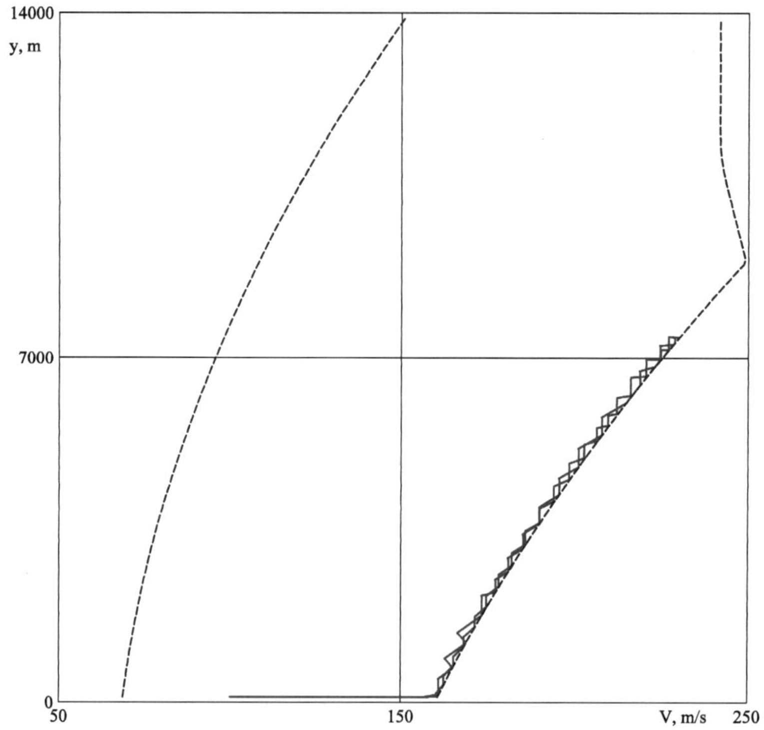

Figure 1. Aircraft movement with the restrictions in terms of velocity and altitude.

3. The algorithm description

We used the standard version of the dynamic programming method and that is why we will say only a few words about it.

For every allowed system state the dynamic programming method requires to find such a preceding state, the passage from which to the allowed state is optimal for the optimization criterion. We defined the state by the range, velocity, and altitude; so we need to save in the computer memory the values of these three functions.

For every range value we have an array of elements. Each element consists of a pair of values: the velocity and the angle. The altitude value corresponds to the number of the element. We will call a set of these three specific values a node. We will call the left layer the array corresponding to the current value of range, and the array corresponding to the next range value will be referred as the right layer. We take an arbitrary admissible node in the right layer by choosing the values of the altitude, velocity, and angle. We then take the left layer node from which it is possible to reach the fixed node using all restrictions. In this way we obtain all possible transitions with the corresponding mass consumption values and we choose the transition with the smallest mass of the airliner at the selected node of the right layer.

At the initial point all three values of the airliner are known due to the initial data. Next, we fill all layers one after another up to the terminal range. There we select the node corresponding to the terminal conditions and we begin the reverse pass-through and reconstruct the entire optimal trajectory.

4. The calculation results

We used the following parameters for calculations: 504 allowable variants of the value y with range of 126 m and a step of 0.25 m, 10,000 possible values for the range using a step equal to50 m,20 allowed angle B values in a diapason about 4.8°,252 allowed velocity values using a step equal to 0.025 m/s in the diapason of 6.3 m/s.

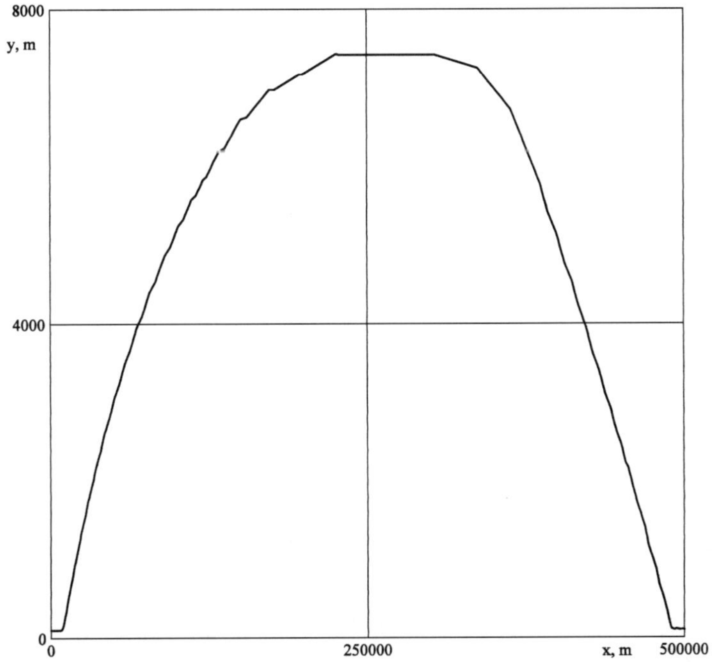

Figure 2. Optimal trajectory.

We see that the optimal trajectory depicted in figure 2 contains rather a short horizontal part in comparison with that one calculated earlier [13].

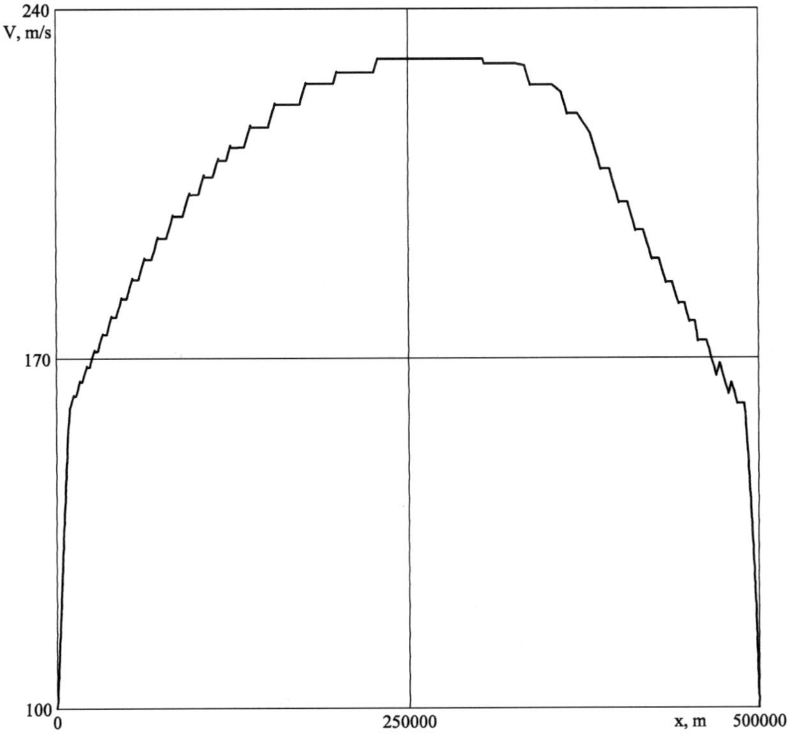

Figure 3. Optimal velocity.

The optimal velocity in figure 3 corresponds to the curve shown by solid line in figure 1.

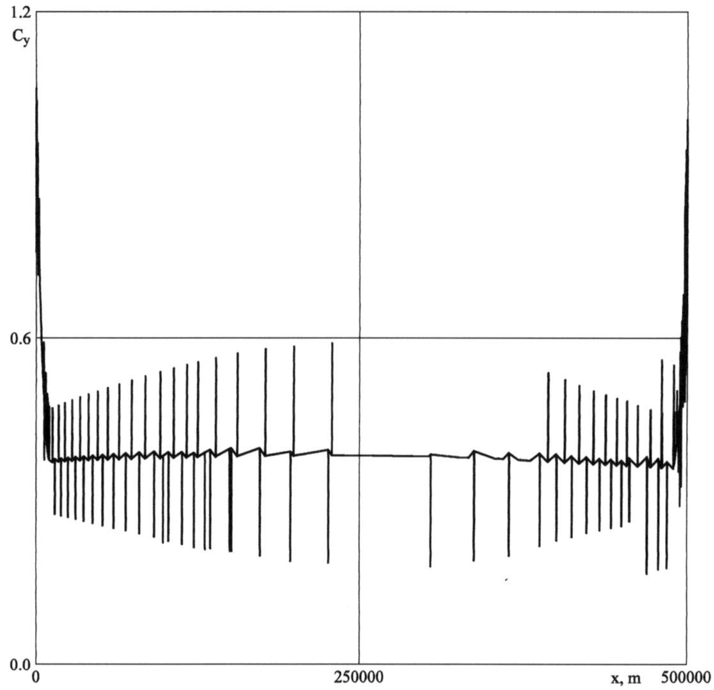

Figure 4. Optimal control C_y.

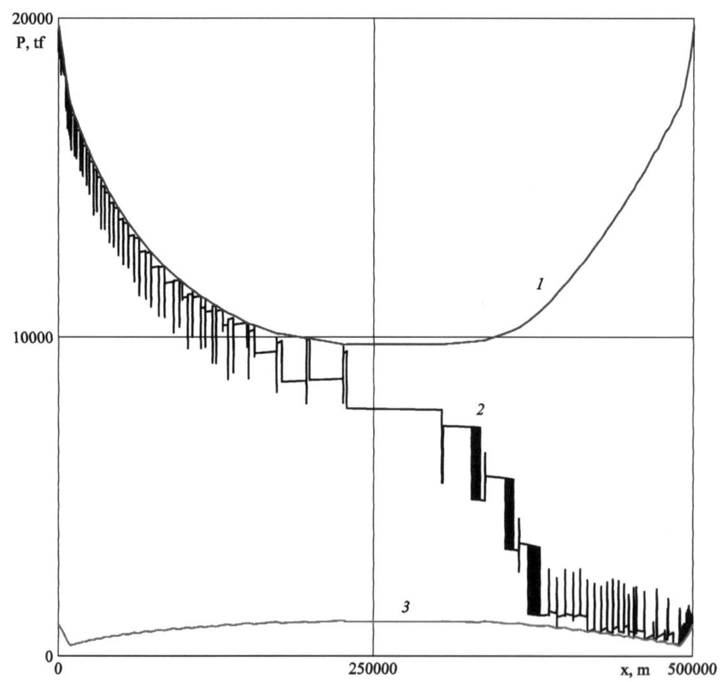

Figure 5.1 – maximal thrust, 2 – optimal thrust, 3 – minimal thrust.

Figures 4 and 5 depict the dependence of the optimal controls C_y and P on the range.

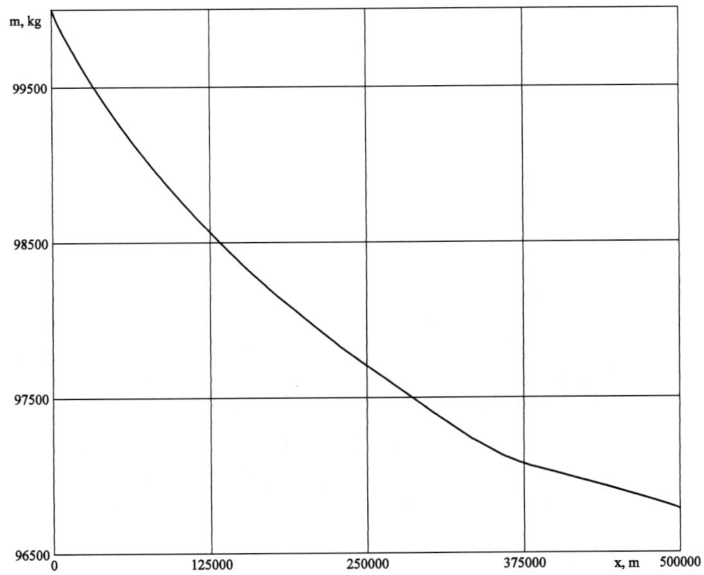

Figure 6. Aircraft mass.

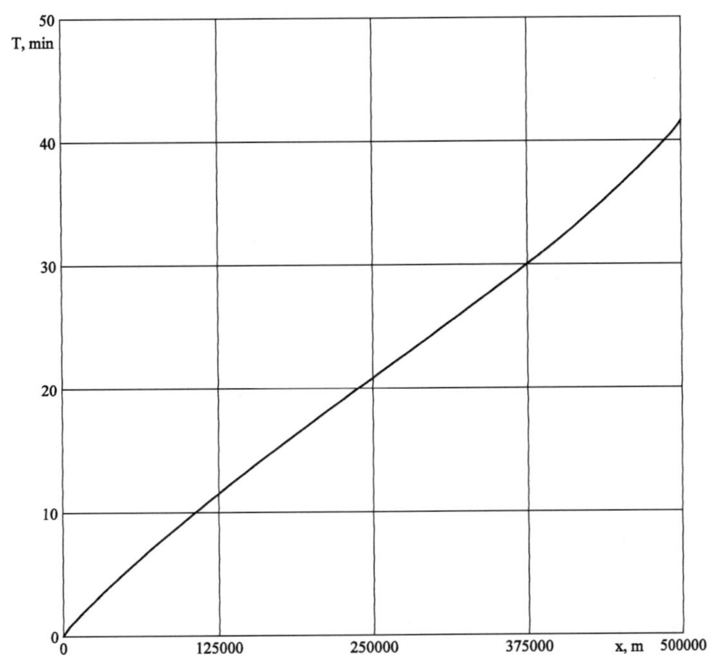

Figure 7. Optimal time in minutes.

Corresponding curves are characterized by significant outliers originating due to inaccuracies in the calculations based on the curves approximations by broken lines.
The dependences of the aircraft mass and the flight time in figures 6 and 7 are smooth because of the low sensitivity of these functions to short-time changes in C_y and P.

References

[1] Pontryagin L S, Boltyanskii V G, Gamkredlidze R V and Mishchenko E F 1963 *The Mathematical Theory of Optimal Processes* (New York: Wiley–Interscience)
[2] Bellman R 1957 *Dynamic Programming* (Princeton, NJ: Princeton Univ. Press)
[3] Bonami P, Olivares A, Soler M and Staffett E 2013 *J. Guid. Control Dyn.***36** pp 1267–77
[4] Li X, Nair P B, Zhang Z, Gao L and Gao C 2014 *J. Aircraft.* **51** pp 1592–603
[5] Dalmauand R and Prats X 2014 *Proc. 6th Int. Conf. Research in Air Transportation*
[6] Mendoz M A and Botez Mr 2014 *Aircraft Noise and Emissions Reduction Symp.*
[7] Mcconnachie D, Bonnefoy P and Belle A 2015 *Eleventh USA/Europe Air Traffic Management Research and Development Seminar*
[8] Franco A and Rivas D 2015 *J. Guid. Control Dyn.* **38** pp 452–67
[9] Soler M, Olivares A and Staffetti E 2015 *J. Aircraft* **52** pp 274–86
[10] Rosenow J, Forster S, Lindner M and Fricke H 2016 Int. Transport. **68** pp 40–43
[11] Rosenow J and Fricke H 2016 *Deutscher Luft-und Raumfahrtkongress*
[12] Gracia-Heras J, Soler M and Saez F J 2016 *J.Aerospace Inform. Syst.* **13** pp 243–65
[13] Grevtsov N M, Kumakshev S A and Shmatkov A M 2017*J. Appl. Math. Mech.* **81**

Acknowledgments
The research was carried out in part within the state assignment (state registration No. AAAA-A17-117021310387-0), by the Presidium of the Russian Academy of Sciences (Program No 30 "Theory and Technologies of Multi-level Decentralized Group Control under Confrontation and Cooperation") and by the Russian Foundation for Basic Research (Grants No. 17-01-00538 and No. 17-08-00742).

IOP Conf. Series: Materials Science and Engineering **468** (2018) 012034 doi:10.1088/1757-899X/468/1/012034

Modeling of compensatory sustainbility on the basis of algebraic formalization of systems

V I Serdyukov[1,3], N A Serdyukova[2] and S I Shishkina[1]

[1]Bauman Moscow State Technical University, Moscow, Russia
[2]Finance and Prices faculty of Plekhanov Russian University of Economics, Moscow, Russia
[3]Laboratory of pedagogical technologies on the basis of means of Informatization and communications of the Federal state scientific institution «Institute of Informatization of education» of the Russian Academy of education, Moscow, Russia
E-mail: shish–bmstu@mail.ru

Abstract. A solution of the problem of modeling the compensatory system's sustainability on the basis of the algebraic formalization of the general concept of the system is proposed. By a compensatory sustainability is meant the internal structure of the system's links, which ensures the interchangeability of its structural resources.

1. Introduction

A solution of the problem of modeling the compensatory system's sustainability on the basis of the algebraic formalization of the general concept of the system is proposed. By a compensatory sustainability is meant the internal structure of the system's links, which ensures the interchangeability of its structural resources.

Formulation of the problem. The problem solved in the article is to answer the following questions:

- on the possibility of mutual replacement of system elements, or factors that determine the system, or the functions of the system to achieve the system's goal,
- what will happen to the system S if in some group of factors $G_{lS}, 1 \le l \le m,$ i.e. in the external or internal system's environment, there will be changes, for example: any of the factors, or several factors, will cease to function; the feedback will stop working, the associativity of the system will be violated.

The relevance of the putted task is explained by its physical meaning and the proposed form of its solution, that is, by the advantages of using modeling on the basis of systems' algebraic formalization. The present paper continues [3–14], in which the methodology of mathematical modeling, based on the algebraic formalization of the general concept of the system, has been proposed.

The advantages of using modeling on the basis of algebraic formalization of systems runs as follows:

- the possibility of using the ready and well–developed language of the theory of algebraic systems developed by A I Maltsev [1],
- the possibility of using the entire powerful apparatus of group theory, that previously has been used in a wide variety of fields of science and beyond mathematics,
- the key difference between modeling on the basis of algebraic formalization of systems and the classical approach used in systems theory [2], is that the basis of algebraic formalization is the concept

Content from this work may be used under the terms of the Creative Commons Attribution 3.0 licence. Any further distribution of this work must maintain attribution to the author(s) and the title of the work, journal citation and DOI.
Published under licence by IOP Publishing Ltd

of a factor that determines a system, rather than an element of the system. This opens up wide possibilities for choosing a model of algebraic methods for studying the system.

The developed concept of algebraic formalization as a tool for studying phenomena, processes and practical results at the theoretical level was applied to the study of problems in systems theory, for which the following results were obtained:

- expressive properties of formalization are investigated and a hierarchy of formalizations is constructed [6], which allows using the feedback between formalization and qualitative analysis to refine previously known concepts in terms of deepening the study of its qualitative properties;
- the axiomatic of the system within the framework of the proposed concept is formalized [7],
- the external and internal attributes of the system are described in the language of algebraic formalization [7];
- the property of the system's integrity is formalized [7];
- an algorithm for studying closed associative systems using group theory has been developed [3–14];
- the number of possible synergetic effects of a closed system has been determined [9,10];
- the concept of an innovation system and its infrastructure has been formalized; within the framework of algebraic formalization, their properties have been studied [9];
- the concept of pseudo–innovation system has been introduced and some comparative properties of innovative and pseudo–innovation systems have been studied [9];
- it is shown that the proposed approach allows to formalize qualitative changes in a closed system and thus to predict their possibility.

For expert systems in education, an algorithm of constructing a database of errors in solving tasks was constructed, a theorem on the description of errors was proved, a classification of errors and their typification were obtained, errors were scaled along their length. An algorithm for compiling a knowledge base has been constructed, a theorem on the description of the correct solutions is proved, a classification of the correct actions has been obtained, correct actions have been scaled by their length. An algorithm for the transition to the recording a solution in the language of NPC (a narrow predicates calculus) from a solution written in the group theory language is proposed, that allows us to consider elementary theories of the group of correct solutions of the problem and of the group of errors [11]. The concept of a dynamical predicate is introduced [11]. Predicate P that is a function with a set of values of two elements $\{0,1\}$, that is, a condition defining a property of the set outlines the static properties of the system if it does not depend upon time or upon changing other factors external to the system. The predicate allows one to characterize the studied properties in a single integrated complex, both numerical indicators and synchronized with them links in dynamics, if they are dynamic predicates, and in static if they are static ones.

The next question when studying the properties of a system and the process of its functioning is the question of how to determine that the system ceases to satisfy some property P, or some complex of properties Π. To answer this question, we introduced the notion of a partial probability measure on the set of one–place predicates defined on the class of groups and closed with respect to taking subgroups and quotient groups [6]. The concept of a partial binomial distribution for realizing the complex of properties $\boldsymbol{P} = \{P_i | i \in I\}$ of the system S in n trials with the probability of successful realization of the complex of properties $\boldsymbol{P} = \{P_i | i \in I\}$ of the system S equal to p is considered. A methodology for modeling a system using a finite group of factors that determine the system has been developed. The connections between the factors of the system are characterized by the interaction between the factors of the system, that is, the binary operation of composition of factors, and the unary operation $*^{-1}$ characterizing the feedback of the system. Thus, an important role in describing the relationships of the system is played by the Cayey table of the group G_S, consisting of the factors describing the system S. It is from these positions the question of the stability of the system is considered in this paper.

2. Basic Concepts and Definitions. Justification of Empirical Results in a Theoretical Way. System's Concept of Formalization

Simulation is, in fact, a representation of a real phenomenon or process, or a functioning system in one or another abstract form. This representation is always carried out according to some rules, which are usually clear, so, the rules are more or less formalized. Therefore, one of the main questions that arises in this connection is the question of the formalization concept development. The history of the problem is presented in [6]. Let's note that the formalization cocept is closely related to logic, and begins with the work of Hilbert, who proposed the formalization of the process of logical inference. The key moment for us here is the connection between logic and algebra. I R Shafarevich, noted that the modern period of the development of science is characterized by the mathematization of science.

Algebra always took leading positions in mathematics. Since the middle of the 19th century, since the appearance of Boolean algebras, the process of algebraization of logic has been occurring (from the preface of the editor of the English edition of R Feys's book Modal Logic). At present moment, this process has become very powerful and branched. Examples of the use in logic of such concepts as a closure operator with certain properties, a topological space, a monoid, a category, and so on, appeared. The reason for this was: the presence of a continual set of logical systems, the direction of studying logical systems as certain constructions, the process of the development of the information society, the needs of insertion modeling, the development of intellectual systems.

By formalization one means the representation of any content area (reasoning, evidence, classification procedures, information retrieval of scientific theories) as a set of characteristics which are typical for it, the use of which allows us to understand it in a more meaningful way. Thus, formalization is a reflection of the results of thinking in precise concepts and statements. In advanced form these signs appear in a mode of a formal system or indicators of the calculus. By formalization we mean a record or representation of some phenomena, processes, practical results, denoted by P and expressed verbally, by means of the FP model, on the special language $\delta = \langle M, C_M \rangle$ with its own semantics and syntax allowing (one–to–one) to display their verbal model FP in the form of symbols or constructions of this language, called the language of formalization. We believe that it is possible to reflect the essence and obtain new profound results in the field of system theory in the most power with the help of the theory of algebraic systems (our concept of a closed system differs from the traditional one and lies in the mainstream of the theory of algebraic systems), A I Mal'tsev, [1], and in the field of closed associative systems – with the help of group theory, which is a well–developed component of the theory of algebraic systems. To confirm, let's recall that A G Kurosh believed that the theory of such algebraic systems as groups cannot be limited to use only in the domains closest to it. We believe that this statement extends not only to mathematical fields and group theory, but also to scientific research in general, and to algebraic systems. The concept of an algebraic system is well formalized, it reflects the fundamental property of things that is symmetry and is therefore connected with the concepts of expediency, proportionality, and optimality. Let us give one example. The application of group theory in physics is possible because the theory of groups is well formalized and allows one to study the invariance principle, according to which a physical law is necessarily invariant in any inertial system and, therefore, retains its shape under all transformations of four–dimensional space–time that transfer inertial systems to inertial systems. The group of all transformations under which Maxwell's equations for the electromagnetic field remain invariant, described by Lorenz, led to the formation of a special theory of relativity. In 1991 – 1993 years on the basis of the study of Yu Ershov's papers there was proposed a method of isolating and studying of pure embeddings (links with special properties that match the internal properties of the subsystem and the system) in groups. This method made it possible to transfer and generalize the well–known results of the theory of purities of abelian groups to the case of arbitrary non–Abelian groups [3]. In 2000 –ers a method for modeling the final states of the system and determining the number of final states using the technique of group theory was developed [4]. In 2008 – 2014 years pure or pure embeddings according to the meaning included in this concept that are connections don't distort the internal relations of subsystem in the system that satisfy the predicate P were used to differentiate links in systems by their nature, i.e. for the classification of links in general closed (with respect to the operation of composition of factors) associative systems. We propose the formalization of the system, based on the use of factors that affect the system, or the factors that determine the system.

P –pure, or P –pure embeddings, (analog of purities) by the meaning of this concept by the sense of this concept, mean attachments that do not distort the internal connections of the system that satisfy the predicate *P*. They will be used by us to distinguish between links in systems by their nature, i.e. to classify relationships in abstract systems. While studying the behavior of a system and its properties, the question of changing the system under the influence of factors that affect it is the key one, so we introduce the following definition of the algebra of factors that determine the system. Factors are conditions, causes, parameters, indicators that influence, operate the process, the phenomenon or result of a process. In the concept of algebraic formalization of the general concept of a system, the following two definitions are fundamental.

Definition 1.1. By the algebra of factors of a general system we shall understand an algebra $\bar{A}=\langle A\,|\,\{f_\alpha^{n_\alpha}\,|\,\alpha\in\Gamma\}\rangle$ with the main set of factors c A and with the set of operations $\{f_\alpha^{n_\alpha}\,|\,\alpha\in\Gamma\}$, that describe interaction of factors, where: n_α – arity of an operation, Γ – the set of symbols.

Actions of operations $\{f_\alpha^{n_\alpha}\,|\,\alpha\in\Gamma\}$ on the set A show the qualitative change in the system under the action of factors affecting it.

Definition 1.2.

A subalgebra $\bar{B}=\langle B\,|\,\{f_\alpha^{n_\alpha}\,|\,\alpha\in\Gamma\}\rangle$ of an algebra $\bar{A}=\langle A\,|\,\{f_\alpha^{n_\alpha}\,|\,\alpha\in\Gamma\}\rangle$ is called a P –pure subalgebra of an algebra $\bar{A}$, if every homomorphism $\bar{B}\xrightarrow{\alpha}\bar{C}$ of a subalgebra $\bar{B}$ of an algebra $\bar{A}$ into an algebra $\bar{C}$ of a signature $\{f_\alpha^{n_\alpha}\,|\,\alpha\in\Gamma\}$, satisfying the predicate *P*, which is closed under taking subalgebras and factor–algebras, can be continued up to a homomorphism of an algebra $\bar{A}=\langle A\,|\,\{f_\alpha^{n_\alpha}\,|\,\alpha\in\Gamma\}\rangle$ into an algebra $\bar{C}=\langle C\,|\,\{f_\alpha^{n_\alpha}\,|\,\alpha\in\Gamma\}\rangle$, in such a way that is the diagram (1) will be a commutative one:

$$0\rightarrow\bar{B}=\langle B\,|\,\{f_\alpha^{n_\alpha}\,|\,\alpha\in\Gamma\}\rangle\xrightarrow{\varphi}\bar{A}=\langle A\,|\,\{f_\alpha^{n_\alpha}\,|\,\alpha\in\Gamma\}\rangle$$

$$\alpha\searrow \qquad \swarrow\beta \qquad (1)$$

$$\bar{C}=\langle C\,|\,\{f_\alpha^{n_\alpha}\,|\,\alpha\in\Gamma\}\rangle$$

that is the equation $\beta\varphi=\alpha$ where φ is an embedding of $\bar{B}=\langle B\,|\,\{f_\alpha^{n_\alpha}\,|\,\alpha\in\Gamma\}\rangle$ into $\bar{A}=\langle A\,|\,\{f_\alpha^{n_\alpha}\,|\,\alpha\in\Gamma\}\rangle$, P – is a predicate on the class of algebras of the signature $\{f_\alpha^{n_\alpha}\,|\,\alpha\in\Gamma\}$, highlighting the class of subalgebras, closed under taking subalgebras and factor–algebras (the same main operations of the same type of algebraic systems will be denoted in the same way in each of the algebras of the same type, is true). Then φ is called a P– pure embedding.

3. Main Results

One of the most important questions in systems theory, for the solution of which the theory of (finite) groups can be used, is the question of the sustainability of the system. In the absence of this quality, the system can not exist. Sustainability combines various properties: resistance to external factors, stability, reliability, etc. As a property, stability shows itself in the responses of the system to disturbances of various kinds. By the stability of the system we shall understand the ability of the system to maintain the current state, under the actions of external and internal effects. If the system S is capable to keep some state $S=S_e$ at instants of time $t\in T$, where T is a continuous or discrete set, under the influence of external and internal effects, that is if $S(t)=S_e$, $t\in T$, then the state S_e will be called the equilibrium state of the system S.

Let S be a close associative system and G_S be a group of factors that define the system S. Let $G_S = \langle G_S, \circ, *^{-1}, e \rangle$ be a group of factors that determine the system S. Let's suppose that G_S is finite and consists of n elements, $|G_S| = n$. Let $G_{1S}, G_{2S}, \ldots, G_{mS}$ be all non isomorphic groups of n elements. The following question arises: what will happen with the system S if in some group of factors $G_{lS}, 1 \leq l \leq m,$ there will be some changes, for example:

– will cease to function any of the factors, or several factors,
– a feedback will stop working,
– an associativity property of the system will be violated,
– the neutral factor will cease to function.

The answers to these questions allow us to analyze the qualitative changes taking place in the system. A partial answer to this question can be obtained with the help of well–known theorems of the theory of finite groups, namely, the following assertions hold.

Theorem 2.1. If one factor from the group of factors G_S of the order $|G_S| = n$, that determine the system S, stops functioning (which is equivalent to removing this factor from the system of factors) the system S loses the closeness property.

The proof is obtained directly from the Lagrange's theorem, since the number of factors that continue to function in the system, is equal to $n-1$, and $n-1$ does not divide n.

Theorem 2.2. If m factors from the group of factors G_S of the order $|G_S| = n$, that determine the system S, stop functioning (which is equivalent to removing these m factors from the system of factors) the system S loses the closeness property.

Proof. In this case $n-m$ does not divide n, and by Lagrange's theorem there is no a subgroup of the order $n-m$ in G_S, so the system S loses the closeness property.

Theorem 2.3. If one of the factors a or a^{-1}, that determine the system S, stops functioning, then the system S loses the feedback property on this factor.

This can also be interpreted in a different way: the signature of the system changes; group signature $\langle \circ, *^{-1}, e \rangle$ exchange onto semigroup signature $\langle \circ, e \rangle$ with unity and instead of a group of factors G_S, that determined the system S, one obtains a semigroup Smg_S of factors that determine the system.

Theorem 2.4. A system with S the group of factors G_S that determine the system keeps the closeness property after stop functioning of the factors $\{a_i | i \in I\}$ if and only if $\langle G_S \setminus \{a_i | i \in I\}, \circ, *^{-1} \rangle$ is a group.

Let us consider the simplest examples illustrating these statements.

Example. Systems simulated by a group of 4 elements

There exist exactly two non–isomorphic groups which consist of four elements: a cyclic group $Z_4 = \langle F_4 \| a_1^4 = e, a_1^2 = a_2, a_1^3 = a_3 \rangle$ and Klein four–group $Z_2^2 = \langle F_4 \| a_1^2 = e, a_2^2 = e, a_3^2 = e, a_4^2 = e \rangle$, which are civen respectively by Cayley tables, Table 1, Table 2:

Table 1. A cyclic group of order 4.

*	e	a_1	a_2	a_3
e	e	a_1	a_2	a_3
a_1	a_1	a_2	a_3	e
a_2	a_2	a_3	e	a_1
a_3	a_3	e	a_1	a_2

Table 2. The Cartesian product of two second order cyclic groups (Klein four group).

*	e	a_1	a_2	a_3
e	e	a_1	a_2	a_3
a_1	a_1	e	a_3	a_2
a_2	a_2	a_3	e	a_1
a_3	a_3	a_2	a_1	e

Let's consider what will happen to the system S, if some of its factors cease to function. Let's introduce following designations:

p_0 – the probability of that the neutral factor e has ceased to function,

p_1 – the probability of that the factor a_1 has ceased to function,

p_2 – the probability of that the factor a_2 has ceased to function,

p_3 – the probability of that the factor a_3 has ceased to function.

Let's suppose that in the system, simulated by the cyclic group Z_4 factor a_1 ceased to function.

It can be illustrated as follows, Table 3 :

Table 3. The Cayley table.

*	e	a_1	a_2	a_3
e	e	a_1	a_2	a_3
a_1	a_1	a_2	a_3	e
a_2	a_2	a_3	e	a_1
a_3	a_3	e	a_1	a_2

Cayley table shows that in this case the system S with the probability p_1 loses closeness property, as for example, $a_2 \cdot a_3 = a_1$. Nevertheless, the remaining set of factors $\{e, a_2, a_3\}$ has a partial feedback on the factor a_2

Let's suppose now that in the system, simulated by the group Z_4, the factor a_2 ceased to function.

It can be illustrated as follows, Table 4:

Table 4. The Cayley table.

*	e	a_1	a_2	a_3
e	e	a_1	a_2	a_3
a_1	a_1	a_2	a_3	e
a_2	a_2	a_3	e	a_1
a_3	a_3	e	a_1	a_2

Cayley table shows that in this case the system S with the probability p_2 loses closeness property, as $a_1 \cdot a_1 = a_2$. The property of feedback on the remaining factors remains.

Let's suppose now that in the system, simulated by the group Z_4, the factor $a_3$ceased to function.

It can be illustrated as follows, Table 5:

Table 5. The Cayley table.

*	e	a_1	a_2	a_3
e	e	a_1	a_2	a_3
a_1	a_1	a_2	a_3	e
a_2	a_2	a_3	e	a_1
a_3	a_3	e	a_1	a_2

Cayley table shows that in this case the system S with the probability p_3 loses closeness property, as $a_1 \cdot a_2 = a_3$. The property of feedback on the remaining factors remains.

Let's suppose now that in the system, simulated by the group Z_4, the factors a_1 и a_3 ceased to function. It can be illustrated as follows, Table 6:

Table 6. The Cayley table.

*	e	a_1	a_2	a_3
e	e	a_1	a_2	a_3
a_1	a_1	a_2	a_3	e
a_2	a_2	a_3	e	a_1
a_3	a_3	e	a_1	a_2

The Cayley table shows that in this case the system S with the probability $(1-p_1)(1-p_2)$ keeps the closeness property and the feedback if factors are independent.

Another important issue arising in connection with the study of the property of stability of a system is the question of the possibility of mutual replacement of the elements of the system, or factors that determine the system, or the functions of the system to achieve the goal of the system. In order to outline possible solutions of this problem, we first consider the relationship between the factors that determine the system and the elements of the system, and then, on this basis, the question of the compensatory properties of the system.

4. Substitution of System's Functions. System's Compensational Possibilities

By replacement (compensation) of (disturbed) functions of the system, we shall mean the adaptation of the system to the changed conditions for the existence and replacement of the nonworking (failed, ineffectively working) elements of the system with respect to the elements of the system that are operating (possibly more efficiently than failed system's elements). Such elements of the system will be called substitutional or compensatory.

Let the system S consists of the following elements:

$S = \{s_\alpha | \alpha \in \Lambda\}$. Let's suppose that the main set of the group of factors determine the system S is $G_S = \{e, a_1, \ldots, a_n\}$, in this case, the elements of the system S, that determine each factor, i.e. there is a one–to–one correspondence are outlined:

$a_i \leftrightarrow S_i = \{s_{\alpha_i} | \alpha_i \in \Lambda_i, i = 1, \ldots, n\} \neq \varnothing$. Let's note to every subset $S_i = \{s_{\alpha_i} | \alpha_i \in \Lambda_i, i = 1, \ldots, n\}$ of the set S one to one corresponds the subset $\{\Lambda_i, i = 1, \ldots, n\}$ of the set Λ.

Let $F_S = \{f_i | i \in I\}$ be functions corresponding to the set of factors $\{a_i | i \in I\}$ that determine the system S. As it was noted above we can assume without loss of generality that $\{a_i | i \in I\}$ coincides with $G_S = \{e, a_1, \ldots, a_n\}$

The following Figure 1 helps to understand the mechanism of compensatory possibilities of the system.

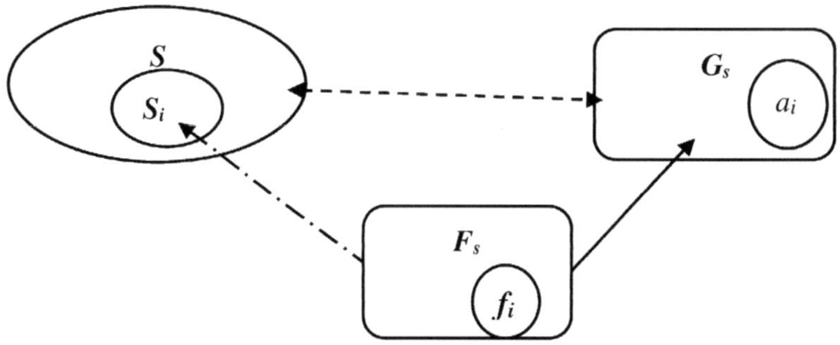

Figure 1. The mechanism of compensatory possibilities of the system.

The Figure 1 can be explained in the following way if to adhere to the classical scheme of studying systems: system – elements – structure. One can consider the factor a_i, $i=1,\ldots,n$, as a cloud of elements $S_i=\{s_{\alpha_i}|\alpha_i\in\Lambda_i, i=1,\ldots,n\}$, and after that to consider the structure of the factors a_i, $i=1,\ldots,n$, or, that is the same the structure of clouds – clusters $S_i=\{s_{\alpha_i}|\alpha_i\in\Lambda_i, i=1,\ldots,n\}$.

5. Compensational Functions of a Factor–Flexible System

Let's consider the set of functions $F_S=\{f_i|i\in I\}$ of the system, which one to one correspond to the set of factors $G_S=\{e, a_1,\ldots,a_n\}=\{a_i, i\in I\}$, that determine the system S, which consists of elements $\{s_\alpha|\alpha\in\Lambda\}$, that is $S=\{s_\alpha|\alpha\in\Lambda\}$.

The system S under the action of the function f_i onto the system S transfers into the state S_i, herewith the model of the system S transformed into the model G_S^i. During this transition, the following can occur:

1) the system S loses the associative property in the case if the action of the function f_i neutralizes the factor a_i, or some several factors together with the factor a_i,

2) the system S loses the associative property,

3) the system S loses the closeness property,

4) in the system S under the impact of the function f_i the influence of the factor a_i will increase and may be the influence of some several factors determine the system S.

We shall consider a special case when after each action of the function $f_i, i\in I$, the model G_S^i is a group. We shall also assume that all the effects of the function f_i onto the system S in the model G_S can be written in the language NPC (Narrow Predicate Calculus) of the signature $\Omega_S=\langle\circ, *^{-1}, e, \{a_i|i\in I\}\rangle$. The set of actions of the function f_i onto the system S in the model G_S will be denoted by $Th(f_i)$ and will be called the elementary theory of the function f_i, that is $Th(f_i)=\{Q_1\ldots Q_n\varphi(x_1,\ldots,x_n)|G_S^i=Q_1\ldots Q_n\varphi(x_1,\ldots,x_n)\}$.

A formula $Q_1\ldots Q_n\varphi(x_1,\ldots,x_n)$ is called a symmetric one if for any substitution $\in S_n$, where S_n is a symmetric group of substitutions of the degree n we have the equivalence $\varphi(x_1,\ldots,x_n)\equiv\varphi(x_{\tau(1)},\ldots,x_{\tau(n)})$. The set of all symmetric formulas of $Th(f_i)$ will be denoted by $SymmTh(f_i)$. The set of all formulas $Q_1\ldots Q_n\varphi(x_1,\ldots,x_n)$ from $Th(f_i)$, such that for transposition $\tau=(i,J)$ the equality имеет $\varphi(x_1,\ldots,x_n)\equiv\varphi(x_{\tau(1)},\ldots,x_{\tau(n)})$ is true will be denoted by $\tau Th(f_i)$, and will be called formulas symmetric by the transposition τ for the function $f_i, i\in I$.

Definition 4.1. Let $\tau = (i, J)$ be a transposition of a degree n. We shall say that the factor a_i replaces the factor a_j, if for the transposition $\tau = (i, J)$ the equality $\varphi(x_1,\ldots,x_n) \equiv \varphi(x_{\tau(1)},\ldots,x_{\tau(n)})$ is true where $Q_1 \ldots Q_n\, \varphi(x_1,\ldots,x_n) \in Th(f_i)$.

Definition 4.2. Let's say that the system S admits the replacements of functions if there exists such transposition $\tau = (i, J)$ of a degree n, that $\tau Th(f_i) \neq \varnothing$.

Hence from we obtain the following theorem.

Theorem 4.3. If for the group of factors G_S that determine the system, the inequality $\tau Th(f_i) \neq \varnothing$ takes place that is if the group G_S has a symmetric formula representation then the system S admits the replacement of functions.

Let's connect the substitution of functions with the sustainability of the system.

Definition 4.4. Let us say that the system S is sustainable by the factors a_i and a_j if $Th(f_i) = Th(f_j)$.

The following theorem is an immediate consequence of the definitions.

Theorem 4.5. In order for the system S to be stable by the factors a_i and a_j it is necessary that it admits the replacement of functions.

6. Conclusions

The advantages of modeling on the basis of algebraic formalization of systems are shown. Answers to the questions posed in the article allow one to analyze the qualitative changes taking place in the system and related to its stability.

References

[1] Maltcev A I 1970 *Algebraic systems* (Moscow: Nauka) 392

[2] Mesarovich M and Takahara Y 1975 *General system theory: mathematical foundations* (Academic Press New York San Francisco London Mathematics in science and engineering vol 113)

[3] Serdyukova N A 1991 On generalizations of purities *Algebra & Logic* 30 **4** pp 432 – 456

[4] Serdyukova N A 2002 *Optimization of tax system of Russia parts I and II* (Budget and TreasuryAcademy, Rostov State Economic University)

[5] Serdyukova N 2014 *The new scheme of a formalization of an expert system in teaching* (RigaICEE/ICIT Proceedings) 032

[6] Serdyukova N A, Serdyukov V I, Uskov A V, Slepov V A and Heinemann C 2017 Algebraic Formalization of Sustainability in Smart University Ranking System *Smart Innovation, Systems and Technologies* vol **75** pp 459 – 474

[7] Serdyukova N, Serdyukov V and Slepov V Formalization of knowledge systems on the basis of system approach *Smart Innovation, Systems and Technologies* (Springer) (ed Uskov V, Robert J H *et al*) vol **41** pp 371 – 381

[8] Uskov A, Serdyukova N A, Serdyukov V I, Heinemann C and Byerly A 2016 Multi objective optimization of VPN design by linear programming with risks models *International Journal of Knowledge–Based and Intelligent Engineering Systems* vol **20** No 3 pp 175–188

[9] Serdyukova N A, Serdyukov V I, Slepov V A, Uskov V L and Ilyin V V 2016 A formal algebraic to modeling smart university as an efficient and innovative *system Smart Innovation Systems and Technologies* vol **59** pp 83–96

[10] Uskov A V, Serdyukova N A, Serdyukov V I, Byerly A and Heinemann C 2016 Optima design of IPSEC–based mobile virtual private networks for secure transfer of multimedia data *Smart Innovation Systems and Technologies* vol **55** pp 51–62

[11] Serdyukova N A, Serdyukov V I and Glukhova L V 2015 Algebraic approach to the system representation of knowledge in the intellectual automated learning and control system *Vector of science of Togliatti State University Series: Pedagogy psychology* vol **3** p 328

[12] Serdyukova N A, Serdyukov V I and Slepov V A 2015 Formalization of knowledge systems on the basis of system approach *Smart Innovation Systems and Technologies* vol **41** pp 371–381
[13] Ilyin V V, Serdyukova N A and Serdyukov V I 2015 Risks of long–term forecasts in the economy of the Russian Federation *Financial analytics: problems and solutions* **44** (278) pp 2–16
[14] Ilyin V V and Serdyukova N A 2008 *System approach to the assessment of financial risks* (Finance **1**) pp 68–72

IOP Conf. Series: Materials Science and Engineering **468** (2018) 012035 doi:10.1088/1757-899X/468/1/012035

Features of numerical simulation of the unsteady vortex flow around aircraft considering airbrake

A S Epikhin and V T Kalugin

Bauman Moscow State Technical University, Moscow, Russia
E-mail: andrey.epikhin@bk.ru

Abstract. The paper presented the specifics of calculating the characteristics of vortex propagation and decay processes which cause aerodynamic tail fin-buffet loads in an aircraft at incompressible subsonic flow. Analysis of numerical diffusion and stability of differencing schemes implemented in OpenFOAM software package has been carried out. Based on the obtained results, differencing schemes are selected and their modification has been done. An algorithm has been presented for combining various approaches for modelling turbulent flows (RANS-LES) by means of zonal isolation and its implementation in OpenFOAM package is carried out. A series of calculations of three-dimensional flow around an aircraft at angles of attack of 0 to 30 degree considering airbrake deflections on 60 degree are conducted. Flow separation at the airbrake side edges and upper edge result in a highly turbulent wake. The corresponding region of vortex flow affects the fin and causes buffet loads.

1. Introduction

Motion of flying vehicles in atmosphere is generally accompanied by flow separation and formation of vortex flows, which result in redistribution of pressure on the aircraft's surface and change in its aerodynamic characteristics. In a subsonic flow around an aircraft's structural elements, zones of separation flow appear behind them where flow parameters have vortex pulsation nature. Studying vortex flows and their interaction with elements of aircraft has become one of the current objectives in the field of aviation and rocket science. Subsonic flow around high-lift system and various versions of external control devices as well as flying at extreme angles of attack result in formation of vortex structures, and unsteady change of their aerodynamic characteristics. On the other hand, the vortex flows generated by various elements of an aircraft such as deflectors or airbrakes may impact control and stabilizing surfaces located in their wake causing their shaking (buffet loads) due to periodic shock loads [1]. At present, most research is focused on studying the processes of tail fin-buffet loads occurrence, in which periodic vortex wake induced by an airbrake can lead to unsteady pulsating loads on the tail fin [1-3].

Nowadays, despite rapid progress in the field of experimental investigation methods, determination of flow structures of maneuverable aircrafts remains a complex task and requires the use of expensive equipment. In this case mathematical modelling of the flow around modern aircraft is an important step of examination and in many instances serves to confirm or disprove a hypothesis describing a physical phenomenon. At present, it is possible to point out a number of current challenges in the field of determining the aerodynamic characteristics at the flow around the aircraft, which are associated with numerical modelling of vortex flows and their interaction with flight control surfaces.

Content from this work may be used under the terms of the Creative Commons Attribution 3.0 licence. Any further distribution of this work must maintain attribution to the author(s) and the title of the work, journal citation and DOI.
Published under licence by IOP Publishing Ltd

IOP Conf. Series: Materials Science and Engineering **468** (2018) 012035 doi:10.1088/1757-899X/468/1/012035

2. Mathematical model

Mathematical model of an incompressible gas includes continuity equation and Navier–Stokes equations:

$$\nabla\cdot\left(\vec{V}\right)=0\ ;$$
$$\rho\frac{\partial\left(\vec{V}\right)}{\partial t}+\rho\vec{V}\nabla V=-\nabla p+\nabla\cdot\tau\ , \tag{1}$$

whereρ is density; $\vec{V}$ is velocity vector; p is pressure; τ is tensor of viscous stresses; t is time.

The numerical solution of the system equations (1) with finite volume method is the most promising approach to calculating aerodynamic characteristics of an aircraft. In the continuity equation, only the velocity vector is included, resulting in the absence of direct link to pressure, which in case of compressed flows is effected through density. In this study, a PISO (Pressure-Implicit with Splitting of Operators) algorithm has been chosen for linking pressure to velocity. These equations are solved under the initial and boundary conditions characteristic for the problems given. After appropriate transformation (discretization), a system of linear algebraic equations is obtained. Values of flow parameters on cell faces necessary for solving conservation equations in algebraic form are determined by means of various differencing schemes.

Since the flow is turbulent and is characterized by the presence of fine and large-scale vortex structures, an eddy-resolving approach has been used for their calculation, combining RANS (Reynolds-averaged Navier–Stokes equations) and LES (Large Eddy Simulation) methods. To close the filtered equations, i.e. to model the interaction of large vortices with fine-scale turbulence, Smagorinskysubgrid model has been used. In this model, the subgrid viscosity is calculated as follows:

$$\nu_{SGS}=\left(C_S\Delta\right)^2\sqrt{2S_{ij}S_{ij}}\ , \tag{2}$$

where Δ is filter size; C_S is an Smagorinsky constant; $S_{ij}$is the resolved rate of strain.

Finite volume method and various approaches to simulation of turbulent flows are implemented in many licensed commercial software packages such as ANSYS CFX, FlowVision, FloWorks, STAR-CCM etc. However, they are closed as well as their source codes, which limits their use in scientific research. An alternative is the development of open source packages such as OpenFOAM, which makes it possible to not only to solve a wide range of tasks of continuum mechanics [4] with the use of standard solvers and utility, but also to improve them.

3. Features of numerical modelling of the unsteady vortex flows

Vortex flows generated by the various elements of an aircraft's structure can act on other stabilizing and control surfaces located in their wake, which causes aerodynamic buffet loads. In order to correctly simulate this phenomenon and to determine pulsating loads, it is necessary to properly calculate the process of propagation and dissipation of vortices. Thus, a criterion of correctness of the vortices' behavior description is Kolmogorov's energetic spectrum. Also for numerical simulation of vortex-breakdown it is necessary to use PISO algorithm and eddy-resolving approaches, for example LES and various hybrid methods combining RANS and LES models. However, an important feature of their application is that accuracy of the vortex flows calculation depends on the differencing schemes of convective flux discretization [5]. The necessary and sufficient condition for the differencing scheme is its stability and the ability to correctly describe dissipation of vortex flows. Therefore, selection of the optimal numerical scheme is one of the main tasks in simulation of vortex flows. On the one hand, the scheme should be low-dissipative i.e. generating the minimum of numerical diffusion and on the other hand, it should provide stable calculation and the absence of oscillations.

3.1. Analysis of numerical diffusionand stability of differencing schemes

The analysis of related works has made it possible to isolate the differencing schemes for applied solutions implemented in OpenFOAM: 1st order upwind scheme (UD), central-differencing linear (CD) and SFCD schemes, blended and limited schemes like "blended", QUICK, LUST, filteredLinear, Gamma and limitedLinear schemes. To investigate the quality of the selected schemes using the OpenFOAM package, two cases have been solved:

- Case of convective transfer of φ scalar with sharp gradients [6] which makes it possible to analyze schemes for stability.
- Case of decaying homogeneous isotropic turbulence [7] which allows for the evaluation of numericaldiffusion.

The problem of convective scalar transfer was solved in a two-dimensional installation with a square calculation area of 1 m^2 in which the scalar is transferred. Boundary and initial conditions are similar to work [6] and are shown in Figure 1. The computational grid consists of 30x30 cells uniformly distributed over the whole area. Stepped profile $\varphi(y)$ is selected for vertical input boundary:

$$\varphi(y)=\begin{cases} 0\ for\ 0\le y<\dfrac{1}{6}, \\ 1\ for\ \dfrac{1}{6}\le y\le 1. \end{cases}$$

Velocity equals to 5 m/s and is directed at a 45 degree angle to grid lines.

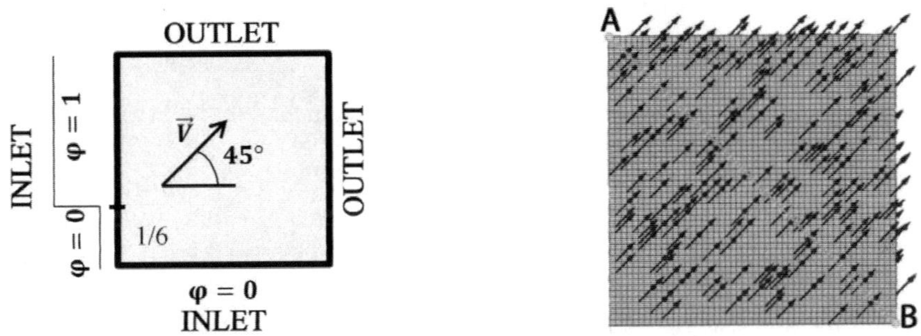

(a) (b)

Figure 1. Case of a scalar's convective transfer: (a) step profile test set-up; (b) computational grid

Figure 2 shows a distinct view of φ field graphs along a diagonal line connecting lower right and upper left corners of the area's geometry (Figure 1 (b), line AB).

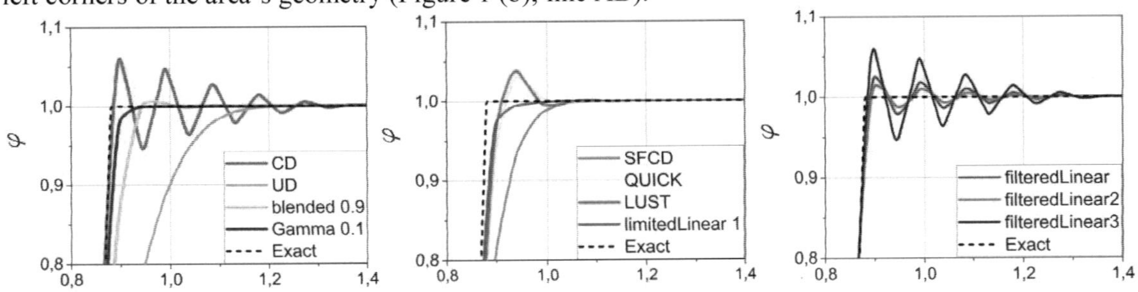

Figure 2. Convection of a step profile

It can be seen that the central-differencing scheme results in numerical oscillations and cannot be widely used in computational simulation of vortex-breakdown. The filteredLinear schemes also cause oscillations at any limiter value. The use of limited schemes, such as Gamma, or mixed schemes, such as blended with ratio of 0.9, eliminates numerical oscillations from the solution.

The differencing schemes should also properly describe the process of vortex dissipation when carrying out the calculations. For this purpose, the problem of decaying isotropic homogeneous turbulence has been solved. This problem helps to find out with high accuracy how dissipative is a certain convective flux discretization scheme [5]. Energy understating in different intervals of the energy spectrum indicates the numerical diffusion of the differencing scheme. This makes it possible to evaluate the suitability of each scheme for tests with the use of eddy-resolving approaches.
Numerical simulation of this task was carried out in a cubic design area with the size of 2π x 2π x 2π and a total number of cells 64^3 (Figure 3 (a)). The "cyclic" periodic boundary condition was used for all three spatial coordinates. Using boxTurb utilities included in the OpenFOAM package, an initial energy spectrum was set which generated a random field of speeds, used as initial condition (Figure 3 (b)). Then calculation was performed, after which the results were stored and transferred back to the energy spectrum which was compared to the experimental data [7]. For calculation, a method of large eddy simulation (LES) and Smagorinskysubgrid model with different constants of C_S were used.

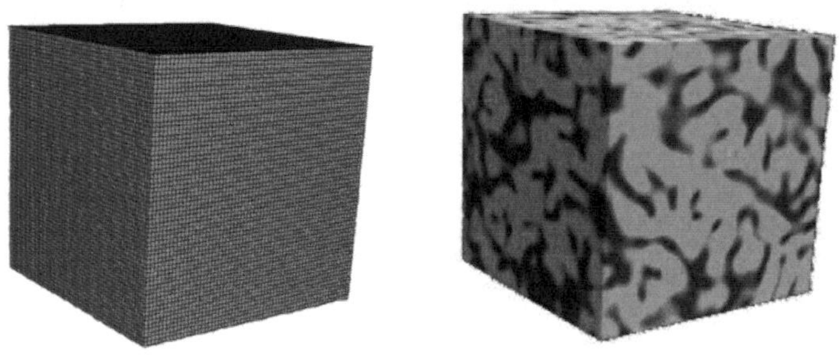

(a) (b)

Figure 3. Case of decaying homogeneous isotropic turbulence: (a) grid; (b) initial speed field
Figure 4 presents the results of calculation with the aid of Smagorinskysubgrid model with a matched C_S constant. When the energy of the turbulent pulsations is reduced in the high-frequency region of the spectrum, the value of the constant C_S decreases and vice versa. In this way, the value of C_S was chosen so that the resulting energy spectrum matched the experimental one. It should be noted that in Smagorinsky model implemented in the OpenFOAM package the default values of $C_S = 0.167$. It can be seen that zeroing the constant C_S does not lead to correct mapping for following schemes: Gamma at $\gamma = 0.1$, SFCD, QUICK, limitedLinear. The least dissipative are central-differencing scheme (CD) at $C_S = 0.2$, schemes of the filteredLinear at $C_S = 0.167$ and blended with coefficient $\beta = 0.9$ at $C_S = 0.1$. These schemes can correctly describe the evolution of vortex structures by using eddy-resolving approaches.

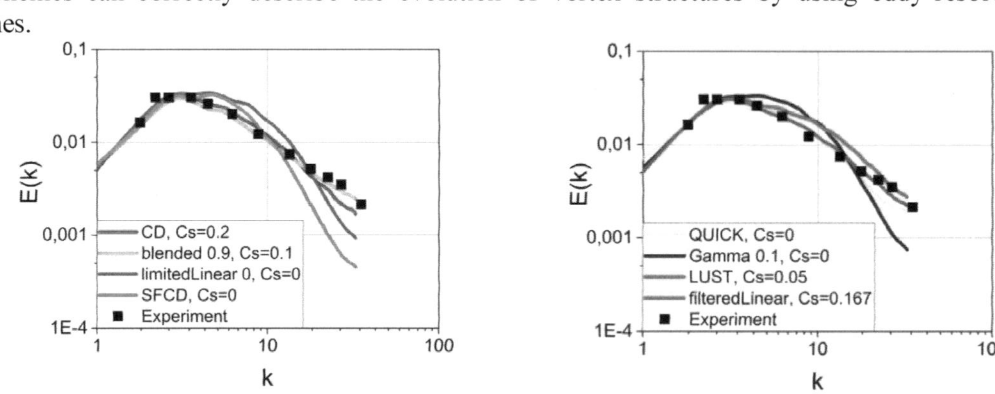

Figure 4. Energy spectra obtained with the aid of Smagorinsky model for various differencing schemes

Thus, the obtained results of the two cases solved allowed to carry out quality analysis of the differencing schemes for simulating the vortex flows which are implemented in the OpenFOAM software package. So, linear scheme correctly describes the evolution of vortex structures at $C_S = 0.2$, although it is unstable. Schemes of filteredLinear correctly map the energy spectrum with a small value understating in the high-frequency region, producing less oscillations than central-differencing scheme. The upwind scheme is stable but it is too dissipative. Use of "blended" scheme with ratio 0.9 makes it possible to increase its stability compared to CD, and at $C_S = 0.1$ it correctly describes the energy spectrum. However, understating viscosity by a factor of 2 leads to poor results in case of near-wall region. The QUICK and LUST schemes do not eliminate oscillation and are quite dissipative, since the correct description of the energy spectrum occurs at zeroing of the subnet viscosity, that is, at $C_S = 0$. Gamma, SFCD, and limitedLinear schemes are stable, but at any values of C_S they show understating of energy in the high-frequency region. It should also be noted that schemes such as upwind, SFCD, blended do not preserve the sharpness of the transfer edge like other schemes. Thus, the analysis of differencing schemes implemented in the OpenFOAM package shows that the schemes based on the central-differencing scheme are unstable and the upwind-differencing schemes are dissipative, therefore it is necessary to improve them in order to eliminate oscillations and maintain an acceptable level of numerical diffusion.

3.2. Modification of differencing schemes

Based on the results obtained above, differencing schemes filteredLinear and Gamma were selected and their improvement were carried out. Modification encompassed the adding of extra free parameters and changing the limiting coefficients in differencing schemes as well as determining their optimal values by solving the cases described above. This made it possible to change the behavior of the initial differencing schemes in the interval where the scheme's data was dissipative or unstable.

The basic idea of most numeric schemes implemented in the OpenFOAM package is that the weight coefficient is determined by each scheme independently and is essentially a function at software level (limiter) which returns its value from 0 to 1. At 0, the scheme is transformed into upwind, at 1 – in linear.

The Gamma scheme is implemented in the OpenFOAM package by the following function which determines weight coefficient:

$$\text{limiter} = \min\left(\max\left(\frac{\varphi_k}{\gamma}, 0\right), 1\right), \tag{3}$$

where $\varphi_k = 1 - (\varphi_N - \varphi_P) / (2\nabla\varphi_P \cdot \vec{d})$ – face value; φ_N and φ_P – the solution of variables at points N and P (cell centers); $\vec{d}$ – vector connecting adjacent cell centers N and P; min and max are functions, which return the minimum and maximum values respectively. According to the equation (3) parameters on face are determined as follows:

$$\begin{cases} \varphi_k \le 0 - \text{limiter} = 0, \text{upwind is used;} \\ \varphi_k \ge 1 - \text{limiter} = 1, \text{linear is used;} \\ \gamma \le \varphi_k < 1 - \text{limiter} = 1, \text{linear is used;} \\ 0 < \varphi_k < \gamma - \text{limiter} = \dfrac{\varphi_k}{\gamma}, \text{blended is used.} \end{cases}$$

It can be seen that the implementation of this scheme is slightly different from the description in [7]. Below there is a modification of this scheme (named GammaM) carried out by adding additional free parameter δ to it, so that it does not become a UD scheme when going beyond the boundary of stability, which makes it possible to reduce its numerical diffusion:

$$\begin{cases} \varphi_k \le 0 - \text{limiter} = \delta, \text{blended is used;} \\ \varphi_k \ge 1 - \text{limiter} = 1, \text{linear is used;} \\ \gamma \le \varphi_k < 1 - \text{limiter} = 1, \text{linear is used;} \\ 0 < \varphi_k < \gamma - \text{limiter} = \dfrac{\varphi_k}{\gamma}, \text{blended is used.} \end{cases}$$

Presented below is filteredLinear scheme implemented in OpenFOAM package:

$$\begin{cases} f = 2 - 0.5\dfrac{\min\left(mag(\varphi_N - \varphi_P - \nabla\varphi_P d), mag(\varphi_N - \varphi_P - \nabla\varphi_N d)\right)}{\max\left(mag(\varphi_N - \varphi_P - \nabla\varphi_P d), mag(\varphi_N - \varphi_P - \nabla\varphi_N d)\right)} \\ \text{limiter} = \max\left(\min\left(f, 1\right), 0.8\right) \end{cases}$$

A modification of this scheme (named filteredLinearM) consists of including additional free parameters α and β and determining their optimal parameters for the purpose of decreasing oscillation:

$$\begin{cases} f = \alpha - 0.5\dfrac{\min\left(mag(\varphi_N - \varphi_P - \nabla\varphi_P d), mag(\varphi_N - \varphi_P - \nabla\varphi_N d)\right)}{\max\left(mag(\varphi_N - \varphi_P - \nabla\varphi_P d), mag(\varphi_N - \varphi_P - \nabla\varphi_N d)\right)} \\ \text{limiter} = \max\left(\min\left(f, 1\right), \beta\right) \end{cases}$$

The following is a study of the modified differencing schemes (GammaM and filteredLinearM), analysis of their free parameters and coefficients for stability and numerical diffusion, and their optimal values obtained by solving the two cases described above with the use of Smagorinsky model with constant $C_S = 0.15$. Figure 5 shows the comparison of modified differencing schemes (GammaM and filteredLinearM) with their original implementation in OpenFOAM software package.

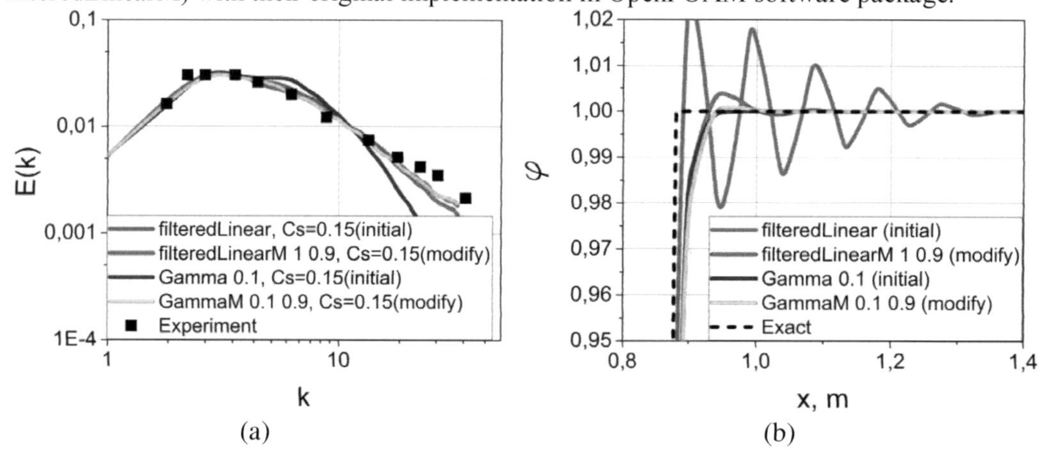

Figure 5. Comparison of modified differencing schemes with their initial realization in OpenFOAM package: (a) energy spectrum; (b) convection of a step profile

Thus, differencing schemes based on the open source OpenFOAM package are proposed and implemented, optimal values of their free parameters and coefficients are determined (γ=0.1,δ=[0.8,0.9] for GammaM and α=1,β=[0.85,0.9] for filteredLinearM) which makes it possible to increase calculation accuracy of propagation and decaying processes for vortex flows. Some validation of the proposed schemes presented in [8].

3.3. Algorithm for combining various approaches for modeling turbulent flows

In case of calculating the flow around complex three-dimensional objects and real aircraft configurations RANS approach is not able to provide plausible accuracy, and LES requires large computational resources while the hybrid DDES method exhibits sufficient accuracy. However, this method imposes certain conditions on the quality of computational grid [9], the construction of which presents considerable difficulties and requires a large amount of time. A possible alternative is the use of zonal RANS-LES approach, the idea of which consists in using a resource-demanding LES approach in areas where accuracy of numerical simulation is to be increased, and RANS in the remaining zones of the computational domain.

Initial file analysis of OpenFOAM package shows that there is no prepared solution for using the zonal approach to determine aerodynamic characteristics. As a result, the algorithm based on combining RANS and LES approaches is proposed and implemented, which makes it possible to use combination of LES and RANS models in different regions of the computational grid. The key idea of this approach is that the majority of turbulence models in OpenFOAM return fields of turbulent viscosity τ and turbulence kinetic energy k as exit values. On the other hand, turbulence in the equation of momentum balance in the OpenFOAM is taken in through approximation of effective stress tensor, the numerical implementation of which is determined independently for each model.

The solution of the problem of combining RANS and LES approaches on software level implies the implementation of an algorithm that makes it possible to obtain common matrix of coefficients from two different models of turbulence, which will then be sent to the equation of momentum conservation in model. In this case, the problem of combining RANS and LES models includes three procedures:

- determination of sub-areas of RANS and LES approaches' application based on flow patterns and study objectives;
- RANS model is used to calculate turbulent viscosity ν_{rans} and LES model is used for calculation of subgrid viscosity ν_{sgs};
- calculation of blended viscosity value ν_t as a function of ν_{rans} and ν_{sgs} using equation proposed J.Fröhlich and von D.Terzi [10].

In order to determine the sub-areas of use of RANS and LES approaches in the OpenFOAM package, a special ZoningArea utility is implemented, which analyzes the calculation grid created and stores the scalar data field in which each cell of the grid corresponds to one of the three values: 1 is the sub-area of RANS, 0.5 - mixed area, 0 - LES sub-area. This field is the initial condition for carrying out numerical calculation. It is worth noting that for RANS approach, in general, stable upwind-differencing schemes are used which are too dissipative for LES. In turn, low dissipative schemes are used for LES, which may be unstable in RANS domain. This fact has also been taken into consideration when implementing this algorithm, which makes it possible to use different numerical schemes for each of the approaches.

The implemented zonal approach combining RANS and LES models allows to carry out numerical modelling of unsteady vortex flows around of aircraft using the open-source utilities of OpenFOAM which increases the accuracy of aerodynamics characteristics determination.

4. Numerical simulation flow around an aircraft considering airbrake

A series of calculations of three-dimensional flow around a maneuverable aircraft using the modified differencing schemes and algorithm described above was carried out. The effect of the airbrake release on the dynamic loads on the tail fin was investigated. The solution was found with the following initial data: velocity of the incoming flow V_∞ = 50 m/s, kinematic viscosity $\nu = 1.5{\cdot}10^{-5}\text{m}^2/\text{s}$, Reynolds number $Re = 6{\cdot}10^5$. Calculation time step was chosen by condition CFL ≤ 0.2. Field of RANS and LES models' use was set by the initial conditions. In this case, it was necessary to accurately model the breakdown of vortices from the canopy and the airbrake of the aircraft at different angles of attack. Therefore, flow in the rest of the region could be obtained by means of RANS model. As RANS model of turbulence, k-ω SST was used, whereas for LES model it was Smagorinsky model with constant $C_S = 0.15$. Figure 6 shows the aircraft design and a grid fragment with cell number equal to 15 million. All calculations were carried out using the equipment of the shared research facilities of HPC computing resources at Lomonosov Moscow State University.

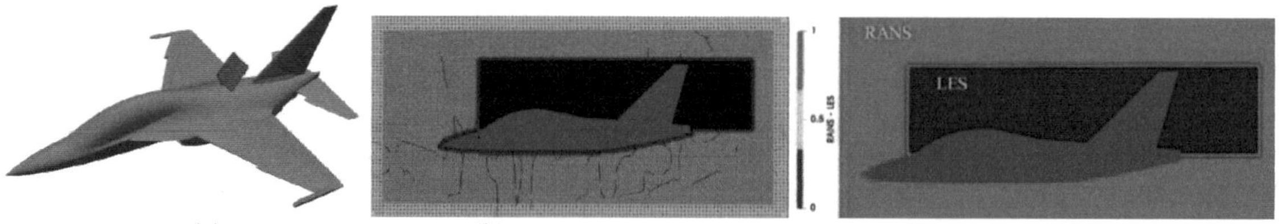

(a) (b) (c)

Figure 6. Case setup of the flow around an aircraft with a airbrake: (a) aircraft design; (b) section of computational grid; (c) field of application for RANS and LES approaches

In Figure 7 and Figure 8 the aerodynamics characteristics of the aircraft with and without the airbrake are presented and compared to experimental data [11].

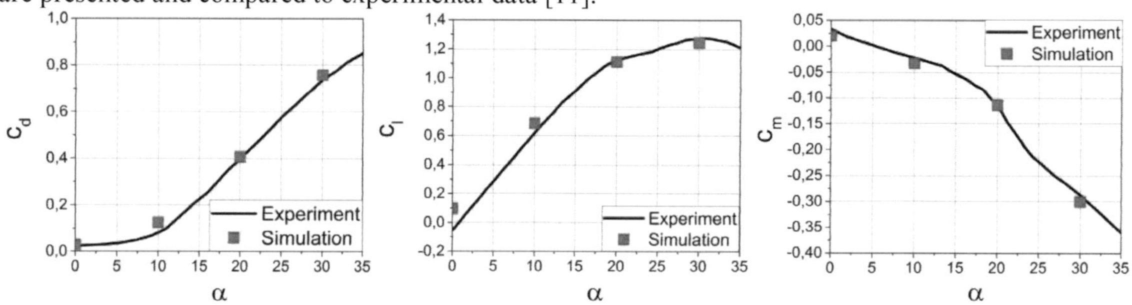

Figure 7. Dependence of the aircraft's aerodynamics characteristics on the angle of attack

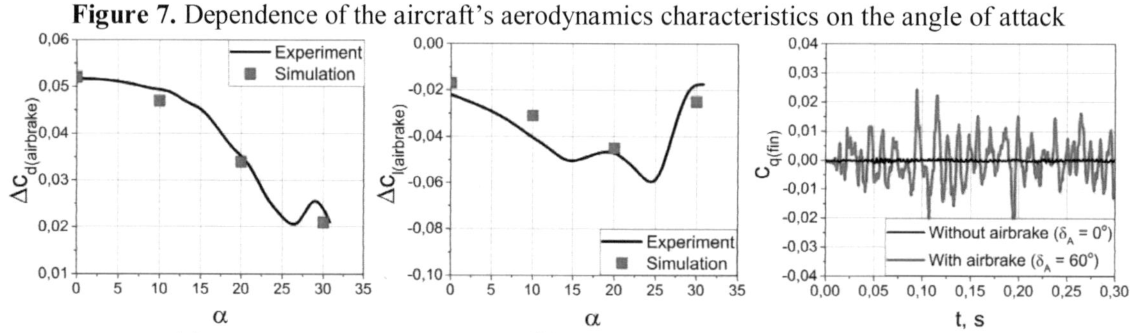

(a) (b) (c)

Figure 8. The influence of the airbrake release on the change of aerodynamic coefficients: (a) drag; (b) lift; (c) lateral force on the tail fin (α= 0°)

It can be seen that the results of numerical simulation using the proposed technique are consistent with experimental data, and the occurrence of aerodynamic buffet loads is confirmed at the airbrake deflection angle of δ_A = 60°. Figure 9 shows flow structures around the aircraft with and without the airbrake at angles of attack α = 0° and 20°.

(a) (b)

Figure 9. Instantaneous vortex structures (iso-surface of Q-criterion) of flow around an aircraft with and without airbrake at angles of attack: (a) α = 0°; (b) α= 20°

5. Conclusion

A technique is proposed for calculating the aerodynamic characteristics of a maneuverable aircraft and its control elements with the use of an improved open source OpenFOAM package, which makes it possible to model the vortex flow around its stabilizing and control surfaces by a subsonic incompressible flow with sufficient accuracy. This makes it possible to carry out the detailed analysis of the processes of aerodynamic buffet loads occurrence at early stages of aircraft design. Implementation of the algorithm for combining RANS and LES approaches and differencing schemes makes it possible to increase reliability of an aircraft's aerodynamic characteristics with the account of the vortex flows. The technique was confirmed by comparison with experimental data.

Modification and improvement of open source OpenFOAM package has been carried out, which makes it possible to use it for research and evaluation of parameters of unsteady vortex flows, and also calculation of force loads on the elements of the aircraft when solving a wide range of scientific tasks in the aircraft industry.

Reference

[1] Garifullin M F 2010 *Bafting* (Moscow: Fizmatlit) p 216.

[2] Breitsamter C and Schmid A 2008 Airbrake-induced fin-buffet loads on fighter aircraft *Journal of Aircraft*. Vol. **45**, № 5. pp 1619-1630.

[3] Chris L, Dansen L, Michael P, Pendleton E 1996 Full-scale wind-tunnel pressure measurements of an F/A-18 tail during buffet *Journal of Aircraft*. Vol. **33**, No. 6, pp 1148-1156.

[4] Epikhin A, Kalugin V, Michkin A, Golubev A 2014 Applicability of the open source package OpenFOAM for numerical modeling separated flow around an aircraft at subsonic and supersonic speeds *Civil Aviation High Technologies*. №**199** pp 23-30.

[5] Kozelkov A, Kurulin V, Tyatyushkina E, Puchkova O, Lashkin S 2013 Convective flux discretization schemes for detached eddy simulations of viscous incompressible turbulent flow *Fundamental research*. № **10** pp 1051-1058.

[6] Jasak H, Weller H G, Gosman A D 1999 High resolution NVD differencing scheme for arbitrarily unstructured meshes *International journal for numerical methods in fluids*. Vol. **31** pp 431-449.

[7] Comte-Bellot G and Corrsin S 1971 Simple Eulerian time correlation of full- and narrowband velocity signals in grid-generated «isotropic» turbulence *Journal of Fluid Mechanics*. Vol.**48** pp 273-337.

[8] Epikhin A, Kalugin V 2017 Computational modelling and methods to reduce fin buffet loads on aircraft *Matem. Mod.* Vol. **29**:10 pp 35–44.

[9] Spalart P R 2001 Young-Person's guide to detached-eddy simulation grids *Tech. Rep. NASA/CR-2001-211032*. NASA (Langley Research Center).

[10] Fröhlich J. and von Terzi D 2008 Hybrid LES/RANS methods for the simulation of turbulent flows *Progress in Aerospace Sciences*. Vol. **44**(5), pp 349–377.

[11] Podobedov V A 2015 *The Yak-130UBS aircraft. Aerodynamics and flight characteristics* (Moscow: Mechanical Engineering) p 348.

A method for attitude control of telescopes making use of an inverted pendulum

H Hanada[1,2], S Tsuruta[1], K Asari[1], H Araki[1,2], K Funazaki[3], A Satoh[3], H Taniguchi[3] and M Kikuchi[3]

1. National Astronomical Observatory, Mizusawa, Oshu, Japan
2. Sokendai (The Graduate University for Advanced Studies), Mitaka, Tokyo, Japan
3. Iwate University, Morioka, Iwate, Japan.

Abstract. We propose an attitude control method of a telescope using an inverted pendulum. A cylindrical telescope tube with a sharp bottom end is put on a motorized XY linear stage. The upper edge of the tube is supported by 4 pressure sensors. When the tube begins to deviate from the vertical, torque arises to try to throw down the tube. Then, the pressure sensors detect the stress and the XY linear stage displaces the bottom of the tube so as to make the stress zero. Sensitivity to the verticality or stress intensity acting on pressure gauges does not depend on the size of the tube but on the mass. This character is advantageous to miniaturization.
Keywords : Inverted pendulum, telescope tube, attitude control, horizontal/vertical reference.

1. Introduction

We have been developing a telescope for in-situ observations of lunar rotation under the background of many experiences in research and observations of the Earth rotations. The main target of the observation of lunar rotation is to make it clear the existence or non-existence of lunar liquid core [1]. Although we have a prospect to attain the expected accuracy from laboratory experiments and preliminary ground observations [2], there remains an issue which may become a problem when it will be onboard a space craft going to the Moon. The telescope uses a mercury pool as a reference of the level surface, and the mercury surface is not very stable for a long term and it has a certain volume and certain weight. If we try to apply such kind of observations to the Mars and other planets, much efforts for lighter weight and miniaturization are essential.

It is necessary to have certain connection between the attitude of a telescope and the reference frame of a target planet if we want to observe motions of the rotation axis of the planet or motions around the axis. There are several devices or materials for this purpose such as a liquid surface, bubble tilt meters and plumb bobs. They each have own advantages and disadvantages. For example, liquid surface is easy to evaporate and vibrate, and mercury surface is especially easy to react with other metals such as copper and aluminum although it almost perfectly follows an equal potential surface and its reflectivity is very high. The plumb bob or a pendulum is affected by friction around a pivot, and is easy to swing although it is simple. Tilt meters such as bubble type which are independent to the telescope has a problem of difficulty in connecting the attitude of the tilt meter and that of the telescope.

Considering the situation mentioned above, we propose a new method to control the attitude of the tube by making it be an inverted pendulum.

Content from this work may be used under the terms of the Creative Commons Attribution 3.0 licence. Any further distribution of this work must maintain attribution to the author(s) and the title of the work, journal citation and DOI.
Published under licence by IOP Publishing Ltd

2. Method

An inverted pendulum is an unstable system whose center of gravity is above the pivot point, and many researches on control method of how to stably keep the pendulum vertical have been made as a typical and classical problem in control engineering. In addition, recently it has become a new trend as a necessary technique for a bipedal walking robot and two-wheeled vehicle [3]. Here, as the first attempt, we propose to utilize the inverted pendulum as a telescope tube which is kept to be vertical. This is a kind of static and precise control although the other applications such as the walking robot are rather dynamic control.

It also has a characteristics of long period pendulum if we make the restoring force be as small as possible. In old days, the Wiechert seismometer was famous as a long period seismometer using an inverted pendulum. The inverted pendulum is also used in a vibration isolation system of gravitational wave detector making the most of its long periodicity [4]. This long periodic character is effective for the static control of the tube of inverted pendulum.

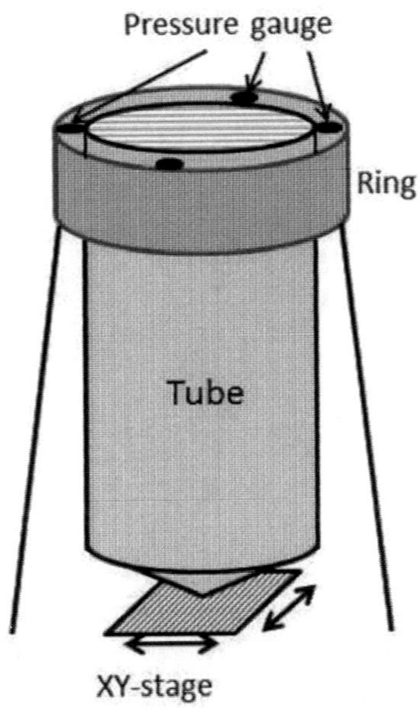

Figure1. Principle of the attitude control of a tube making use of an inverted pendulum.

A tube supported by a single point at the bottom looks unstable, but it is possible to control the attitude with high sensitivity because it tends to throw down even if it just slightly deviates from the vertical. This method gives little constraint upon optical system of the telescope since there is no need to incorporate the reference of the horizontal plane into the optical system and therefore there is no need to place a structure such as a supporting mechanism on the way of the incident rays unlike the case of plumb bob. Therefore, there is some possibilities of realizing a small sized telescope using this method.

We put a tube with a sharp bottom end on a XY stage, and surround the top of it by a ring with 4 pressure gauges inside of it (Figure 1). If the tube deviates from the vertical direction, force acts on the pressure gauges. Then we move the bottom of the tube horizontally until the force becomes zero, and the tube is kept to be vertical. This attitude control does not restrict the optical system of the telescope because any optical element as the horizontal reference plane like a mercury pool is not necessary, nor nothing comes in the field of view.

3. Advantages of the inverted pendulum

The inverted pendulum is advantageous in possibility of high sensitivity compared with normal pendulum. Normal pendulum has a pivot and the motion of the pendulum around the pivot is controlled by gravitational torque and friction momentum. The gravitational torque N and the friction momentum M are respectively expressed as

$$N = mgl \sin\theta \tag{1}$$

$$M = \mu Pd / 2 \tag{2}$$

where m is mass of the weight, g gravitational acceleration, l arm length of the pendulum, θ deviation angle from the vertical, μ friction coefficient, P load, and d diameter of the ball pivot as shown in Fig. 2. The gravitational torque and the friction momentum are balanced when the deviation angle θ is about 5 arc second, if m = 1 kg, l = 0.1 m, μ= 0.001 and d = 0.005 m. This means that the normal pendulum has a limit in the verticality due to the friction moment at the pivot. The inverted pendulum, on the other hand, has no limit in principle.

When an inverted pendulum with mass m (kg) deviates from the vertical direction by angle θ, the force P acting horizontally is represented as

$$P = mg \sin\theta \tag{3}$$

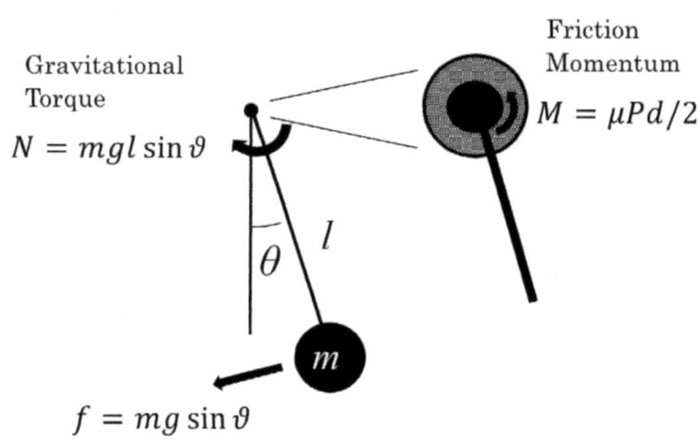

Figure 2. Gravitational torque and friction momentum in a normal pendulum.

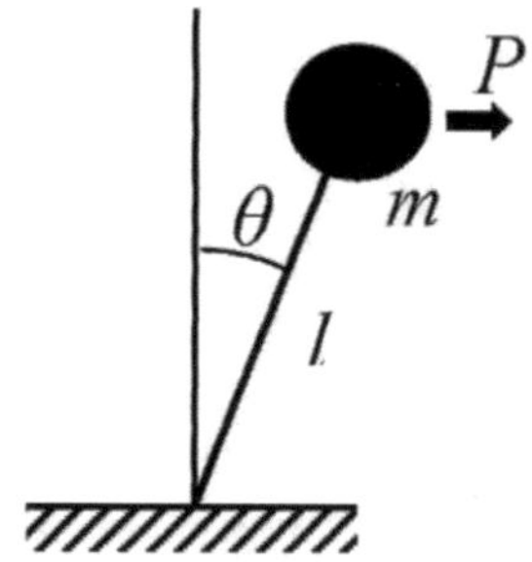

Figure 3. Gravitational torque in an inverted pendulum.

If we suppose m = 1 kg, θ = 1 arc second (4.8×10^{-6} rad), P becomes about 5×10^{-5} N (50 μN). We can detect the force of 0.005μN which is about 1/10,000 of the force mentioned above if we use the most sensitive pressure gauge (FT-S100, FemtoTools AG). This means that we can control the attitude of a tube with the sensitivity of 0.1 milli-arc second.

On the other hand, it has the dynamic range of about 20,000 times as large as the resolution, thus the most sensitive sensor has the range of 100 μN which corresponds to the force generated by the pendulum tilted by 2 arc seconds. Therefore, we must keep the tube within 2 arc seconds from the vertical direction by some other way.

The sensitivity of verticality does not depend on height or width but on the mass. As shown in Figure 4, even if the height or the width of a pendulum is different, if the mass is the same, the horizontal component of the force applied to the frame, or the sensitivity to the tilted angle from the vertical is the same. This is also understood from the fact that the height of the center of gravity is not included in equation (3). This is advantageous for miniaturization. The accuracy of the horizontal position control of the bottom end, however, depends on the height to the center of gravity. In other words, smaller the size becomes, more precise control is required. For example, if we attempt to control a pendulum of 10 cm height of gravity with 1 milli-arc-sec accuracy, resolution of an XY linear stage should be as small as 5×10^{-10} m (0.5 nm). This resolution is not impossible if we utilize a certain reduction mechanism or if we use sub nanometer resolution actuator using piezoelectric elements [5].

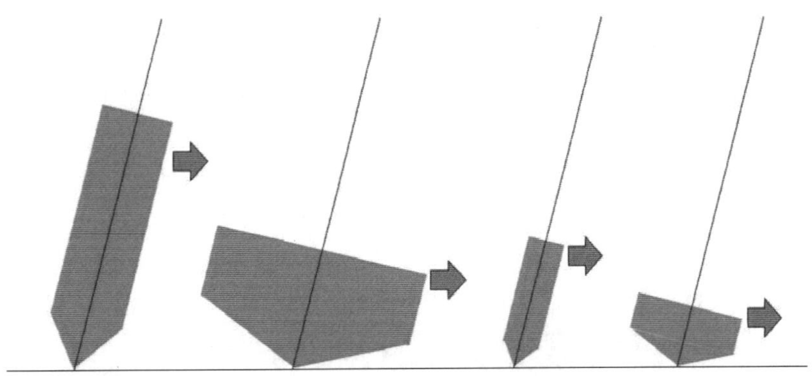

Figure 4. Horizontal forces generated by pendulums tilted by a certain angle does not depend on the height nor the dimension but on the mass.

4. A concrete method of controlling the inverted pendulum

It seems to be simple to directly put pressure gauges between the tube and the frame as shown in Figure1. However, the more sensitive to the vertical, the more tight the control system becomes. For example, if there is a gap between the pressure gauge and the inverted pendulum (or the frame) even if it is narrow, the pendulum allows tilt caused by the gap. On the contrary, if we put the pressure gauge with no gap, the sensor detects a pressure even if the pendulum is in the vertical direction.

We propose to put a spring between the pendulum and the frame, and we detect the pressure by measuring extension or contraction of the spring. Many pressure gauges detect a pressure as deformation of an elastic material. We can make the pendulum vertical by adjusting the horizontal position of the bottom end so as to the extension or contraction of the spring be the same as the initial value. The spring constant is selected so that the restoring force should be as small as possible. Then the proper period of the pendulum becomes long which makes the control easier. We use a simple model of a two dimensional inverted pendulum which has freedom of motion in one direction.

IOP Conf. Series: Materials Science and Engineering **468** (2018) 012036 doi:10.1088/1757-899X/468/1/012036

Motion of such a pendulum has been theoretically investigated [6]. The equation of motion of the inverted pendulum which is supported by two springs is written as:

$$I\frac{d^2\theta}{dt^2} = mgl\sin\theta + ml\cos\theta\frac{d^2x}{dt^2} - kh^2\theta \tag{4}$$

where I is the moment of inertia of the pendulum, θ tilt angle, t time, m the mass, g gravitational acceleration, l the distance between the pivot and the center of gravity of the pendulum, x position of the pivot, k spring constant and h the distance between the pivot to the supporting point of the spring as shown in Figure5. We set the positive direction of x-axis be rightward in the figure, and the tilt of the pendulum toward positive x-axis is represented by the angle of positive θ. The 2nd term of the right hand side of eq. (4) represents the motion of the pivot, which is used for controlling of the pendulum. The spring constant k is for the combined spring composed with two springs.

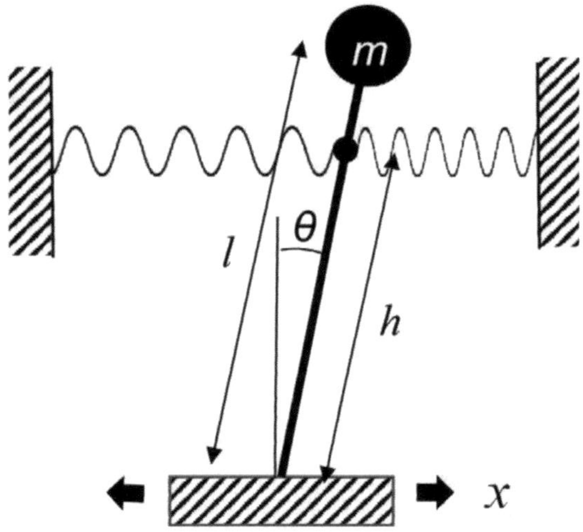

Figure 5. A two dimensional model of inverted pendulum with two spring.

We made an experimental device of two dimensional pendulum as shown in Figure 6 in order to check if this method is effective for the attitude control of a telescope tube, although it is not aiming at the highest sensitive experiments. An electronic level (Talyvel 4, Taylor Hobson Ltd) was put on the pendulum for monitoring of the tilt of it. The motion of the pendulum was simulated at first by using parameters listed in Table 1. The results are shown in Figure 7 for the cases of natural motion (upper) and controlled motion (lower). This shows that it is easy to make the natural period of the pendulum be as long as about 10 s, and that the pendulum is set to be vertical by moving the bottom end to the opposite direction as 150 times as large as the amount of extension or contraction of the spring. This characteristics was also checked by experiments using the experimental device.

We have a prospect that this method is useful for attitude control of the telescope tube, although we have not attained the best accuracy of attitude control of the inverted pendulum.

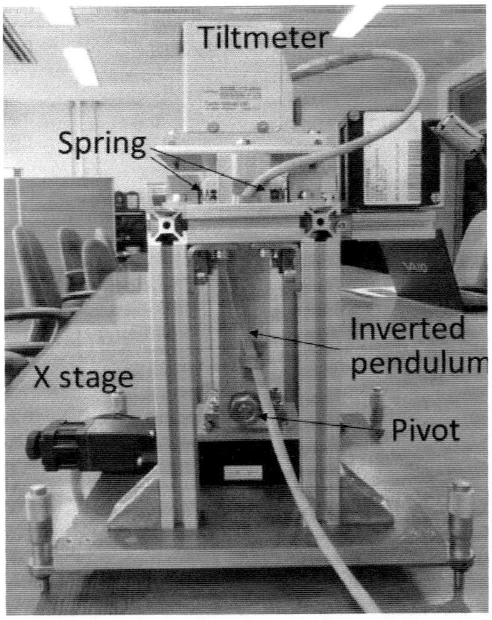

Figure 6. An experimental device of two dimensional inverted pendulum.

Table 1. Parameters used for the simulations.

Parameters	Values	Unit
I	0.0262	kgm^2
m	1.665	kg
g	9.8	ms^{-2}
l	0.116	m
k	101	N/m
h	0.13	m

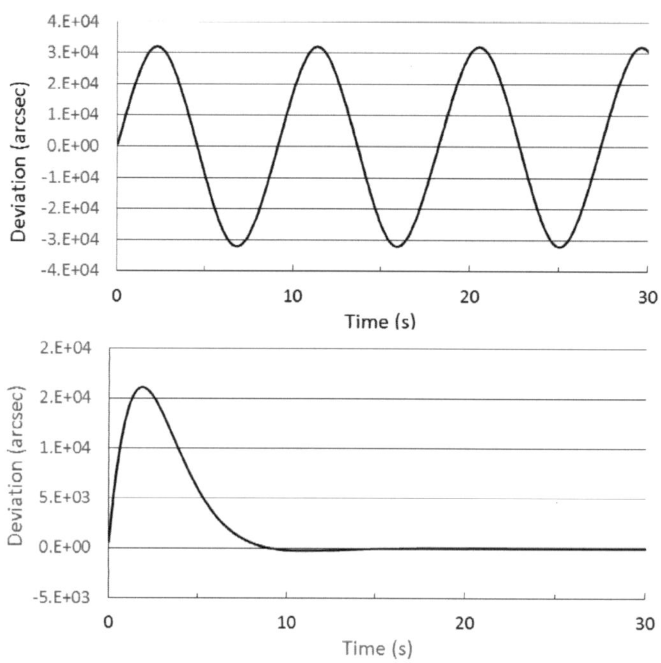

Figure 7. Simulation results of the motion of the inverted pendulum with springs. Natural motion (upper) and a motion controlled by moving the pivot by as 150 times as large as extension of the length of the springs (lower).

5. Concluding remarks

We propose a method of attitude control using an inverted pendulum which can be applied for a telescope tube. It is the first attempt to make use of the inverted pendulum for attitude control of a telescope tube, although the control of the inverted pendulum itself has been investigated for many years. The new method has a characteristics in not using a mercury pool nor a normal pendulum nor a independent tilt meter. This method is free from the difficult treatment of the mercury pool, limit of vertical sensitivity in the normal pendulum and the independency of the tilt meter to the tube itself..

We found that this method can be used for the attitude control and we will pursue the accuracy of the control by extending to a two dimensional model in the future. This method can open the new way in the future mission with a small and light telescope for observation of rotation on the Moon or on the planet. This method also can contribute to establishment of a new reference of a horizontal plane instead of mercury pool. This work is supported by JSPS KAKENHI Grant Number 16K05545.

References:

[1] Hanada H, Araki H, Tazawa S, Tsuruta S, Noda H, Asari K, Sasaki S, Funazaki K, Satoh A, Taniguchi H, Kikuchi M, Takahashi T, Yamazaki A, Ping J, Kawano N, Petrova N, Gouda N, Yano T, Yamada Y, Niwa Y, Kono Y and Iwata T 2012 *Science China* **55** pp 723-32 [DOI: 10.1007/s11433-012-4673-1]

[2] Hanada H, Tsuruta S, Asari K, Araki H, Noda H, Tazawa S, Kashima S, Funazaki K, Satoh A, Taniguchi H, Kato H, Kikuchi M, Sasaki H, Hasegawa T, Yano T, Gouda N, Kobayashi Y, Yamada Y, Iwata T and Gusev A 2017 *J. Gyroscopy Navigation* (in press)

[3] Felix G, D'Arrigo A, Colombi S and Rufer AC 2002 *IEEE Trans. Industrial Electronics* **49** pp 107-14
[4] Takamori A, Raffai P, Marka S, DeSalvo R, Sannibale V, Tariq H, Bertolini A, Cella G, Viboud N, Numata K, Takahashi R and Fukushima M 2007 *Nuclear Inst. Method Phys. Res.* A **582** pp 683-92 [DOI: 10.1016/j.nima.2007.08.161]
[5] Scire F E and Teague E C 1978 *Rev. Sci. Inst.* **49** pp 1735-40 [DOI: org/10.1063/1.1135327]
[6] Hauser J, Saccon A and Frezza R 2005 *Proc 4th IEEE Conf. Decision Control and European Control Conf. 2005* pp 6176-80

IOP Conf. Series: Materials Science and Engineering **468** (2018) 012037 doi:10.1088/1757-899X/468/1/012037

Constraint stabilization of two-wheeled sleigh

I Kaspirovich

Peoples' Friendship University of Russia, (RUDN University), Moscow, Russia
E-mail: aspirovich.ivan@mail.ru

Abstract. In this paper a problem of constraint stabilization of a two-wheeled sleigh is considered. This problem is solved with the help of the Chaplygin's approach, in which Lagrange equations of the second kind are modified with respect to the nonholonomic constraints. For the obtained equations we define the functions of reactions forces of constraints with respect to their stabilization. During the numerical integration some of the stabilization parameters are defining at each step of the summation. This gives an advantage in comparison with the classical stabilization approach.

1. Introduction

The main problem of the classical mechanics is to find a solution of motion equations that are often can be written as a system of ordinary differential equations. But there are only few systems that we can solve by obtaining an analytic solution. In general the methods of numerical integration are used to obtain a specific result. Most of them are realized with the help of the summation of deference schemes. But when we have a constraint mechanical system deviations from the constraints will accumulate and lead to the numerical solution instability. So J. Baumgarte suggested to modify the functions of reaction forces by equating the derivatives of the constraints to the linear form of themselves [1]. The optimal range of values of linear form coefficients allow us to control the accumulation of the deviations. The problem of determining this range was considered in [2–4].
The constraint stabilization method is applied to the specific types of nonholonomic systems, called Chaplygin's. When the structure of constraint equations and a Lagrange function allow us to separate independent variables from dependent ones Lagrange equations of the second kind can be modified so they can include constraint terms in them [5]. In this paper we consider a problem of defining linear form stabilization coefficients for the Chaplygin nonholonomic system.

2. Motion Equations

2.1. Chaplygin systems

Let us consider the motion of a nonholonomic system defined by the set of generalized coordinates $(q^1, \ldots, q^n)$ and generalized velocities $(v^1 = dq^1/dt = \dot{q}^1, \ldots, v^n = \dot{q}^1)$. If the structure of nonholonomic constraint equations allows us to divide a set of generalized velocities into dependent and independent parts and the Lagrange function and external forces do not depend on the dependent coordinates then a nonholonomic system is called a Chaplygin system. For example, for the constraint equation

$$\dot{q}^i = b_a^i(q^1, \ldots, q^p)\dot{q}^a, \qquad (1)$$

Content from this work may be used under the terms of the Creative Commons Attribution 3.0 licence. Any further distribution of this work must maintain attribution to the author(s) and the title of the work, journal citation and DOI.
Published under licence by IOP Publishing Ltd

the nonholonomic system can be considered as Chaplygin's. Here and elsewhere we consider the summation under the repeating indexes. Latin letters $i, j, k, \ldots$ stand for the summation under the dependent variables $(q^{p+1}, \ldots, q^n)$ and letters $a, b, c, \ldots$ for the summations under independent ones $(q^1, \ldots, q^p)$. So there are only p independent variables due to the fact that there are $n-p$ constraint equations (1).
It has been shown that the motion equations for the Chaplygin systems can be written in the following form:

$$\frac{\mathrm{d}}{\mathrm{d}t}\frac{\partial L^*}{\partial \dot{q}^a} - \frac{\partial L^*}{\partial q^a} + \left(\frac{\partial L}{\partial \dot{q}^i}\right)^* \left(b_c^i q^a - b_a^i q^c\right)\dot{q}^c = 0, \tag{2}$$

where the asterisk means the operation of defining the dependent velocities through the independent ones due to the equation (1).
Let us consider that the Lagrange function of the mechanical system with constraint equations (1) takes form

$$L = \frac{1}{2}\sum_{r,s=1}^{n} g_{rs}(q^1, \ldots, q^p)\dot{q}^r\dot{q}^s - U(q^1, \ldots, q^p). \tag{3}$$

So it is just represented as a sum of the kinetic energy T and a potential energy U. As we can see both of these functions do not depend on the coordinates $(q^{p+1}, \ldots, q^n)$ then such nonholonomic system can be treated Chaplygin's.
After applying the asterisk operation to the Lagrange function (3) we will obtain a modified function

$$L^* = \frac{1}{2} g_{ab}^*(q^1, \ldots, q^p)\dot{q}^a\dot{q}^b - U(q^1, \ldots, q^p), \tag{4}$$

where the bilinear coefficients g_{ab}^* are determined as follows:

$$g_{ab}^* = g_{ab} + g_{ib}b_a^i + g_{ia}b_b^i + g_{ij}b_a^i b_b^j.$$

So we can see that these bilinear coefficients that can be interpreted as metric tensor for Chaplygin system define through the linear coefficients b_b^j of the nonholonomic constraint equations (1). Motion equations (2) for the Lagrange function (4) take form

$$\begin{cases} g_{ac}^*\ddot{q}^a + \tilde{\Gamma}_{c,ab}\dot{q}^a\dot{q}^b = 0, \\ \tilde{\Gamma}_{c,ab} = \Gamma_{c,ab}^* + \left(b_b^i q^c - b_c^i q^b\right)\left(g_{ia} + g_{ij}b_a^j\right), \\ \Gamma_{c,ab}^* = \frac{1}{2}\left(g_{ac}^* q^b + g_{bc}^* q^a - g_{ab}^* q^c\right). \end{cases} \tag{5}$$

If a determinant of the matrix $\langle g_{ac}^* \rangle$ is nonzero then we can raise an index of the tensor $\tilde{\Gamma}_{ab}^d = g^{*cd}\tilde{\Gamma}_{c,ab}$. The obtained tensor is an analog of the affine connection in a nonholonomic basis. The torsion tensor $T_{ab}^d = \tilde{\Gamma}_{ab}^d - \tilde{\Gamma}_{ba}^d$ is nonzero for the Chaplygin systems due to equation (5). So this fact defines the properties of a nonholonomic motion.

2.2. Reaction forces

To solve the problem of the constraint stabilization of a Chaplygin system we have to introduce an extra set of holonomic and nonholonomic constraints:

$$\begin{cases} f_\mu(q^1, \ldots, q^p) = 0, \\ h_\alpha(q^1, \ldots, q^p, v^1, \ldots, v^p) = 0, \end{cases} \tag{6}$$

where Greek indexes $\mu, \nu = 1, \ldots, m_1$, $\alpha, \beta = 1, \ldots, m_2$, $m_1 + m_2 = r < p$ stand for the summation under constraint terms.
To take into account the set of constraints (6) we have to introduce extra terms of the external generalized forces in (5) which are called the reaction forces. These forces can be defined with the help of the Lagrange multipliers $(\lambda_1, \ldots, \lambda_r)$. So we consider them as follows:

$$R_a = \lambda_\mu \frac{\partial f_\mu}{\partial q^a} + \lambda_{m_1+\alpha}\frac{\partial h_\alpha}{\partial v^a}.$$

Now we have to construct a system of an algebraic equations to determine the Lagrange multipliers as functions of generalized coordinates and velocities so the reaction forces will be defined. For this we must consider second full time derivative of holonomic constraints f_μ and first derivative of nonholonomic constraints h_α. The obtained equations combined with the motion equations (5) compose a system of algebraic equations with respect to generalized accelerations and Lagrange multipliers:

$$\begin{cases} \ddot{f}_\mu(q^1, \dots, q^p) = 0, \\ \dot{h}_\alpha(q^1, \dots, q^p, v^1, \dots, v^p) = 0, \\ \ddot{q}^a = Q^a + R_a, \end{cases} \tag{7}$$

where $Q^a = -\tilde{\Gamma}^a_{bc}\dot{q}^b\dot{q}^c$.

Due to system (7) a solution for Lagrange multipliers can be found from the following system of $r \times r$ algebraic equations:

$$\begin{cases} \frac{\partial f_\mu}{\partial q^a}\lambda_\nu\frac{\partial f_\nu}{\partial q^a} + \frac{\partial f_\mu}{\partial q^a}\lambda_{m_1+\alpha}\frac{\partial h_\alpha}{\partial q^a} - \frac{\partial f_\mu}{\partial q^a}Q^a + \frac{\partial^2 f_\mu}{\partial q^a \partial q^b}\dot{q}^a\dot{q}^b = 0, \\ \frac{\partial h_\alpha}{\partial q^a}\lambda_\nu\frac{\partial f_\nu}{\partial q^a} + \frac{\partial f_\mu}{\partial q^a}\lambda_{m_1+\beta}\frac{\partial h_\beta}{\partial q^a} - \frac{\partial h_\alpha}{\partial q^a}Q^a = 0. \end{cases} \tag{8}$$

If this system is nondegenerate then Lagrange multipliers can be uniquely defined as functions of generalized coordinates and velocities which determines the functions of the reaction forces.

2.3. Constraint stabilization

After determining the reaction forces we consider the following system of equations:

$$\begin{cases} \ddot{q}^a = Q^a + R_a, \\ f_\mu(q^1, \dots, q^p) = 0, \\ h_\alpha(q^1, \dots, q^p, v^1, \dots, v^p) = 0. \end{cases} \tag{9}$$

To find its solution we can apply some methods of a numerical integration.

A deference scheme of the numerical integration can be realized with the help of the program cycle. But we have to be careful, because when we deal with the multiple repeating operations of the summation, deviations from the constraints, caused mostly by rounding errors, accumulate. And it leads to the numerical solution instability.

A method of constraint stabilization can be applied to control the accumulation of these deviations. According to it the second and the first derivatives in (7) should be equated to the linear form of the constraints themselves. Such as $\ddot{f}_\mu = -k^\nu_\mu\dot{f}_\nu - l^\nu_\mu f_\nu$ and $\dot{h}_\alpha = -m^\beta_\alpha h_\beta$. These equations are called the equations of perturbation and the linear coefficients - perturbation parameters. The main problem of the constraint stabilization is to determine the optimal range of the values of the perturbation parameters so the numerical solution will be stable in relation to the constraint equations.

Due to the method of constraint stabilization the equations (8) can be rewritten in the form

$$\begin{cases} \frac{\partial f_\mu}{\partial q^a}\lambda_\nu\frac{\partial f_\nu}{\partial q^a} + \frac{\partial f_\mu}{\partial q^a}\lambda_{m_1+\alpha}\frac{\partial h_\alpha}{\partial q^a} - \frac{\partial f_\mu}{\partial q^a}Q^a + \frac{\partial^2 f_\mu}{\partial q^a \partial q^b}\dot{q}^a\dot{q}^b = -k^\nu_\mu\dot{f}_\nu - l^\nu_\mu f_\nu, \\ \frac{\partial h_\alpha}{\partial q^a}\lambda_\nu\frac{\partial f_\nu}{\partial q^a} + \frac{\partial f_\mu}{\partial q^a}\lambda_{m_1+\beta}\frac{\partial h_\beta}{\partial q^a} - \frac{\partial h_\alpha}{\partial q^a}Q^a = -m^\beta_\alpha h_\beta. \end{cases}$$

The reaction forces now also depend on perturbation parameters like the Lagrange multipliers. So at each step of the numerical summation stabilization term of the reaction forces should control values of the deviations.

Let us consider Euler's first order difference scheme. For the system (8) it can be written in the form

$$\begin{aligned} q^a_{(n+1)} &= q^a_{(n)} + hv^a_{(n)}, \\ v^a_{(n+1)} &= v^a_{(n)} + h(R_{a(n)} - Q^a_{(n)}), \\ t_{(n+1)} &= t_{(n)} + h, \end{aligned} \tag{10}$$

where indexes in the round brackets stand for the number of the step of the summation.

To avoid some bulky expressions let's write the perturbation equations in the matrix form:

$$\dot{\mathbf{f}} = t\begin{pmatrix} f_1 \\ \vdots \\ f_{m_1} \\ g_1 \\ \vdots \\ g_{m_1} \\ h_1 \\ \vdots \\ h_{m_2} \end{pmatrix} = \mathbf{Kf},$$

where $g_\mu = \dot{f}_\mu$ and elements of the matrix $\mathbf{K}$ represent the perturbation parameters.
Let us expand the vector function $\mathbf{f}(q_{(n+1)}, v_{(n+1)})$ to the Taylor series of the powers h:

$$\mathbf{f}(q_{(n+1)}, v_{(n+1)}) = \mathbf{f}(q_{(n)}, v_{(n)}) + h\frac{\mathrm{d}\mathbf{f}(q_{(n+1)}, v_{(n+1)})}{dh}\bigg|_{h=0} + \frac{h^2}{2}\mathbf{f}^{(2)},$$

where $\mathbf{f}^{(2)} = \frac{\mathrm{d}^2\mathbf{f}(q_{(n+1)}, v_{(n+1)})}{dh^2}\bigg|_{\xi\in[0,h]}$ is a vector of integral reminder terms in the form of a mean value theorem.

Such as $\mathrm{d}\mathbf{f}(q_{(n+1)}, v_{(n+1)})/\mathrm{d}h\big|_{h=0} = \mathbf{f}(q, v)t|_{(q,v)=(q_{(n)}, v_{(n)})} = \mathbf{Kf}(q_{(n)}, v_{(n)})$. We can estimate the value of the deviations at $(n+1)$ step if we consider that at n step the following inequality takes place: $|f_{(n)}| < \varepsilon$, where ε is chosen small control parameter and $|f_{(n)}|$ is a norm of a vector in Euclidean space. So if we consider triangle inequality to our expansion we can obtain the following result:

$$|f_{(n+1)}| \le |\mathbf{I_{2m_1+m_2}} + \mathbf{hK}|\varepsilon + \frac{h^2}{2}|\mathbf{f}^{(2)}|, \tag{11}$$

where $I_{2m_1+m_2}$ is a unit matrix, and $|\mathbf{I_{2m_1+m_2}} + \mathbf{hK}|$ is an $L_{2,1}$ matrix norm.
If we want the norm of deviation vector to be also less than ε the following equality should take place:

$$|\mathbf{I_{2m_1+m_2}} + \mathbf{hK}| \le 1 - \frac{h^2}{2\varepsilon}|\mathbf{f}^{(2)}|. \tag{12}$$

Formula (12) defines the range of the values of perturbation parameters at which the values of constraint deviations do not increase.

3. Stabilization of two-wheeled sleigh

As an example we consider the well-known mechanical problem of two-wheeled sleigh on an inclined plane. Lagrange function and nonholonomic constraints are as follows:

$$\begin{aligned} &L = \frac{m}{2}(\dot{x}^2 + \dot{y}^2) + m_0 l\dot{\theta}(\dot{y}\cos\theta - \dot{x}\sin\theta) + \frac{J}{2}\dot{\theta}^2 + \frac{C}{2}(\dot{\varphi}_1{}^2 + \dot{\varphi}_2{}^2), \\ &\dot{x}\cos\theta + \dot{y}\sin\theta + b\dot{\theta} + a\dot{\varphi}_1 = 0, \\ &\dot{x}\cos\theta + \dot{y}\sin\theta - b\dot{\theta} + a\dot{\varphi}_2 = 0, \\ &\dot{y}\cos\theta - \dot{x}\sin\theta = 0, \end{aligned} \tag{13}$$

where $m = m_0 + 2m_I$, m_0 – mass of the body, m_I – mass of the wheel, (x, y) are the coordinates of the intersection point $M(x, y)$ between the symmetry axis and the wheel axle, θ is an angle between the symmetry axis and the projection of Ox axis on the inclined plane, φ_1 and φ_2 are rotation angle of wheels correspondingly, l is a distance between $M(x, y)$ and a mass center of the sleigh, $J = m_0 k_0^2 + 2m_I b^b + 2A$ – moment of inertia about the vertical axis passing through the point $M(x, y)$, k_0 – radius of inertia of the wheel also about the vertical axis, A – moment of inertia of a wheel about the diameter axis, C – axial moment of inertia of the wheel, a – radius of the wheel, b – its length of a semiaxis.
Let us consider that sleigh should follow to a particular trajectory - an extra holonomic constraint:

$$y = \gamma(x) \quad \text{or} \quad f(x, \theta) = \tan\theta - \frac{\mathrm{d}\gamma}{\mathrm{d}x} = 0. \tag{14}$$

The structure of the Lagrange function (13) and the constraint equations (14) allow us to separate independent coordinates (x, θ) and dependent ones $(y, \varphi_1, \varphi_2)$. So the problem can be solved with the help of Chaplygin's approach. The motion equations (9) with respect to the constraint stabilization take form

$$\begin{cases} \frac{\tilde{m}\ddot{x}}{\cos^2\theta} + \frac{\tilde{m}\dot{x}\dot{\theta}\tan\theta}{\cos^2\theta} - \frac{m_0 l\dot{\theta}^2}{\cos\theta} = -\lambda\frac{d\gamma}{dx}, \\ \tilde{J}\ddot{\theta} + \frac{\tilde{m}\dot{x}^2\tan\theta}{\cos^2\theta} - \frac{m_0 l\dot{\theta}\dot{x}}{\cos\theta} = \frac{\lambda}{\cos^2\theta}, \\ \ddot{f} + k_2\dot{f} + k_1 f = 0, \end{cases} \tag{15}$$

where $\tilde{m} = m + \frac{2C}{a^2}$, $\tilde{J} = J + 2C\frac{b^2}{a^2}$. The dependent variables are determined through the constraint equations in (13).

We consider deference first order scheme (10) to obtain a stable numerical solution. To control the value of constraint deviations we also consider the stabilization terms in (15). There are only two perturbation parameters in them so we can specifically chose, for example, the value of k_2 and determine the value of k_1 with the help of the inequality (12). It be written in the form:

$$(1 + h^2 k_{1(n)}^2 + (1 - hk_2)^2)^{1/2} = 1 - \frac{h^2}{2\varepsilon}|\mathbf{f}_{(n+1)}^{(2)}| + \delta, \tag{16}$$

where δ is a chosen parameter. So at each step of the numerical the value of the perturbation parameter $k_{1(n)}$ can be considered self-defining.

If $\gamma(x) = A\sin kx$ and an initial data is chosen corresponding to the constraints, numerical solution will lead to the following result

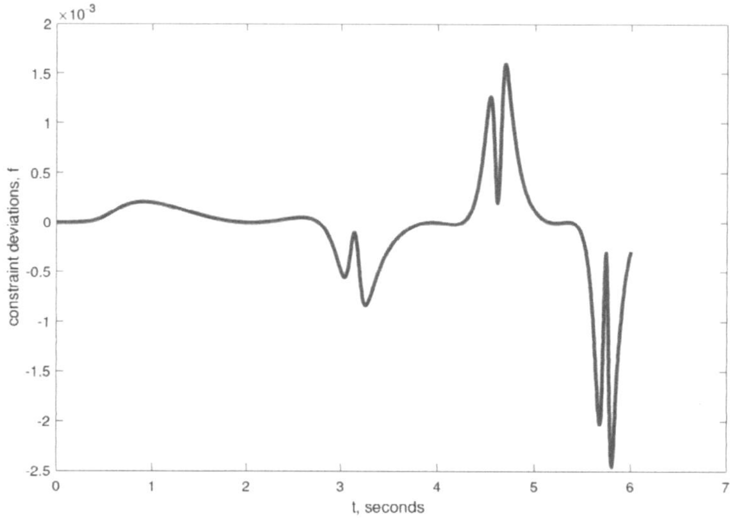

Figure 1. Value of deviation dependence on time.

Basically, we are not interested in the trajectory curve because we have established it as sinusoid. But the plot of deviation value dependence on time can be quite illustrative for this problem. So the self-defining value of a particular perturbation parameter allows us to provide the minimum value of deviations.

References

[1] Baumgarte J 1972 *Computer Methods in Applied Mechanics and Engineering* **1** pp 1–16
[2] Mukharlyamov R G 2015 *J. Comput. Syst. Sci. Int.* **54** pp 13–26
[3] Kaspirovich I E 2016 - 2016 Application of constraint stabilization to nonholonomic mechanics *2016 2nd International Conference on Industrial Engineering, Applications and Manufacturing (ICIEAM) 2016 2nd International Conference on Industrial Engineering,*

Applications and Manufacturing (ICIEAM) (Chelyabinsk, Russia, 19.05.2016 - 20.05.2016) (IEEE) pp 1–4
[4] Abramov N V, Mukharlyamov R G and Kirgizbaev Z K 2013 *Control dynamics of systems with program constraints* (Nizhnevartovsk: Nizhnevartovsk State University)
[5] Chaplygin S A 1949 *Research of nonholonomic system dynamics* (Moscow: Gostekhizdat)

Acknowledgments
This work is supported by RFBR, project 16-08-00558 A.

IOP Conf. Series: Materials Science and Engineering **468** (2018) 012038 doi:10.1088/1757-899X/468/1/012038

To development of general theory of gearings synthesis on the basis of conical wheels formed by a tool of rack and disk type

O N Tsukanov

Bauman Moscow state technical University, Moscow, Russia
E-mail: Tsonzz@mail.ru

Abstract. Generalized scheme and special cases of schemes gears, which are synthesized on the basis of conical shaped blanks of cogwheels, formed by a tool of rack and disk type, are presented. The advantages of such gears in comparison with traditional cylindrical and conical gears first of all, layout advantages and ability to eliminate backlash in the gearing both at manufacturing stage and at stage of operation of designed mechanisms are described. The main hypothesis used in the development of general theory of synthesis of gears under consideration involved is formulated. The notion of a general type initial link is given and essence of synthesis of gears in generalizing coordinates is disclosed, which allows to maximally reveal geometric-kinematic possibilities of gearing. The main provisions of general theory of synthesis of considered gears in a limiting area existence of gearing by means of a change in the shape and coordinates of points local areas, corresponding to one or another set of qualitative indicators are formulated.

1. Introduction

One of topical problems of machine mechanics is the task of creating high-resource and high-precision mechanisms for various, including extreme, operating conditions in a wide range of loads and speeds. Wide possibilities for solving this problem are opened by use of gears synthesized on the basis of conical shaped blanks of cogwheels with crossed, intersecting and parallel geometric axes, tooth profiles of which are formed by a rack and disk type tool. Accordingly, in the first case we obtain a conical wheel with an involute tooth profile, the so-called involute-conical wheel [1], and in the second case – with a not involute tooth profile (not involute conical wheel) [2]. Comparatively simple and high-performance technology for shaping of teeth a worm cutter and a worm grinding wheel, relating to tools of rack type, provides high profile accuracy [2, 3]. As disc-type tool for cutting teeth a dolbyak is used, and for final processing – a disc shaver [4].

2. Schemes and advantages of gears under consideration

In figure 1 shown the basis scheme of a conical gear, in figure 2 shows possible schemes, so-called, cylinder-conical gears [4], obtained when one of wheels of a conical gear degenerates into a cylindrical wheel, and in figure 3 shown scheme, so-called, a flat-wheel gear, obtained when a conical wheel of a cylinder-conical gear degenerates into a flat one.

In figure 1: Σ – the inter axial angle; $xOyw$ – the fixed coordinate system with the origin O in initial cross-section of initial link; $x_kO_ky_kw_k$ – the mobile coordinate system with the beginning of O_k in initial cross-section of envelope link; φ_k – the current angle of rotation of envelope link; k – the index of parameters of envelope link. Parameters of initial link the index are not assigned to.

Content from this work may be used under the terms of the Creative Commons Attribution 3.0 licence. Any further distribution of this work must maintain attribution to the author(s) and the title of the work, journal citation and DOI.
Published under licence by IOP Publishing Ltd

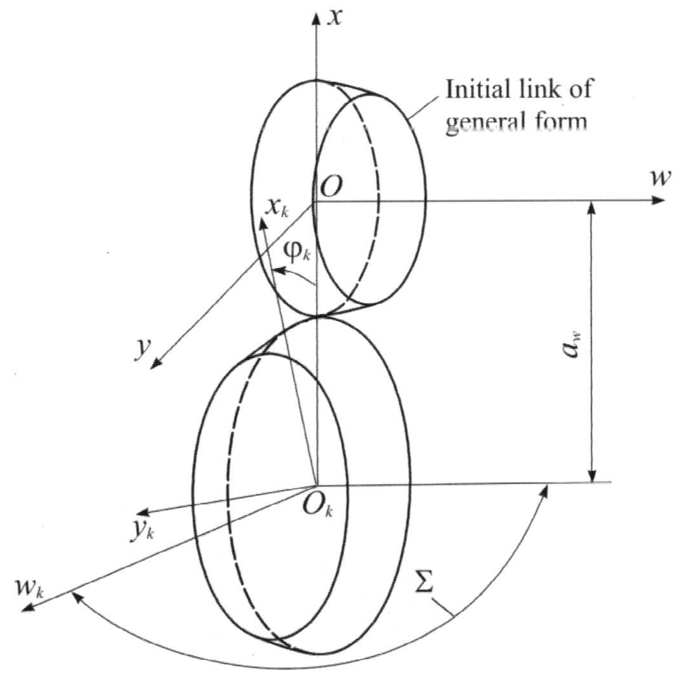

Figure 1. The basis scheme of a conical gear

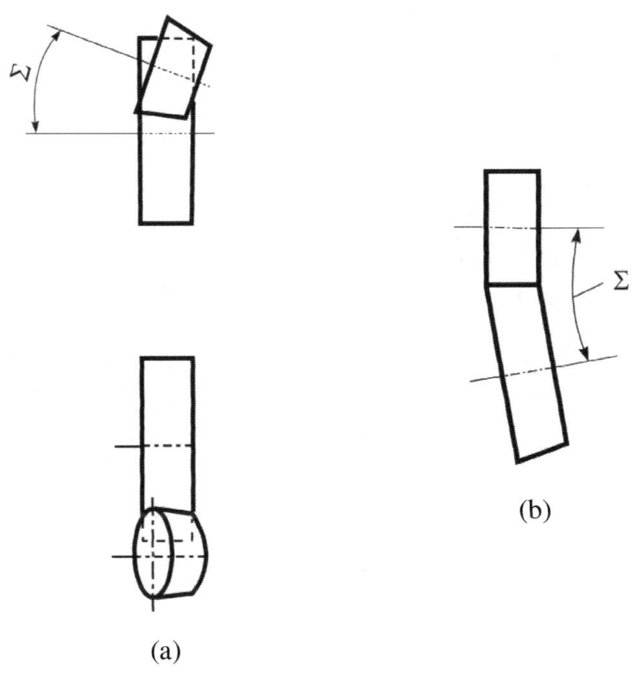

Figure 2. Schemes of cylinder-conical gears: (a) – with crossed geometric axes of wheels; (b) – with intersecting geometric axes of wheels

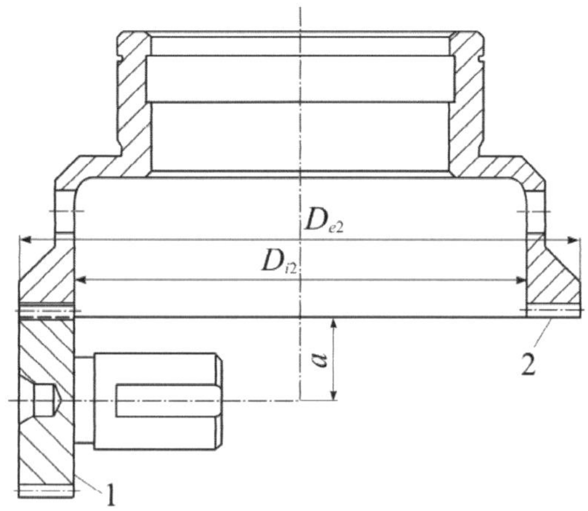

Figure 3. Constructional scheme of the flat-wheel gear: 1 – the cylindrical gear wheel; 2 – the flat wheel

In figure 3: D_{i2} – an inner diameter of the flat wheel; D_{e2} – an outer diameter of a flat wheel; a – a distance from the pinion axis to the plane of the vertices of the teeth of the wheel.

One of main advantages of considered gears is possibility of eliminating a backlash of an output shafts and compensating for a displacement of teeth contact spot during assembly and during operation (as working surfaces of teeth are wear out) by means corresponding displacement of a conical wheel [3, 4].

Another important advantage is the possibility of obtaining such a layout of drives that not canon be realized with traditional gears (cylindrical and conical) [4]. For example, a hypoid cylinder-conical gear can be realized at an arbitrarily small inter axial distance, which is impossible when using a helical gear. A not hypoid cylinder-conical gear can be realized at small inter axial angles (5°...15°), which is impossible when using conventional conical gear due to the lack of equipment for making wheels with a small cone angle. The most compact cylinder-conical gears are gears internal gearing, in which the initial link is an involute cylindrical wheel with internal teeth, and the envelope link is a not involute conical wheel with external teeth.

In figure 4 the scheme of precession planetary gearbox with the cylinder-conical gear of type $2k$-h is shown [4].

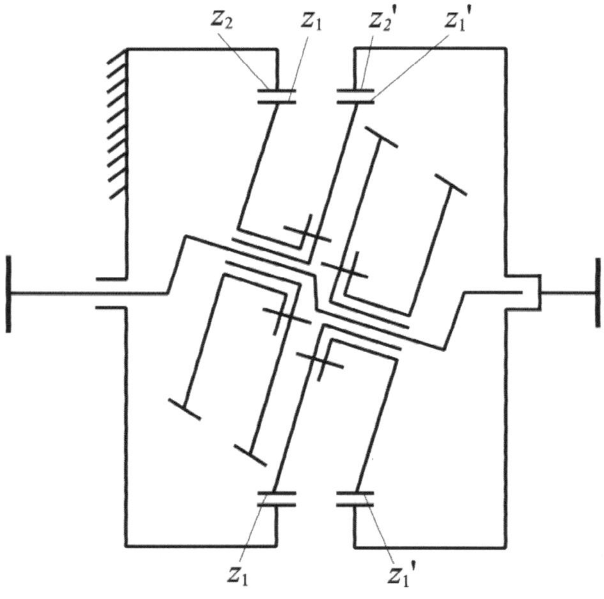

Figure 4. The scheme of precession planetary gearbox with the cylinder-conical gear of type $2k$-h

In figure 4: z_1 – is number of teeth of conical sattelites, engaged with the unmovable central wheel; z_2 – is number of teeth of the stationary central wheel; z_1' – is number of teeth of conical sattelites, engaged with the moving central wheel; z_2' – is number of teeth of the moving central wheel.
Gear has a many pairs of contacting teeth when a small difference in the number of teeth of central wheels and satellites, self-inhibition an output shaft, a wide range of gear ratios (maximally about 2000). Compared with a typical single-satellite cylindrical planetary gear, this transmission can transfer torque in 2,5 ... 3 times more when same dimensions, gear ratio and load of teeth. By load capacity, weight and dimensions such gears are competitive with wave gears, but much more durable than them, since it is possible to obtain high hardness of teeth working surfaces.
In this gear are two cylindrical the initial links. Accordingly teeth of conical gears must be cut gear-cutting tool with internal teeth. Due to absence of such tool of a conical wheels are forming by a worm gear-cutting tool, an axis of which moves along a curved trajectory [4]. The task of the researcher-designer consists in the justified choice of this trajectory and, accordingly, the function of changing the displacement factor of the forming rack in depending of the koordinate of cross-section of the cut wheel. The main criterion is the quality of the localized contact of teeth, which eliminates their edge contact due to errors in manufacture and installation of gears, reduces likelihood of interference due to deformation of elements of the mechanism. In particular, any required degree of contact localization can be obtained by longitudinal modification of teeth of the conical gear by a certain choice of the tool displacement coefficients along its edges [5].
On basis of the presented planetary gearbox the high-precision and high-resource electromechanical drive (service life up to 10 years) for mechanisms of angular rotation of objects of space vehicles (antennas, scanning devices, solar batteries, etc.) was created. He contains also electric motor of new type and a coupling sleeve, was created [6].

3. Prerequisites for development of general theory of gearings synthesis

3.1. Basic concepts of general theory of gearings synthesis

In basis for the development of general theory of synthesis of the considered gears the hypothesis that the main features, formerly the others determining the geometry and geometric-kinematic possibilities of gearing, are the shape of a vertex surface and type of teeth (teeth without transition surfaces) of initial link, and method synthesis of gearings in generalizing coordinates (independently of the parameters of the initial forming contour) allows to fully reveal these possibilities.
The gear wheel with not involute profile and curvilinear shape of the generatrix of the vertex surface of teeth, obtained during the process of their formed by a rack with a curved profile of teeth in cross-section moving along a certain curvilinear trajectory, is suggested be called the initial link of general form. By according to the shape of the lateral surface of teeth, it is not involute and the initial surface can have any shape. Blanks for the initial link of a general view, it is expedient to choose a technological conical and, in a particularistic case, when the cone angle is zero, of a cylindrical shape. If a contour of forming rack has a rectilinear profile of the teeth, and the axis of the worm cutter moves along a straight path, we get an involute conical wheel [4].
The profile of the tooth of the initial link in general case can theoretically be formed by rolling along a certain (main) circle of a segment of some curve of the line $f(x, y) = 0$ (figure 5), in particular, arcs of a circle, branches of a parabola, hyperbolas, etc., but practically by rolling method of workpiece forming rack with the corresponding profile of teeth. A rolling curve can consist of several sections of different shapes, and in most general case can be specified by a set of points with a certain step of discreteness. The corresponding profile of teeth can practically be realized on metal cutting machines with numerical program control.
With given values of independent parameters of a gear scheme (an inter axial angle, angles of an initial cones, a radius of an initial circle, and an angle of inclination of a tooth on an initial surface of the initial link), the geometry of a spatial gearing in general case is completely determined by values of generalizing coordinates of points of a tooth surface of an initial link in thes cross-sections: of the

gearing angle α_{tw}, which determines the angular position of the normal n and the tangent τ to profile of the tooth in the xOy coordinate system; of the profile angles α_x, α_y, which determine the distances r_n and r_τ from the axis of the initial link to the normal and tangent, respectively (look up figure 5), and the relative thickness of the tooth on the circumference of the vertices m_a in fractions of a diameter of a largest base circle $d_{b\,\max}$ – of scale factor of the gear.

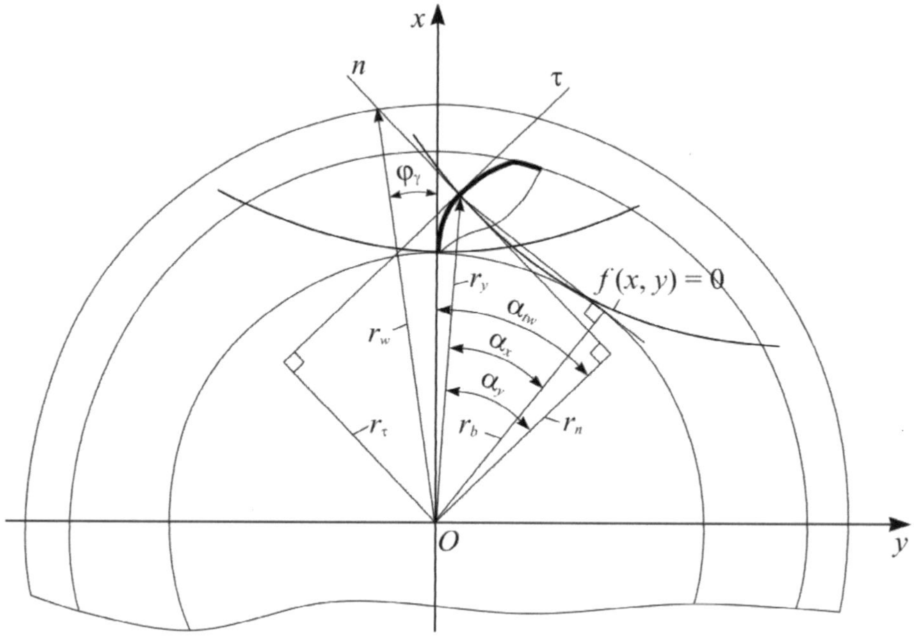

Figure 5. To formation of the tooth profile in a cross-section of an initial link

The radius of an initial circle r_w is suggested to be chosen more or equal to the radius of the circumference of the vertices of teeth, that is the area of their profiling in each cross-section of an initial link, is limited by an initial circle, and in space by an initial surface.

To ensure continuity of developed theories the module of the radius-vector of the current profile point r_y as the function of the radius of the current base circle r_b is definite.

The type of the function $f(\alpha_x, \alpha_y) = 0$ in a cross-section of the initial link depends on the type of function of line that is rolled along base circle. If this function is selected by some criterion of quality gearing, then on the contrary, it determines the function of the line rolling along the base circle. If the workpiece is cylindrical, then we use the base cylinder, and if conical, then is base cone. In the second case the radii of base circles of teeth cross-sections will depend on the angle of the cone.

The function of the surface of the vertices of the teeth of the initial link and the associated with her function of changing the displacement coefficient depending on coordinate the cross-section is proposed to be chosen as a result of analyzing of zones of inertial touch of the producing and the reproducible surfaces at the stage of computer designing of various types of gears.

If in any cross-section of an initial link $\alpha_x = \alpha_y$, then we obtain a not involute gearing with involute, in particular, with a cylindrical initial link, and the theory of synthesis of such gearing in generalizing parameters [7] developed by the author of this article can be considered as a special case of the general theory of gearings. In other words, synthesis of gearings on the basis of an involute initial link is performed in the system of two generalizing coordinates (α_{tw}, α_y), and synthesis of gearings on the basis of the not involute initial link – in the system of three generalizing coordinates (α_{tw}, α_y, α_x).

In the general case the surface of gearing exists in some limiting region, called the generalized area of existence of the gearing (GAEG), which is determined by the set of values of independent parameters of a gear scheme, the number of teeth of the initial link and their relative thickness on circle of vertex in the system of generalizing coordinates α_{tw}, α_x, α_y.

For the values of generalizing coordinates $\alpha_{tw\ max}$, α_x, α_y, given in the generalized area existence of gearing, one of the extreme points of the surface of gearing on the vertex of the tooth of the initial link determines the local area existence of gearing (LAEG).

Proceeding from the results of the studies described in [7], it can be assumed that the GAEG and LAEG in the indicated coordinates will have the form shown in figure 6.

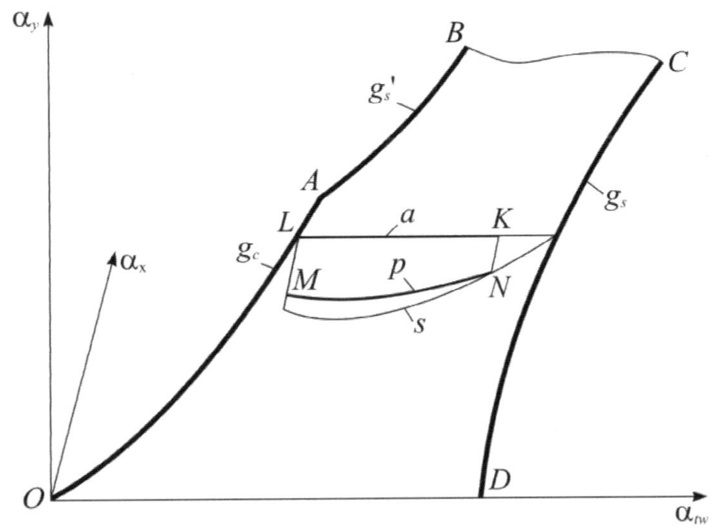

Figure 6. To the definition of generalized and local areas of existence (*OABCD* and *KLMN*) of spatial gearings in generalizing coordinates

GAEG is limited by limiting lines of sharpening (g_s, g_s') and interference (g_c) of teeth of an envelope link.

LAEG is limited to a straight line segment *a* corresponding to the specified angles α_x and α_y on the surface vertex of teeth of the initial link, the local curve of sharpening of the tooth of the envelope link *s* corresponding to these values, and the line *p*, whose shape depends on the shape of the surface vertex of teeth of this link.

The surface of teeth of the forming rack is reproduced by translational movement of the line of contact with the surface of the teeth of the original link (characteristics). The points of the characteristic are found by turning the points of teeth cross-profile by the angle φ_γ on the projection plane *xOy* (look up figure 5) and moving to the distance $w = p_\gamma\varphi_\gamma$ from this plane, where p_γ – is the screw parameter.

3.1.1. Basic provisions of the general theory of gearings synthesis. Taking into account the above, the author establishes the following basic positions of general theory of synthesis of the considered gears.

1. The function of the surface of teeth vertices of an initial link and associated with her dependence of displacement coefficient of the forming rack from coordinate of a cross-section of a conical wheel are determined depending on the purpose of the gear train, working conditions, and the results of computer analysis of inertial zones of touching of a producing and the reproducible surfaces.

2. The computer synthesis of gearings is performed in the system of three generalizing coordinates, whose values in the characteristic (usually extreme) points of the gearing surface are selected from the generalized area their existence as result of analysis of whole complex of geometric-kinematical and load-bearing indicators of gearing.

3. Control qualitative indicators of gearings at the stage of their computer analysis in the generalized area existence of gearing is accomplished by changing the shape and position of the local

area existence of gearing corresponding to one or another set of indicators, at which parameters of control are determined with considering the geometry of the vertex surface and the lateral surface of teeth of an initial link.

4. In process of computer analysis of gearings a rational part generalized area thes existence is defined as a set of isolines of qualitative indicators within which the most favorable complex of these indices should be sought based on general laws governing the influence of parameters of control on them under various criteria working capacity of gears.

References

[1] Gavrilenko V A and Bezrukov V I 1976 The geometrical calculation of gears, composed of involute-conical wheels *Herald of engineering* № 9 pp 40–44

[2] Lopatin B A and Tsukanov O N 1998 Methods of forming working surfaces of gears with small inter axial angle *Gears and transmissions* № 1 pp 38–49

[3] Lopatin B A, Bezrukov V I and Kazartsev D N 1989 Design of gears from involute-conical wheels without backlash *CAD of gears and of gearboxes: theses of reports of the scientific and technical seminar* (Izhevsk: Publishing house IzhSTU) pp 100–102

[4] Lopatin B A and Tsukanov O N 2005 *Cylinder-conical gears* (Chelyabinsk: Publishing house SUSU)

[5] Poluektov Ye A and Lopatin S D 2011 Modification of tooth profiles in cylindro-conical gear of internal gearing *Science SUSU: materials of scientific conference. Series of technical sciences* (Chelyabinsk: Publing center SUSU) pp 160–164

[6] Tsukanov O N, Lopatin D B and Poluektov Ye A 2009 Electromechanical drive for the mechanisms of the angular rotation of space objects *Herald of engineering* № 2 pp 14–16

[7] Tsukanov O N 2011 *Basics of synthesis not involute gearings in generalizing parameters* (Chelyabinsk: Publing center SUSU)

IOP Conf. Series: Materials Science and Engineering **468** (2018) 012039 doi:10.1088/1757-899X/468/1/012039

Hybrid algorithms for optimization and diagnostics of hydromechanical systems used in food production biotechnology

I G Blagoveshchenskii[1,2], M M Blagoveshchenskay[2], V D Sulimov[1] and P M Shkapov[1,3]

[1] Bauman Moscow State Technical University, Moscow, Russia
[2] Moscow State University of Food Production, Moscow, Russia
E-mail: spm@bmstu.ru

Abstract. Reliable computer modelling in hydromechanical system dynamics is very important for studies related to optimizing food manufacturing and biotechnology processes. Searching best parameters of a computational model of a system and its diagnosing are formulated as inverse problems. The current error function is built on base of comparison of corresponding parts of the computed eigenspectrum and the measured one. The incompleteness of the registered data and the presence of multiple eigenfrequencies lead to the fact that the error function to be minimized is nonconvex and not everywhere differentiable. Since the function has many local minima, it is necessary to use global optimization methods. Two novel hybrid global optimization algorithms for solving the inverse problems are described. The hybrid algorithms combine the up-to-date stochastic algorithm for scanning the search space and deterministic derivative-free procedures for local descent. Numerical results on benchmarking the hybrid algorithm are presented to illustrate the efficiency of the approach.

1. Introduction

Profitable and safe operation of the technological equipment in contemporary food production biotechnology is highly related to problems of ensuring the stability of hydromechanical systems used and their individual circuits as coupled dynamic systems. Relevance of these problems has recently been acquired in connection with the following factors: the growth of the energy intensity of the equipment; the supercritical regimes of processing food masses; the inclusion of new, autonomously processing and independently operating systems and assemblies into existing technological schemes and production infrastructure; the introduction of innovative technologies using equipment vibrations and flow pulsations; the formation of two-phase gas-liquid mixtures; the application of automated control systems, regulation and production management. For these reasons, one should take into account the following important characteristics of the hydromechanical system: spectra of natural frequencies, vibration modes, amplitude and phase characteristics, a stability margin and a quality of transient processes [1]. When solving the above issues, information technologies for investigations of hydromechanical systems used in food production are fully utilized [2, 3]. The product properties are sufficiently dependent on process variables and operating conditions. Usually the following two types of optimization problems are under consideration. Namely, designing the integrated biotechnology process (finding the process variables, the operating conditions, etc.) and updating parameters of the mathematical model of the hydrosystem (the inverse problem: searching the parameters of the nonlinear dynamic model giving the best fitness to a given set of measured data). On the basis of the

Content from this work may be used under the terms of the Creative Commons Attribution 3.0 licence. Any further distribution of this work must maintain attribution to the author(s) and the title of the work, journal citation and DOI.
Published under licence by IOP Publishing Ltd

same approach, the problems of diagnosing the operating hydrosystem or identifying the properties of the flow of food masses and detecting anomalies in equipment units with parameters directly determining the dynamic characteristics, as well as optimizing the spectrum of natural frequencies of the hydromechanical system can be solved.

Numerical analysis of the hydromechanical system requires adequate computational models representing the physical behaviour of the system. In actual practice in the field of developing dynamic models, numerical estimates generally do not sufficiently match the measured data obtained from vibration experiments. The up-to-date methodology on reducing the above discrepancies is to update the assumptions made for physical idealizations and parameters of the computational model until the agreement between numerical predictions and experimental results satisfy practical requirements. In particular, substantial effort has been made in the development of special-purpose procedures for updating parameters of computational models using data extracted from vibration tests. The most used approach to computational model updating is the inverse sensitivity technique where residuals between numerical and measured eigenvalues and/or mode shapes are minimized in order to improve user-selected parameters. In the gradient-based inverse sensitivity methods of the model updating it is critically important to define an appropriate error function between the computed and measured modal data. The estimated parameters can be obtained by minimizing the error function, which is generally a non-linear function of the updating parameters. Moreover, in defining the error function the correct pairing of computed modal data with the measured modal data is essential. The reason is that the pairing of computed and experimental modal data based on the sequential order of mode numbers may not be correct at all times. With the aim of establishing the correct modal correspondence the modal assurance criterion is generally used. One further important stage in model updating is the selection of the parameters to be updated. In particular, the computed natural frequencies and corresponding mode shapes of the computational model should be sensitive to the selected updating parameters. Quite often only the measured eigenvalues are used for model updating. This is because firstly the measured mode shapes contain more noise than the natural frequencies and secondly the mode shapes are not very sensitive to parameter changes. Nevertheless, the mode shapes are necessary to ensure that the modes under consideration are paired correctly. The error function in the minimization problem is the weighted sum of squares of the differences in the eigenvalues. Generally the problem is non-linear and requires an appropriate iterative solution, and in turn this requires the formulation and computation of the sensitivity matrix of the function to be minimized with respect to the updating parameters. The eigenvalue derivatives with respect to the updating parameters can be evaluated by the differentiation of the characteristic dynamic equation. Possible availability of repeated or very close eigenvalues one should take into account. As the minimization problem is usually ill-posed, then various regularization techniques should be applied [4].

It is common practice to reformulate the above problems to nonlinear programming problems subject to preset constraints. In the framework of the implicit methodology the inverse problem is formulated as an optimization problem on minimizing a criterion (error) function. Incompleteness of the experimental data and presence of repeated eigenvalues result in the error function being non-convex and non-differentiable. Since the error function has numerous local minima, it is necessary to use global optimization methods. Properties of deterministic global optimization algorithms are well-studied. It is common knowledge that the performance of deterministic algorithms essentially depends on the problem dimensionality. Alternatively, stochastic algorithms are capable of scanning efficiently the search space, but they are computationally expensive and their convergence properties are not guaranteed [5]. Developing hybrid global optimization algorithms, which combine both stochastic and deterministic techniques, is represented the promising approach in reactor equipment diagnostics and optimizing food production and bioengineering processes [6–8]. In the subsequent text certain of the subjects concerning structures of hybrid algorithms are discussed.

The plan of the remainder of this paper is as follows. The section 2 contains statement of the inverse eigenvalue problem under consideration. Section 3 provides brief description of hybrid global optimization algorithms. In section 4 some results of successful computational experiments on benchmarking the hybrid algorithm are presented to illustrate peculiarities of the proposed approach. Section 5 gives conclusions and discussion on further work.

2 Formulation of the problem

It is supposed that a set of performance index values associated with a computational model to be updated is defined by a set of control variables. Experimental spectral data registered by permanent instrumentation may be incomplete. So the goal is to determine vectors of control variables using only measured data on natural frequencies of the object. The standard approach is to set the inverse spectral problem and then to solve the corresponding least squares problem

$$\min_{x \in X \subset R^n} F(x),$$

where $F(x) = \sum_{i=1}^{N} w_i \left(\zeta_i(x) - \zeta_i^*\right)^2$; x, X – the vector of control variables and its feasible domain of the error function $F(x)$ respectively; the w_i stand for weighting factors that reflect the confidence level in the measurements; N is the number of eigenvalues under consideration; $\zeta_i(x)$ and ζ_i^* denote the eigenvalues that correspond to computed (from solutions of the direct problem) and to measured natural frequencies respectively.

As practical observations show, the error function in the considered regularized problem is often multiextremal. Therefore, it is necessary to turn to methods of global optimization. It is clear that if the measured spectral data exactly match to the computational model then the solution of the minimization problem will cause error function to take its global minimum value of zero. Let us suppose that there is a unique solution of the ill-posed inverse spectral problem and that this corresponds to the global minimum of the error function. However, the fact is that the theoretical question of the uniqueness of solutions of the problem may not be relevant to practical applications in which there is the additional complication of accuracy of experimental measurements. Furthermore, some complications may arise due to incompleteness of measured spectral data, influence of the two-phase interference on the flow dynamics, the presence of noise, etc. Within the scope of this work we take it as a convenient and reasonable assumption that global minimization of the error function in the above inverse problem will yield correct and sufficiently accurate model updating for objects under consideration.

In generalization of the statements of extremal problems the following global optimization problem is formulated: to find

$$f(x^*) = \min_{x \in X \subset R^n} f(x), \quad (1)$$

where

$$X = \{x \in R^n : a_j \le x_j \le b_j,\ j \in J\}, \quad (2)$$

$f(x)$ is the criterion function; $J = \{1, \dots, n\}$ is the bounded set of indexes; x^* is the global solution. The criterion function $f(x)$ of the problem (1)–(2) is supposed to be continuous Lipschitzian function. Also it is supposed that the real-valued function $f : R^n \to R$ is multiextremal, not everywhere differentiable, and there is a numerical procedure for calculating values of the function in points of the supposed area.

3. Hybrid algorithms

The problem at hand is to find global extrema of not everywhere differentiable criterion function. When the number of variables is large they use stochastic global optimization algorithms. Two currently available powerful stochastic algorithms PCA (Particle Collision Algorithm) and M-PCA (Multi-Particle Collision Algorithm) were introduced in [9] and [10] respectively. The PCA algorithm is inspired by the physics of a nuclear particle traveling inside a nuclear reactor (the scattering and the absorption phenomena). The original (canonical) PCA has some essential advantages in relation to well known heuristic global optimization algorithms such as the Genetic Algorithms, Simulated Annealing, Fast Simulated Annealing, etc. Specifically, the PCA does not require any additional parameters other than the number of iterations; the algorithm is extremely easy to implement and can be applied to both continuous and discrete optimization problems. The algorithm works as follows. First an initial configuration is chosen, then a modification of the old configuration into a new one is implemented. The qualities of the two configurations are compared. A decision then is made on whether the new configuration is acceptable. If it is, the current configuration acts as the old configuration for the next step. If it is not acceptable, the algorithm proceeds with a new change of the old configuration. The acceptance of current trial solution with certain probability makes it possible to avoid the convergence to local optima. In spite of its advantages over well-known meta-heuristics in solving test problems, practical application of the PCA is restricted because of solutions remain too expensive. As possible development, the local search procedure in the algorithm could be improved. The Multi-Particle Collision Algorithm is based on the canonical PCA, but a new characteristic is introduced: the use of several particles, instead of only one particle to act over the search space. So, the new outer loop for the particle control has been added to the basic global optimization algorithm. Thanks to use of several particles the M-PCA can better explore the variable space, avoiding convergence to a local solution. Coordination between the particles was achieved throw a blackboard strategy, where the current information is shared among all the particles in the process [10]. The M-PCA (similar to the PCA) also has only one parameter to be determined, the number of iterations. The algorithms are claimed to be robust for problems with presence of discontinuities or where the criterion function values are noisy. Usually stochastic algorithms yield too computationally expensive solutions, so this drawback restricts their applications. However, the PCA and the M-PCA are in their early stages. In spite of its advantages over Genetic Algorithms and Simulated Annealing in solving test problems, practical application of the PCA is restricted because of solutions remain too expensive. As possible development, the local search procedure in the algorithm could be improved. Developing hybrid algorithms that combine a robust stochastic algorithm for scanning the variable space with deterministic local search method is a promising way [11, 12].

A hybridization of the PCA and the deterministic (standard) Nelder-Mead simplex algorithm was introduced in [13]. The Nelder-Mead simplex method belongs to the class of derivative-free methods, so it is relatively insensitive to numerical noise and does not depend on convexity or smoothness of the criterion function. In practical cases, when highly accurate solutions of optimization problems may not be required due to inaccurate computational modelling, the method is an applicable tool. The hybrid algorithm (named NMPCA) was successfully applied to a nuclear reactor core design optimization problem. During processing two procedures are performing: a wide search in the solution space using the stochastic PCA and a local search in the promising areas with the deterministic simplex algorithm. The algorithm works well with irregular criterion functions and it can overcome kinks, local solutions, and discontinuities in functions being minimized. The local search procedure is performed iteratively until a certain number of fitness function evaluations being reached. It was established that the NMPCA performs better than the Metropolis PCA and such well known metaheuristics as Genetic Algorithms and Particle Swarm Optimization. Nevertheless, a standard implementation of the Nelder-Mead Simplex algorithm during the local search can be expensive for problems with a large number of variables or with costly criterion functions. It is pertinent to note that the convergence theory for Nelder-Mead simplex method is rather far from completion; so the method can fail to converge or converge to non-stationary points [14]. By this is meant that the question of reliability of the algorithm as a whole is an open question. Another hybrid algorithm HJPCA that integrates the PCA based on the Metropolis algorithm and the popular deterministic Hooke–Jeeves method for the local search was proposed in [15]. The HJPCA implements a wide search in the space of variables using the stochastic algorithm and then perform the local search by the derivative-free method. The Hooke–Jeeves derivative-free algorithm performs two types of search: an exploratory search and a pattern search. A simple search strategy and small volume of memory required are the advantages of the method. In implementation of the hybrid HJPCA the local search procedure is performed until a predetermined stopping criterion (a sufficiently small step size) is reached.

Two novel hybrid global optimization algorithms combining the Metropolis-based stochastic algorithm M-PCA with the deterministic derivative-free methods for local search are presented in [16, 17]. The first local search method implements a variant of the modified Nelder-Mead Simplex algorithm [18]. The second approach to the local search is the Hooke–Jeeves method. The resulting hybrid algorithms M-PCAMNM and M-PCAHJ belong to a class of algorithms which do not require derivatives. Moreover, the hybrid algorithm M-PCAHJ may be considered as a modification of the earlier algorithm HJPCA. According to published data on benchmarking, the HJPCA outperforms two state-of-the-art algorithms C-GRASP and GLOBAL [15]. As the M-PCA is more efficient than canonical PCA, so the more recent hybrid algorithm M-PCAHJ outperforms the HJPCA.
A new derivative-free stochastic Nelder-Mead (SNM) Simplex algorithm for continuous simulation optimization is proposed in [19]. It is proven that the SNM is a globally convergent algorithm. The essential drawback of the SNM is that its performance can deteriorate as the problem dimensionality is getting large. It is proven that the expansion and contraction steps of the standard Nelder-Mead Simplex algorithm possess a descent property when the criterion function is uniformly convex [20]. This property offers some new insights on the behaviour of the Nelder-Mead algorithm. The adaptive Nelder-Mead Simplex (ANMS) method is proposed whose expansion, contraction and shrink parameters depend on the dimension of the optimization problem. The presented numerical results have shown that the ANMS method outperforms the standard Nelder-Mead method for high dimensional problems.
Recently the substantial progress of parallel computing systems and programming techniques has stimulated the research on parallel global optimization algorithms [21]. The main reason is that solving complex global optimization problems involves intensive computations and evaluation of criterion functions may be rather expensive. Certain of the parallel schemes implementing the Nelder-Mead simplex method were presented in [22]. The advantage of the approach is that the parallel simplex algorithm is generic and can be applied to any optimization problem to which the Nelder-Mead simplex is applicable. An implementation of the Nelder-Mead simplex method in parallel using a distributed memory is described in [23]. It is pointed out that in simulations the proposed implementation exhibits large speedups and is scalable to large problem sizes. Taking in consideration that parallel versions of the M-PCA were discussed in [10], it may be concluded that developing a parallel hybrid algorithm on base the M-PCA and parallel schemes of the Nelder-Mead simplex method is a promising way.

4. Numerical example

The benchmark is n -dimensional Levy–Montalvo 1 problem (LM1) [24]

$$f(x) = \frac{\pi}{n}\Big\{k\sin^2(\pi y_1) +$$

$$+\sum_{i=1}^{n-1}\Big[(y_i - A)^2\big(1 + k\sin^2(\pi y_{i+1})\big)\Big] + (y_n - A)^2\Big\},$$

where $y_i = 1 + 0.25(x_i - 1)$, $-10 \le x_i \le 10$, $i = 1,2,\ldots,n$, $k = 10$, $A = 1$. The global minimum of LM1 is known to be $f\left(x^*\right) = 0$ with $x_i^* = 1$, $i = 1,2,\ldots,n$.
The hybrid global optimization algorithm M-PCAHJ is used. It is supposed $n = 15$. The stopping criterion is $f(x) < \varepsilon = 10^{-6}$. When current scanning the search space is done the set of initial trial points for the local search is determined. The finishing iteration history of the problem variables x_i , $i = 1,15$, versus the number of iterations N_{iter} is depicted in Figures 1–3. The corresponding history for the criterion function $f(x)$ is demonstrated in Figure 4. After 20 iteration it have been obtained: $f(x^{20}) \approx 0{,}6232 \cdot 10^{-6}$.

IOP Conf. Series: Materials Science and Engineering **468** (2018) 012039 doi:10.1088/1757-899X/468/1/012039

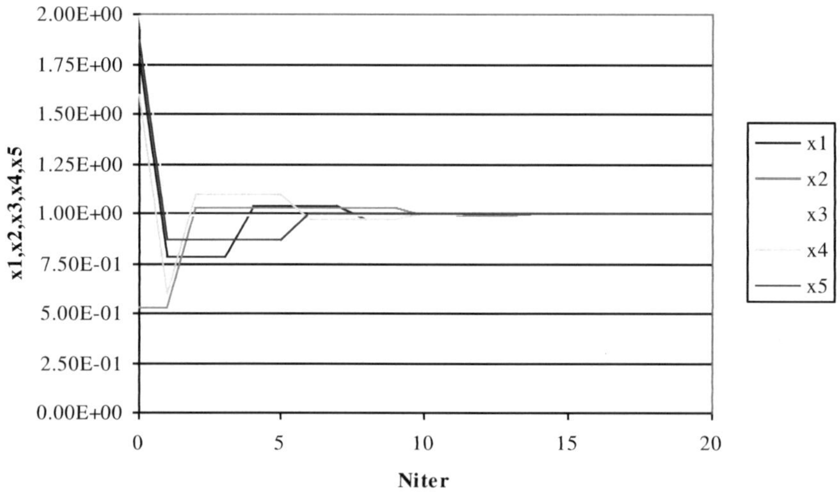

Figure 1. Number of final iterations N_{iter} of the M-PCAHJ vs. control variables x_i, $i = \overline{1,5}$

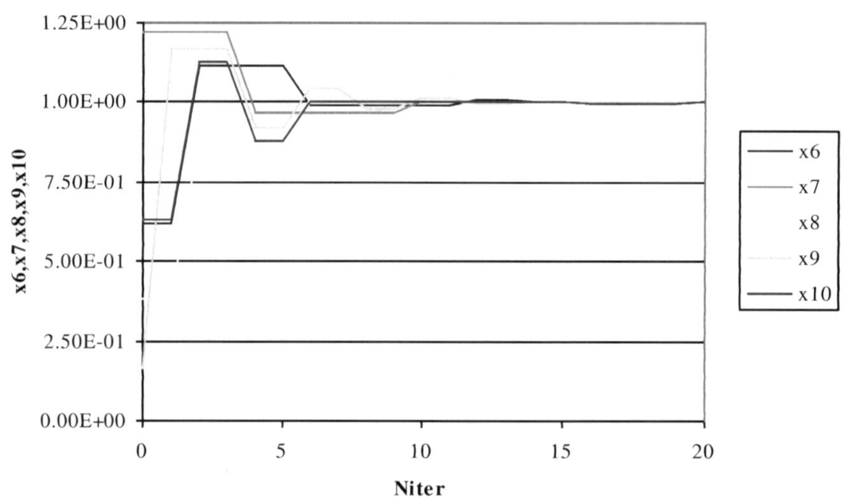

Figure 2. Number of final iterations N_{iter} of the M-PCAHJ vs. control variables x_i, $i = \overline{6,10}$

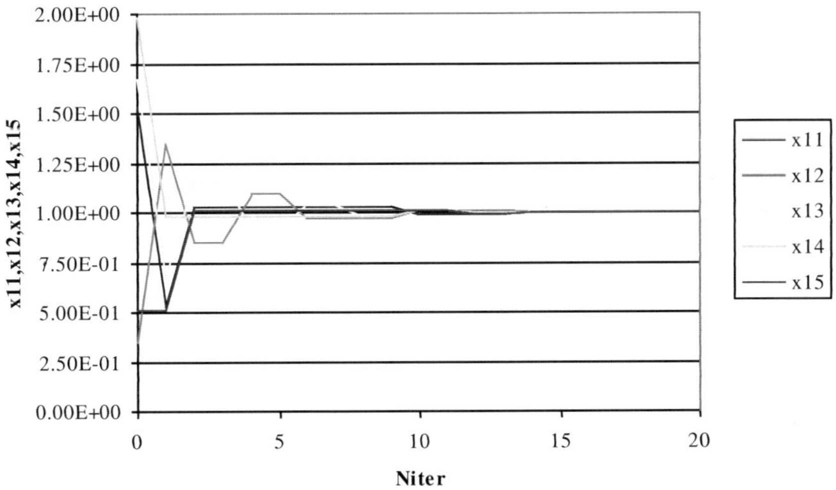

Figure 3. Number of final iterations N_{iter} of the M-PCAHJ vs. control variables x_i, $i = \overline{11,15}$

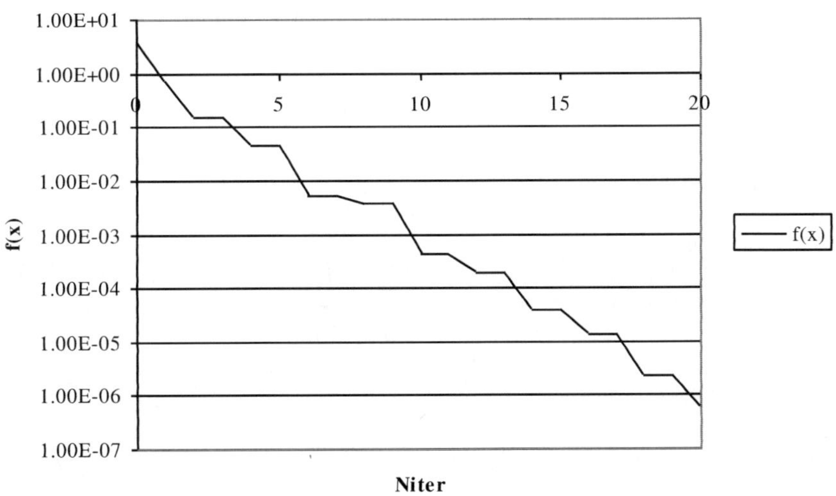

Figure 4. Number of final iterations N_{iter} of the M-PCAHJ vs. the error function $f(x)$

5. Conclusions

Developing hybrid global optimization algorithms that combine stochastic algorithms for scanning the variable space with deterministic local search methods is a promising way. Novel hybrid algorithms M-PCAMNM and M-PCAHJ that integrate the Metropolis-based M-PCA and derivative-free methods for the local search are described. Computational experiments show the principal applicability of the algorithms for solving inverse problems for hydromechanical systems in the field of food production and bioengineering. Some results on benchmarking the M-PCAHJ are presented. The future work will be devoted to increasing the computational efficiency of the tools with applications to inverse problems.

References

[1] Kostin A M, Yablokov A Ye, Blagoveshchenskaya M M 2015 Perspektivnyye metody tekhnicheskoy diagnostiki oborudovaniya pishchevykh proizvodstv *Voprosy Nauki [Questions of Science]* **7** pp 4-10 (in Russian).

[2] Blagoveshchenskaya M M, Sulimov V D, Shkapov P M 2010 Metodologiya razrabotki osnov modelirovaniya i diagnostiki gidromekhanicheskikh sistem pishchevykh proizvodstv po ikh dinamicheskim kharakteristikam *Vysokiye intellektual'nyye tekhnologii i innovatsii v obrazovanii i nauke: Materialy XVII Mezhdunarodnoy nauchno-metod. Konf. 11-12 Fevralya 2010 g., S.-Peterburg.(SPb.: Izd-vo Politekhn. Univer.)* **2** pp 95-98 (in Russian).

[3] Blagoveshchenskaya M M, Zlobin L A 2005 *Informatsionnyye tekhnologii sistem upravleniya tekhnologicheskimi protsessami* (M.: Vyssh. Shk.) (in Russian).

[4] Kirsch A 2011 *An Introduction to the Mathematical Theory of Inverse Problems 2nd edition* (Springer, New York).

[5] Žilinskas A, Zhigljavsky A 2016 Stochastic global optimization: A review on the occasion of 25 years of Informatica *Informatika* **27** (2) pp 229-256.

[6] Kinelev V G, Shkapov P M, Sulimov V D 2003 Application of global optimization to VVER-1000 reactor diagnostics *Progress in Nuclear Energy* **43** (1-4) pp 51-56.

[7] Sulimov V D, Shkapov P M 2015 Hybrid methods of computer diagnosis of two-phase flow in the circulation loop *Mathematical Modelling and Numerical Methods*, **3** pp 68-88 (in Russian).

[8] Tumuluru J S, McCulloch R 2016 Application of hybrid Genetic Algorithm routine in optimizing food and bioengineering processes *Foods*, **5** (4) pp 76-87.

[9] Sacco W F, de Oliveira C R E 2005 A new stochastic optimization algorithm based on particle collisions *Proceedings of the 2005 ANS Annual Meeting. Transactions of the American Nuclear Society*, **92** pp 657-659.

[10] Luz E F P, Becceneri J C, de Campos Velho H F 2008 A new multi-particle collision algorithm for optimization in a high performance environment *J. of Computational Interdisciplinary Sciences*, **1** pp 3-10.

[11] Voglis C, Parsopoulos K E, Papageorgiou D G, Lagaris I E, Vrahatis M N 2012 MEMPSODE: A global optimization software based on hybridization of population-based algorithms and local searches *Computer Physics Communications*, **183** (2) pp 1139-1154.

[12] Lara P C S, Portugal R, Lavor C 2014 A new hybrid classical-quantum algorithm for continuous global optimization problems *J. of Global Optimization*, **60** pp 317-331.

[13] Sacco W F, Filho H A, Henderson N, de Oliveira C R E A 2008 Metropolis algorithm combined with Nelder–Mead Simplex applied to nuclear reactor core design *Annals of Nuclear Energy*, **35** pp 861-867.

[14] McKinnon K I M 1998 Convergence of the Nelder-Mead simplex method to a non-stationary point *SIAM J. Optimization*, **9** pp 148–158.

[15] Rios-Coelho A C, Sacco W F, Henderson N A 2010 Metropolis algorithm combined with Hooke-Jeeves local search method applied to global optimization *Applied Mathematics and Computation* **217** (2) pp 843-85.

[16] Sulimov V D, Shkapov P M, Sulimov A V 2017 Jacobi stability and computational diagnostics of dynamical systems using hybrid algorithms *Fundamental and Applied Problems of Engineering and Technology*, **4-2 (324)** pp 61-71 (in Russian).

[17] Sulimov V D, Shkapov P M, Nosachev S K 2014 Hooke–Jeeves Method-based local search in a hybrid global optimization algorithm *Science&Education*, **6** pp 107-123 (in Russian).

[18] Xiao H F, Duan J A 2014 Multi-direction-based Nelder–Mead method *Optimization: A Journal of Mathematical Programming and Operations Research*, **63** pp 1005-1026.

[19] Chang K H 2010 Stochastic Nelder-Mead simplex method – a new globally convergent direct search method for simulation optimization *SIAM J. Optimization* **9** pp 148–158.

[20] Gao F, Han L 2012 Implementing the Nelder–Mead simplex algorithm with adaptive parameters *Computational Optimization with Applications* **51** (1) pp 259–277.

[21] Voglis C, Hadjidoukas P E, Papageorgiou D G, Lagaris I E 2013 A parallel hybrid optimization algorithm for fitting interatomic potentials *Applied Soft Computing* **13** pp 4461-4492.

[22] Lee D, Wiswall M 2007 A parallel implementation of the simplex function minimization routine *Computational Economics* **30** (2) pp 171-187.

[23] Klein K, Neira J 2014 Nelder–Mead simplex optimization routine for large-scale problems: a distributed memory approach *Computational Economics*, **43** (4) pp 447-461.

[24] Montaz Ali M, Khompatraporn C, Zabinsky Z B 2005 A numerical evaluation of several stochastic algorithms on selected continuous global optimization test problems *J. of Global Optimization*, **31** (4) pp 635-672.

Jacobi stability and updating parameters of dynamical systems using hybrid algorithms

V D Sulimov[1], **P M Shkapov**[1] **and A V Sulimov**[2]

[1] Bauman Moscow State Technical University, Moscow, Russia
E-mail:spm@bmstu.ru

Abstract. Continuously time evolving dynamical systems are widely used as effective theoretical tools for the mathematical modelling of various processes. The global stability of dynamical systems is described by the theory of Lyapunov stability. An alternative approach is represented by the local (Jacobi) stability analysis. Consideration is being given to Jacobi stability of dynamical systems in the framework of the Kosambi–Cartan–Chern theory. In this way it is common practice to describe the time evolution of a dynamical system in geometric terms, by presenting it as a geodesic in the Finsler space. Geometric invariants of the system under investigation are obtained. Eigenvalues of the second invariant determine the Jacobi stability of the system. The inverse problem on updating parameters of the system on base of given eigenvalues of the invariant is stated. Criterion functions are supposed being continuous, Lipschitzian, multiextremal, not necessary everywhere differentiable. Global solutions are searched for by use of novel hybrid algorithms that combine the up-to-date stochastic Quasi Reflective Multi-Particle Collision Algorithm for scanning a search space and deterministic methods for local descend. Two-parametric smoothing approximations of criterion functions are inserted for local searching in the first algorithm. The second algorithm implements the local search procedure using numerical approximations of space-filling curves to reduce the original multi-dimensional problem to an univariate one satisfying the Hölder condition. A numerical example of updating parameters and Jacobi stability analysis of the Lorenz system is presented.

1. Introduction

Continuously time evolving dynamical systems represent powerful theoretical tools for the mathematical modelling of the evolution of natural phenomena. They are widely used in practice, and they provide extremely useful descriptions of various processes. In most of the applications the usefulness of the dynamical systems is determined mainly by their predictive capabilities. It is well known that the predictive potential of the dynamical systems depends on the stability of their solutions. In realistic investigations some uncertainties in the measured initial conditions in physical, engineering or biological systems do always appear, so that they must be taken into account when formulating the mathematical equations of the processes under consideration. In practical applications, a dynamical system is required to provide some essential information, not only on the time evolution of the system, but also on the possible deviations of the trajectories of the dynamical system with respect to a given reference trajectory. When studying a dynamical system it is necessary to obtain both the local description of the stability, and the global description of the late-time deviations of the trajectories. The global stability of the solutions of the dynamical systems is described by the well established theory of the Lyapunov stability. In this approach to the problem of stability analysis the fundamental mathematical quantities of interest are the Lyapunov exponents. The Lyapunov exponents measure the exponential deviations of the trajectories of a dynamical system from a given reference trajectory [1]. An alternative approach to the study of the dynamical systems is represented by the Kosambi–Cartan–Chern

Content from this work may be used under the terms of the Creative Commons Attribution 3.0 licence. Any further distribution of this work must maintain attribution to the author(s) and the title of the work, journal citation and DOI.
Published under licence by IOP Publishing Ltd

(KCC) theory [2, 3]. Principal formulations of the KCC theory are based on a Finsler geometry and a theory of a tangent bundle [4–6]. A fundamental idea is to consider that there is a one to one correspondence between a second-order dynamical system and geodesic equations in an associated Finsler space (a recent survey of the theory is given in [7]). The KCC theory is a differential geometric theory of the variational equations for the deviations of the whole trajectory to nearby ones. In this geometric description of dynamical systems one associates a nonlinear connection, and a Berwald type connection to the differential system, and five geometrical invariants are obtained. The second invariant (also called the curvature deviation tensor) gives the Jacobi stability of the given dynamical system [8, 9]. Geometric methods can be employed successfully to formulate a general geometric theory of dynamical systems, or, more particularly, of systems of second order ordinary differential equations. The KCC theory has been applied to the dynamical systems in such fields of research as biology, geophysics, and cosmology [10–12]. The Lorenz system is one more example of successful application of the approach [13, 14].
Further the following inverse problem for a dynamical system depending on free parameters is stated. In the framework of the KCC theory, it is suggested that the eigenvalues of the curvature deviation tensor are given approximately (from indirect measuring and computer processing). The objective is to reconstruct the free parameters of the system from given data and to draw the conclusion on the Jacobi stability of the system. The optimization approach has been applied to solve the regularized inverse eigenvalue problem. Generally, criterion functions of the problem are supposed being continuous, Lipschitzian, multiextremal, not necessary everywhere differentiable. Consequently, global nondifferentiable optimization methods should be implemented. Two novel hybrid optimization algorithms combining the up-to-date stochastic algorithm (for scanning the search space) and deterministic methods for the local descent are applied. Two-parametric smoothing approximations of the criterion function are inserted during the local search in the first algorithm. The second algorithm implements the local search procedure by use of numerical approximations of space-filling curves.
The present paper is organized as follows. The basics of the KCC theory to be used in the sequel are sketched in Sec. 2. Some results on application of the theory to dynamical systems are given [2, 3]. The Lorenz system is reformulated as a set of two second-order differential equations; then the curvature deviation tensor and its eigenvalues are obtained explicitly [14]. Sec. 3 contains the statement of the regularized inverse problem. Brief description of two novel hybrid global optimization algorithms is presented in Sec. 4. Numerical example of solving the inverse problem for the Lorenz system is given in Sec. 5. The paper will be concluded in Sec. 6.

2. Jacobi stability of dynamical systems

In the following it is assumed that the dynamical variables x^i, $i=1,2,\dots,n$, are defined on a real, smooth n-dimensional manifold M; the tangent bundle of M is denoted by TM [3]. Sometimes M is thought as R^n, $M=\mathrm{R}^n$, so that $TM=T\mathrm{R}^n=\mathrm{R}^n$. A subset Ω of the Euclidian $(2n+1)$-dimensional space $\mathrm{R}^n\times\mathrm{R}^n\times\mathrm{R}^1$ is considered. On Ω it can be introduced a $(2n+1)$-dimensional coordinate system (x^i, y^i, t), $i=1,2,\dots,n$, where $(y^i)=(y^1,y^2,\dots,y^n)$, and t stands for the usual time coordinate. The coordinates y^i are defined as

$$y^i=\left(\frac{dx^1}{dt},\ \frac{dx^2}{dt},\ \dots,\ \frac{dx^n}{dt}\right).$$

In many cases of interest the equations of motion of a dynamical system can be derived from a Lagrangian function $L:TM\to\mathrm{R}$ using the Euler–Lagrange equations

$$\frac{d}{dt}\frac{\partial L}{\partial y^i}-\frac{\partial L}{\partial x^i}=F_i,\quad i=1,2,\dots,n, \tag{1}$$

where in the case of mechanical systems F_i, $i=1,2,\dots,n$, represents the external force. If the Lagrangian L is regular, one can show that the Euler–Lagrange equations Eq. (1) are equivalent to a system of second-order ordinary differential equations [3]

$$\frac{d^2x^i}{dt^2} + 2G^i(x^i, y^i, t) = 0,\ i \in \{1, 2, \dots, n\};$$

where each function $G^i(x^i, y^i, t)$ is C^∞ in a neighborhood of some initial conditions (x_0, y_0, t_0) in Ω; functions $G^i(x^i, y^i, t)$, $i \in \{1, 2, \dots, n\}$, are defined locally on TM.
In order to study the geometrical properties associated to the dynamical system defined by the previous equations, on the base manifold M one introduces first the nonlinear connection N, with coefficients $N^i{}_j$, defined as [3]

$$N^i{}_j = \frac{\partial G^i}{\partial y^j}.$$

By using the KCC-covariant derivative, it is possible to obtain the deviation equation in the covariant form as

$$\frac{D^2\xi^i}{\partial t^2} = P^i{}_j \xi^j,$$

where ξ^j are the components of a contravariant vector field defined along the path $x^i(t)$; then it has been denoted

$$P^i{}_j = -2\frac{\partial G^i}{\partial x^j} - 2G^l G^i{}_{jl} + y^l \frac{\partial N^i{}_j}{\partial x^l} + N^i{}_l N^l{}_j + \frac{\partial N^i{}_j}{\partial t};$$

where it has been introduced the tensor $G^i{}_{jl}$, defined as [2]

$$G^i{}_{jl} = \frac{\partial N^i{}_j}{\partial y^l},$$

and which is called the Berwald type connection. The tensor $P^i{}_j$ is the second KCC-invariant, and it may be called alternatively as the deviation curvature tensor. The deviation equation represents the well known Jacobi equation. In either Riemann or Finsler geometry, when the system of dynamic equations describes the geodesic motion, then the current equation is the Jacobi field equation corresponding to the given geometry.
A scalar quantity which can be constructed from $P^i{}_j$ is the trace P of the curvature deviation tensor, which can be obtained from the relation

$$P = P^i{}_j = -2\frac{\partial G^i}{\partial x^i} - 2G^l G^i{}_{il} + y^l \frac{\partial N^i{}_i}{\partial x^l} + N^i{}_l N^l{}_i + \frac{\partial N^i{}_i}{\partial t}.$$

Further, the application of the KCC theory to the set of differential equations connected with the fluid mechanics is considered. In particular, the Lorenz system of three nonlinear ordinary differential equations is given by [14]

$$\frac{1}{\sigma}\frac{dX}{dt} = -X + Y, \tag{2}$$

$$\frac{dY}{dt} = -XZ + \rho X - Y, \tag{3}$$

$$\frac{dZ}{dt} = XY - \beta Z, \tag{4}$$

where σ, ρ, β stand for some free parameters; $X, Y,$ and Z are functions of time alone. From a mathematical point of view the set of equations (2)–(4) represents an approximation to a system of partial differential equations describing finite amplitude convection in a fluid layer heated from below. From a physical point of view the parameters $\sigma, \rho,$ and β can be interpreted as the Prandtl number, the normalized Rayleigh number, and the wave length number, respectively. Although the Lorenz set of equations is deterministic, its solution shows chaotic behavior for $\rho>\sigma(\sigma+\beta+3)/(\sigma-\beta-1)$ and $\sigma>\beta+1$ [14].

From Eq. (2), it takes place

$$Y = X + \frac{1}{\sigma}\frac{dX}{dt}. \tag{5}$$

By substituting Eq. (5) into Eq. (3), one obtains

$$\ddot{X} + (1+\sigma)\dot{X} + \sigma XZ + \sigma(1-\rho)X = 0 \tag{6}$$

By taking the derivative of Eq. (4) with respect to the time we find

$$\ddot{Z} = \dot{X}Y + X\dot{Y} - \beta\dot{Z} = \dot{X}\left(X + \frac{1}{\sigma}\dot{X}\right) + X\left(\dot{X} + \frac{1}{\sigma}\ddot{X}\right) - \beta\left[X\left(X + \frac{1}{\sigma}\dot{X}\right) - \beta Z\right]. \tag{7}$$

By use of Eq. (6) and Eq. (7) it could be obtained

$$\ddot{Z} + \left(\frac{1+\sigma}{\sigma} + \frac{\beta}{\sigma} - 2\right)X\dot{X} - \frac{1}{\sigma}\dot{X}^2 + (1-\rho+\beta)X^2 + X^2Z - \beta^2 Z = 0.$$

After changing the notation so that

$$X = X^1,\quad \dot{X} = Y^1,\quad Z = X^2,\quad \dot{Z} = Y^2,\quad Y = X^3,$$

the Lorenz system is equivalent with the following system of second-order equations

$$\frac{d^2X^1}{dt^2} + (1+\sigma)Y^1 + \sigma X^1X^2 + \sigma(1-\rho)X^1 = 0, \tag{8}$$

$$\frac{d^2X^2}{dt^2} + \left(\frac{1+\sigma}{\sigma} + \frac{\beta}{\sigma} - 2\right)X^1Y^1 - \frac{1}{\sigma}\left(Y^1\right)^2 + (1-\rho+\beta)\left(X^1\right)^2 + \left(X^1\right)^2X^2 - \beta^2X^2 = 0. \tag{9}$$

Once X^1 and X^2 are known, the variable X^3 can be obtained as

$$X^3 = X^1 + \frac{1}{\sigma}Y^1.$$

So, the Lorenz system can be represented by the set of two differential equation as follows

$$\frac{d^2X^i}{dt^2} + 2G^i(X^i, Y^i) = 0,\quad i = 1, 2.$$

From Eq. (8) and Eq. (9) one can get

$$G^1(X^1, X^2, Y^1) = \frac{1}{2}[(1+\sigma)Y^1 + \sigma X^1X^2 + \sigma(1-\rho)X^1],$$

$$G^2(X^1, X^2, Y^1) = \frac{1}{2}\left[\left(\frac{1+\sigma}{\sigma} + \frac{\beta}{\sigma} - 2\right)X^1Y^1 - \frac{1}{\sigma}\left(Y^1\right)^2 + (1-\rho+\beta)\left(X^1\right)^2 + \left(X^1\right)^2X^2 - \beta^2X^2\right].$$

Now the components of the nonlinear connection and the Berwald type connection can be found; after that the components of the curvature deviation tensor of the Lorenz system can be defined as follows [14]

$$P_1^1 = -\sigma X^2 - \sigma(1-\rho) + \frac{1}{4}(1+\sigma)^2,$$

$$P_2^1 = -\sigma X^1,$$

$$P_1^2 = \left(1 - \frac{\beta}{2\sigma}\right)Y^1 + \left[\frac{1-\sigma^2 - 7\beta\sigma + \beta + 4(\rho-1)\sigma}{4\sigma}\right]X^1 - X^1X^2,$$

$$P_2^2 = -\left(X^1\right)^2 + \beta^2.$$

In the context of the Jacoby stability analysis the following three equilibrium points of the Lorenz system are defined [14]: $S_0(0,0,0)$, if $\rho\le 1$, $S_+\left[\sqrt{\beta(\rho-1)}, \sqrt{\beta(\rho-1)}, \rho-1\right]$ and $S_-\left[-\sqrt{\beta(\rho-1)}, -\sqrt{\beta(\rho-1)}, \rho-1\right]$, $\rho>1$, respectively. Further the stability of the point S_+, $\rho>1$, is under consideration. Consequently, the components of the curvature deviation tensor can be represented in the following form

$$P_1^1(S_+) = \frac{1}{4}(1+\sigma)^2,$$

$$P_2^1(S_+) = -\sigma\sqrt{\beta(\rho-1)},$$

$$P_1^2(S_+) = \frac{\sqrt{\beta(\rho-1)}[(-7\sigma+1)\beta-\sigma^2+1)]}{4\sigma},$$

$$P_2^2(S_+) = \beta^2 - \beta(\rho-1).$$

Analytic expressions for the eigenvalues of the curvature deviation tensor in the considered equilibrium point are as follows [14]

$$\lambda_+(S_+) = \frac{1}{2}\left\{\sqrt{(\beta(\beta-\rho+1)+\frac{1}{4}(\sigma+1)^2+\beta(\rho-1)[\beta(7\sigma-1)+\sigma^2-1]} + \right.$$

$$\left. +\beta(\beta-\rho+1+\frac{1}{4}(\sigma+1)^2\right\},$$

$$\lambda_-(S_+) = \frac{1}{2}\left\{-\sqrt{(\beta(\beta-\rho+1)+\frac{1}{4}(\sigma+1)^2+\beta(\rho-1)[\beta(7\sigma-1)+\sigma^2-1]} + \right.$$

$$\left. +\beta(\beta-\rho+1+\frac{1}{4}(\sigma+1)^2\right\}.$$

Convenient results for two other equilibrium points can be obtained in an analogous manner. Now one can formulate the following theorem, giving the Jacoby properties of the equilibrium points of the Lorenz system.

Theorem [14]. *(a) The equilibrium point* $S_0(0,0,0)$ *of the Lorenz system is Jacobi unstable.*

(b) If the free parameters β, $\rho > 1$ *and* σ *of the Lorenz system satisfy simultaneously the conditions*

$$\beta(\beta-\rho+1)+\frac{1}{4}(\sigma+1)^2 < 0$$

and

$$\frac{1}{4}\beta\{\beta[-7\rho\sigma+\rho+\sigma(\sigma+9)]-2\sigma(\rho-1)(\sigma+1)\} > 0,$$

respectively, then the equilibrium points $S_+\left[\sqrt{\beta(\rho-1)},\ \rho-1\right]$ *and* $S_-\left[-\sqrt{\beta(\rho-1)},\ \rho-1\right]$ *of the Lorenz system are Jacobi stable, and Jacobi unstable otherwise.*

3. Inverse problems

From now, a new system of notation will be used. It is considered an inverse problem governed to the operator equation in the framework of the structured mathematical model

$$Ax = y,\quad x\in X,\quad y\in Y,$$

where X, Y are Hilbert spaces; A stands for the compact linear operator, acting from X into Y. The right part of the confused operator equation represents given data y^δ, that are known approximately from the experiment. It is supposed that the input data error δ is established and take place the inequality $\left\| y^\delta - y \right\| \le \delta$. Stable approximate solutions are needed when the data y^δ are given approximately. It should be noted that in many applications inverse problems are ill-posed. Further the approach on base of the Tikhonov regularization method is implemented [15, 16].

The approximate solution of the considered inverse problem is connected with the search of the Tikhonov's functional minimum

$$x_\alpha^\delta = \arg\min_{x\in X} J_\alpha(x),\quad \alpha>0;$$

where x_α^δ is the regularized solution of the equation $Ax = y^\delta$ with the regularization parameter α; moreover, the functional to be minimized is defined as follows

$$J_\alpha(x) = \left\| Ax - y^\delta \right\|_Y^2 + \alpha \| x \|_X^2,$$

where $\left\| Ax - y^\delta \right\|_Y^2$ is the error functional (the squared norm in the space Y); $\alpha\| x \|_X^2$ is the stabilizing functional.

The method to be implemented is called as the converged regularization method if the following condition takes place

$$\sup_{\left\| y^\delta - y \right\| \le \delta} \left\| R_\alpha y^\delta - x \right\| \to 0 \text{ when } \delta \to 0;$$

here a family of linear restricted operators $R_\alpha : Y \to X$ with the regularization parameter $\alpha > 0$ such that

$$\lim_{\alpha \to 0} R_\alpha Ax = x \quad \forall x \in X .$$

are introduced. When using data $x \in X_1 \subset X$, given *a priori* the solution error in worth case is determined in the form [15]

$$\Delta(\delta, X_1, R_\alpha) = \sup\{ \left\| R_\alpha y^\delta - x \right\| : x \in X_1, y \in Y, \left\| y^\delta - y \right\| \le \delta\} .$$

The inverse problem for eigenvalues of the deviation curvature tensor of the dynamical system is stated as follows. For eigenvalues compared pairwise it is possible to construct the following numerical set of the error criteria

$$f_i(x) = \left| \lambda_i(x) - \lambda_i^*(x) \right|, x \in X \subset \mathrm{R}^n, i \in J,$$

where $\lambda_i(x), \lambda_i^*(x)$ are eigenvalues from the current and the given spectra, respectively; x is the vector of the control variables; X stands for the feasible area; n is the problem dimension; $J = \{1, \dots, n\}$; R^n is the n-dimensional real linear space. The inverse problem is formulated as follows: to find the vector of the control variables $x \in X$ such that minimizes the maximal value of the error criterion

$$\min_{x \in X \subset \mathrm{R}^n} \max_{i \in I} \{ f_i(x) \}. \tag{P}$$

The solution of the problem (P) is represented by the admissible vector $x^* = \left(x_1^*, \dots, x_n^* \right)^T$, such that the scalar error function $f(x) = \max \{ f_1(x), \dots, f_N(x) \}$ takes the minimal value. Further the regularized problem $(\mathrm{P})_\alpha$ with the multiextremal not everywhere differentiable criterion function $f(x)$ and the regularization parameter $\alpha > 0$ is considered.

In generalization of the statements of extremal problems the following global optimization problem is formulated: to find

$$f\left(x^* \right) = \min_{x \in X \subset \mathrm{R}^n} f(x), \tag{10}$$

where

$$X = \{ x \in D : g_i(x) \le 0, i \in I \}, \tag{11}$$

$$D = \left\{ x \in \mathrm{R}^n : a_j \le x_j \le b_j, j \in J \right\}, \tag{12}$$

$f(x)$ is the criterion function; $g_i(x)$ stands for the constraint function, $i \in I$; $I = \{1, \dots, m\}$ is the bounded set of indexes; m is the number of constraints; D is the search area; x^* is the global solution. The functions $f(x)$, $g_i(x)$, $i \in I$, of the problem (10)–(12) are supposed to be continuous Lipschitzian functions. Also it is supposed that the real-valued function $f:\mathrm{R}^n \to \mathrm{R}$ is multiextremal, not everywhere differentiable, and there is a numerical procedure for calculating values of the function in points of the supposed area.

4. Hybrid algorithms

A review of popular global optimization methods is given in [17]. The currently available stochastic Multi-Particle Collision Algorithm (M-PCA) [18] has some essential advantages in relation to well known global optimization algorithms such as the Genetic Algorithms, Simulated Annealing, Fast Simulated Annealing, etc. Specifically, the M-PCA does not require any additional parameters other than the number of iterations; the algorithm is extremely easy to implement and can be applied to both continuous and discrete optimization problems. The M-PCA performs using the analogy with nuclear particle collision reactions, in particular scattering and absorption. So, a particle that hits a high-fitness "nucleus" would be "absorbed" and would explore the boundaries. Otherwise, a particle that hits a low-fitness region would be scattered to another region. This reasoning makes it possible to simulate the exploration of the search space and the exploitation of the most promising areas of the fitness landscape throw successive scattering and absorption collision events. The original M-PCA works as follows. First an initial configuration is chosen, then a modification of the old configuration into a new one is implemented. The qualities of the two configurations are compared. A decision then is made on whether the new configuration is acceptable. If it is, the current configuration acts as the old configuration for the next step. If it is not acceptable, the algorithm proceeds with a new change of the old configuration. It is pertinent to note that acceptance of current trial solution with certain probability may avoid the convergence to local optima.

However, the M-PCA is in its early stages. In spite of its advantages over Genetic Algorithms and Simulated Annealing in solving benchmark problems, practical application of the M-PCA is restricted because of solutions remain too expensive. As possible development, the local search procedure in the algorithm could be improved. It seems promising to use gradient methods for local minimization of the error function. But in so doing the problem of non-differentiability of the function should be taken into account. A hybridization of the stochastic M-PCA and deterministic methods was introduced in [19]. During processing two procedures are performing: a wide search in the solution space using the M-PCA and a local search in the promising areas with the deterministic techniques. The local search procedure is performed iteratively until a certain number of criterion function evaluations being reached.

New versions for the M-PCA meta-heuristics are presented in [20]. In order to provide more effective trial solutions, the concept of opposition and reflection is introduced to improve the capacity of the search space for the M-PCA. One of the new versions of the M-PCA is the Quasi Reflective M-PCA (QRM-PCA). This new version of the global optimization algorithm makes it possible to realize a more intensive exploration of the search space. Two novel hybrid global optimization algorithms integrating the stochastic QRM-PCA and deterministic methods for local minimization are presented. The first hybrid algorithm (named here QRM-PCALMSI) combines the QRM-PCA and deterministic gradient techniques [21]. In this new version of the hybrid algorithm the local search procedure is a standard deterministic Linearization Method. Two-parametric smoothing approximations with iteratively refinement (FunctionLMSI()) are introduced during the local search that makes it possible to expand the approach on the class of non-differentiable problems. Solutions of numerous benchmarks show the computational efficiency of the hybrid algorithm. The pseudo code of the hybrid algorithm QRM-PCALMSI is presented in compacto as follows.

for $j \leftarrow 1,\ n_{particles}$ **do**

Generate an initial solution $OldConfig_j$

Determinate quasi-reflective solution $QOldConfig_j$

if Fitness($QOldConfig_j$) < Fitness($OldConfig_j$) **then**

$OldConfig_j \leftarrow QOldConfig_j$

end if

```
end for
UpdateBlackboard()
wile Stopping criteria not yet met do
    for j ← 1, n_particles do
        Perturbation()
        if Fitness( NewConfig_j )<Fitness( OldConfig_j ) then
            if Fitness( NewConfig_j )<BestFitness then
                BestFitness=Fitness( NewConfig_j )
                BestConfig ← NewConfig_j
            end if
            OldConfig_j ← NewConfig_j
            FunctionLMSI()
        else
            Scattering()
        end if
        if It is time to get the reflected population (each n_reflection iterations) then
            Determine reflective solution QOldConfig_j
            if Fitness( QOldConfig_j )<Fitness( OldConfig_j ) then
                OldConfig_j ← QOldConfig_j
            end if
        end if
    end for
    if It is time to update the blackboard (each n_blackboard iterations) then
        UpdateBlackboard()
    end if
end while
BestConfig ← UpdateBlackboard()
return BestConfig
```

In distinction to the stochastic QRM-PCA, the hybrid QRM-PCALMSI explores the search space and when a better-than-previous solution is found, it is used as an initial point for the LMSI-method. Thus, function Exploration() in the QRM-PCA is replaced by the FunctionLMSI() in the first hybrid algorithm. The second hybrid algorithm (named here QRM-PCASFC) combines the QRM-PCA and deterministic space-filling curve method [22]. Some powerful algorithms for multi-extremal non-convex optimization problems are based on reducing the initial multi-dimensional problem to the equivalent Hölder problem of one dimension. This reduction can be executed by applying Peano-type space-filling curves mapping a unit interval on the real axis onto a multi-dimensional hypercube. The Hilbert technique is used for building the development of the Peano space-filling curve depending on the free parameter that stands for the number of subdivision levels. The approach needs not any derivatives of the function to be minimized with updating parameters. Some disadvantage of this approach is in the fact that one-dimensional problem obtained by the above reduction leaks information on the closeness of iteration points in the initial multi-dimensional space. Solutions of some benchmarks show the computational efficiency of the hybrid algorithms.

5. Numerical example

It is supposed that from the analysis of experimental data for the Lorenz system the following approximate eigenvalues of the deviation curvature tensor are established: $\lambda_{+}^{*} \approx 68.5$; $\lambda_{-}^{*} \approx -103.0$. These indirect data have been obtained by the modelling the Lorenz system under the standard free values. The relative error of the given data is less than $0{,}1\%$. The problem is to reconstruct values of the free

IOP Conf. Series: Materials Science and Engineering **468** (2018) 012040 doi:10.1088/1757-899X/468/1/012040

parameters of the Lorenz system from given λ_+^* и λ_-^*. The control variables are represented by the relative variables x_1, x_2, x_3, (determined in %), which correspond to the required parameters σ, ρ, β. The criterion function of the considered problem is defined as follows

$$F(x) = \sum_{i=1}^{2} \gamma_i f_i^2(x) + \alpha \| x \|_2^2,$$

where $\gamma_i, f_i(x)$ are the weighting coefficient and the particular criterion that correspond to the i th eigenvalue λ_i: $\lambda_1(x) = \lambda_+(x)$, $\lambda_2(x) = \lambda_-(x)$; $f_i(x) = (\lambda_i^* - \lambda_i(x))$, $i = 1, 2$; α is the regularization parameter; $x \in R^3$.
The hybrid global optimization algorithm QRM-PCALMSI is used. The iteration history of x_i, $i = 1, 3$, when changing the number of iterations is depicted in Figure 1. The corresponding history for the criterion function $F(x)$ and the norm of the search direction $Nr(w)$ is demonstrated in Figure 2. After 22 iterations the following solution is obtained: $F(x^{22}) \approx 0{,}5879$; $x_1^{22} \approx 70{,}27\%$, $x_2^{22} \approx 71{,}14\%$, $x_3^{22} \approx 66{,}47\%$. The reconstructed approximate free parameters of the Lorenz system are as follows: $\sigma \approx 9{,}9698$; $\rho \approx 27{,}805$; $\beta \approx 2{,}6607$. The maximal relative error of the solution is less than $0{,}7\%$. It should be mentioned that there is a good correspondence between the accuracy of the approximate solution of the inverse problem and the given input data.

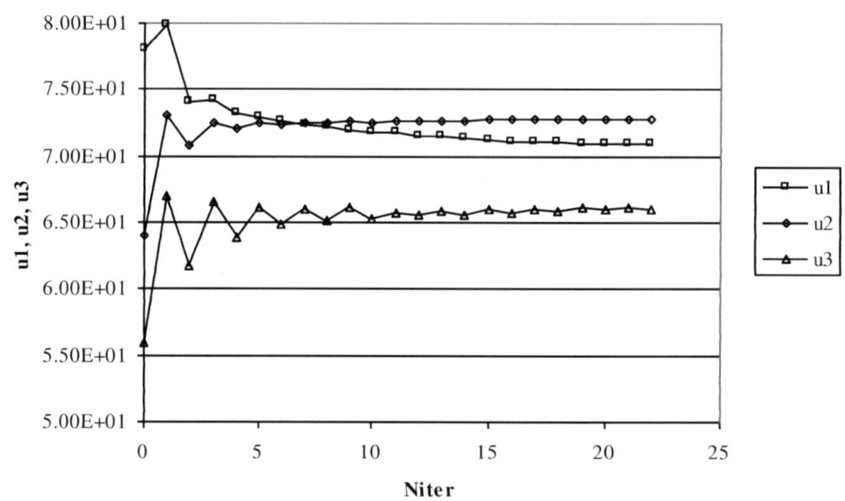

Figure 1. Number of final iterations N_{iter} of the QRM-PCALMSI vs. control variables x_i, $i = \overline{1,5}$

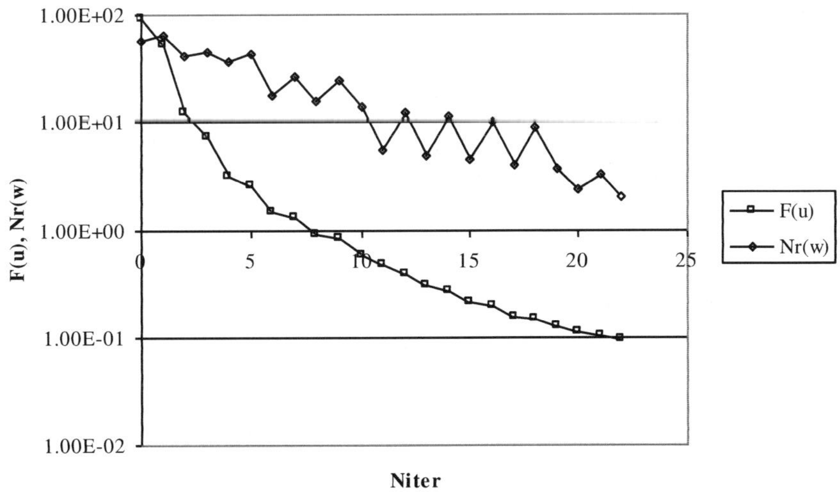

Figure 2 .Number of final iterations N_{iter} of the QRM-PCALMSI vs. the error function $f(x)$

So, the free parameters of the Lorenz system have been obtained, and they are close to the standard ones: $\sigma * = 10$, $\rho * = 28$, $\beta * = 2,(6)$. It should be noted that in this case all the equilibrium points of the considered system are Jacobi unstable [14].

6. Conclusion

Some questions on the Jacobi stability of the dynamical system based on determining the second geometric invariant in the framework of the Kosambi–Cartan–Chern theory are considered. The technique on reconstructing free parameters of the systems from eigenvalues of the curvature deviation tensor is introduced. The optimization approach is used for solving the inverse problem. Two novel hybrid global optimization algorithms QRM-PCALMSI and QRM-PCASFC being realized in programs are presented. The model updating of the free parameters of the Lorenz system is implemented. The accuracy of the approximate solution is in a good correspondence with the accuracy of the given indirect experimental input data.

References

[1] Giesl P, Hafstein S 2015 Review on computational methods for Lyapunov functions *Discrere & Continuous Dynamical Systems B* **20** (8) pp 2291-2331.
[2] Udrişte C, Nicola I R 2007 Jacobi stability and geometric dynamics *J. of Dynamic Systems and Geometrical Theories* **5**(1) pp 85-95.
[3] Harko T, Pantaragphong P, Sabau S V 2016 Kosambi-Cartan-Chern (KCC) theory for higher order dynamical *systems International J. of Geometric Methods in Modern Physics* **13** (2), pp 1656014.
[4] Bao D, Chern S-S, Shen Z 2000 *An Introduction to the Riemann–Finsler Geometry* (Springer-Verlag, New York)
[5] Antonelli P L, Ingarden R S, Matsumoto M 1993 *The Theory of Sprays and Finsler Spaces with Applications in Physics and Biology* (Kluwer Academic, Dodrecht)
[6] Bucataru I, Miron R 2007 *Finsler–Lagrange Geometry. Applications to Dynamical Systems.* (Editura Academiei Române, Bucureşti)
[7] Böhmer C G, Harko T, Sabau S V 2012 Jacobi stability analysis of dynamical systems – applications in gravitation and cosmology *Advances in Theoretical and Mathematical Physics* **16** (4) pp 1145-1196.

[8] Punzi R, Wohlfarth M N R 2009 Geometry and stability of dynamical systems *Physical Review E* **79** (4) pp 046606.

[9] Sabau S V 2005 Systems biology and deviation curvature tensor *Nonlinear Analysis: Real World Applications* **6** (3) pp 563-587.

[10] Yamasaki K, Yajima T 2013 Lotka–Volterra system and KCC theory. Differential geometric structure of competitions and predations *Nonlinear Analysis RWA* **14** pp 1845-1853.

[11] Gupta T 2016 Jacobi stability analysis of Rikitake *system International J. of Geometric Methods in Modern Physics* **13** (7), pp 1650098.

[12] Lake M J, Harko T 2016 Dynamical behavior and Jacobi stability analysis of wound strings *European Physical J. C* **76** (6), pp 331.

[13] Lorenz E N 1963 Deterministic nonperiodic flow *J. of the Atmospheric Sciences* **20** (1), pp. 130-141.

[14] Harko T, Ho C Y, Leung C S, Yip S 2015 Jacobi stability analysis of the Lorenz system *International J. of Geometric Methods in Modern Physics* **12** (7) pp 1550081.

[15] Kirsch A 2011 *An Introduction to the Mathematical Theory of Inverse Problems.* 2nd edition. (Springer, New York).

[16] Goncharsky A V 2013 Supercomputer technologies in inverse problems of ultrasound tomography *Inverse Problems* **29** (7) pp 1-22.

[17] Floudas C A, Gounaris C E 2009 A review of recent advances in global optimization *J. of Global Optimization* **45** (1) pp 3-38.

[18] Luz E F P, Becceneri J C, de Campos Velho H F 2008 A new multi-particle collision algorithm for optimization in a high performance environment *J. of Computational Interdisciplinary Sciences* **1** pp 3-10.

[19] Sulimov V D, Shkapov P M 2012 Application of hybrid algorithms to computational diagnostic problems for hydromechanical systems. *J. of Mechanics Engineering and Automation* **2** (12), pp 734-741.

[20] Torres R H, da Luz E F P, de Campos Velho H F 2015 Multi-Particle Collision Algorithm with reflected points *Proceeding Series of the Brazilian Society of Applied and Computational Mathematics* **3** (1), pp 010433

[21] Sulimov V D, Shkapov P M, Sulimov A V 2016 Optimization of singular values of parameter dependent matrices using hybrid algorithms *Herald of the Bauman Moscow State Tech. Univ., Nat. Sci.* **5** (68) pp 46-66 (in Russian).

[22] Lera D, Sergeev Ya D 2015 Deterministic global optimization using space-filling curves and multiple estimates of Lipschitz and Hölder constants *Computations in Nonlinear Science and Numerical Simulations* **23** (1-3), pp 326-342

Modelling of dynamics of mechanical systems with regard for constraint stabilization

R G Mukharlyamov

Institute of Physical Research and Technologies, RUDN University, Moscow, Russia
E-mail: robgar@mail.ru

Abstract. The main purpose of dynamical processes modelling is to formulate the motion equations of the system with regard for active forces and constraints restricted its movement. Desirable properties of system's motion, which are provided by the influence of additional forces and by the variation of inertial system's properties, can be specified by the constraint equations. Nikolay Zhukovskiy studied two main problems on constructing motion equations: defining the force function, that determines a set of motion trajectories, and analysing its stability. The representation of constraint equations as partial integrals of motion equations allows to provide an asymptotic stability of the corresponding integral manifold and to solve the problem of constraint stabilization at numerical solution of dynamics equations.

1. Introduction

The basis of the mathematical model of mechanical system's dynamics is the system of motion equations. If the set of constraint equations allows to uniquely represent the kinematic state of a system using generalized coordinates and velocities, then the system's behavior is determined with some particular level of accuracy that depends on applicable methods. Lagrange multipliers can be applied for accounting the constraint influence in the case of impossibility of generalized coordinates and velocities introduction. At the same time, constraint reactions can be considered as control forces providing the realization of constraint equations.

The problem of defining additional control forces, which allow the motion of a mechanical system to have appropriate properties, is related to the inverse dynamical problems [1]. So, based on the properties of planetary motion sir Isaak Newton [2] established the form of the gravity force and later it was discovered [3] that the motion of a material point on a conic section is the effect of a central force, depending on the point's position [4], [5]. The problem of determining a force function, corresponding to a holonomic system with given integrals, was considered by Gavriil Suslov [6]. Also, Nikolay Zhukovskiy established the method of determining a force function, based on a given set of trajectories of a material point on a curved surface, and gave a solution to the problem of motion strength of a representation point [8], using indicators of the kinetic energy system's force function. General Lyapunov's theory of motion stability [9] allowed to formulate the stability criteria of a bunch of trajectories [10] and to develop some new methods of constructing the systems of differential equations, having a given stable integral manifold [11].

Determining the Lagrange multipliers constraint equations are usually considered as the first integrals of motion equations so that the initial data corresponds to them. Deviations of initial data and an application of some approximation methods cause the disruption of constraint equations. The problem

Content from this work may be used under the terms of the Creative Commons Attribution 3.0 licence. Any further distribution of this work must maintain attribution to the author(s) and the title of the work, journal citation and DOI.
Published under licence by IOP Publishing Ltd

of a constraint stabilization was firstly mentioned by Joachim Baumgarte in his paper [12].An asymptotic stability of corresponding to the constraint equations integral manifold of a system of dynamics equations is a necessary condition for solving this problem. The solution of this problem can be obtained by introducing additional forces or by changing the inertial properties of a system [13], [14].

The problem of a constraint stabilization leads to the problem of constructing the system of differential equations admitting constraint equations as partial integrals and defining asymptotic stable invariant set [16] or integral manifold [17] of this system. It is quite possible to construct a system of motion equations with required accuracy [18] of constraints deviations using a general approach of solving inverse dynamical problems. Some relevant dynamical analogies allow us to apply methods and equations of classical mechanics to solve problems of modeling and dynamical control of systems of different nature.

2. Problem statement

Let the state of a mechanical system be determined by generalized coordinates $\mathbf{q}=\left(q^1,...,q^n\right)$, velocities $\mathbf{v}=\left(v^1,...,v^n\right)$, $dq^i/dt=v^i$, $i=1,...,n$, Lagrange function $L=L(\mathbf{q},\mathbf{v},t)$ and non-potential generalized forces, interacting with a mechanical system, $\mathbf{Q}=(Q_1,...,Q_n)$, $Q_i=Q_i(\mathbf{q},\mathbf{v},t)$. Let's consider that the system has both holonomic and nonholonomic constraints

$$\mathbf{f}(\mathbf{q},t)=0\,,\ \mathbf{f}=\left(f^1,...,f^m\right), \tag{1}$$

$$\mathbf{f}'(\mathbf{q},\mathbf{v},t)=0\,,\ \mathbf{f}'=\left(f^{m+1},...,f^s\right),\ s\le n\,. \tag{2}$$

So the problem is to determine dynamical equations of this mechanical system that provide the constraint stabilization during the numerical integration.

3. Dynamical equations of an extended system

Let's introduce new variables: $\breve{\mathbf{q}}=(q^{n+1},...,q^{n+m})$, $\hat{\mathbf{q}}=(q^{n+m+1},...,q^{n+s})$, $\breve{\mathbf{v}}=(v^{n+1},...,v^{n+m})$, $\hat{\mathbf{v}}=(v^{n+m+1},...,v^{n+s})$, $\tilde{\mathbf{v}}=(v^{n+1},...,v^{n+s})$, such as

$$\breve{\mathbf{q}}-\mathbf{f}(\mathbf{q},t)=0\,, \tag{3}$$

$$\tilde{\mathbf{v}}-\mathbf{g}(\mathbf{q},\mathbf{v},t)=0\,, \tag{4}$$

$$\mathbf{g}=(\breve{\mathbf{g}},\hat{\mathbf{g}})\,,\ \breve{\mathbf{g}}=\frac{\partial\mathbf{f}}{\partial\mathbf{q}}\mathbf{v}+\frac{\partial\mathbf{f}}{\partial t},\hat{\mathbf{g}}=\mathbf{f}'\,.$$

The system of equations that defines the virtual displacements can be constructed based on the equation (4) and it takes form

$$\mathbf{G}\delta\mathbf{q}=\delta\tilde{\mathbf{v}}\,,\mathbf{G}=\frac{\partial\mathbf{g}}{\partial\mathbf{v}}\,. \tag{5}$$

A vector of virtual displacements $\delta\mathbf{q}$ can be defined due to the system (5) considering the vector $\delta\tilde{\mathbf{v}}$ to be arbitrary. If the columns of the matrix $\mathbf{G}$ are linear independent, then the solution of the system (5) $\delta\mathbf{q}=(\delta\mathbf{q})^\tau+(\delta\mathbf{q})^n$ consists of a general solution $(\delta\mathbf{q})^\tau=\delta l[\mathbf{GC}]$ of the homogeneous equation and a partial solution $(\delta\mathbf{q})^n=\mathbf{G}^+\delta\tilde{\mathbf{v}}$ of the nonhomogeneous equation

$$\delta\mathbf{q}=\delta l[\mathbf{GC}]+\mathbf{G}^+\delta\tilde{\mathbf{v}}\,. \tag{6}$$

Here δl is an arbitrary scalar value, $[\mathbf{GC}]$ is a cross product of the vectors $\mathbf{g}^\sigma=\left(g_1^\sigma,...,g_n^\sigma\right)$, $\sigma=1,...,s$, composing the columns of the matrix $\mathbf{G}$ and arbitrary vectors $\mathbf{c}^\gamma=\left(c_1^\gamma,...,c_n^\gamma\right)$, $\gamma=s+1,...,n-1$, $\mathbf{G}^+=\mathbf{G}^T\left(\mathbf{G}\mathbf{G}^T\right)^{-1}$.

Let's consider that $L = L(\mathbf{q}, \mathbf{v}, t)$ is a Lagrange function, $\mathbf{Q} = (Q_1, ..., Q_n)$, $Q_i = Q_i(\mathbf{q}, \mathbf{v}, t)$ are non-potential generalized forces, interacting with a mechanical system, $\mathbf{R} = (R_1, ..., R_n)$, R_i are components of the vector of constraint reaction. Let's define the functions $\tilde{L} = \tilde{L}(\mathbf{q}, \mathbf{v}, \breve{\mathbf{q}}, \tilde{\mathbf{v}}, t)$ and $\tilde{D} = \tilde{D}(\mathbf{q}, \mathbf{v}, \breve{\mathbf{q}}, \tilde{\mathbf{v}}, t)$ that satisfy the following conditions: $\tilde{L}(\mathbf{q}, \mathbf{v}, 0, 0, t) = L(\mathbf{q}, \mathbf{v}, t)$, $\tilde{D}(\mathbf{q}, \mathbf{v}, 0, 0, t) = 0$ and $\tilde{D}(\mathbf{q}, \mathbf{v}, \breve{\mathbf{q}}, \tilde{\mathbf{v}}, t) \geq D(\breve{\mathbf{q}}, \tilde{\mathbf{v}}) > 0$ if $\breve{\mathbf{q}}, \tilde{\mathbf{v}}$ do not go to zero simultaneously and $D(0,0) = 0$. If $\mathbf{R} = (R_1, ..., R_n)$, R_i are components of the vector of constraint reaction, then D'Alembert's principle for the extended system with the Lagrange function $\tilde{L}$ takes form

$$\mathbf{E}(\mathbf{q}, \mathbf{v}) - \mathbf{Q} - \mathbf{R} = 0, \tag{7}$$

$$\mathbf{E}(\tilde{\mathbf{q}}, \tilde{\mathbf{v}}) + \frac{\partial \tilde{D}}{\partial \tilde{\mathbf{v}}} = 0, \tag{8}$$

$$\mathrm{E}(\mathbf{q}, \mathbf{v}) = \frac{d}{dt}\frac{\partial \tilde{L}}{\partial \mathbf{v}} - \frac{\partial \tilde{L}}{\partial \mathbf{q}}, \ \tilde{\mathbf{q}} = (\breve{\mathbf{q}}, \hat{\mathbf{q}}).$$

We can define the sum of elementary works by scalar multiplying equations (7) and (8) by $\delta\mathbf{q}$ and $\delta\tilde{\mathbf{v}}$ correspondingly

$$(\mathbf{E}(\mathbf{q}, \mathbf{v}) - \mathbf{Q} - \mathbf{R})^T \delta\mathbf{q} + \left(\mathbf{E}(\tilde{\mathbf{q}}, \tilde{\mathbf{v}}) + \frac{\partial \tilde{D}}{\partial \tilde{\mathbf{v}}}\right)^T \delta\tilde{\mathbf{v}} = 0, \tag{9}$$

then we can rewrite it with regard for (6):

$$(\mathbf{E}(\mathbf{q}, \mathbf{v}) - \mathbf{Q} - \mathbf{R})^T [\mathbf{GC}]\delta l + \left((\mathbf{E}(\mathbf{q}, \mathbf{v}) - \mathbf{Q} - \mathbf{R})^T \mathbf{G}^+ + \left(\mathbf{E}(\tilde{\mathbf{q}}, \tilde{\mathbf{v}}) + \frac{\partial \tilde{D}}{\partial \tilde{\mathbf{v}}}\right)^T\right)\delta\tilde{\mathbf{v}} = 0. \tag{10}$$

The equality (10) can be accomplished only if the following conditions are satisfied

$$(\mathbf{E}(\mathbf{q}, \mathbf{v}) - \mathbf{Q} - \mathbf{R})^T [\mathbf{GC}] = 0, \tag{11}$$

$$(\mathbf{E}(\mathbf{q}, \mathbf{v}) - \mathbf{Q} - \mathbf{R})^T \mathbf{G}^+ + \left(\mathbf{E}(\tilde{\mathbf{q}}, \tilde{\mathbf{v}}) + \frac{\partial \tilde{D}}{\partial \tilde{\mathbf{v}}}\right)^T \tag{12}$$

Let's choose the vector $\mathbf{R}$, so that the elementary work of constraint reactions due to the displacements $(\delta\mathbf{q})^T$ will be equal to zero: $\mathbf{R}^T[\mathbf{GC}] = 0$. This fact denotes the correspondence to the ideal constraints of the initial system $\mathbf{R} = \mathbf{G}^T\boldsymbol{\lambda}$, $\boldsymbol{\lambda} = (\lambda_1, ..., \lambda_s)$. From the identity (11) follows the equation, describing the variation of generalized coordinates of the system. Identity (12) can be reduced to the equation of the constraints' perturbations and, as a result, with regard for kinematic equations, identities (3) – (4) and initial conditions we can obtain the following system of differential-algebraic equations for $\mathbf{q}, \mathbf{v}, \tilde{\mathbf{q}}, \tilde{\mathbf{v}}, \boldsymbol{\lambda}$:

$$\frac{d\mathbf{q}}{dt} = \mathbf{v}, \ \frac{d}{dt}\frac{\partial \tilde{L}}{\partial \mathbf{v}} - \frac{\partial \tilde{L}}{\partial \mathbf{q}} = \mathbf{Q} + \mathbf{G}^T\boldsymbol{\lambda}, \tag{13}$$

$$\frac{d\tilde{\mathbf{q}}}{dt} = \tilde{\mathbf{v}}, \ \frac{d}{dt}\frac{\partial \tilde{L}}{\partial \tilde{\mathbf{v}}} - \frac{\partial \tilde{L}}{\partial \tilde{\mathbf{q}}} = -\frac{\partial \tilde{D}}{\partial \tilde{\mathbf{v}}}, \tag{14}$$

$$\breve{\mathbf{q}} = \mathbf{f}(\mathbf{q}, t), \ \tilde{\mathbf{q}} = (\breve{\mathbf{q}}, \hat{\mathbf{q}}), \ \tilde{\mathbf{v}} = \mathbf{g}(\mathbf{q}, \mathbf{v}, t), \tag{15}$$

$$\mathbf{q}(t_0) = \mathbf{q}_0, \ \breve{\mathbf{q}}(t_0) = \mathbf{f}(\mathbf{q}_0, t_0), \ \hat{\mathbf{q}}(t_0) = \hat{\mathbf{q}}_0, \ \mathbf{v}(t_0) = \mathbf{v}_0, \ \tilde{\mathbf{v}}(t_0) = \mathbf{g}(\mathbf{q}_0, \mathbf{v}_0, t_0). \tag{16}$$

It is necessary to complete the right sides of differential equations (13) – (14) to solve the systems (13) – (16). The values of the forces of constraint reactions are determined by defining the multipliers $\boldsymbol{\lambda}$ that ensure the equalities (15). If we assume that the values of the deviations from the solution of the system (13), (14) are defined with the help of extra variables $\breve{\mathbf{q}}, \tilde{\mathbf{v}}$, then the solution $\breve{\mathbf{q}} = 0$, $\tilde{\mathbf{v}} = 0$ of this system corresponds to the constraint equations and its stability depends on the choice of the Lagrange function $\tilde{L}$ and dissipative function $\tilde{D}$.Let's consider that functions $\tilde{L}$ and $\tilde{D}$ take form

$$\tilde{L}=\tilde{T}-\tilde{P},$$
$$2\tilde{T}=\mathbf{v}^T\mathbf{A}(\mathbf{q})\mathbf{v}+\tilde{\mathbf{v}}^T\tilde{\mathbf{A}}(\mathbf{q},\tilde{\mathbf{q}})\tilde{\mathbf{v}},\ 2\tilde{P}=\tilde{\mathbf{q}}^T\tilde{\mathbf{H}}(\mathbf{q})\tilde{\mathbf{q}},\ 2\tilde{D}=\tilde{\mathbf{v}}^T\mathbf{B}(\mathbf{q},\tilde{\mathbf{q}})\tilde{\mathbf{v}}, \tag{17}$$
$$\tilde{\mathbf{H}}(\mathbf{q})=\begin{pmatrix}\breve{\mathbf{H}}(\mathbf{q}) & 0\\ 0 & 0\end{pmatrix}.$$

The values of $\boldsymbol{\lambda}$ are determined from the equations (13) – (15):

$$\boldsymbol{\lambda}=\mathbf{M}^{-1}(\mathbf{q},\mathbf{v},t)\mathbf{h}(\mathbf{q},\mathbf{v},t),\ \mathbf{M}=\mathbf{G}\mathbf{A}^{-1}(\mathbf{q})\mathbf{G}^T,$$
$$\mathbf{h}=\tilde{\mathbf{A}}^{-1}(\mathbf{q},\tilde{\mathbf{q}})\left(\frac{\partial\tilde{L}}{\partial\tilde{\mathbf{q}}}-\frac{d\tilde{\mathbf{A}}(\mathbf{q},\tilde{\mathbf{q}})}{dt}\tilde{\mathbf{v}}-\mathbf{B}(\mathbf{q},\tilde{\mathbf{q}})\tilde{\mathbf{v}}\right)+\mathbf{G}\mathbf{A}^{-1}(\mathbf{q})\left(\frac{d\mathbf{A}(\mathbf{q})}{dt}\mathbf{v}-\frac{\partial\tilde{L}}{\partial\mathbf{q}}\right)-\frac{\partial\mathbf{g}}{\partial\mathbf{q}}\mathbf{v}-\frac{\partial\mathbf{g}}{\partial t},$$
$$\tilde{\mathbf{q}}=\mathbf{f}(\mathbf{q},t),\ \tilde{\mathbf{v}}=\mathbf{g}(\mathbf{q},\mathbf{v},t).$$

Dynamical equations (13) with regard for the values $\tilde{L}$, $\tilde{\mathbf{q}}$, $\tilde{\mathbf{v}}$, $\boldsymbol{\lambda}$ as functions of the variables $\mathbf{q},\mathbf{v},t$ are reduced to the following system of differential equations:

$$\frac{d\mathbf{q}}{dt}=\mathbf{v},\ \frac{d\mathbf{v}}{dt}=\mathbf{A}^{-1}(\mathbf{q})\left(\frac{\partial\tilde{L}}{\partial\mathbf{q}}-\frac{d\mathbf{A}(\mathbf{q})}{dt}\mathbf{v}+\mathbf{Q}(\mathbf{q},\mathbf{v},t)+\mathbf{G}^T(\mathbf{q},\mathbf{v},t)\boldsymbol{\lambda}(\mathbf{q},\mathbf{v},t)\right), \tag{18}$$
$$\tilde{L}=\tilde{L}(\mathbf{q},\mathbf{v},\mathbf{f}(\mathbf{q},t),\mathbf{g}(\mathbf{q},\mathbf{v},t),t),$$

that has the partial integrals:

$$\mathbf{f}(\mathbf{q},t)=0,\ \mathbf{g}(\mathbf{q},\mathbf{v},t)=0. \tag{19}$$

4. Stability of integral manifold

The constraint (1), (2) stabilization requires the asymptotic stability of an integral manifold of the system (18) given by the equalities (19). This stability of an integral manifold can be defined with the help of the following terms.

An integral manifold of the system (18), given by the equalities (19), is stable if for any $\varepsilon>0$ there exists such $\delta>0$, so the solution $\mathbf{q}=\mathbf{q}(t)$, $\mathbf{v}=\mathbf{v}(t)$ of the system, corresponding to the initial conditions $\mathbf{q}(t_0)=\mathbf{q}_0$, $\mathbf{v}(t_0)=\mathbf{v}_0$: $\|\mathbf{f}(\mathbf{q}_0,t_0)\|\le\delta$, $\|\mathbf{g}(\mathbf{q}_0,\mathbf{v}_0,t_0)\|\le\delta$, for all $t\ge t_0$ will satisfy the inequalities $\|\mathbf{f}(\mathbf{q}(t),t)\|\le\varepsilon$, $\|\mathbf{g}(\mathbf{q}(t),\mathbf{v}(t),t)\|\le\varepsilon$.

An integral manifold of the system (18), given by the equalities (19), is asymptotically stable, if it is stable and the following conditions are satisfied: $\lim\limits_{t\to\infty}\|\mathbf{f}(\mathbf{q}(t),t)\|=0$, $\lim\limits_{t\to\infty}\|\mathbf{g}(\mathbf{q}(t),\mathbf{v}(t),t)\|=0$.

It is obvious that the stability of an integral manifold is determined by the corresponding stability of the trivial solution $\tilde{\mathbf{q}}(t)=0$, $\tilde{\mathbf{v}}(t)=0$ of the system of the perturbation constraint equations (14). With the regard for values of $\tilde{L},\tilde{D}$ (17) the system (14) can be written in the form:

$$\frac{d\tilde{\mathbf{q}}}{dt}=\tilde{\mathbf{v}},\ \frac{d\tilde{\mathbf{v}}}{dt}=-\tilde{\mathbf{S}}(\mathbf{q},\tilde{\mathbf{q}})\tilde{\mathbf{q}}-\tilde{\mathbf{K}}(\mathbf{q},\tilde{\mathbf{q}})\tilde{\mathbf{v}}+\tilde{\mathbf{A}}^{-1}(\mathbf{q},\tilde{\mathbf{q}})\left(\tilde{\mathbf{v}}^T\frac{\partial\tilde{\mathbf{A}}(\mathbf{q},\tilde{\mathbf{q}})}{\partial\tilde{\mathbf{q}}}\tilde{\mathbf{v}}\right),$$
$$\tilde{\mathbf{S}}(\mathbf{q},\tilde{\mathbf{q}})=\tilde{\mathbf{A}}^{-1}(\mathbf{q},\tilde{\mathbf{q}})\tilde{\mathbf{H}}(\mathbf{q}),\ \ \tilde{\mathbf{K}}(\mathbf{q},\tilde{\mathbf{q}})=\tilde{\mathbf{A}}^{-1}(\mathbf{q},\tilde{\mathbf{q}})\left(\frac{d\tilde{\mathbf{A}}(\mathbf{q},\tilde{\mathbf{q}})}{dt}+\mathbf{B}(\mathbf{q},\tilde{\mathbf{q}})\right),$$

or after expanding the matrices $\tilde{\mathbf{S}}(\mathbf{q},\tilde{\mathbf{q}})$ and $\tilde{\mathbf{K}}(\mathbf{q},\tilde{\mathbf{q}})$ in a series in powers of $\tilde{\mathbf{q}}$:

$$\tilde{\mathbf{S}}(\mathbf{q},\tilde{\mathbf{q}})=\tilde{\mathbf{S}}(\mathbf{q},0)+\frac{\partial\tilde{\mathbf{S}}(\mathbf{q},0)}{\partial\tilde{\mathbf{q}}}\tilde{\mathbf{q}}+\tilde{\mathbf{S}}^{(2)},\ \tilde{\mathbf{K}}(\mathbf{q},\tilde{\mathbf{q}})=\tilde{\mathbf{K}}(\mathbf{q},0)+\frac{\partial\tilde{\mathbf{K}}(\mathbf{q},0)}{\partial\tilde{\mathbf{q}}}\tilde{\mathbf{q}}+\tilde{\mathbf{K}}^{(2)},$$

it can take the following form:

$$\frac{d\breve{\mathbf{q}}}{dt}=\tilde{\mathbf{v}},\ \frac{d\tilde{\mathbf{v}}}{dt}=-\mathbf{S}(\mathbf{q})\breve{\mathbf{q}}-\mathbf{K}(\mathbf{q})\tilde{\mathbf{v}}+\tilde{\mathbf{V}}^{(2)}, \tag{20}$$

$$\mathbf{S}(\mathbf{q})=\tilde{\mathbf{S}}(\mathbf{q},0),\ \mathbf{K}=\tilde{\mathbf{K}}(\mathbf{q},0),$$

$$\tilde{\mathbf{V}}^{(2)}=\left(\frac{\partial\tilde{\mathbf{S}}(\mathbf{q},0)}{\partial\breve{\mathbf{q}}}\breve{\mathbf{q}}+\tilde{\mathbf{S}}^{(2)}\right)\breve{\mathbf{q}}+\left(\frac{\partial\tilde{\mathbf{K}}(\mathbf{q},0)}{\partial\breve{\mathbf{q}}}\breve{\mathbf{q}}+\tilde{\mathbf{K}}^{(2)}\right)\tilde{\mathbf{v}}+\tilde{\mathbf{A}}^{-1}(\mathbf{q},\breve{\mathbf{q}})\left(\tilde{\mathbf{v}}^{T}\frac{\partial\tilde{\mathbf{A}}(\mathbf{q},\breve{\mathbf{q}})}{\partial\breve{\mathbf{q}}}\tilde{\mathbf{v}}\right).$$

Let's introduce the following notation:

$$\mathbf{y}=(\breve{\mathbf{q}},\tilde{\mathbf{v}}),\ \mathbf{W}(\mathbf{q})=\begin{pmatrix}0 & \mathbf{E}\\ -\mathbf{S}(\mathbf{q}) & -\mathbf{K}(\mathbf{q})\end{pmatrix},\ \mathbf{Y}^{(2)}=\begin{pmatrix}0\\ \tilde{\mathbf{V}}^{(2)}\end{pmatrix},$$

and rewrite the system (20) in an abbreviated form:

$$\frac{d\mathbf{y}}{dt}=\mathbf{W}(\mathbf{q})\mathbf{y}+\mathbf{Y}^{(2)}. \tag{21}$$

If all of the roots of the characteristic polynomial of the matrix $\mathbf{W}(\mathbf{q})$ at all of the possible values of the generalized coordinates $q_1,...,q_n$ at their domain Ω_q have negative real parts then a trivial solution of the equation with primary approximation:

$$\frac{d\mathbf{y}}{dt}=\mathbf{W}(\mathbf{q})\mathbf{y}$$

is asymptotic stable. The problem the holonomic constraint stabilization is studied in the paper [12] using the equations of perturbed constraints with the constant matrix

$$\mathbf{W}=\begin{pmatrix}0 & \mathbf{E}\\ -\omega^2\mathbf{E} & -\gamma\mathbf{E}\end{pmatrix}.$$

An algorithm of solving the problem of stabilization with the matrix $\mathbf{W}=\mathbf{W}(\mathbf{q},\mathbf{v})$, determined with the help of the matrix $\mathbf{G}$ and its derivative, is established in the paper [19].

In general the method of Lyapunov's function is applied [11] to define the sufficient conditions of the stability of a trivial solution (21). If the constraints are scleronomic

$$\mathbf{f}(\mathbf{q})=0,\ \mathbf{g}(\mathbf{q},\mathbf{v})=0,$$

then we can take as a Lyapunov's function a positive definite quadratic form with the constraint matrix with coefficients $2V=\mathbf{y}^T\mathbf{U}\mathbf{y}$. If the derivative of this function

$$\frac{dV}{dt}=\mathbf{y}^T\mathbf{U}\mathbf{W}(\mathbf{q})\mathbf{y}+\mathbf{Y}^{(3)}$$

is negative definite then the trivial solution $\mathbf{y}=0$ of the equation (21) is asymptotic stable.

5. Numerical solution

If the perturbation constraint equations have asymptotic stable trivial solution then we can limit our choice with the simplest numerical methods of solving the dynamical equations (18). So the application of the finite-difference scheme

$$\mathbf{x}_{k+1}=\mathbf{x}_k+\tau\mathbf{X}_k,\ \mathbf{x}_k=\mathbf{x}(t_k),\ \tau=t_{k+1}-t_k,\ x(t_0)=x_0,$$

$$\mathbf{x}=\begin{pmatrix}\mathbf{q}\\ \mathbf{v}\end{pmatrix},\ \mathbf{X}=\begin{pmatrix}\mathbf{v}\\ \mathbf{F}(\mathbf{q},\mathbf{v},t)\end{pmatrix},$$

$$\mathbf{F}(\mathbf{q},\mathbf{v},t)=\mathbf{A}^{-1}(\mathbf{q})\left(\frac{\partial\tilde{L}}{\partial\mathbf{q}}-\frac{d\mathbf{A}(\mathbf{q})}{dt}\mathbf{v}+\mathbf{Q}(\mathbf{q},\mathbf{v},t)+\mathbf{G}^T(\mathbf{q},\mathbf{v},t)\boldsymbol{\lambda}(\mathbf{q},\mathbf{v},t)\right)$$

with regard for (21) leads to the inequality

$$\|\mathbf{y}_{k+1}\| \le \|\mathbf{E} - \tau \mathbf{W}_k\| \|\mathbf{y}_k\| + \Upsilon_k^{(2)},$$

where $\Upsilon_k^{(2)}$ is a reminder. From inequality (23) follows the estimation $\|\mathbf{y}_{k+1}\| \le \varepsilon$ if for all $k = 0,1,2,...,N$ the following conditions: $\|\mathbf{y}_k\| \le \varepsilon$, $\|\mathbf{E} - \tau \mathbf{W}_k\| \le \alpha \le 1$, $\|\Upsilon_k^{(2)}\| \le 2\varepsilon(1-\alpha)$ are satisfied.
If solving equations (18) we use the finite difference scheme

$$\mathbf{x}_{k+1} = \mathbf{x}_k + \Delta \mathbf{x}_k, \ \ \Delta \mathbf{x}_k = \tau(1-\sigma)\mathbf{X}_k + \tau\sigma\hat{\mathbf{X}}_k,$$

$$\hat{\mathbf{X}}_k = \mathbf{X}_k(\tilde{\mathbf{x}}_k, t_k + \alpha\tau), \ \tilde{\mathbf{x}}_k = \mathbf{x}_k + \alpha\tau\mathbf{X}_k, \ \sigma > 0, \ \alpha > 0, \ k = 0,1,2,...,N,$$

where $\alpha, \sigma - const$ and for all $k = 0,1,2,...,N$ the following conditions are satisfied

$$2\alpha\sigma = 1, \ \|\mathbf{y}_k\| \le \varepsilon, \ \|\Upsilon_k^{(3)}\| \le 6\varepsilon(1-\beta), \ \left\| \mathbf{I}_{m+r} + \tau\mathbf{W}_k + \frac{1}{2}\tau^2\left(\mathbf{W}_k^2 + \left(\frac{d\mathbf{W}}{dt}\right)_k\right)\right\| \le \beta \le 1,$$

then we have the estimation $\|\mathbf{y}_{k+l}\| \le \varepsilon$. The conditions of the constraint stabilization were obtained by the Runge-Kutta method in the paper [21].

6. Example

Our goal is to determine the law of the variation of the force $\mathbf{F}$ providing the stable motion of the rocket on the trajectory $f(x,y) = 0$. The rocket is considered as a material point that has coordinates and velocities

$$q = (x,y), \ \mathbf{v} = (v_x, v_y), v_x = \frac{dx}{dt}, \ v_y = \frac{dy}{dt}$$

and it interacts with the force of gravity $m\mathbf{g}$ directed vertically downwards. The deviation of the point from its trajectory and its derivative are denoted as

$$\breve{q} = f(x,y), \ \breve{v} = \frac{\partial f}{\partial x}v_x + \frac{\partial f}{\partial y}v_y. \tag{22}$$

Let's introduce Lagrange and dissipative functions

$$2\tilde{L} = m\mathbf{v}^2 - 2mgy + \breve{v}^2 - c\breve{q}^2, \ 2D = k\breve{v}^2, \tag{23}$$

$$\mathbf{v}^2 = v_x^2 + v_y^2, \ c,k,g - const.$$

From the equalities (22) follows the equation:

$$\frac{\partial f}{\partial x}\delta x + \frac{\partial f}{\partial y}\delta y = \delta\breve{v},$$

that determines virtual displacements of a point depending on the arbitrary values δs and $\delta\breve{v}$:

$$\delta x = -\frac{\partial f}{\partial y}\delta s + \frac{\partial f}{\partial x}\left(\left(\frac{\partial f}{\partial x}\right)^2 + \left(\frac{\partial f}{\partial y}\right)^2\right)^{-1}\delta\breve{v}, \ \delta y = \frac{\partial f}{\partial x}\delta s + \frac{\partial f}{\partial y}\left(\left(\frac{\partial f}{\partial x}\right)^2 + \left(\frac{\partial f}{\partial y}\right)^2\right)^{-1}\delta\breve{v}.$$

Using D'Alembert-Lagrange principle

$$\left(\frac{d}{dt}\frac{\partial\tilde{L}}{\partial v_x} - \frac{\partial\tilde{L}}{\partial x}\right)\delta x + \left(\frac{d}{dt}\frac{\partial\tilde{L}}{\partial v_y} - \frac{\partial\tilde{L}}{\partial y}\right)\delta y + \left(\frac{d}{dt}\frac{\partial\tilde{L}}{\partial\breve{v}} - \frac{\partial\tilde{L}}{\partial\breve{q}} - \frac{\partial D}{\partial\breve{v}}\right)\delta\breve{v} = 0$$

let's construct the dynamical equations of the rocket

$$\frac{dy}{dt} = v_y, \ \frac{d}{dt}(mv_x) = \lambda\frac{\partial f}{\partial x}, \ \frac{d}{dt}(mv_y) = -mg + \lambda\frac{\partial f}{\partial y}, \tag{24}$$

and the perturbation constraint equations

$$\frac{d\breve{q}}{dx} = \breve{v}, \ \frac{d\breve{v}}{dt} = -c\breve{q} - k\breve{v}. \tag{25}$$

Let's introduce μ as a velocity of a particle separation from the rocket and considering the equality

$$\frac{d}{dt}(m\mathbf{v}) = m\frac{d\mathbf{v}}{dt} - \mu\left(\frac{dm}{dt}\right)\frac{\mathbf{v}}{|\mathbf{v}|},$$

we can rewrite (24) in the following form:

$$\frac{dx}{dt} = v_x, m\frac{dv_x}{dt} = T_x + \lambda\frac{\partial f}{\partial x}, \frac{dy}{dt} = v_y, m\frac{dv_y}{dt} = T_y - mg + \lambda\frac{\partial f}{\partial y}, \tag{26}$$

$$T_x = \frac{\mu v_x}{\sqrt{v_x^2 + v_y^2}}\frac{dm}{dt}, T_y = \frac{\mu v_y}{\sqrt{v_x^2 + v_y^2}}\frac{dm}{dt}.$$

Right parts of the equations (26) contain the traction force $\mathbf{T} = (T_x, T_y)$ directed along the tangent to the motion trajectory and the constraint reaction $\mathbf{R} = \lambda(\partial f / \partial x, \partial f / \partial y)$ directed along the normal. These two forces form the unknown force $\mathbf{F}$. From the equations (22), (25), (26) we can determine the Lagrange multipliers

$$\lambda = -\frac{m}{N^2}\left(\frac{\partial^2 f}{\partial x^2}v_x^2 + 2\frac{\partial^2 f}{\partial x \partial y}v_x v_y + \frac{\partial^2 f}{\partial y^2}v_y^2 + cf(x,y) + k\left(\frac{\partial f}{\partial x}v_x + \frac{\partial f}{\partial y}v_y\right) + g\frac{\partial f}{\partial y}\right) - \frac{1}{N^2}\frac{\mu}{\sqrt{v_x^2 + v_y^2}}\left(\frac{\partial f}{\partial x}v_x + \frac{\partial f}{\partial y}v_y\right)\frac{dm}{dt}, \tag{27}$$

$$N^2 = \left(\frac{\partial f}{\partial x}\right)^2 + \left(\frac{\partial f}{\partial y}\right)^2.$$

The motion on the trajectory will be asymptotic stable if the roots the characteristic equation of the system (25) $\kappa^2 + k\kappa + c = 0$ have negative real parts. Numerical solution of the system (26), (27)

$$x(t_k) = x_k,\ x_{k+1} = x_k + \tau v_{xk},\ y_{k+1} = y_k + \tau v_{yk},$$

$$(v_x)_{k+1} = (v_x)_k + \frac{\tau}{m_k}\left(T_x + \lambda\frac{\partial f}{\partial x}\right)_k, (v_y)_{k+1} = (v_y)_k + \frac{\tau}{m_k}\left(T_y - mg + \lambda\frac{\partial f}{\partial y}\right)_k,$$

will satisfy the equality $\|q_k\| \le \varepsilon$ for all $k = 0,1,2,...,K$ if the conditions $\|q_0\| \le \varepsilon$, $\|\mathbf{E} - \tau\mathbf{W}\| \le \alpha \le 1$, $\|\Upsilon_k^{(2)}\| \le 2\varepsilon(1-\alpha)$ are satisfied, where $\mathbf{W}$ is a matrix of the coefficients of the system (25), $\Upsilon_k^{(2)}$ is a reminder of the function's $\breve{q} = f(x,y)$ expansion.

7. Conclusions

The methods of solving inverse dynamical problems and the conditions of the stability of a set of trajectories based on the Zhukovskiy's papers allow us to develop an algorithm od solving dynamical problems of mechanical systems and problems of dynamical processes control in the systems with different nature.

This work is supported by RFBR, project 16-08-00558 A.

Reference

[1] Galiullin A S 1986 *Methods of solving inverse dynamical problems* (Moscow: Nauka)
[2] Newton I 1989 *Philosophiæ Naturalis Principia Mathematica. Translation from Latin by A.N. Krylov* (Moscow: Nauka)
[3] Bertrand M G 1873 *Compterendus* **77** pp 849–853
[4] Imshenetskiy V G 1879 *Bulletin of Kharkov Mathematical Society* pp 1–11
[5] Darboux M G 1877 *Compterendus* **84** pp 760–768
[6] Suslov G K 1890 *On force function with given integrals* (Kiew: Kiew University)

[7] Zhukovskiy N E 1937 *Collection of Papers* **1** pp 293–308
[8] Zhukovskiy N E 1937 *Collection of Papers* **1** pp 110–205
[9] Lyapunov A M 1956 *General problem of motion stability (Collection of papers* vol 2*)* (Moscow: Academy of Science, USSR)
[10] Zubov VI 1957 *Methods of A.M. Lyapunov and its application* (Leningrad: LSU)
[11] Mukharlyamov R G 1969 *Differential Equations* **5** pp 688–699
[12] Baumgarte J 1972 *Comp. Math. Appl. Mech. Eng* **1** pp 1–16
[13] Kozlov V V 2015 *Regul. Chaot. Dyn.* **20** pp 205–224
[14] Kozlov V V 2015 *Regul. Chaot. Dyn.* **20** pp 401–427
[15] Erugin N P 1979 *Book for reading of general course of differential equations* (Minsk: NaukaiTekhnika)
[16] Levi-Chivita T and Amaldi U 1952 *Course of theoretical mechanics* II (Moscow: IL)
[17] Mukharlyamov R G 1971 *Differential Equations* **7** pp 1825–34
[18] Mukharlyamov R G 2006 *Applied Mathematics and Mechanics* **70** pp 236–49
[19] Ascher U M, Chin H, Petzold L R and Reich S 1995 *Mechanics of Structures and Machines* **23** pp 135–57
[20] Mukharlyamov R G 2015 *Izvestiya RAN* 15–28
[21] Mukharlyamov R G and Assaye W B 2013 *Bulletin of Peoples' Friendship University of Russia. Series: Mathematics, Informatics and Physics* **3** pp 81–92

Considering the Motion of the Elastic Lunar Pole

M Yu Barkin and V A Pankratov

Bauman Moscow State Technical University, Moscow, Russia
E-mail: barkin@yandex.ru

Abstract. In connection with the development of the Japanese project ILOM on the study of the rotational motion of the Moon using a telescope mounted on the surface of the moon (near the north pole), there are increased demands on the accuracy of the analytical and numerical description of its physical librations. In particular, studies of the influence of the elastic properties of the Moon on its rotational motion with the high-precision description of its orbital motion and taking into account the resonant nature of motion are of a significant importance. In this work it is shown that the elastic properties of the Moon result in extending of the oscillations period of its pole by 8.0 days.

1. Introduction

The studies of the rotational and translational-rotational motion of the viscous-elastic Earth and the Moon are devoted to the work of many authors [1], [2]. Basically, these works were devoted to the study of the evolution of the Earth-Moon system. For modern studies of the Moon, it is also important to study the manifestations of the elastic properties of the Moon in its physical libration. In connection with the development of the Japanese project ILOM on the study of the rotational motion of the Moon using a telescope mounted on the surface of the Moon (near its North Pole), there are increased demands on the accuracy of the analytical and numerical description of its physical librations [3], [4]. In particular, studies of the influence of the liquid core of the Moon, its solid core, the elasticity of the mantle of the Moon on its rotational motion with a highly accurate description of the orbital motion of the Moon and taking into account the resonant nature of motion are of a significant interest. In this paper, we explore the dynamic role of the elasticity of the Moon in the movement of the pole of its axis of rotation. The influence of the liquid and solid core of the Moon on its libration is not taken into account in this work. To describe the rotational motion of the Moon the equations of the Liouville problem are in Andoyer variables [5] and the perturbation theory [6] is used. The significant effects that must be taken into account when interpreting observational data using the lunar telescope are found. In this work it is shown that the elastic properties of the Moon lead to an extension in the oscillations period of its pole by 7-8 days.

2. Statement of the problem. The canonical equations of rotational motion of a deformable body in Andoyer variables.

We will consider the Moon as a weakly deformed body with deformations due to its own rotation and under the influence of the Earth gravitational force. We assume that its particles weakly deviate from their original positions during the motion process or are displaced in a known manner with respect of time with a small speed. It can also be assumed that the body has an internal solid shell, and we associate a Cartesian coordinate system $C\xi\eta\zeta$ with it (body axes), and an external deformable shell.

Content from this work may be used under the terms of the Creative Commons Attribution 3.0 licence. Any further distribution of this work must maintain attribution to the author(s) and the title of the work, journal citation and DOI.
Published under licence by IOP Publishing Ltd

Let the Cartesian coordinate system $Cxyz$ with the origin at the Moon's center of mass maintain a constant orientation in space.
We introduce the Andoyer variables that are associated with the vector of the angular momentum **G** of the Moon:

$$G,\ \theta,\ \rho,\ l,\ g,\ h \tag{1}$$

Let $CG_1G_2G_3$ be the intermediate coordinate system associated with the vector $\mathbf{G}$. The axis CG_3 is directed along the vector $\mathbf{G}$, and the axis CG_1 is located in the Cxy plane of the main coordinate system and is directed along the line of intersection of the planes CG_1G_2 and Cxy in the direction of the ascending node of the plane CG_1G_2 (Figure 1). Let $G = |\mathbf{G}|$ be the module of the angular momentum vector, ρ and h are angles that uniquely determine the orientation of the coordinate system $CG_1G_2G_3$ axes of the with respect to the coordinate system $Cxyz$: ρ is an angle between the axis Cz and the angular momentum vector $\mathbf{G}$, h is an angle between the positive directions of the coordinate axes Cx and CG_1 (h is the longitude of the ascending node of the intermediate plane CG_1G_2).
The orientation of the axes $C\xi\eta\zeta$ with respect to the intermediate coordinate system $CG_1G_2G_3$ is determined by the Euler angles l, g, θ (Figure 1). Nutation angle θ is the angle between the positive directions of the axes CG_3 and $C\zeta$. The precession angle g is the angle between the axis CG_1 and the line of intersection of the coordinate planes CG_1G_2 and $C\xi\eta$. The angle of self rotation l is the angle between the positive direction to the ascending node of the $C\xi\eta$ plane and the $C\xi$ axis.
Thus, if the angles h and ρ set the orientation of the intermediate coordinate system $CG_1G_2G_3$ with respect to the main coordinate system, and the Euler angles l, g, θ set the orientation of the connected coordinate axes with respect to the intermediate coordinate system $CG_1G_2G_3$.

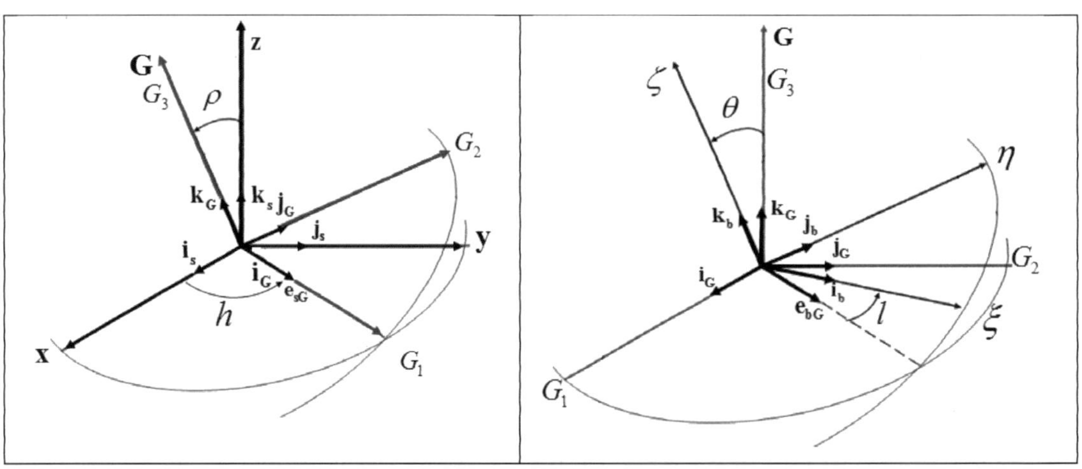

Figure 1. Basic coordinate systems and Andoyer variables.

IOP Conf. Series: Materials Science and Engineering **468** (2018) 012042 doi:10.1088/1757-899X/468/1/012042

In the figure 1 we denote the unit vectors of the corresponding coordinate systems $Cxyz$, $CG_1G_2G_3$ and $C\xi\eta\zeta$ respectively:

$$\mathbf{i}_s, \mathbf{j}_s, \mathbf{k}_s; \quad \mathbf{i}_G, \mathbf{j}_G, \mathbf{k}_G; \quad \mathbf{i}_b, \mathbf{j}_b, \mathbf{k}_b, \tag{2}$$

and also the unit vectors $\mathbf{e}_{bs}$, $\mathbf{e}_{Gs}$ and $\mathbf{e}_{bG}$ aredirected along the intersection lines of the corresponding coordinate planes Cxy, $C\xi\eta$, Cxy, CG_1G_2 and CG_1G_2, $C\xi\eta$ (figure 1). In other words Andoyer variables determine the magnitude and orientation of the angular momentum vector $\mathbf{G}$ of the body (Moon) in two coordinate systems: in the basic ecliptic coordinate system $Cxyz$, in the intermediate coordinate system $CG_1G_2G_3$ associated with the vector $\mathbf{G}$ and the axes $C\xi\eta\zeta$ associated with the celestial body (see Figure 1) [6], [7].

We define three more Andoyer variables: L, G, H. L is the projection of the vector $\mathbf{G}$ onto the polar axis of the body $C\zeta$, H is the projection of the vector $\mathbf{G}$ onto the axis Cz and G is the magnitude of the vector $\mathbf{G}$.

Obviously,

$$L = \mathbf{G} \cdot \mathbf{k}_b, \quad G = \mathbf{G} \cdot \mathbf{k}_G, \quad H = \mathbf{G} \cdot \mathbf{k}_s \tag{3}$$

or

$$L = G\cos\theta, \quad G = |\mathbf{G}|, \quad H = G\cos\rho. \tag{4}$$

The vector of the angular momentum $\mathbf{G}$ of the rotational motion of the body in the coordinate system $Cxyz$ is determined by the formula

$$\mathbf{G} = G_\xi \mathbf{i}_b + G_\eta \mathbf{j}_b + G_\zeta \mathbf{k}_b, \tag{5}$$

and its projections of the vector on the coordinate axes of the body are:

$$\begin{aligned} G_\xi &= Ap - Fq - Er + P, \\ G_\eta &= -Fp + Bq - Dr + Q, \\ G_\zeta &= -Ep - Dq + Cr + R. \end{aligned} \tag{6}$$

Here A, B, C and D, E, F are the axial and centrifugal moments of inertia of the body (by assumption, they are some known functions of time); P, Q and R are the projections of the angular momentum vector of the relative motion of the particles of the body on the coordinate axes $C\xi$, $C\eta$ and $C\zeta$. According to the assumptions of the problem these 9 dynamic characteristics are given functions of time and, in general, they can be represented as the sum of two terms:

$$\begin{aligned} &A = A_0 + \delta A(t), \quad B = B_0 + \delta B(t), \quad C = C_0 + \delta C, \\ &D = D_0 + \delta D(t), \quad E = E_0 + \delta E(t), \quad F = F_0 + \delta F(t), \end{aligned} \tag{7}$$

where A_0, B_0, C_0, D_0, E_0, F_0 are some constant (initial or unperturbed) values of the components of the body inertia tensor, for example, corresponding to the non-deformed state of the body, and

$$\delta A(t), \ \delta B(t), \ \delta C(t), \ \delta D(t), \ \delta E(t), \ \delta F(t) \tag{8}$$

are the perturbing additives to them due to the displacement of body particles and the corresponding density variations. We consider that the terms in (7) and the components of the angular momentum of the particles' relative motions P, Q and R are the given functions of time. They can be obtained sometimes from the satellite observations of variations of the Earth gravitational field or other celestial bodies.

In Andoyer variables, the components of the angular momentum (5)-(6) are determined by simple formulas [6]:

$$\begin{aligned} G_\xi &= \mathbf{G}\mathbf{k}_b = G\mathbf{k}_G\mathbf{i}_b = G\sin\theta\sin l = \sqrt{G^2 - L^2}\sin l, \\ G_\eta &= \mathbf{G}\mathbf{j}_b = G\mathbf{k}_G\mathbf{j}_b = G\sin\theta\cos l = \sqrt{G^2 - L^2}\cos l, \\ G_\zeta &= \mathbf{G}\mathbf{k}_b = G\mathbf{k}_G\mathbf{k}_b = G\cos\theta = L. \end{aligned} \tag{9}$$

To solve practical problems of studying forced variations in the rotation of specific celestial bodies (Earth, Moon, etc.) instead of the moments of inertia, it is sometimes convenient to use the second harmonic coefficients of the gravitational potential. Their standard non-normalized values are related to the moments of inertia (7) by simple relations [8]:

$$J_2 = -C_{20} = \frac{2C - A - B}{2mr_0^2}, \; C_{22} = \frac{B - A}{4mr_0^2}, \tag{10}$$

$$S_{22} = \frac{F}{2mr_0^2}, \qquad C_{21} = \frac{E}{mr_0^2}, \qquad S_{21} = \frac{D}{mr_0^2}$$

Accordingly, the differences of the axial moments of inertia of the planet and the product of inertia are expressed in terms of the geopotential coefficients by the formulas:

$$A - B = -4C_{22}mr_0^2, \; A - C = (C_{20} - 2C_{22})mr_0^2, \; B - C = (C_{20} + 2C_{22})mr_0^2, \tag{11}$$

$$F = 2S_{22}mr_0^2, \qquad E = C_{21}mr_0^2, \qquad D = S_{21}mr_0^2$$

If we introduce the polar dimensionless moment of inertia I by the formula $C = I \cdot mr_0^2$, then from formulas (11) we get the expressions for the moments of inertia

$$A = (I + C_{20} - 2C_{22})mr_0^2, \; B = (I + C_{20} + 2C_{22})mr_0^2, \quad C = I \cdot mr_0^2. \tag{12}$$

Let the axes of inertia that are associated with the body are the main ones in the absence of temporal variations of the moments of inertia (in the undeformed state). In this case, the constant parameters of the problem are related by the following relations:

$$J_2^{(0)} = -C_{20}^{(0)} = \frac{2C_0 - A_0 - B_0}{2mr_0^2}, \quad C_{22}^{(0)} = \frac{B_0 - A_0}{4mr_0^2}, \tag{13}$$

$$S_{22}^{(0)} = \frac{F_0}{2mr_0^2} = 0, \quad C_{21}^{(0)} = \frac{E_0}{mr_0^2} = 0, \qquad S_{21}^{(0)} = \frac{D_0}{mr_0^2} = 0.$$

Due to the deformations of the planet and, in general, due to any variations of its mass geometry, the geopotential parameters (10) experience temporal variations, for which similar relations take place:

IOP Conf. Series: Materials Science and Engineering **468** (2018) 012042 doi:10.1088/1757-899X/468/1/012042

$$\delta J_2 = -\delta C_{20} = \frac{2\delta C - \delta A - \delta B}{2mr_0^2}, \qquad \delta C_{22} = \frac{\delta B - \delta A}{4mr_0^2}, \tag{14}$$

$$\delta S_{22} = \frac{\delta F}{2mr_0^2}, \qquad \delta C_{21} = \frac{\delta E}{mr_0^2}, \qquad \delta S_{21} = \frac{\delta D}{mr_0^2}$$

Near-surface changes and redistributions of masses (for example, fluid masses for the Earth) make the greatest contribution to the temporal variations of the moments of inertia and coefficients of the geopotential. In these cases, in practice, an additional relation is used between variations of the moments of inertia,

$$\delta A + \delta B + \delta C = 0. \tag{15}$$

The relation (15) follows in fact from the definition of the spherical moment of inertia about the center of the planet, which does not change with arbitrary surface redistributions of masses in a thin spherical layer. From formulas (14) and (15) it is easy to get expressions for variations of the moments of inertia - components of the inertia tensor:

$$\frac{\delta A}{C} = -\frac{2}{J}\delta C_{22} - \frac{1}{3J}\delta J_2, \qquad \frac{\delta B}{C} = \frac{2}{J}\delta C_{22} - \frac{1}{3J}\delta J_2, \qquad \frac{\delta C}{C} = \frac{2}{3J}\delta J_2,$$

$$\frac{\delta F}{C} = \frac{2}{J}\delta S_{22}, \qquad \frac{\delta E}{C} = \frac{1}{J}\delta C_{21}, \qquad \frac{\delta D}{C} = \frac{1}{J}\delta S_{21}. \tag{16}$$

To describe the rotational motion of a weakly deformable moon in a potential force field, we use canonical equations in Andoyer variables. A detailed derivation of these equations is given in [6]. They have the form:

$$\frac{dl}{dt} = \frac{\partial K}{\partial L}, \quad \frac{dL}{dt} = -\frac{\partial K}{\partial l},$$

$$\frac{dg}{dt} = \frac{\partial K}{\partial G}, \quad \frac{dG}{dt} = -\frac{\partial K}{\partial g}, \tag{17}$$

$$\frac{dh}{dt} = \frac{\partial K}{\partial H}, \quad \frac{dH}{dt} = -\frac{\partial K}{\partial h},$$

$$K = \frac{1}{2}G^2\left\{\left(a\sin^2 l + b\cos^2 l - f\sin 2l\right)\sin^2\theta + c\cos^2\theta - \sin 2\theta\left(e\sin l + d\cos l\right)\right\} -$$

$$-G\left[\left(\Omega_\xi \sin l + \Omega_\eta \cos l\right)\sin\theta + \Omega_\zeta\cos\theta\right] - U\left(L, G, H, l, g, h, t\right). \tag{18}$$

Here entered the notation:

$$a = \frac{BC - D^2}{\Delta}, \quad b = \frac{AC - E^2}{\Delta}, \quad c = \frac{AB - F^2}{\Delta},$$

$$d = -\frac{ED + FC}{\Delta}, \quad e = -\frac{DF + BE}{\Delta}, \quad f = -\frac{FE + AD}{\Delta}, \tag{19}$$

$$\Delta = ABC - AD^2 - BE^2 - CF^2 - 2DEF.$$

The resulting formulas (19) are exact and, with the above assumptions, can be replaced by fairly simple approximate formulas.

The force function U in (18) should be represented as a function of canonical variables (1) and time. The latter problem is usually solved using known representations of the direction cosines a_{ij} of the coordinate axes $C\xi\eta\zeta$ of the body in the main coordinate system $Cxyz$: $a_{ij}(\theta,\rho,l,g,h)$. In the formulas for the Hamiltonian (18) and direction cosines we have:

$$\sin\theta=\frac{\sqrt{G^2-L^2}}{G},\quad \cos\theta=\frac{L}{G},\quad \sin\rho=\frac{\sqrt{G^2-H^2}}{G},\quad \cos\rho=\frac{H}{G}. \tag{20}$$

3. Elongation of the period of free movement of the pole of the moon due to its elasticity.

To estimate the period of oscillations of the pole of a weakly deformable Moon due to its rotational deformation, we will use the method proposed in [6], [9] to estimate the periods of Chandler oscillations of the poles of the Earth and Mars. According to this method, the rotational motion of a solid moon and a weakly deformable moon can be approximately described by the same Hamiltonian in shape (written in Andoyer variables), if we change the main moments of inertia of a solid model to close values with a special correction. If in the first case (rigid Moon) the Hamiltonian of the rotational motion has the form:

$$H=\frac{G^2-L^2}{2}\left(\frac{\sin^2 l}{A}+\frac{\cos^2 l}{B}\right)+\frac{L^2}{2C}-U, \tag{21}$$

then in the case of a weakly deformable Moon, the Hamiltonian retains its overall shape,

$$\mathrm{H}_0=\frac{G^2-L^2}{2}\left(\frac{\sin^2 l}{\mathrm{A}}+\frac{\cos^2 l}{\mathrm{B}}\right)+\frac{L^2}{2\mathrm{C}}, \tag{22}$$

where

$$\mathrm{A}=A+3D_\omega,\quad \mathrm{B}=B+3D_\omega,\quad \mathrm{C}=C \text{ и } D_\omega=k_2\frac{a_e^5\omega_0^2}{9f} \tag{23}$$

In expression (23) a_e is the equatorial radius of the Moon, ω_0 is the unperturbed value of the angular velocity of rotation of the Moon, k_2 is the Love number, f is the gravitational constant. In other words in the second case the axial moments of inertia of the body experience certain constant increments. Thus in the coordinate system $C\omega_1\omega_2\omega_3$ associated with the vector of angular velocity (the axis $C\omega_3$ is directed along the vector $\boldsymbol{\omega}$), the components of the inertia tensor of the equatorial thickenings of the Earth, caused by its rotational deformation, are determined by the formulas [10]: $A_\omega=B_\omega=D_\omega$, $C_\omega=2D_\omega$. To increment the polar moment of inertia of the Moon, due to its rotational deformation, we can use the relation

$$2D_\omega=\Delta C_M=\frac{2}{9}k_2\frac{a_e^5\omega_0^2}{f}=\frac{2}{9}k_2\frac{m_M r_M^2}{m_M f}\cdot\frac{a_e^2}{r_M^2}\omega_0^2 a_e^3. \tag{24}$$

According to modern observations [11], [12] we have the following values of the parameters of the Moon:

$$k_2 = 0.024, \quad C = 0.873486 \cdot 10^{42}\ g \cdot \text{cm}^2, \tag{25}$$

$$\frac{1}{m_M f}\omega_0^2 a_e^3 = 9.377115 \cdot 10^{-6}, \quad \frac{a_e^2}{r_M^2} = 1.00171619.$$

Here a_e is the average equatorial radius of the Moon. The angular velocity is comparable to its unperturbed value $\omega_0 = n_F$, n_F is the angular velocity of the change in the argument of lunar orbital motion. The corresponding period and the indicated rotation correspond to the motion of the Moon according to the laws of Cassini. Now, using formulas (24), (25), we find the variation of the polar moment of inertia of the Moon due to its rotational deformation:

$$\Delta C_M = \frac{2}{9} k_2 C_M \frac{1.00171619}{0.39349} \frac{1}{m_M f}\omega_0^2 a_e^3 = 1.352754 \cdot 10^{-7} C_M = 1.181612 \cdot 10^{35}\ \text{g} \cdot \text{cm}^2. \tag{26}$$

It can be said that the elastic Moon, deformed by its own rotation, rotates as some fictitious (additional) solid, but with changed moments of inertia.

So the elastic Moon (deformed by its own rotation) rotates according to the Euler- Poinsot laws, but like some other free solid body whose moments of inertia slightly differ from the main moments of inertia of the Moon itself A, B and C. We omit the intermediate calculations and give the final expressions for the Chandler period of motion of the pole of a weakly deformable Moon without taking into account the gravitational attraction of the central celestial body (Earth), and taking into account the gravitational attraction of the Earth taking into account the resonant properties of the perturbed orbital motion of the Moon.

$$T_q^* = \frac{T_F}{\sqrt{\left(\frac{\mathrm{B}}{\mathrm{A}} - 1\right)\left(\frac{\mathrm{B}}{\mathrm{C}} - 1\right)}}, \quad T_q = \frac{T_F}{\sqrt{\left(\frac{\mathrm{B}}{\mathrm{A}} - 1 - \frac{1}{B\Omega^2} \cdot \frac{\partial^2 \langle \mathbf{U} \rangle}{\partial l^2}\right)\left(\frac{\mathrm{B}}{\mathrm{C}} - 1 - B\frac{\partial^2 \langle \mathbf{U} \rangle}{\partial L^2}\right)}}. \tag{27}$$

Here T_F is the period of the daily rotation of a celestial body (around its axis). The first of these formulas in (27) is obtained for a free celestial body, i.e. not subject to gravitational influence from external celestial bodies. The second formula takes into account the gravitational influence of the central external body (for the Moon it is primarily the Earth, for the Earth and Mars it is the Sun) by means of the averaged value of the force function $\langle \mathbf{U} \rangle$ of its Newtonian interaction with the planet in question. In this case, disturbances in the orbital motion of the perturbing body are taken into account. The values of the gravitational components in the expression of the period of motion of the pole were calculated for the high-precision orbit of the Moon and they are according to our estimates:

$$\frac{1}{B\Omega^2} \cdot \frac{\partial^2 \langle U \rangle}{\partial l^2} = -1830.440218 \cdot 10^{-6} \qquad B\frac{\partial^2 \langle U \rangle}{\partial L^2} = 0.184727 \cdot 10^{-6}. \tag{28}$$

When calculating the dimensionless quantities (28), the coefficients of the second harmonic of the selenium potential were used according to its modern model obtained during the implementation of the Japanese project Selenium [13].

Table 1 compares the Chandler periods calculated by the formulas of our theory for the Earth, Mars and the Moon for comparison. The Euler period $\mathbf{T_{Eu}}$ and the Chandler period $\mathbf{T_{Ch}}$ for the Earth, Mars and the Moon (first, second and third row) are calculated using the first formula from (27) for the observed inertia moments *A, B* and *C*, and for reduced inertia moments *A, B* and *C* respectively. In these calculations, the gravitational influence of the central body on the movement of the pole is not taken into account. The fourth line contains the values of the periods $\mathbf{T}_{\mathbf{Eu}}$ and $\mathbf{T}_{\mathbf{Ch}}$ and already taking into account the gravitational contribution to the motion of the pole. The values of the indicated periods are given in days, the differences between the periods in days and in percentages are given in relation to the values $\mathbf{T}_{\mathbf{Eu}}$ and $\mathbf{T}_{\mathbf{Ch}}$ of the periods for the inelastic planet (or the Moon).

Table 1. Influence of elasticity on the period of motion of the pole of the Moon and planets.

Planet	$\mathbf{T}_{\mathbf{Eu}}$ (day)	$\mathbf{T}_{\mathbf{Ch}}$ (day)	$\Delta\mathbf{T}_{\mathbf{Ch}}$ (day)	$\Delta\mathbf{T}_{\mathbf{Ch}}$ %
Earth	**304.300**	**447.086**	**142.786**	$46.92\%\mathbf{T}_{\mathbf{Eu}}$
Mars	**191.816**	**218.049**	**26.233**	$13.68\%\mathbf{T}_{\mathbf{Eu}}$
Moon*	53922.70	53944.95	22.70	$0.042\%\mathbf{T}_{Eu}$
Moon	**27267.65**	**27275.62**	**7.97**	$0.029\%\mathbf{T}_{q}$

It is important to emphasize that in the case of the Moon, the formula for the period T_q^* in (27) is not at all applicable for estimating the period of free movement of the Moon's pole. Since the additional terms in the expressions of the period T_q (27) are comparable in magnitude with the first terms and actually lead to a decrease in the period of the pole movement by about 2 times [14]. This is well illustrated by the values of the period of the pole oscillations without taking into account the force interaction (line 3) and taking that into account (line 4). For illustration, here are the values of the periods (for solid and elastic models) for the Earth and Mars. They increase, respectively, by 47% and 14%. The latter values are obtained by the first formula of (27), in which the attraction of the central body is not taken into account. Thus, the elasticity of the planet most significantly affects the period of motion of the pole of the axis of rotation, and in the case of the Moon (synchronous satellite), this influence is small, but significant for a highly accurate description of its rotation, which is significant and significant for the implementation of the project for studying the orientation of the Moon using a telescope mounted on the surface of the Moon.

References

[1] Vilke V G, Markov Yu G 1988 Evolution of translator-rotational motion of an viscoelastic planet in the gravitational field *Astr. Journ.*, Vol. **65** №4 pp 861–867 [in Russian]

[2] Vilke V G, Shatina A V 2001 Motion evolution of a double planet *Cosmic Research* Vol. **39**. No 3. pp 295–305.

[3] Hanada H et al. 2005 Application of a PZT telescope to In situ Lunar orientation Measurements (ILOM). *Proc. Int. Ass. Geod.* (Springer) pp 163-168.

[4] Barkin Yu, Matsumoto K, Hanada H, Sasaki S, Petrova N, Barkin M 2012 The influence of elastic properties of the Moon on its pole motion *Book of abstracts of the 118th Meeting of the Geodetic Society of Japan, Sendai, Japan, (29 October – 2 November 2012, Sendai)* pp 149-150.

[5] Barkin Yu V, Dyomin V G 1984 Dynamics of the system of rigid and variable celestial bodies *Book of abstracts of the 8th Republic interuniversity mathematics and mechanics conference (4-6 of September 1984) Part III. Theoretical and applied mechanics, Alma-Ata* pp 10. [in Russian]

IOP Conf. Series: Materials Science and Engineering **468** (2018) 012042 doi:10.1088/1757-899X/468/1/012042

[6] Barkin Yu V 2000 Perturbated rotational motion of weakly deformable celestial bodies *Astronomical and Astrophysical Transactions* Vol.**19**. Issue 1, pp 19-65. doi: 10.1080/10556790008241350
[7] Arkhangelsky Yu A 1977 *Analytical dynamics of a rigid body* (Moscow: Nauka) p 328 [in Russian]
[8] Aksyonov E P 1977 *Theory of artificial satellites' motion* (Moscow: Nauka) p 360. [in Russian]
[9] Barkin Yu, Ferrandiz J 2010 Elliptical Chandler pole motions of the Earth and Mars *EGU General Assembly 2010, held 2-7 May, 2010 in Vienna, Austria*, p. 2936
[10] Munk W, MacDonald G 1960 *The rotation of the Earth* (Cambridge University Press), p 348.

Acknowledgments
The work was fulfilled with the financial support of the Russian Science Foundation (Project No. 17-71-10254).

IOP Conf. Series: Materials Science and Engineering **468** (2018) 012043 doi:10.1088/1757-899X/468/1/012043

Coupled Multi-Physics Modelling in Continuum Mechanics

Murashkin E.V.[1,2] and Radayev Y.N.[2]

[1]Bauman Moscow State Technical University, Moscow, Russia
[2]Ishlinsky Institute for Problems in Mechanics of the Russian Academy of Sciences, Moscow, Russia
E-mail: evmurashkin@gmail.com, radayev@ipmnet.ru

Abstract. Contemporary models of solids of thermoelasticity requires to include multiphysics coupling and employ non classical e.g. elastic behavior. The permanent generalization of elastic model is the Cosseratt micropolar model. Now this model is to be applied to growing solids, biomaterials, granular media, concrete. The basic concepts of the continuum mechanics is considered in connection with the difference in the dimensions of the continuum and external space. New field variables are introduced representing complex continuum properties. A generalization of the model of the micropolar continuum has been proposed. The action and the action density for the complex continuum model are discussed. A new field-theoretic model of a nonlinearly elastic continuum is developed assuming existence of an isometric immersing into an external plane space.

1. Introduction

Modern mathematical continuum models are based on the notion of *differentiable manifold* (or *differentiable space*) [1, 2]. Such models are best applicable to the mechanical behavior of solids, for example solids with microstructure. The continuum is assumed as of a differentiable manifold of a given dimension M ; $M = 3$ in applied problem.

It is often necessary to introduce additional structures on a differentiable manifold. For example, a Riemannian structure on a manifold allows us to define a differentiable spaces class [3], [4] playing an important role. The theory of finite deformations known from many discussions can be readily developed for a continuum with a Riemannian metric structure.

Historically, the theory of manifolds takes its origin from the geometry of curved surfaces in three-dimensional Euclidean space. Development of the theory of differentiable manifolds came to rejection of an outer (external) space notion. Internal geometry of manifolds as a special part of the theory has been developed. However, the external space with immersed solids is a permanent attribute of the continuum mechanics.

For a long time, such views do not cause any objections. Along with affine-metric structures of various complexity levels have been used to modeling continua with defects and damages. The best example of a complex continuum is a growing solid manufacturing by a specific technology or synthesized by a complex physicochemical process. The question about the possibility of immersion of a continuum in a 3D plane space has been never discussed previously. It is well known that continua with a complex affine-metric structure can not be immersed in a three-dimensional external plane space. Therefore it is impossible give a reason to deformation of such a continuum from the view point of three-dimensional representations and forms. Fundamentally new assumption of this study is the rejection of the external space concept as a 3D plane space with an immersed solid.

Content from this work may be used under the terms of the Creative Commons Attribution 3.0 licence. Any further distribution of this work must maintain attribution to the author(s) and the title of the work, journal citation and DOI.
Published under licence by IOP Publishing Ltd

The present study deals with the foundations of the new mechanics of a complex continuum of an arbitrary dimension M admitting an existence of a plane external space of dimension $N \geq M$. We are going to incorporate the formalism of the physical field theory [5, 6] into theory of differentiable manifold.

2. Mechanics of a complex continuum

A vicinity of each differentiable manifold point can be mapped to a domain of an affine space of dimension M. In this sense, the manifold admits local comparisons with domains of an affine space. The number of maps is not more than countable. Images of the manifold on intersecting maps are to be consistent. The concept of manifold (and consequently continuum) does not require any determination of the outer space into the continuum can be immersed. Individual points of the continuum are represented by a special variables denoted by ξ identifying by the map coordinates ξ^{α}. Thus, there is a correspondence (coordinate mapping)

$$\xi \to \xi^{\alpha}, \tag{1}$$

representing by ξ^{α} a continuum point ξ on a map.

The idea of an outer space is explicitly or implicitly accepted in continuum mechanics. The solids are assumed to be immersed into an external space, providing an opportunity for the observer to discriminate the positions of material point in space, to measure distances and angles, and to register the changes of lengths and distortions of the angles of the linear material elements. From this point we able to introduce the concept of deformation. The external space is tacitly assumed to be 3D with the natural Euclidean metric. Another important assumption is that a continuum must always allow an embedding in the same external space during deforming.

Since the continuum is modeled by some differentiable manifold, the question immediately arises of the possibility of its immersion into an external space. It is well known that such immersion in a 3D plane space is not always possible. For instance, a Riemannian manifold of dimension M in the general case is immersed in a plane space of dimension $N = \frac{1}{2}M(M+1)$. In any case, the following inequality is fulfilled

$$N \geq M. \tag{2}$$

Let us elucidate the notion of a plane space. The plane space is usually defined as a special class of Riemannian spaces (see in details [3]). A Riemannian space is to be plane one if there is a coordinate system with constant components of the metric tensor. In a plane space there is always a coordinate system y^j ($j = 1, \ldots, N$) with metrics determined by the differential form

$$ds^2 = \sum_j c_j dy^j dy^j \qquad (j = 1, \ldots, N),$$

where constants $c_j = \pm 1$ depend on space type. The coordinate system y^j is called *Cartesian coordinate system*. It is clear that for an external space the Cartesian coordinate system is preferable.

We can define *embedding mapping (or emorsion)* for a continuum that allows immersion in an external space establishing a correspondence between the points of the continuum ξ and positions (places) in external space. If the continuum admits an immersion into some external plane space of dimension N then every point ξ of the continuum takes some position X. The position X in external space in the continuum mechanics is called referential one. By virtue of the embedding map, the referential position X must be bijectively connected with the variable ξ:

$$\xi \to X. \tag{3}$$

It is possible taking into account the equations (1) and (3) to parameterize the referential position of

IOP Conf. Series: Materials Science and Engineering **468** (2018) 012043 doi:10.1088/1757-899X/468/1/012043

the continuum in the external space by the coordinates ξ^{α} :

$$X \to \xi^{\alpha}. \tag{4}$$

The points of the continuum change their positions in the external space during deformation process. The spatial position of individual material point ξ is denoted by x. In the general form, the deformation of the continuum can be expressed by mapping the individual points of the continuum into actual positions in the external space

$$\xi \to x. \tag{5}$$

Transformation (also called deformation) of the position variables in the external space according to equations (3) and (5) reads

$$X \to x. \tag{6}$$

Variables X and x can be represented by curvilinear coordinates X^{α} ($\alpha = 1, \ldots, M$), x^j ($j = 1, \ldots, N$). The coordinates ξ^{α} can be equally used instead of the coordinates X^{α}. The (Lagrangian) metrics $\grave{g}_{\alpha\beta}$ is directly inherited from the external space. The spatial (Eulerian) metrics g_{ij} is the one of the external space. The convective metrics is characterized by the metric tensor $g_{\alpha\beta}$ and is defined by deformation (6) contrary to the metrics $\grave{g}_{\alpha\beta}$ and g_{ij}.

Modeling of the mechanical behavior of complex continua may require the introduction of additional degrees of freedom. Additional degrees of freedom (for example, rotational ones) are introduced into the theory in the form of tensor variables with contravarianrt indices, just as it does for translational freedom degrees related to the external space. These tensor variables are called d-variables and they are denoted by d^k. d-variables in the case of a micropolar continuum are a set of unit vectors (directors). The generalization of the micropolar continuum model immersed in a space of greater dimension is achieved due to the increase in the number of linearly independent directors associated with the infinitesimal element.

Furthermore, additional field variables can be used in the mechanics of a complex continuum. They are denoted also in the vector form by contravariant components φ^k. As an example, we note that in the hyperbolic thermomechanical continuum, the temperature displacement acts as an additional field variable.[1]

Deformation gradient

$$\partial_{\alpha} x^j \qquad (j = 1, \ldots, N; \quad \alpha = 1, \ldots, M), \tag{7}$$

where ∂_{α} denotes the partial differentiation with respect to the referential variable characterizes the elementary affine deformation. The deformation gradient $\partial_{\alpha} x^j$ transforms the referential linear element dX^{α} to its actual position in the external space dx^j according to

$$dx^j = \left(dX^{\alpha}\right)\partial_{\alpha} x^j. \tag{8}$$

The convective metrics $g_{\alpha\beta}$ is obtained by a deformation gradient and the Eulerian metrics g_{ij}

$$g_{\alpha\beta} = g_{ij}\left(\partial_{\alpha} x^i\right)\left(\partial_{\beta} x^j\right). \tag{9}$$

In the case $M = N$, a complete study of the finite deformation geometry can be given regardless of

[1] The hyperbolic coupled thermomechanical continuum model is developed in monograph [6]. There thermal field variable is denoted by ϑ.

the mathematical dimension (see, for example, [7]). The study is based on the possibility of a polar decomposition of an arbitrary tensor of the second rank [8]. The polar decomposition of deformation gradient $\partial_\alpha x^j$ ($j=1, \ldots, N$; $\alpha=1, \ldots, M$) allows to present it in the form of a multiplicative composition of pure deformation (stretching relative to mutually orthogonal principal deformation axes in the referential position) and rotation in the external space of the referential polyhedron of the principal deformation axes to its actual position. Nothing similar can be said in the case when

$$N > M.$$

Further development of the theory is carried out within the frameworks of physical field theory. Therefore, we need to introduce the action integral and its density called also as Lagrangian. The functional arguments of the Lagrangian are field variables and their gradients. In the case of field theory on a P–dimension manifold of the action is a P–dimension integral variantional functional:

$$\mathrm{I} = \int \mathrm{L}\left(\varphi^k\ \partial_\alpha \varphi^k, \partial_\gamma \partial_\alpha \varphi^k, \ldots, X^\beta\right) d^P X, \tag{10}$$

where the integration extends to an unspecified domain of the manifold
L denotes natural Lagrangian density (action density);
φ^k are physical field variables;
X^β ($\beta=1, \ldots, P$) are space–time coordinates;
$d^P X$ are natural elemental volume (product of the coordinate differentials).

In the case of a complex continuum immersed in an external space the coordinates of the external space x^j are treated as physical fields. The same is true for the system of d–variables and additional field variables. So the action density L can be given by the following functional form

$$\mathrm{L} = \mathrm{L}\left(x^j, d^j, \varphi^j, \dot{x}^j, \dot{d}^j, \dot{\varphi}^j, \partial_\alpha x^j, \partial_\alpha d^j, \partial_\alpha \varphi^j, X^\beta\right). \tag{11}$$

The functional form (11) plays a significant role in the mathematical model of a complex continuum. In addition to the differential field equations and the constitutive equations this form allow us to construct the objective deformation tensors.

The first question in a study of complex continua is the strain tensors being an objective one and their complete systems sufficient for the mathematical representation of the deformed state. Objectivity is understood as the invariance of the strain tensors with respect to rotations of the preferred coordinate systems of the external space. The same can be said on the translations of coordinate systems.

The action density L is an objective function. It must invariant under translations of the coordinates x^j. As a result, we can conclude that the functional form (11) is clearly not dependent on external spatial coordinates (in this case, on the coordinates y^j). Therefore, it is reduced to the following form (we confine ourselves to static states):

$$\mathrm{L} = \mathrm{L}\left(d^j, \partial_\alpha y^j, \partial_\alpha d^j\right). \tag{12}$$

For the same reason, the action density L should not depend on rotations of the preferred coordinate system polyhedron in the external space. The invariance of the action density with respect to the rotations of the coordinate frame is a manifestation of the isotropy of the external space, i.e. absence of preferred directions in this space. For a functional form of the equation (12) this means that it actually functionally depends only on such algebraic combinations of contravariant vectors of the external space

$$d^j,\ \partial_\alpha y^j,\ \partial_\alpha d^j, \tag{13}$$

which invariant with respect to orthogonal transformations of the external coordinate system y^j.

The problem of constructing rational combinations of the contravariant vectors 13 that would not be sensitive to rotations of the preferred external coordinate system is solved by using the algebraic theory of rational invariants of tensors and tensor systems [9]. In particular, all irreducible rational invariant forms for the system (13) can be easily enumerated: they are inner and skew products of the space vectors (13). The mentioned predicts form a system of independent objective strain tensors. Consider a nonlinearly elastic continuum of dimension M whose metrics admits its isometric embedding into an external plane space of the same dimension. The functional form (12) in this case reverts to

$$\mathrm{L} = \mathrm{L}\left(\partial_\alpha y^j\right), \tag{14}$$

where the functional arguments are the M contravariant vectors of the external space

$$y^j. \tag{15}$$

The form (14) of irreducible invariant functional arguments of the Lagrangian L is furnished by

$$\partial_\alpha y^j \partial_\beta y^j, \qquad \det\left(\partial_\alpha y^j\right). \tag{16}$$

The first combination in 16 is the convective metrics of the deformed continuum $g_{\alpha\beta}$; the second one is the Jacobian of the deformation related to the convective metrics by the rational equation

$$(\det\left(\partial_\alpha y^j\right))^2 = \det\left(g_{\alpha\beta}\right). \tag{17}$$

The latter equation is considered from the algebraic point of view as a *syzygy* for the system of invariants (16), and from mechanical viewpoint it is the continuity equation in Lagrange variables. If we remain within the frameworks of such a scheme, which involves operating only with rational invariants, then the syzygy (17) should be taken into account as a kinematic constraint, and the Lagrangian (14) in accordance with the rule of Lagrange multipliers should be replaced by a new one

$$\mathrm{L}^{å}\left(\partial_\alpha y^j \partial_\beta y^j, \det\left(\partial_\alpha y^j\right)\right) = \mathrm{L}\left(\partial_\alpha y^j \partial_\beta y^j, \det\left(\partial_\alpha y^j\right)\right) - \lambda\left[(\det\left(\partial_\alpha y^j\right))^2 - \det\left(g_{\alpha\beta}\right)\right],. \tag{18}$$

where the functional arguments (16) should be considered as kinematically independent, λ denotes indefinite Lagrange multiplier.

3. Conclusions

- The basic concepts of the continuum mechanics have been elucidated in connection with the difference in the dimensions of the continuum and external space.
- New field variables have been introduced representing complex continuum properties. A generalization of the model of the micropolar continuum has been proposed.
- The action and the action density for the complex continuum model have been discussed.
- A new field-theoretic model of a nonlinearly elastic continuum has been developed assuming existence of an isometric immersing into an external plane space.

References

[1] Arnol'd V I 1989 *Mathematical Methods of Classical Mechanics*. (NY: Springer-Verlag)
[2] Mishchenko A and Fomenko A 1988 *A Course of Differential Geometry and Topology.* (Moscow: MIR Publishers)
[3] Eisenhart L P 1968 *Riemannian Geometry.* (Princeton University Press)
[4] Rashevskii P K 1967 *Riemannian Geometry and Tensor Analysis.* (Moscow: Nauka)
[5] Kovalev V A and Radayev Yu N 2009 *Wave Problems of the Field Theory and Cross-Coupled Thermoelastic Waves.* (Moscow: Fizmatlit)
[6] Kovalev V A and Radayev Y N 2010 *Wave Problems of the Field Theory and Thermomechanics.* (Saratov University Press)
[7] Sedov L I 1997 *Mechanics of Continuous Media.* (Singapore: World Scientific Pub Co Inc)
[8] Horn R A and Johnson C R 2012 *Matrix Analysis (2nd Edition).* (Cambridge University Press)
[9] Gurevich G B 1964 *Foundations of the theory of algebraic invariants.* (Groningen: P. Noordhoff)

Acknowledgements

The present study is financially supported by the Ministry of Science and Higher Education of the Russian Federation (State Registration Number AAAA-A17-117021310381-8) and by the Russian Foundation for Basic Research (project No. 18-01-00844).

Precise measuring-computational predictive monitoring – the conditional of self-organization of the life cycle

M I Kiselev

Bauman Moscow State Technical University, Moscow, Russia
E-mail: vip-u@yandex.ru

Abstract. Formation of aviation which was laid down by the fundamental works of N.E. Zhkovsky and its further development at the forefront of scientific and technological progress occurs in the constant overcoming of the information deficit. The biosphere gives samples of successful provision of reliable functioning, but the perfection of living organisms is achieved through natural selection, which lasts for thousands of years. Improving the objects of the technosphere requires an increase in the level of information and metrological support for the production and operation of facilities. At present, the parameters of the basic structural materials are known with relative error at the level of percent or its fractions. The metrological level of the control and diagnostic equipment is the same. The lack of information is compensated by long and costly experimental and production work-up, as well as by the system of preventive maintenance and inspections, but is not a complete guarantee against accidents and catastrophes. The existing information barrier can be overcome by the transition from amplitude to phase measurement methods. The phase-chronometric version of this approach provides effective monitoring of the cyclic systems - from the clock mechanism to the powerful turbo-aggregate with a relative error of at least $5 \cdot 10^{-4}$ % at the industrial frequency. The combination of a precise phase-chronometric system with a mathematical model of the object that is constantly refined from the received measurement data makes it possible to realize a measurement and computational monitoring of its technical state and emergency protection. Simultaneously, the problem of information and metrological support of the life cycle of objects included into the self-organizing system of the Research Institute - Design ofiice - Plant - operated facility is being solved.
Keywords: phase-chronome1tric method, monitoring, cyclic system, the amplitude-phase method, biosphere, technosphere.

1. Introduction

A transition from aeronautics including a balloon with a motor – a dirigible, to the flight of "heavier than air" apparatus - gliders and aircrafts equipped with engines is associated with the name of N.E. Zhukovsky. An invaluable "consultant" on the most complicated path of scientific and design search was the Nature itself. Thus, the cross sectional profile of a subsonic airplane wing (the famous "profile of the NEZh"), which provides the formation of the lift force in flight, actually repeats a similar profile of the deployed wing of the soaring bird. But, moreover, it became clear over time that the mode of unconstrained soar of a kite spreaded its wings is provided not only by using the ascending air currents that are sensitively captured by the bird, but, that is especially remarkable, by the operative and fine stabilizing correction of the spatial structure of the incoming air flow directly

Content from this work may be used under the terms of the Creative Commons Attribution 3.0 licence. Any further distribution of this work must maintain attribution to the author(s) and the title of the work, journal citation and DOI.
Published under licence by IOP Publishing Ltd

adjacent to the bird's plumage. This correction is provided by the information obtained from the nerve (sensory) endings distributed along birds' body surface, which are transmitted through the nerve network to the muscles acting as the drive for the associated with them feathers - elementary executive organs.

This perfect adaptive system is the product of evolutionary development of the biosphere that lasts thousands of years and a natural selection is playing a decisive role in its improvement (according to Charles Darwin). [1]

Improving the facilities of the technosphere is also stimulated by the processes taking place in the social sphere, where profitability, effectiveness of combat use, reliability are decisive in the competitive struggle of goods. At the same time, the specific characteristics of the biosphere objects are still often remaining an inspiring example for the creators of new technology.

2. The main obstacles to the development of aviation and cosmonautics

Aviation and cosmonautics are the cutting edge of scientific and technological progress, forced to develop on the boundary between knowledge and ignorance. Corkscrew, shimmy, flutter, vibrational burning, sparging is not a complete list of the fundamental obstacles that have stood in the way of development of these technology branches.

Many of these phenomena, primarily flutter, have a threshold character inherent in self-oscillating systems, and are now sufficiently well studied. The incentive to this was the application of aviation technology achievements in power engineering and transport (for example, gas turbine drives of electric generators and pumps for gas transportation, ground and water hovercraft, gliding, underwater wings).

However, besides the whole range of stability and mass dimensional optimization problems arising here, the problem of time-factor influence for a fleet of actively operated equipment becomes more and more urgent. This problem significantly differs from the moral aging one, gaining its value in the conditions of a modern market economy with its mechanisms of competition.

The current situation is that if earlier the main efforts of science and production were aimed at improving the efficiency of the work cycle, then now ensuring the efficiency of a life cycle gets an increasing importance at all its stages. The situation is aggravated by the growing need for a multifactorial optimization of the interaction of the technosphere, the biosphere and the socio-economic sphere.

3. The tasks of ensuring reliability and durability, estimating the residual re-source and obtaining an informed forecast of failure time

At the same time, the tasks of ensuring reliability and durability, estimating the residual resource and obtaining an informed forecast of failure time -an accident or a catastrophe-are at the foreground.

To solve these problems, it is necessary to provide operational precision measurement control (monitoring) not only for the deviation of the parameters of the structural elements of the functioning technical facilities, but also for the physical and mechanical characteristics of the structural materials themselves.

It is obvious that the stability of the parameters of the used information-measuring system should be as high as possible.

It is known that the technical tools of chronometry possess the greatest stability, and time and frequency are physical quantities measured with minimal error. The situation most objectively characterizes the fact that the physical properties of structural materials - metals, alloys, various plastics and composites are known, as a rule, with accuracy up to the third or fourth decimal place, which corresponds to (0.1-1.0)% of the nominal value.

If we take into account that the relative errors in measuring the parameters characterizing the technical state of functioning objects with modern control and diagnostic instruments based mainly on vibro-acoustic principles have practically the same accuracy level, we can also judge the level of metrological support of the technosphere as a whole.

IOP Conf. Series: Materials Science and Engineering **468** (2018) 012044 doi:10.1088/1757-899X/468/1/012044

As a result, the maintenance of the working capacity of national equipment is largely ensured by the established system of preventive maintenance and preventive inspections and by a thorough and long-term multi-stage experimental development of the created equipment, starting with laboratory studies of the first models and makets to pilot-industrial tests of the head samples of the series.

4. Systematic monitoring of measurement of technical condition of cyclic aggregates

Studies have shown that it is expedient to maintain precise systematic measurement control of the technical state of cyclic, primarily rotary machines and mechanisms, recording the passage moments of the boundaries of the working cycle phases intervals. It is most rational to ensure the equality of these intervals, and to make their boundaries like contrasting marks applied to one of the elements of the running gear performing periodic (oscillatory or rotational) motions. It is established that the variations of the working cycle phase passage duration, formed from cycle to cycle, contain information about the technical state of the measured object and its evolution.

Indeed, such an approach, both to miniaturized clockwork and to powerful (up to 300 MW) turbine units has opened fundamentally new opportunities for their study directly in the process of functioning. [2]

Thus, for a clock mechanism, a long-distance and contactless (optoelectronic) recording of the oscillation periods of its balanced wheel made it possible to study the response of the mechanism to all possible perturbing effects such as shock, vibration, temperature variations, etc. Obtaining essentially new measurement information was achieved by reducing the errors in measuring the oscillation period of the balance to 1 μs, using the "quartz" time. Traditional watch calibration devices, previously used in the watch industry, were based on the vibro-acoustic principle and had a greater error. It guaranteed only the level of qualitative research.

Attempts at interpreting the results of optoelectronic measuring control of clockwork mechanisms functioning led to the need to refine the mathematical model of their functioning. The result was the substantiation of the principle possibility of creating not only a system for automatic designing of watch movements, but also the rationale for the intellectualization of their production without the participation of personnel. [3] However, the practical implementation of the current concept in the watch industry did not take place due to the loss of relevance of their mass industrial production in our country.

Nevertheless, the developed approach attracted the attention of specialists in the field of electric power engineering in connection with the emerging problems of self-excitation of emergency-dangerous torsional oscillations of the running gear of the turbine unit and the physical aging of the power generating equipment. The seriousness of the problem was evidenced by the well-known accidents that occurred at Kashirskaya (2002) and Reftinskaya (2006) electrostations. In connection with this the indispensable, windy systems of systematic precise measurement control of the technical condition of turbo-aggregates were required. Such systems were realized using the phase-chronometric approach. Significantly, in the transition from miniature clock mechanisms to massive and large-sized industrial turbo-aggregates (power ~ 200-300 MW), the absolute error in measuring the period and its multiple fractions of the turn of the running gear was 10^{-7} s, which corresponds to a relative error of $\sim 5 \cdot 10^{-4}$at the industrial frequency. [4]

The efficiency of the application of such a phase-chronometric system is characterized by the possibility of providing a detailed registration of the rotational mode of the turbo-generator shaft at a level of 19 (one corner minute). Due to this, constant torsional oscillations of the rotating undercarriage of the turbine unit in a stationary mode of operation and bursts of torsional vibrations of the rotating undercarriage of the turbine unit, excited by incoming current and voltage impulses coming to the stator windings of the generator, coming from the network where they were excited by a lightning discharge were detected (Figure 1).

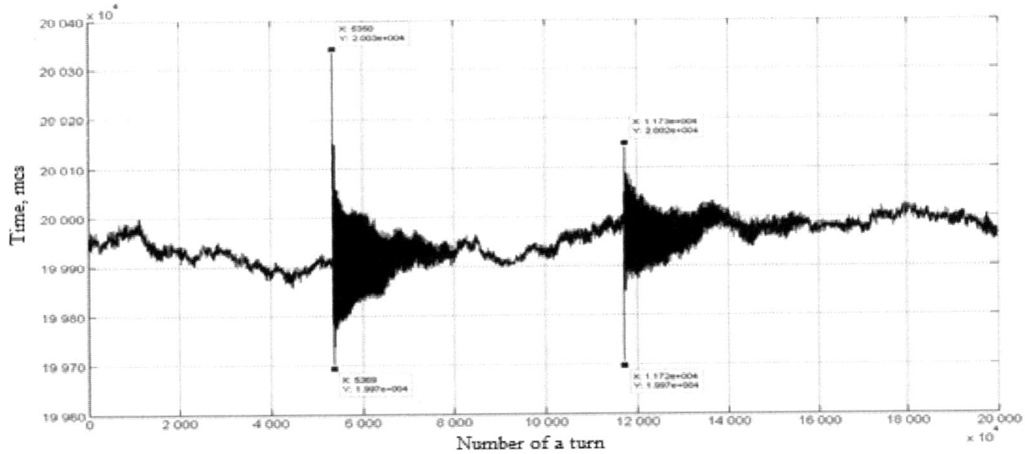

Figure 1. Reaction (surge of torsional vibrations) of a rotating multi-ton turbine unit to a lightning discharge

The effectiveness of the approach is also evidenced by the spectra of torsional oscillations of the shafting line, obtained as a result of mathematical processing of the regular periodograms (Figure 2) for the case of the normal functioning mode (Figure 3) and in the presence of defects (Figure 4).

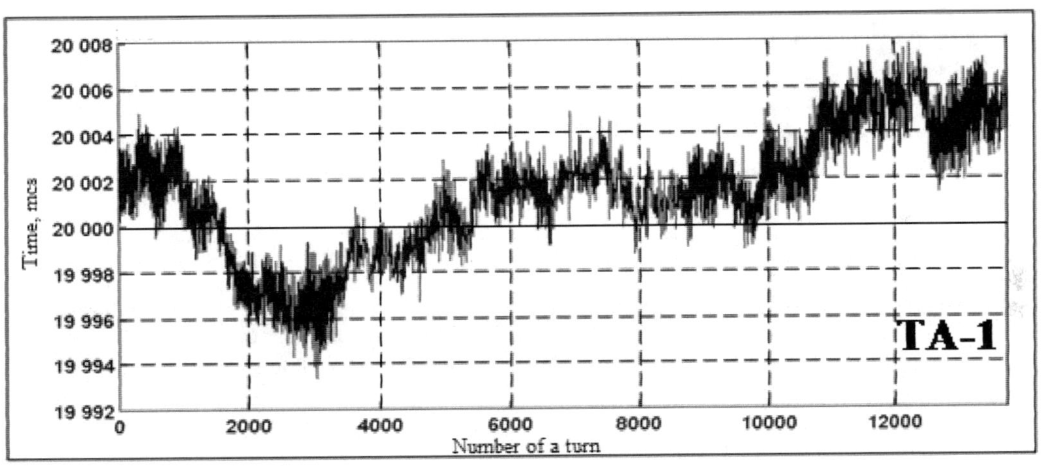

Figure 2. Periodogram of the turbine unit shaft rotation, nominal mode

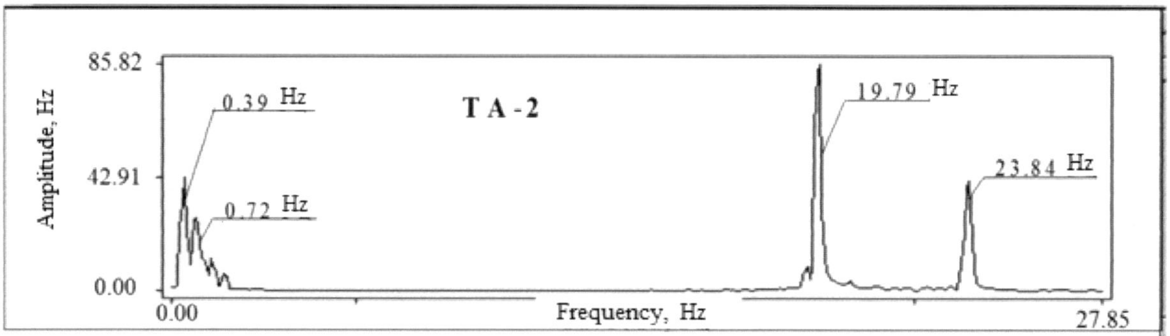

Figure 3. Spectrum of torsional oscillations of the shaft of the turbine unit in normal conditions

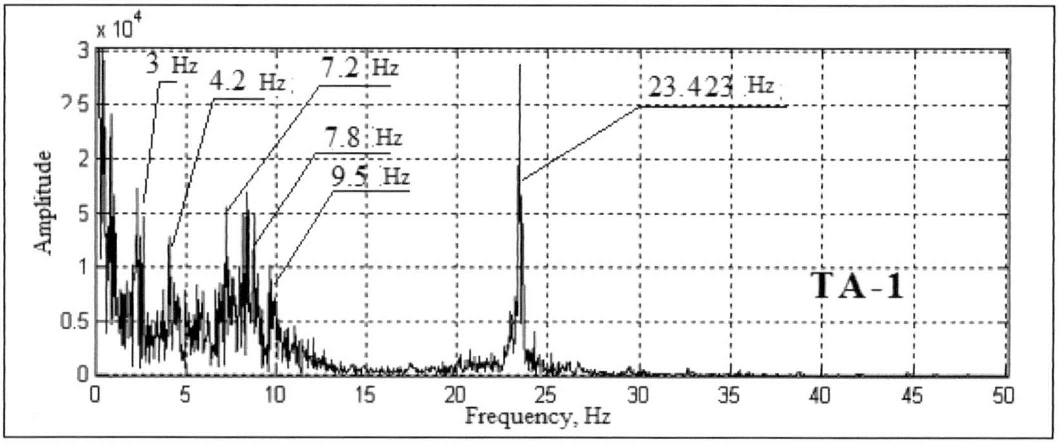

Figure 4. Spectrum of torsional oscillations of the shaft of the turbine unit in abnormal conditions

These results served to justify the scheme for obtaining the frequency characteristics of a turbo-aggregate as a multimass system formed by a set of running parts of a high-pressure turbine, two medium pressure turbines, a low-pressure turbine, an electric generator and an exciter connected by sections of a single shafting line.

The achieved accuracy makes it possible to use the accumulated detailed information about the current technical state of the turbo unit, its evolution and to build its forecast (estimate of the residual resource).

The effectiveness of these works will undoubtedly increase if they are accompanied by the development of precise materials science, which will ensure the achievement of the necessary accuracy of the values of physicomechanical parameters of structural materials. Their substantial refinement can be achieved by conducting studies using

- self-oscillators with resonators, made from the studied materials in the form of strings, shells, tuning forks, etc.,
- the apparatus of the thermodynamics of nonequilibrium processes and, if necessary, mesomechanics.

The achieved level of development testifies to the reality and expediency of applying the phase-chronometric approach to cyclical machines and mechanisms in transport, in the processing and extractive industries, up to the housing and communal sphere. [5] - [9]

Special mention should be made of the reality and prospects of the developed approach to information and metrological support of the life cycle of industrial products in the mode of its self-organization. [10]

It seems real not so much the expediency of such approaches as their inevitability in connection with the entry of our civilization in the era of the fourth industrial revolution ("Industry 4.0" or "Internet-things"), when unoccupied industrial production on the basis of intellectual technology with its mathematical support using Neural networks are more efficient and competitive than its traditional forms. [11]

An example of active interest in "Industry 4.0" at the level of industrial enterprises and departments is demonstrated by the Republic of Kazakhstan. [12]

It is known that an increase in the accuracy of measurements of the functioning objects parameters of the technosphere provides not only an improvement in the quality of the assessment of their technical condition, but also broadens the horizon of its forecast. At the same time, the problem of emergency mathematical analysis of a fast working cycle (for example, for a gas turbine engine) becomes more urgent with the aim of making a decision promptly.

5. Conclusion

But what is more, modern science demonstrates undoubted achievements in reliable mathematical support of scientific forecasts in the socio-historical sphere. [13], [14].

References

[1] Kiselev M I 2013 Why do we need this accuracy? *Metrology* **7** pp 4-7

[2] Kiselev M Iand Pronyakin V I 2001 Phase method of investigation of cyclic machines and mechanisms based on the chronometric approach *Measuring technique* **9** pp 15-18

[3] Kiselev M I and Pronyakin V I 2003 Measuring-computational creation of watch mechanisms *Measuring technique* **5** pp 22-28

[4] Kiselev M I, Zroychikov N A, Pronyakin V I and Chivilev Ya V 2006 Precision study of turbo-aggregate operation by optoelectronic means *Heat power engineering* **11** pp 10-13

[5] Kiselev M I2016 Phase chronometry: problems and prospects *Instruments* **10** pp 51-54

[6] Kiselev M I and Komshin A S2012 Features of dynamics of diesel-generator sets of diesel locomotives *Bulletin of UNN. N.I. Lobachevsky* **5 (2)** pp 107-112

[7] Komshin A S, Kiselev M I and Syrytsky A B 2015 Introduction of measuring and computing systems for maintaining the life cycle of metalworking equipment and tools based on the phase-chronometric method *Stankoinstrument* **1** pp 89-96

[8] Syrytsky A B and Boldasov D D2015 Phase-chronometric monitoring system for wear of cutting tools *Metalworking* **5** pp 2-10

[9] Potapov K G and Syrytskiy A B 2014 Implementation of the measuring phase-chronometric system for diagnosing the technical condition of lathes*Instruments* **5** pp 18-22

[10] Kiselev M I2014 Features of information support of the life cycle of engineering objects in connection with the tightening of the requirements for their quality *All materials Encyclopedic reference book* **6** pp 2-9

[11] Kiselev M I and Novikov S V "Industry 4.0": some problematic issues*Stankoinstrument* **2** 42-46.

[12] "Industry 4.0" and measurement in industry *Experience of Kazakhstan*(https // kazinmetr.intopress / news / 2/176 /)

[13] Turchin P V 2007 *Historical Dynamics: Towards a Theoretical History*(Moscow: URSS) p 368

[14] Korotaev A V and Malinetsky G G 2009 *Problems of Mathematical History: Historical Reconstruction, Forecasting, Methodology*(Moscow: URSS) p 248

IOP Conf. Series: Materials Science and Engineering **468** (2018) 012045 doi:10.1088/1757-899X/468/1/012045

Its own spectrum band – a property of mechanical systems

I N Ovchinnikov

Bauman Moscow State Technical University, Moscow, Russia
E-mail: iovchin@bmstu.ru

Abstract. In experimentally and modelling, the results have been obtained, revealing the existence of the hardes tmode of broadband random vibration in function of the spectral width when it impacts on the mechanical system. During broadband random vibration there are extremums of tension in the material and vibration speed during the capture of «n» original frequencies of the system (n≥2). This fact indicates on presence of the "own bands" of the mechanical system, similar to the natural frequencies, but which appears at random loading.

1. Introduction

It is known that more than 70% of malfunctions in technics are caused by a vibration, which results in a fatigue failure of parts of a construction, transported loads, also in breakages and halting of electronic devices. Obtaining of reasonable compromise between the maximum of reliability and minimum of weight is laid in the development of methods of adequate investigation of materials' behavior in constructions with given resource and reliability, and that shows the importance of vibration trials. At present trials on broadband random vibrations (BRV) are used most often.

International Electrotechnical Commission (IEC) sets two main requirements for vibration trials on its' result's veracity and reproducibility in different laboratories, which are especially important in conduction of acceptance trials. Reproducibility of trials' results cannot be reached with low veracity. Therefore, the reliability of the test results is given special attention (a new sensor deformation). It is also necessary that ranges of vibration load, which are formed, take into account the dynamics of the trials' object [1], what is done during harmonic loadings. At present ISO series standards ignore that [2]. Long attempts to replace the random loading to an equivalent determinate loading have not been successful [3,4,5 etc]. Only in [6] the problem was solved: such a replacement is incorrect and unacceptable.

Hardest mode BRV, mentioned in [7], but determined by dynamics of trials' object, is necessary for raising levels of reproducibility of trials' results, standardization of modes of vibration and fatigue trials, forming modes of accelerated trials, in technological processes.

In [8] is theoretically shown than extremums of parameters of vibration loads exists in any mechanical system. They correspond to extremums of power of mechanical forces, appearing in system during broadband random vibration. In other words, there is such width of the range that the ratio (in probabilistic aspect) of amplitudes of components of oscillatory process on its own frequencies optimally. Existence of such components of oscillatory process out of original frequencies, which is usual for BRV, just lowers lifespan of the objects.

Here are presented the results of the experiment and simulation of the oscillations of the cantilever beam.

Content from this work may be used under the terms of the Creative Commons Attribution 3.0 licence. Any further distribution of this work must maintain attribution to the author(s) and the title of the work, journal citation and DOI.
Published under licence by IOP Publishing Ltd

2. Experiment

Forming of hardest condition in its essence is a problem of optimization vibration loading with given power. It comes to discovering the class of effects f(x,t), which creates extreme conditions of object load. Mathematically it can be expressed by relation:

$$extr\ F[y,\dot{y},\ddot{y},...y^{n}(t),x,t,f(x,t)], \tag{1}$$

for $f(x,t)\in\Phi$,

where F – functional, characterizing vibrational movement of the object.

In general case the problem (1) is difficult to solve, but in practical realization of the trials there are limitations on kind of effect (power, number of harmonics), on control coordinates for object movement (dangerous intersection) and so on. That is why one can investigate particular problems:

$$\max_{f(x,t)} \dot{y}_{x=u},\ t\in T \tag{2}$$

$$\max_{f(x,t)} D_{y_{x=u}},\ t\in T \tag{3}$$

$$\min_{f(x,t)} t_r,\ t\in T, \text{where } \frac{1}{T}\int_0^T f^2\,dt < C \tag{4}$$

In other words, it is necessary, by changing the range of the spectrum, to find in which one the vibration speed (2) or dispersion of vibration speed (3) in the intersection during the time t are maximal, and time before fracture of object t_r (4) is minimal for given limited power of tension.

In order to raise veracity and reliability of trials' results a series of indestructible contactless capacitive sensors, which allow to measure almost limitless by value flexural deformations for indefinite time, was developed. Capacitive sensors, on the opposite of resistive-strain sensors, do not deform with an object, do not accumulate damage, that is why they keep their metrological characteristics constant. Capacitive sensor of deformation EED [9] has been developed in order to measure deformation of flat and cylindrical surfaces with zero gauss curvature in radial direction. It consists of metal lining 1 with cylindrical working surface, which through the layer of dielectric 2 is pushed towards (or placed close to) surface under study 3, which is the second lining of capacitor (Figure 1).

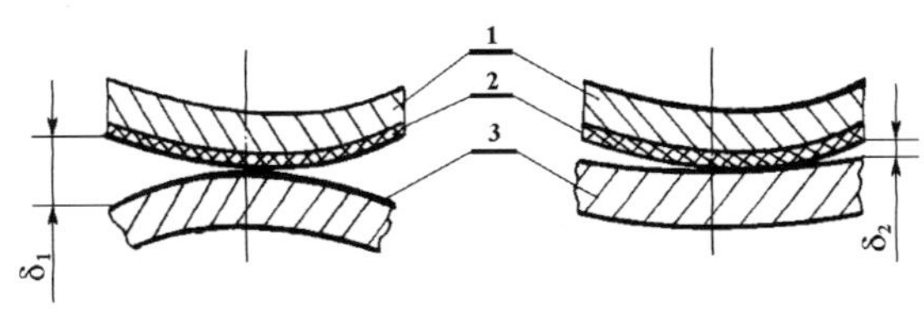

Figure 1. Capacitive sensor of deformations for flat and cylindrical surfaces (EED1)

With the bending of an object the distance δ between linings of the capacitor is changed and electric signal of sensor is proportional to the change of curvature and curvature deformation of an object. Area of linear characteristic of capacitive sensor in statics and dynamics is higher than it is of resistive-strain sensors. Its range covers values of deformation in the given experiment including low-cycle area of fatigue. Sensibility of the sensor is defined by curvature of measuring lining and can vary in a wide range. Two linings from two sides of an object raise sensibility by 2 times. Measuring lining does not deform with an object, and consequently its lifespan almost unlimited.

Experimental investigation of lifespan. Investigation of effect of vibration spectrum width Δf on vibration loading (vibration speed) and lifespan was conducted on cantilevered fixed beam in kinematic excitation (Figure 2).

Investigation of effect of vibration spectrum width Δf on vibration and lifespan was conducted on cantilevered fixed beam in kinematic excitation (Figure 3).

Figure 2. Three first forms of original oscillations of testing object (beam)

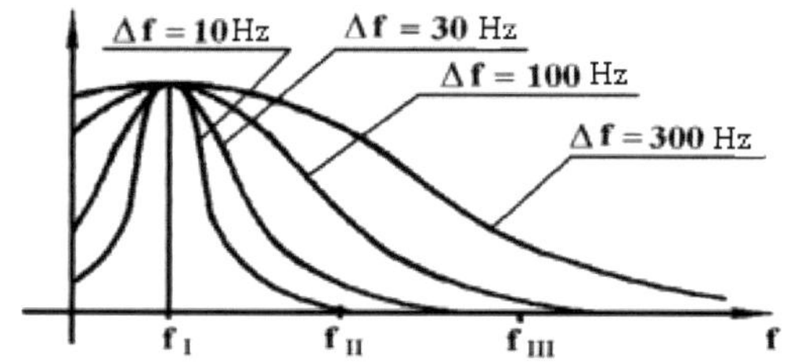

Figure 3. Spectrums of tests

The trial was conducted with random modes for following values of Δf: 10 Hz (mode №5); 30 Hz (mode №6); 100 Hz (mode №7); 300 Hz (mode №8) (Figure 3). Numbers of testing modes are given with correspondence to full series of testing modes [10]. The results of trials in harmonic mode№1 are introduced only for qualitative comparison.

Experiment on fatigue was conducted according to method, which does not simplify any random process, because replacement of random vibration loading with harmonic mode is not correct [6]. All measurements and, most importantly deformations were applied from the start of trial until the fracture of the object.

As it common for lifespan investigations the results were introduced in a form of fatigue curves – dependence of logarithm of time till the fracture on average tension in the material (Figure 4). It is shown that with the widening of load range lifespan of the material of the object is falling in the beginning and then raising.

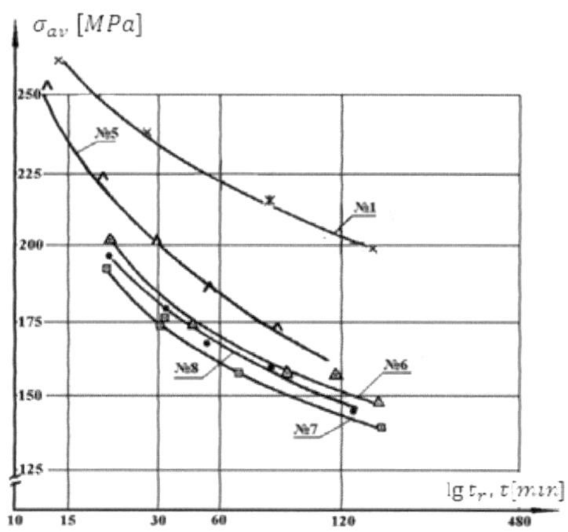

Figure 4. Fatigue curves of modes №№ 1(harmonic), 5-8(random)

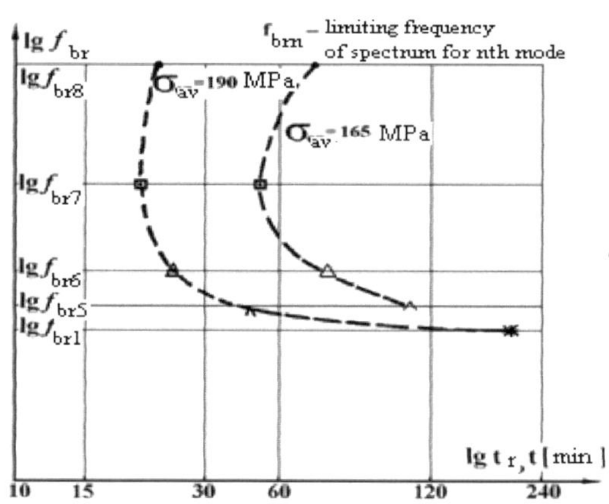

Figure 5. Dependence of the object lifespan from width of vibration load spectrum with constant average stress

Existence of extremum of lifespan depending on limiting frequency of spectrum of vibration loading during constant tension in the objects is clearly shown in picture 5, which has been obtained by processing results of fatigue trials. As a limiting frequency of spectrum for harmonic mode (mode №1 [10]) logarithm of first original frequency of the object f_I=27 Hz, and for random mode with continuous spectrum – logarithm of sum of average spectrum frequency (first original frequency of the object) and half of spectrum width for chosen mode (Figure 3): $\lg f_{br} = \lg\left(f_1 + \frac{1}{2}\Delta f \right)$.

During random oscillations the object was significantly being exited on the second and the third original frequencies (f_2=177 Hz, f_3=475 Hz).
In order to raise informational content of the results of trials "curves of the vibration loading", which show the dependence of logarithm of time until fracture on vibration speed in sample's dangerous cross-section (Figure 6), have been drawn first time in practice.

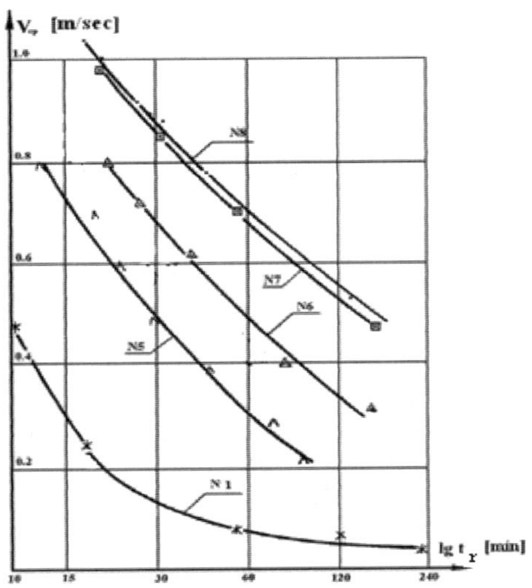

Figure 6. Depending t_r of the samples of V_{AV} in a dangerous cross-section of the beam ("curves of the vibration loading")

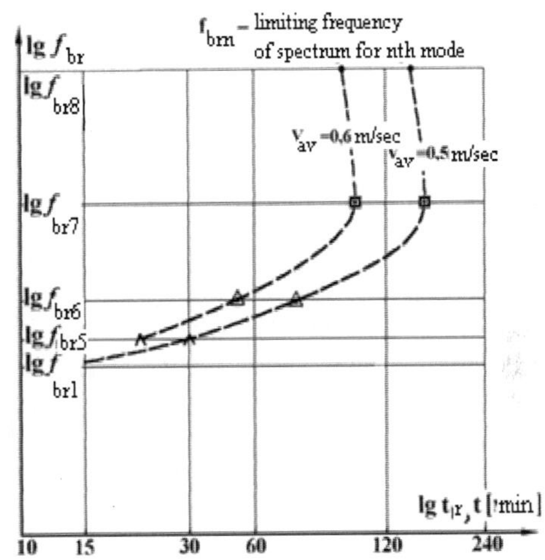

Figure 7. Dependence of material samples lifespan on width of vibration loading spectrum with constant average vibration speed.

Similar existence of extremum of lifespan depending on limiting width of spectrum $f_{br,}$ but with constant average vibration speed in dangerous intersection of the object, obtained by transforming "curves of the vibration loading", is shown on Figure 7.
Despite the fact that curvature of graphs 5 and 7 in given coordinates is relatively small, there is no doubt in the existence of extremums, because the points on the curves correspond to high amount of fractured test objects. The veracity of the test results is high enough thanks to application more perfect methods of trials and deformation sensors.
The experimental results are agreed well with the simulation results.
Similar existence of extremum of lifespan depending on limiting width of vibration loading spectrum $f_{br,}$ but with constant average vibration loading in dangerous intersection of the object, obtained by transforming "vibration loading curves" [10], is shown on Figure 7.
Despite the fact that curvature of graphs 5 and 7 in given coordinates is relatively small, there is no doubt in the existence of extremums, because the points on the curves correspond to high amount of fractured test objects. The veracity of the test results is high enough thanks to application more perfect methods of trials and deformation sensors.

3. Simulating

With the purpose of decision of these tasks for the console fixed beam were investigated by digital simulation the dependences of average square of vibration speed meaning$< \dot{S}^2 >$ of object in dangerous section at given average square meaning of tensions $< \dot{\sigma}^2 >$from width of a spectrum and also average square meaning tensions at constant average square meaning of vibration speed [11].
As a generator of random excitations it was used a standard subroutine. For formation of a loading spectrum of "white noise" type the filters giving signals with a uniform spectrum on displacement <S> were applied; f_{1bd} bottom boundary frequency of a spectrum, f_{2bd} - top boundary frequency of a spectrum.
The real experiment on durability was carried out to a cinematically excited beam, console fixed on shaking table, where S(t) – displacement of the shaking table. Proceeding from this, the vibrating influence on a beam is proportional to acceleration $\ddot{S}(t)$.
The imitating model was under construction on the basis of the equation of bending fluctuations of a beam.

$$m_o \frac{\partial^2 y(x,t)}{\partial t^2} + A \frac{\partial^4 y(x,t)}{\partial x^2} + \alpha \frac{\partial y(x,t)}{\partial t} = f(x,t)\,. \tag{5}$$

Under boundary conditions:

$y(x) = 0;\quad y'(x) = 0 \qquad$ at $x = 0$;

$y''(x) = 0;\quad y'''(x) = 0 \qquad$ at $x = l$;

$y(x,t)$ – deflection of a beam;

$m_o = \rho F$ – weight of unit of length of a beam;

F – area of cross section;

l – length of a beam;

A=EI – bending tough;

E – module of elasticity of a material of a beam;

I – inertial moment of section;

α – factor of damping;

f(x, t) – external loading.

At cinematic excitation external loading is described by the following expression:

$f(x,t) = -m_o \ddot{S}(t)$,

where S(t) – displacement of a table of the vibrator (jammed end of a beam).
Let's proceed from the equation (5) to the dimensionless form of record, having accepted $x = \varepsilon \cdot l$, $y = \upsilon \cdot l$:

$$\frac{\partial^2 \upsilon}{\partial t^2} + \frac{EI}{m_0 l^4} \cdot \frac{\partial^4 \upsilon}{\partial \varepsilon^4} + \frac{\alpha}{m_0} \frac{\partial \upsilon}{\partial t} = -\frac{\ddot{S}}{l} \tag{6}$$

On the basis of Galerckin method the transition from equations (6) was carried out to the system of linear differential equations.
The research of accuracy of modeling was carried out with 16 members. There was shown that it is possible to use only three members in Galerckin approximation.
Obtained system of equations was numerically integrated by a Rhounge-Kutta method at random tension $\ddot{S}(t)$ formed, as it was specified, with generator of a random signal and filters.
In result the dependences of average square tension meaning were received.

IOP Conf. Series: Materials Science and Engineering **468** (2018) 012045 doi:10.1088/1757-899X/468/1/012045

$$<\sigma^2>=\frac{1}{T}\int_0^T \sigma^2 dt\text{, where } \sigma=\frac{hE}{2}\frac{\partial^2 y(x,t)}{\partial x^2}=\frac{hE}{2l}\sum_{k=1}^{3} f_k(t)\ddot{y}_k\,.$$

In function of boundary frequency of vibrating excitation f_{2bd} (6) (Figure 8) at constant root-mean-square meaning of vibration speed

$$<\dot{S}^2>=\frac{1}{T}\int_0^T S^2 dt = const\,.$$

Dependence of average square meaning of vibration speed $<\dot{S}^2>$ also was received from boundary frequency of entrance influence f_{2bd} at given constant meaning of average square meaning of a tension $<\sigma^2>=$ constant (Figure 9).

The received dependences (Figure 8 and Figure 9) have extreme at boundary frequency 120 Hz<f_{2bd}<160 Hz. It is possible to accept f_{extr} about 140 Hz. The resonant frequencies for model (1) have meanings: f_1=28 Hz, f_2=175 Hz, f_3=485 Hz.

The carried out research allows to consider, that at constant average square of a tension (vibration speed) of a beam the dependence of an average square of vibration speed (tension) from boundary frequency of vibration has extremes, laying in the field of meanings between resonant frequencies of a beam, but is significantly closer to n-th own frequency.

The result, received by modeling, will well be coordinated with the fatigue experiment [10,11], in which vibration was formed with 1/3- octave filters with average frequency f_{av}=f_1 and width of a spectrum Δf=10, 30, 100 and 300 Hz.

Extreme meanings $\dot{S}^2$ and σ^2 are received for a mode with Δf=100 Hz, boundary frequency f_{bd} which with account flatness of characteristic of the filter is close to the meaning f_{2bd}, appropriate calculated extremes. On the mode with Δf=100 Hz the time before destruction of a beam was the least, i.e. the specified mode has appeared the hardest one.

At the first stages of revealing extremes in the described dependences it not so important the accuracy of concurrence of experimental and calculated results, how the fact of confirmation both in experiment, and in account of the existence of effective width of a spectrum, which correspond extreme parameters of loading.

Modeling of influence of a broadband random spectrum on the console fixed beam by the described method was carried out in a range to sixteen first own frequencies of a beam. At capture by a loading spectrum of every one following own frequency of a beam vibration speed and tension in dangerous section accepted the extreme meanings. Appropriate width of a spectrum was named as an "own strip of a spectrum".

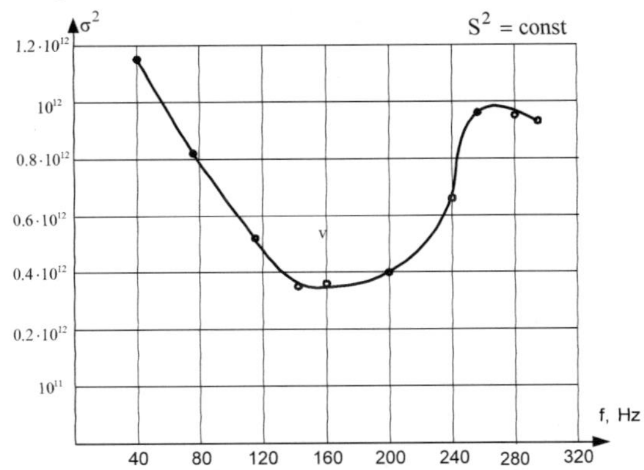

Figure 8. Extreme of root-mean-square meaning of a tension in function of boundary frequency of vibrating action

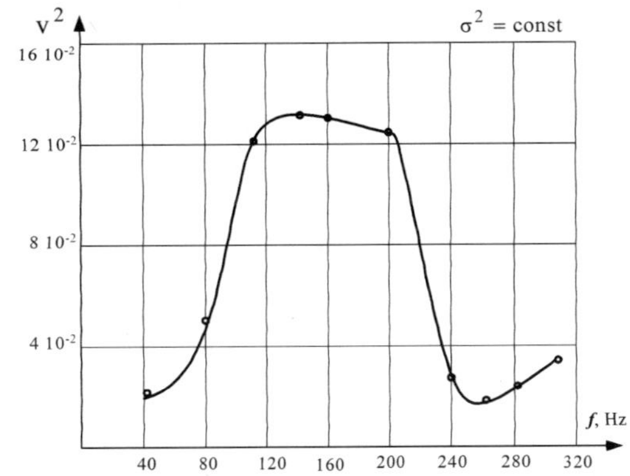

Figure 9. Extreme of root-mean-square meaning of vibration speed in function of boundary frequency

IOP Conf. Series: Materials Science and Engineering **468** (2018) 012045 doi:10.1088/1757-899X/468/1/012045

The received result can be treated as existence in mechanical systems "of own strips of a spectrum" (pass bands), in sense sensitivity to vibration, similar to own frequencies, but shown at broadband random vibration. The received result is similar submitted in [8], where the theory of dynamic systems was used.

4. Discussion

The modeling of impact of broadband random spectrum on cantilevered beam by described method was held in a range of first sixteen natural frequencies of the beam; the results are similar to those shown in Figure 8 and Figure 9 and for the first three natural frequencies are illustrated in Figure 10a, 10b, 10c.

In Figure 10a the amplitude frequency response of cantilever beam in the area of first three natural frequencies is given. The Figure 10b presents by the three spectrum of equal power slightly different by the width and, of course, amplitude very intensively exciting the beam in the area of first two and first three natural frequencies generated by the vibrator control system in place where the beam is fixed. The Figure10c shows the changes of average voltage and average vibration velocity in the rigid fixing of the beam depending on the width of the spectrum of external influence. At some width of this spectrum (for clarity, this is average spectrum, Figure 10b) the vibration velocity has a maximum and voltage – a minimum. Such a picture repeats with the coverage by the spectrum of external vibration exposure of the first two, three, etc. natural frequencies of the beam. The curves in Figure 10c are in fact a kind of analog amplitude frequency response of the beam, which characterize her reaction on changing by the width the random vibration loading.

The obtained result can be interpreted as the existence of "own spectrum bands" (passbands) of mechanical systems, in terms of sensitivity to vibration exposure similar to natural frequencies, but manifested in broadband random vibration loading.

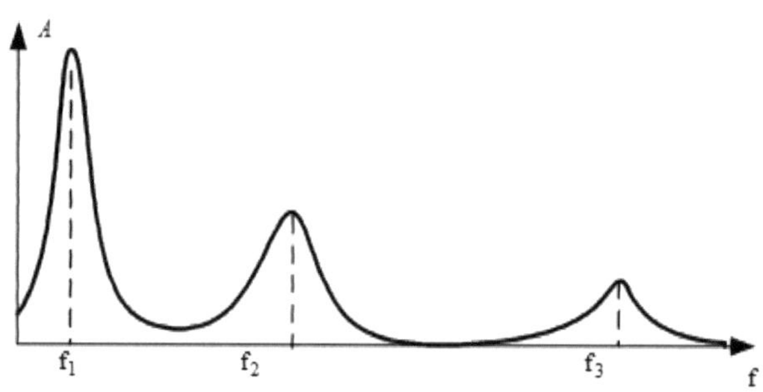

Figure 10a. Amplitude frequency response of testing object

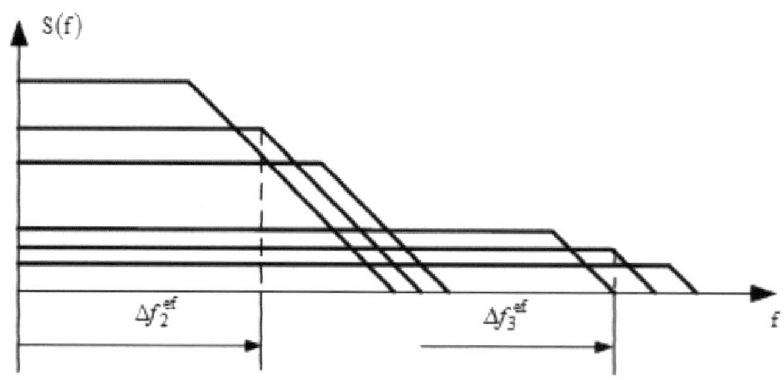

Figure 10b. The spectral density of external Influence

IOP Conf. Series: Materials Science and Engineering **468** (2018) 012045 doi:10.1088/1757-899X/468/1/012045

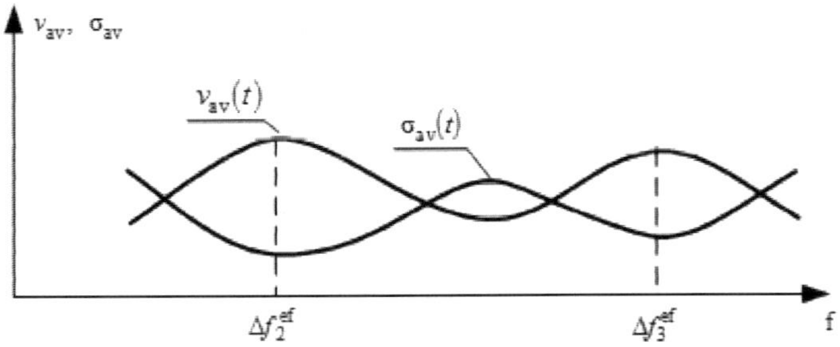

Figure 10c. The Figure reaction parameters of testing object

5. Conclusions

1. The existence of extremums in the dependencies of lifespan, tension and vibration speed, which are invariant toward the width of the vibration loading spectrum, has been noticed first time. The width of the spectrum of the continuous vibrational load corresponding to these extremes was called the previously effective spectral width [11,12, 13] (Figure 10b, 10c), and after [14] - the "own bands of the spectrum": Δf_n^{own}, where $n \geq 2$, determines the number of natural frequencies of the object that are taken into account when vibrating.
2. The hardest BRV condition of mechanical system is defined by its characteristics, similar to natural frequencies of the system.
3. Knowledge about the hardest BRV condition is useful not only for conducting accelerated trials but also for standardization of trials, for evaluation of level of danger of operating vibrations, for technological processes, acoustics and probably for resistivity to earthquakes.
4. The developing of the automated system for definition of hardest mode of vibration loading is necessary, which would determine the own spectrum bands of mechanical system, as determining the natural frequencies of this system.

References

[1] Krendell S 1967 *Random fluctuations* (Moscow: Peace) p 356
[2] IEC60068-2-64-93 international standard *Tests on the effects of random vibration using the digital broadband vibration test control system* (GOST R 51502–99)
[3] Kolovskiy M Z 1963 About the replacing random vibration exposure by the harmonic process *Proc. of the Ac. of Sc. of the USSR, Dep. of tech. Sc., Machine sc.* **2** pp 93-101
[4] Hu JM 1997 Correlation of a Sinusoidal Sweep Test to Field Random Vibration *J. of the Ins. of Environmental Sc.* **40** (6) pp 35-41
[5] McNeill S I 2008 Implementing the Fatigue Damage Spectrum and Fatigue Damage Equivalent Vibration Testing *79 th Shock and Vibration Symp.* pp 1-20
[6] Ovchinnikov I N, Stepnev VA and Brancevich PY 2012 Authenticity of the equivalent vibration tests *Proc. 19 Int. Cong. on Sound and Vibration*
[7] Chelomey VN 1981 *Vibration in technique* **V5** Genkin M D Measuring and testing (Moscow Publishing house «Engineering») p 496
[8] Ovchinnikov I N 2005 The usage of “transparency frequency bands” of mechanic systems for standardization of tests to random vibration *Proc. 12 Int. Cong. on Sound and Vibration*
[9] Ovchinnikov I N 1995 Capacitive sensor for measuring flexural deformations *Devices and control system* **3** p 25
[10] Arutyunov S K, Kolesnikov K S and Ovchinnikov I N 1985*The regularities of fatigue failure in random vibration loading* Engineeringpp. 81-86

[11] Ovchinnikov I N 2004 *Vibration test, diagnosis and prediction of fatigue destruction* (Moscow: BMSTU) p 128
[12] Ovchinnikov I N 2008 Simulation of oscillations of a beam at random vibration loading *Proc. 15 Int. Cong. on Sound and Vibration*
[13] Ovchinnikov I N and Brancevich P Y 2014 Experimental determination oft she hardest condition of broad band random vibration *Proc. 21 Int. Cong. on Sound and Vibration*
[14] Ovchinnikov I N 2015 A new property of the mechanical systems – own spectrum bands *Proc. 22 Int. Cong.s on Sound and Vibration*

Acknowledgments
The work is executed at financial support of Russian Foundation for basic research, grant No. 18-08-01512.